A MANUAL

OF THE

GEOLOGY OF INDIA

A MANUAL

OF THE

GEOLOGY OF INDIA.

CHIEFLY COMPILED FROM THE OBSERVATIONS
OF THE GEOLOGICAL SURVEY.

STRATIGRAPHICAL AND STRUCTURAL GEOLOGY.

BY

R. D. OLDHAM, A.R.S.M.

GYAN

Published by

Gyan Publishing House
5, Ansari Road
Daryaganj, New Delhi-110002
Phone: 011-47034999, 9811692060
E-mail: books@gyanbooks.com

Distribution Network
gyanbooks.com
India, USA, Canada, UK, Australia

ISBN: 978-93-6208-011-0 (PB)
First Published, 1879

2nd Impression 2024

Printed at: Gyan Press, Delhi.

A MANUAL OF THE GEOLOGY OF INDIA
Author: R. D. OLDHAM

Photo etching from a Photograph by Mess.rs Johnston & Hoffmann.

Survey of India Offices, Calcutta, July 1892.

SIMVO - 22,300 Ft. - FROM THE NORTH WEST (EASTERN SPUR OF KÁNCHANJANGA)

A MANUAL

OF THE

GEOLOGY OF INDIA.

CHIEFLY COMPILED FROM THE OBSERVATIONS
OF THE GEOLOGICAL SURVEY.

STRATIGRAPHICAL AND STRUCTURAL GEOLOGY.

SECOND EDITION REVISED AND LARGELY REWRITTEN BY

R. D. OLDHAM, A.R.S.M.,
SUPERINTENDENT, GEOLOGICAL SURVEY OF INDIA.

Published by order of the Government of India.

CALCUTTA:
SOLD AT
OFFICE OF THE SUPERINTENDENT OF GOVERNMENT PRINTING;
GEOLOGICAL SURVEY OFFICE; AND BY ALL BOOKSELLERS;
LONDON: TRUBNER & CO.

MDCCCXCIII.

PREFATORY NOTICE.

IN the beginning of 1887 my predecessor, Mr. Medlicott, wrote as follows in his Annual Report of the Geological Survey:—"The two first parts of the Manual of the Geology of India, issued in 1879, have been out of print for some time, and the question of re-writing it has been much upon my mind. Parts of it would require abridgment, leaving local information to be sought for in the special Memoirs; and parts of it would need alteration and addition in view of extended information. The greater part of the two volumes was written by Mr. Blanford, who was for the time relieved of other work. To re-write the whole while carrying on the manifold current duties of the Survey has been more than I could attempt in India with any justice to either."

The directing of the Survey since Mr. Medlicott's retirement is even fuller of current duties, not the least of which has been a considerably increased system of frequent tours over the length and breadth of the land; so that, however pressing it may also have been on my mind, I have been unable even to venture on the elaboration of a revised form of Messrs. Medlicott and Blanford's most excellent work; and I therefore gladly accepted Mr. R. D. Oldham's offer to prepare a fresh issue accordant with our progressive survey of the Empire.

Mr. Oldham had had a varied experience of survey work over widely separated tracts in India where he had opportunities of studying most of our representative formations in their peninsular and extra-peninsular development: while of his own motion he devoted his first period of well-earned leave to a comparative study of our Gondwána representatives in Australia. His close acquaintance with the literature, as evidenced in the careful *Bibliography of Indian Geology*, compiled by him in 1888, had already indeed predisposed me in favour of a possible ultimate placing of a second issue of the Manual in his hands; and in now authorising that issue I would fain hope that my choice may be justified.

WILLIAM KING,

Director, Geological Survey of India.

PREFACE TO THE FIRST EDITION.

THE want of a general account of Indian Geology has been felt for some years. The regular Geological Survey of India may be considered to have commenced in 1851; and but few of those who took part in the work during the earlier years now remain in the service. It is desirable, before all the older members of the Survey pass away, that some record of the early observations, many of which are unpublished, should be rescued from oblivion, for the benefit of future explorers. The published Memoirs and Records of the Survey, moreover, have now become too numerous and bulky for general use; and it is difficult for any one, without much study, to gather the more important observations on the geology of the country from amidst the mass of local details. Many papers on Indian geology are also scattered through various Indian and European periodicals. As a guide to all who have occasion to acquire a knowledge of Indian geology, or who desire information from a love of the science, some compendium of the observations hitherto collected has become absolutely necessary; and the present Manual has been drawn up, by direction of the Government of India, to supply the deficiency.

It was originally desired by the Government that this work should be prepared by the late Dr. Oldham, or that the compilation should have the advantage of his supervision. As Dr. Oldham was the first Superintendent of the Survey, and remained at the head of the Department from its commencement in 1851 to 1876, he would, unquestionably, have been admirably qualified to carry out the work; and it was his own desire to do so, as the completion of his labours in India. Failing health, however, and the pressure of other duties, prevented him from even commencing the task; and when, at length, he was unable any longer to remain in the country, the duty of preparing a Summary of Indian Geology was left to his successor. At this time the only preparation that had been made for the work was the partial compilation of a general Geological Map of the Peninsula.

The double authorship was not entirely a matter of choice; although undertaken, and carried out, most willingly by both the writers. Both have been engaged in the work of the Survey almost from the commencement; and as each has, in the course of his service, examined very large areas of the country, the combination secures the description and discussion, from personal knowledge, of a much larger portion of India. At the same time the advantages of wider experience and thought may not be found an adequate compensation for want of uniformity and occasional discrepancies—the natural results of divided authorship. To secure, so far as possible, the responsibility of each author for the facts and opinions stated, the initials of each are affixed in the Table of Contents to the chapters contributed by him. Every such chapter has been read and revised by the other writer; but the alterations have in no case been of more than trivial importance; so that each chapter may be practically taken as an individual contribution. The number of subjects is so large, and the connection between them, in many cases, so slight, that the lack of uniformity will not, it is hoped, seriously detract from the usefulness of the Manual.

In addition to the subjects discussed in the present work, it was, at first, proposed to add an account of the Economical Geology, and to treat in a special chapter of the known Mineral Resources of India. But the length to which the Manual has already extended has rendered it advisable to postpone this very important subject, and to reserve it for a separate volume.

Although many of the details in the work now issued have not previously been published, and although the discussion of the observations involves several new deductions and suggestions, the book is, in the main, a compilation; and it is quite possible that, especially in treating of areas and formations of which the authors have no personal knowledge, full justice has not always been done to the views of original observers. It has, in several instances, been thought more important to point out possible causes of error than to endorse opinions which, although very possibly correct, are not sufficiently supported by published data to be accepted as conclusive. In all such cases full references to previous publications have been furnished; and an examination of the details given in the latter will, it is hoped, serve to correct any errors of interpretation on the part of the authors of the present work.

The numerous and large areas left blank in the annexed Map show, at once, how far the present publication falls short of completeness, and how imperfectly the promise implied in the title is fulfilled. A note upon the Map further explains that large portions of it have been coloured from very imperfect information, from sketch surveys or rapid traverses affording no sufficient opportunity for a proper study of the formations. It had, however, become imperative, as a duty to the public, for reasons already mentioned, to bring together a summary of the work accomplished since the commencement of the Survey; and it was equally essential, for the Survey itself, that some general record of the results obtained up to date should be compiled. These objects could only be attained by attempting a general Map and Review of the Geology of India; but the reader must not forget that the present attempt is more of the nature of a progress report than of a finished work.

The Map, it is feared, will be found defective in several other respects. Under the circumstances it was impossible to prepare a special reduction of the topography; and, amongst the Maps of India available in the Surveyor General's Office, there was, practically, no choice but to accept that on the scale of 64 miles to the inch, then well advanced towards completion, as a basis for the geological details. The scale is inconveniently small for all parts of the country that have been geologically mapped in any detail, and the mountain ranges have not been inserted; so that many features discussed in the text are not indicated. But the most serious drawback is in the names of places. Many towns of importance are omitted, owing to the small scale; and other names of interest, for purposes of geological description, such as those of fossil-localities, or of villages near important sections, are wanting. Nor is this all. The spelling of Oriental names is a well-known cause of perplexity; and the confusion has been increased by the unfortunate circumstance that, while one system has been adopted by the Great Trigonometrical Survey, and employed in all the maps, including those of the detailed Topographical Surveys, issued by the Department, an entirely distinct system has been employed by the Revenue Survey, by whom the maps of all the best known parts of the country have been prepared. Under the first system, each letter in the Indian language is represented by a corresponding letter in the Roman character; diacritical marks and accents being employed to distinguish such consonants or vowels in the latter as are required to represent two or more sounds, and the Italian or German sounds of the vowels being used, instead of the English. Under the second system, an attempt is made to represent the original sound by English spelling; double vowels being largely used, but no diacritical marks. The imperfection of the latter plan is manifest; because, in the first place, the sounds, of the vowels especially, in English, are variable, and incapable, in many cases, of representing those of Oriental languages; and, secondly, the representation of the true names by supposed equivalents is arbitrary, depending chiefly on the ear, often very imperfectly trained, of the transcriber. When maps of large areas, as in the present case, are compiled, the mixture of names, spelt according to two different systems, is inevitable. The attempt at a general revision of the nomenclature, however desirable, would have involved serious delay.

Of late, the Government has adopted a compromise in the question of spelling, and lists of the principal places in each province have been issued; the familiar and well-known names being spelt in the manner that has become customary by usage, whilst transliteration is employed in all other cases, with the exception that no diacritical marks are used for consonants. This system is obligatory for all official publications; and it has, consequently, been adopted in the present work. In some cases, however, the lists for particular provinces have not been published in time to be available; and in the following pages it is not unfrequently necessary to mention places not contained in the lists, and the proper vernacular pronunciation of which is unknown to the writers. In such cases, an attempt has been made to spell the name according to the recognised system; but it is only fair to warn the reader that no dependence can be placed on many names of places, specially upon those in the south of India, when taken from old maps.

In the preparation of the Map a large share has been taken by various Officers of the Geological Survey, all of whom have contributed. The colouring and printing have been carried out at the Surveyor General's Office, under the superintendence of Captain Riddell, R.E., to whom the authors beg to express their obligations for the labour he has given to the work, and for the assistance he has afforded to them personally.

In the plates of fossil plants and animals at the end of the work some of the most common and characteristic forms of organic remains found in India are represented. The plants have been selected and arranged by Dr. Feistmantel, and the tertiary Mammalia by Mr. Lydekker. All the plates are lithographed by Mr. Schaumburg, whose work will answer for itself. The majority of the figures are from original drawings, or from the "Palæontologia Indica;" the remainder are copied from other works; but these copies have, in many cases, been compared with specimens.

PREFACE TO THE SECOND EDITION.

THE first edition of the Manual of the Geology of India was found to supply a want so much felt that it soon went out of print, and for several years has only been procurable from the dealers in second-hand books. It had, besides, become out of date in many parts, and the Government of India decided that the time had come for the preparation of a new edition.

So great have been the strides made in our knowledge of Indian Geology in the last fifteen years that it has been found possible to entirely change the arrangement of the book and to adopt the more scientific and orderly course of describing the rocks in chronological order, instead of breaking the book up into a series of descriptions of separate districts, as was found to be inevitable in the preparation of the first edition. There is not, unfortunately, everywhere the same certainty regarding the proper position in a chronological arrangement of particular groups of beds. Frequently the true homotaxis is unknown, and there is a conspicuous instance of the mistakes which may then be made, in certain rocks of the Arakan Yoma, which were supposed to be triassic at the date when this second edition was written, but have been shown to be eocene as the pages were passing through the press, and consequently rocks which should have been classed with the tertiary were described in the chapter devoted to the carboniferous and trias.

Many districts have remained untouched since the publication of the first edition, and in all such cases, and wherever indeed no serious modifications of the original text have been necessary, it has been allowed to stand practically as in the original publication. All these passages have been carefully revised and generally more or less condensed, while many minor alterations needful to adapt them to the altered scope and arrangement of the work have been made, as well as such alterations as have been necessary on account of the advance of our knowledge. In the table of contents the portions in which the first edition has been taken as the basis of the text are distinguished by a different type from that which refers to the portions which are new or have been entirely re-written.

As there is now a special volume devoted to the economic aspects of the geology of India, not to mention the Dictionary of Economic Products and the Handbooks of Commercial Products of the Imperial Institute, it has been decided to exclude all references to economic geology in this work. The references would necessarily have to be too brief for commercial purposes, while they would have confused the stratigraphical descriptions and increased the bulk of the volume. It is, therefore, better that these subjects should be relegated to the books specially devoted to them.

For the rest, the remarks in the preface to the first edition, relative to the geographical basis of the Geological Map of India and the spelling of place names, still hold good. As regards the latter, the Imperial Gazetteer of India has been adopted as a standard by the Government of India and the Royal Geographical Society. The spelling in that work has consequently been followed, and as regards places not mentioned in that work the same system of spelling has been adopted so far as possible. The scale of the Geological Map adopted for this edition is smaller than that of the map issued with the first edition. It is hoped that the alteration will make the map more convenient to consult, while the amount of detail that can be exhibited remains practically the same, and the impossibility of finding a map showing all the places mentioned in the text has been rectified, so far as possible by an index of place names giving their geographical co-ordinates.

The date of the map will be seen to differ from that of the title-page of the book; this is due to the fact that it was necessary to go to press with the map before the text was completed. The map consequently represents the state of information available in the office of the Geological Survey on the 31st December 1891, while the text dates about four months later, and subsequent additions and corrections have been made in footnotes up to the beginning of this year.

Finally, I have to express my obligations for the help so readily rendered in the preparation of this work, specially to Prof. Suess, for the loan of the illustration block on p. 202, and to Dr. W. T. Blanford, who, in addition to other assistance, has been good enough to read the proofs of the passages referring to the fossil tertiary and pleistocene mammalia.

TABLE OF CONTENTS.

(Ordinary type refers to matter which has been adapted from the first edition.)

Heavy type refers to matter which is new or entirely rewritten in the present edition.]

CHAPTER I.

PHYSICAL GEOGRAPHY.

CHAPTER II.

METAMORPHIC AND CRYSTALLINE ROCKS.

CHAPTER III.

TRANSITION SYSTEMS.

CHAPTER IV.

OLDER PALÆOZOIC (CUDDAPAH AND VINDHYAN) SYSTEMS OF THE PENINSULA.

CHAPTER V.

OLDER PALÆOZOIC SYSTEMS OF THE EXTRA-PENINSULAR AREA.

CHAPTER VI.

CARBONIFEROUS AND TRIASSIC ROCKS OF EXTRA-PENINSULAR INDIA.

CHAPTER VII.

THE GONDWÁNA SYSTEM.

CHAPTER VIII.

HOMOTAXIS OF THE GONDWÁNA SYSTEM.

CHAPTER IX.

MARINE JURASSIC ROCKS.

CHAPTER X.

Marine cretaceous rocks of the Indian Peninsula.

CHAPTER XI.

Deccan Trap.

CHAPTER XII.

CRETACEOUS ROCKS OF THE EXTRA PENINSULAR AREA.

CHAPTER XIII.

TERTIARY DEPOSITS (*excluding those of the Himálayas*).

CHAPTER XIV.

TERTIARIES OF THE HIMÁLAYAS (*including the North-Western Punjab*).

CHAPTER XV.

LATERITE.

CHAPTER XVI.

Pleistocene and recent deposits (*exclusive of the Indo-Gangetic alluvium*).

CHAPTER XVII.

The Indo-Gangetic plain.

CHAPTER XVIII.

THE AGE AND ORIGIN OF THE HIMÁLAYAS.

CHAPTER XIX.

GEOLOGICAL HISTORY OF THE INDIAN PENINSULA.

LIST OF ILLUSTRATIONS.

WOODCUTS IN TEXT.

A MANUAL

OF

THE GEOLOGY OF INDIA

CHAPTER I.

PHYSICAL GEOGRAPHY.

Scope of the work—Threefold division of British India—Contrast between extra-peninsular and peninsular areas—Mountain ranges of the Peninsula—Extra-peninsular mountain ranges—Drainage of extra-peninsular ranges—River system of the Peninsula—Evidence of changes of level in the Peninsula—Changes of coast line—Glacial epoch in India—Volcanoes—Doubtful cases of volcanic action—Salses, or mud volcanoes.

THE limits of the area, whose geology is treated of in this book, coincide with the limits of the jurisdiction of the Governor-General of India. In some few cases references to the geology of adjoining countries will be found, but such are not many, and are all imperfect as they depend on observations which were made during hurried traverses and under circumstances precluding the idea of detailed geological work. In spite of these drawbacks the results have often been important, interesting and impossible to ignore, but there is generally less reason to congratulate ourselves on the knowledge obtained of countries across the border than to deplore our ignorance of large tracts within it.

The general shape and principal features of British India, the great triangular promontory with the pear-shaped island of Ceylon south-east of its extremity, the great range of the Himálayas to the north, and the large area of Burma to the east, running down into the narrow strip of Tenasserim along the east coast of the Bay of Bengal, are well enough known, as well as the principal political divisions of the empire. But, for geological purposes, the important point to be noticed is the threefold division of this area into, *1st*, the great Indo-Gangetic alluvial plain, comprising the Punjab and Hindustán proper, with Bengal and the eastern prolongations up the valleys of the Brahmaputra and the Bárak; *2nd*, the triangular area of the Peninsula, lying to the south; and *3rd*, the extra-peninsular area, comprising the hilly country west, north, and east of the Indo-Gangetic plain.

Nor is this division an arbitrary one. The geological history of the peninsular and extra-peninsular areas has been radically different. Since the latter end of the palæozoic era the former appears to have been an area of dry land; no sedimentary formations of marine origin have been found except near the present sea coasts, and there they thin out against the older rocks on which they rest, in a manner suggesting that the shore line cannot have been very far removed from the present position of the coast when they were being deposited. In the extra-peninsular area, on the other hand, marine deposits range through the palæozoic and mezozoic eras, and only in the latter part of the tertiary period is there any great development of deposits formed on dry land.

Structurally too the two areas differ greatly. The Peninsula has undergone no great compression since the close of the palæozoic era, and the beds all lie at low angles of dip. In the extra-peninsular area the conditions are totally different; the rocks have everywhere undergone great compression and disturbance since the commencement of the tertiary period, a disturbance which ranges in degree from the comparatively regular, though high dipping, folds of the Balúchistán and Punjab hills, to the complicated overfolds and thrust faults of the Himálayas.

This difference in geological history finds its expression in the difference of the present contours of the two areas. In the extra-peninsular area we have mountain ranges which coincide with regions of special elevation, that is, the courses of the principal chains, and often of the minor ridges, are governed by their structure and are the direct result of the compression, and consequent disturbance and elevation they have undergone. As a result of this, the valleys are deep, narrow, and steep-sided, the rivers and streams rapid and torrential in their nature, and, as a rule, evidently actively at work in deepening their valleys. In the peninsular area, on the other hand, the mountains are all remnants of large table-lands, out of which the valleys and low lands have been carved. The valleys, with a few local exceptions, are broad and open, the gradients of the rivers low, and the whole surface of the country presents the gently undulating aspect characteristic of an ancient land surface.

Such, broadly speaking and subject to some minor exceptions, are the contrasting characteristics of the two areas. In the country lying west of the Arávallis, between them and the Indus, there is a tract of geographically debateable ground, which exhibits a combination of the characteristics of the two areas. The rocks exposed are very largely secondary and tertiary beds of marine origin, agreeing in this with those of the extra-peninsular area, while in their low undulating dips and absence of any marked degree of disturbance, they approach the type of the peninsular area. On the north-east again beds, belonging to formations which are characteristically peninsular, are found in the Himálayas of Sikkim and north of the Assam valley and in the hills intervening between the Brahmaputra and Bárak rivers. We will find the explanation of these exceptions to the geological contrast between the two areas in the great structural disturbances which took place during the tertiary period, and profoundly modified the outlines of that ancient land surface of which the Peninsula proper is but a remnant.

The nomenclature of Indian mountain ranges is still a difficulty, it being a rare exception that any definite term is applied to a mountain chain, throughout its extent, by the people of the country. In many parts of India peaks and passes have names, but the ranges have none and, even if names exist, their application is not unfrequently vague. Thus, the ancient name of 'Vindhya,' applied to the hills separating Hindustán proper or the Gangetic country from the Deccan (Dakshin or south), has now, by common consent, been restricted to the hills north of the Narbadá, but it appears almost certain that the term originally applied also to the ranges now known as Sátpura, south of the river, and it is very probable that the latter hills were more especially indicated by the term "Vindhya" than the former. The term "Sátpura" again was of very indefinite application and probably included other ranges besides that to which it is now restricted. The names here applied are those employed by the latest writers on Indian geography, but some of them are by no means generally adopted on maps.

The most important mountain ranges of the Peninsula are the Sahyádri, or Western Gháts, running along the western coast from the Tápti river to Cape Comorin, at the southern extremity of the Peninsula; the Sátpura, running east and west on the south side of the Narbadá valley, and dividing it from the drainage areas of the Tápti to the westward, and the Godávari to the eastward; and the Arávalli, striking nearly south-west to north-east, in Rájputána. The so-called Vindhyan range, north of the Narbadá, and the eastern continuation of the same north of the Son valley, known as the Káimur range, are merely the southern scarps of the Vindhyan plateau comprising Indore, Bhopál, Bundelkhand, etc. The plateaux of Hazáribágh and Chutiá Nágpur (Chota Nagpore) in south-western Bengal appear to form a continuation to the eastward of the Sátpura range, but there is no real connection between these elevations and the Sátpura chain. They are formed of different rocks and there is no similarity in the geological history of the two areas, so far as it is known. In many maps a range of mountains is shown along the eastern coast of the Peninsula, and called the Eastern Gháts. This chain has not the same unity of structure or outline as the Western Gháts. It is composed to the southward of the south-eastern scarp of the south Mysore plateau, on the east of the Yellakonda range along the eastern margin of the Cuddapah transition basin, and further north of the south-eastern scarp of the Bastár-Jaipur plateau, north-west of Vizagapatam, and of several short isolated ridges of metamorphic rocks, separated from each other by broad plains and having in reality but little connection with each other. There are also several minor ranges, such as the Rájmahál hills in western Bengal, the Indhyádri between the Tápti and Godávari, the Nallamalai (Nullamullay) near Cuddapah, north-west of Madras, and the little metamorphic plateaux, such as the Shevaroys, Pachamalai, etc., scattered over the low country of the Carnatic, south-west of Madras.

The peculiarity of all the main dividing ranges of India is that they are merely plateaux, or portions of plateaux, which have escaped denudation. There is not throughout the length and breadth of the Peninsula, with the possible exception of the Arávalli, a single great range of mountains that coincides with a definite axis of elevation, not one, with the exception quoted, is along an anticlinal or synclinal ridge. Peninsular India is, in fact, a table-land worn away by sub-aerial denudation, perhaps to a minor extent on its margins by the sea, and the mountain chains are merely the dividing lines, left undenuded between different drainage areas. The Sahyádri range, the most important of all, consists to the northward of horizontal or nearly horizontal strata of basalt and similar rocks, cut into a steep scarp on the western side by denudation, and similarly eroded, though less abruptly, to the eastward. The highest summits, such as Mahábaleshwar (4,540 feet) are perfectly flat-topped, and are clearly undenuded remnants of a great elevated plain. South of about 16° north latitude, the horizontal igneous rocks disappear, the range is composed of ancient metamorphic strata, and here there is, in some places, a distinct connection between the strike of the foliation and the direction of the hills, but still the connection is only local and the dividing range consists either of the western scarp of the Mysore plateau, or of isolated hill groups, apparently owing their form to denudation. Where the rocks are so ancient as are those that form all the southern portion of the Sahyádri, it is almost impossible to say how far the original direction of the range is due to axes of disturbance; but the fact that all the principal elevations, such as the Nílgiris (Neilgherries), Palnís (Pulneys), etc., some peaks of which rise to over 8,000 feet, are plateaux, and not ridges, tends to show that denudation has played the principal share in determining their contour.

The southern portion of the Sahyádri range is entirely separated from the remainder by a broad gap, through which the railway from Madras to Beypur passes west of Coimbatore. The Anamalai, Palní, and Travancore hills, south of this gap, and the Shevaroy and many other hill groups scattered over the Carnatic, may be remnants of a table-land once united to the Mysore plateau, but separated from it and from each other by ancient marine denudation. Except the peculiar form of the hills, there is but little in favour of this view, but on the other hand there is nothing to indicate that the hill groups of the Carnatic and Travancore are areas of special elevation.

The whole of the western Sátpuras, from their western termination in the Rájpipla hills to Asírgarh, consist of basaltic traps, like the Sahyádri. It is true the bedding is not horizontal, but the dips are low and irregular, and have no marked connection with the direction of the range. The central Sátpuras, comprising the Pachmarhí or Mahádeva hills, from the gap in the range at Asírgarh to near Narsinghpur, are composed chiefly of horizontal, or nearly horizontal, traps, but partly of sandstones and of metamorphic rocks, and there is here again, as in the southern Sahyádri, some connection between the strike of the foliation in the latter and the direction of the ranges. The highest peaks, however—those of Pachmarhí (4,380 feet)—are of horizontal mesozoic sandstones. Farther east still the Sátpuras consist entirely of horizontal traps, terminating in the plateau of Amarkantak, east of Mandlá. East of this plateau there is, north of Biláspur, a broad expanse of undulating ground at a lower level, and farther to the eastward again rises the metamorphic plateau of Chutiá Nágpur, capped in places by masses of horizontal trap and laterite. These formations were apparently once continuous, across the low ground near Biláspur, with the same strata on an equal elevation at Amarkantak. Similar outliers occur on the Bundelkhand plateau, north of the Narbadá, all tending to the same conclusion—that the low valleys of central India are merely denudation hollows, cut by rain and rivers out of the original plateau of the Peninsula. The chief exceptions to this law—the instances in which the strike and dip of the rocks appear to have produced important effects on the contour of the country—are to be found amongst the metamorphic and transition formations.

It is true that some small ridges are formed of azoic and mesozoic sandstones, in places where the beds of these systems have been disturbed, but the only important lines of disturbance in either appear to be due to older axes of metamorphic foliation, and it is a rare case to find that the strike of the sandstones appears to have much effect upon the directions of the hills and valleys. A possible exception occurs in the Dámodar valley in Bengal, but even this is a disputed case, and the subject will be discussed in the chapters relating to the Gondwána system.

This remarkable absence in the Indian Peninsula of any evidence of disturbance in late geological times—a feature which abruptly distinguishes the whole area from the remainder of Asia—will be further noticed in the sequel; at present it is sufficient to remember that the principal mountain chains of the Indian Peninsula are, with one exception, not coincident with axes of disturbance or elevation, and to note the contrast in the extra peninsular area.

The Arávalli differs from the other great ranges of India in being entirely composed of disturbed rocks, with the axes of disturbance corresponding with the direction of the chain. The formations found in the Arávalli range belong to the transition rocks, and are of great antiquity; for the most part they are much altered, they are quite unfossiliferous, and there is evidence which renders it probable that the elevation of the range dates from a period anterior to the deposition of the Vindhyan rocks, themselves of unknown age but almost certainly not of later date than carboniferous, whilst the fact that these Vindhyan rocks are found almost horizontal in the neighbourhood of the Arávalli range, on both sides of the chain, shows that here, as elsewhere in the Peninsula, the forces which have affected the extra-peninsular area in later geological epochs have not been felt.

Passing to the other side of the Indo-Gangetic plain—no matter whether the region reached be to the westward in Sind and the Punjab, to the northward in the Himálayas, or to the eastward in Chittagong and Burma—the mountain ranges, with the exception of the Salt-range and the Assam range, are everywhere composed of disturbed and contorted beds, and the disturbance has invariably affected rocks of late geological age. The amount of alteration may be small or great, the hills may consist of simple anticlinal folds as in Sind, or of the most complicated inversions as in parts of the Himálayas, the strike of the bedding may vary from east and west to north and south, but two characters are constant—great disturbance affecting all the formations, and the coincidence of the direction of the ranges with synclinal and anticlinal axes.

The nomenclature of extra-peninsular mountain ranges is compassed with the same difficulty as those of the Peninsula, owing to the absence of local names except for individual ridges, peaks or passes, and has been further complicated by a want of unanimity among geographers as to the true limit of the term 'mountain range.' Different geographers have recognised from two to seven distinct ranges in that great system of mountains, collectively known as the Himálayas, which rises to the north of the Indo-Gangetic plain, and the opinions regarding the true western limit of the principal range have varied from that which regards it as ending in the Simla spur to that which looks on it as continuous with the Hindu Kush.

These contradictory opinions are all more or less correct, according to the limited point of view of the individual author, but if we look below the accidents of surface contour to the underground structure of this great mass of hills, we find in their geological structure and composition that they owe their elevation to a great series of earth movements, which must be regarded as a single and continuous system of disturbance, and this structural unity is now generally held to unite the separate chains into a single system of mountains to which the term of 'range' is inapplicable, unless we give it a wider application than is usual. The details of the physical geography of this great system of mountains, which stretches from the Indus to the Brahmaputra, have so important a bearing on the history of its elevation that they will be deferred to a subsequent chapter.

At its north-western extremity the great snowy range of the Himálayas bends round into the Hindu Kush, which runs south-westwards along the southern side of the upper Oxus valley. On their southern boundary the strike bends round to southwards at the valley of the Jehlam, and from this termination of the Himálayas proper there extends through Afghánistán and Balúchistán a complicated series of hill ranges of whose detailed geography very little is known.

The proper nomenclature and classification of these hill ranges is a difficult matter to determine, and, in the present state of our knowledge, both geographical and geological, it is impossible to arrive at a satisfactory result. An attempt has been made to classify these ranges according to the "system of disturbance" they belong to, which would hardly need mention were it not for the eminence of its author and the nature of the conclusions arrived at. In his great work "Das Antlitz der Erde," Prof. Suess has regarded the Salt-range of the Punjab as forming part of the foot hills of the Hindu Kush, and has united all the ranges between it and what is known to geographers as the Hindu Kush into one range, on the ground that they belong to the "Hindu Kush system of disturbance." This is still a matter for proof. We have no knowledge that there is a unity in the disturbance of the rocks composing these hills, comparable to that seen in the Himálayas, and until such knowledge is forthcoming it would be useless to adopt a system of nomenclature so opposed to that ordinarily current, and so certain to lead to needless confusion. As regards the Salt-range we shall shortly see that it ought to be classed by itself, and the individuality of the Safed Koh range, as depicted on our maps, is so marked that it is difficult to believe that it has not also an individuality of structure.

In one sense the whole of the ranges west of the Indus may be classed together, for there can be little doubt that they were contemporaneously elevated, and that the greater part of this upheaval, if not the whole, took place within the tertiary era. But, structurally, they may be divided into two classes, whose strike is about north and south and east and west, respectively, and these two alternate with, and pass into, each other in a manner that is at present not understood.

The most northerly range is the Safed Koh which runs eastwards from the neighbourhood of Kábul, forming the south side of the valley of the Kábul river, till it ends in British territory. Further south the Suláimán range is found striking north and south along the western frontier, and at its southern limit this bends round into the east and west running hills of the Bugti country, a strike which extends to near Quetta. The hills again take a southerly bend, and the Brahuik and Kirthar ranges run north and south. Further west the strike again changes, and in western Balúchistán the ranges run east and west.

At their eastern extremity the Himálayas are met, in a manner that has not been worked out, by a series of hills which at first strike south-westwards, afterwards bending round to a more southerly direction in the Nágá hills, where the principal ridge is known as the Patkoi. They run through the Manipur country southwards, till they are continuous with the range of hills, known as the Arakan Yoma, which lies between the Irawadi valley and the Bay of Bengal.

Besides these principal systems of hill ranges there are two minor ones to be noticed which, despite a total difference of the rocks they are composed of, show a considerable similarity in their structure and geographical position, subtending as they do the angles between the southern margin of the Himálayas and the hills which meet them at either extremity. The first of these is the Salt-range in the Punjab, the second that set of hills, called for convenience the Assam range, inhabited by the Gáro, Khási and Jaintia tribes. In both cases the hills are composed of a plateau with a steeply scarped face to the south, along which there is an axis of abrupt folding, accompanied by more or less faulting. This similarity of structure and position, in spite of great difference in the rocks of which they are composed, would seem to show that their elevation is a direct result of the same great series of movements of the earth's crust which resulted in the elevation of the hill ranges forming the extra-peninsular limits of the Indo-Gangetic alluvium, but they cannot be regarded as belonging to any of the mountain systems whose re-entering angles they subtend.

To the east of the Irawadi valley there is a great series of mountain chains, stretching southwards through the Shan states till it terminates in the Malay peninsula, of whose geography and geology even less is known than of the hills beyond the western frontier of India.

The drainage system of the extra-peninsular hills everywhere shows the peculiarity that, though the valleys often run along the strike of the hills for long distances, the streams and rivers always sooner or later break across the axes of maximum elevation. This peculiarity is very noticeable in the case of the Himálayas, whose river system must be deferred to a future chapter, but is in some respects more conspicuous in the case of the hills west of the Indus.

Here the ridges are mostly formed by the hard limestone cores of anticlinal folds through which the streams flow in narrow precipitous-sided

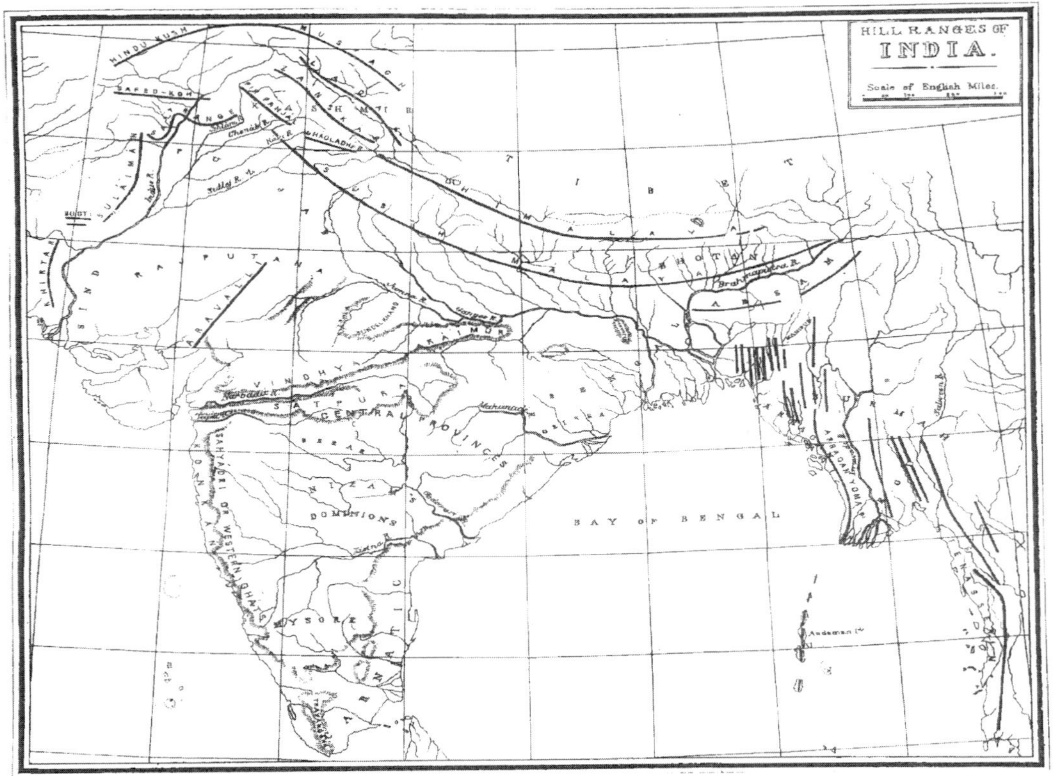
HILL RANGES OF
INDIA.
Scale of English Miles.
HINDU KUSH
SAFED-KOH
SULAIMAN
SALT RANGE
PIR PANJAL
DHAOLADHAR
KHIRTAR
SIND
RAJPUTANA
ARAVALLI
TIBET
HIMALAYA
BHOTAN
ASSAM
Brahmaputra R.
Ganges R.
Jumna R.
Chenab R.
Indus R.
Sutlej R.
KAIMUR
VINDHYA
SATPURA
CENTRAL PROVINCES
BERAR
NIZAM'S DOMINIONS
SAHYADRI OR WESTERN GHATS
KONKAN
MYSORE
CARNATIC
BENGAL
ORISSA
Mahanadi R.
Godavari R.
Kistna R.
BAY OF BENGAL
ARRACAN YOMA
BURMAH
Salween R.
TENASSERIM
Andaman Is.

gorges, locally known as 'tangi.' They form a very characteristic feature on the scenery of the hill country west of the Indus, and are commonly attributed to some disruptive force, which opened a way for the stream to flow through the mountain. Such is not, however, the true explanation, not only can it be disproved in many cases by the continuity of the solid rock across the stream bed, but it is inadequate, as it fails to account for the broad open valleys invariably found above the gorges. The fact is that they have been gradually cut down by the streams that flow through them, and the same absence of vegetation which increases the erosive power of rain wash on soft clays enabling broad valleys to be formed where they are exposed, diminishes its action on the hard limestones, at the same time it reduces chemical action to a minimum, and the absence of moisture almost entirely deprives the night frosts of their power to disintegrate the rock. As a consequence, the steep sides of the ravines cut by the streams, where they meet with compact rock, remain standing almost perpendicular, while in the intervening stretches of soft clay the valley widens out.

It must not be supposed that the whole country was elevated to the height of the crests of the ridges through which these "tangis" were cut. It will be shown in the sequel that the compression, contortion and consequent elevation of the hills was taking place at the same time that the valleys were being excavated, but sometimes the rate of elevation was too great for the streams, and areas of closed drainage were formed in which extensive alluvial and æolian deposits have been accumulated. These are particularly common in Balúchistán, where they are usually occupied by a broad expanse of wind-blown loess.

In the eastern hills the same features are to be seen, but, owing to the greater rainfall and dense vegetation, as well as to the different type of rock forming the hills, there is not the same abrupt alternation of broad open valley with deep and narrow gorges as on the west. The chemical action of the humic acids developed in the jungle-clad soil has smoothed off the steepness of the sides of the gorges, while the vegetation has protected the softer clays from being so easily washed away. As another result of the more rapid erosion of the stream beds there are no areas of closed drainage, but there are broad alluvial valley plains, such as those of Manipur and the Kubo valley, where differential movements of elevation of the beds of the streams have checked their velocity and compelled them to deposit their solid burden.

The river system of the Peninsula, omitting the drainage into the Ganges and small streams flowing to the west coast, is nearly all taken by six large rivers, of which two, the Narbadá and Tápti, drain the north-western portion and escape into the Gulf of Cambay, while the drainage of all the rest of the Peninsula, even from the crests of the Western Gháts within sight of the sea, flows eastwards by four great deltaic rivers, the Mahánadí, Godávari, Kistna, and Cauvery,—the only other streams of any importance being the northern and southern Penner.

This easterly trend of the drainage is probably of very ancient date, as there are patches of littoral marine deposits along the east coast, ranging as far back as the close of the jurassic period, which show that, since that period at least, the eastern coast of the Peninsula has maintained very much its present position. On the west coast no marine sediments older than the upper tertiary are known, if we except the cretaceous beds of the lower Narbadá valley. At the close of the Deccan trap period, that is the commencement of the tertiary era, dry land must have extended considerably west of the coast line; south of the trap area the evidence is only negative, but the absence of any large valleys draining in this direction suggests that the present position of the shore line is of more recent origin than that of the east coast, and that the earth movements which gave rise to it were either too slow, or more probably not of a nature, to change the easterly course of the drainage.

There can be no doubt that, beyond the limits of the Peninsula, there have been very great changes in the distribution of land and sea since the commencement of the tertiary era and, even in the latest part of it, the great disturbances which the rocks have undergone must have been accompanied by great changes of shore line. But when we come to the post-tertiary period and enquire whether, on the whole, there has been elevation or subsidence, the evidence is contradictory. In the alluvium of the Gangetic delta, and near Pondicherri, beds of peat, at various levels below the surface of the ground, show that there has been subsidence, but this is the usual, if not invariable, condition in a delta, and it is more than probable that all the large deltas along the coast are being gradually depressed.

Along the non-deltaic portions of the coast evidences of sub-recent elevation are found in coral reefs and marine deposits raised above the present level of the sea. The low level laterite of the east coast lies on a gentle slope of the older rocks, unaffected by subaerial erosion, such as is formed by the sea, and must have been deposited either before or shortly after this was raised above sea-level.

The escarpment of the Sahyádri range—a remarkable feature of the hills parallel to the western coast of the Peninsula—has frequently been noticed as furnishing evidence of a rise of land. Throughout the trap country of the Bombay presidency, the Western Gháts rise from the Konkan in an almost unbroken wall, varying in height from 2,000 to 4,000 feet, cut back in places by streams, projecting here and there into long promontories, but preserving throughout a singular resemblance to sea cliffs. This resemblance, however, ceases to a great extent to the southward, where the metamorphic rocks replace the horizontal basaltic traps. The escarpments of the Málwá plateau, north of the Narbadá, and of the Deccan plateau, south of

Khándesh, although far inferior in elevation to the scarp of the Sahyádri, resemble the latter too closely in appearance to justify the assumption, without further evidence, that the cliffs of the Western Gháts are of marine origin. The parallelism of the Sahyádri escarpment to the sea-coast is suggestive of a connection between the two, and this connection is strengthened by the facts that a thickness of at least 4,000 feet of bedded trap has been removed from the surface of the Bombay Konkan, and that the plane of marine denudation, already mentioned as supporting the low level laterite, extends in places nearly to the foot of the scarp. The circumstance that the hills of the Sahyádri are inhabited by certain fresh-water mollusca belonging to the genus *Cremnoconchus*, which is unknown elsewhere and is so closely allied to Indian forms of the littoral marine genus *Littorina* as to render it probable that both are descended from the same ancestors, also tends to strengthen the view that the Sahyádri mountains were formerly washed by the sea. But it is certain that great denudation has taken place since the scarp was a sea cliff, and it is far from improbable that, if the sea ever extended to the base of the Western Gháts, the epoch belonged rather to tertiary than post-tertiary times. It is also possible that the isolation of the different hill ranges of Southern India, and the denudation of the Pálghát Gap, south of the Nílgiri plateau, are due in part to ancient marine action of the same date as the formation of the Sahyádri escarpment. In this case, as in so many others connected with Indian geology, all that is now possible is to suggest probable interpretations of phenomena, and to leave them for future exploration to confirm or contradict.

On the other hand, a sudden deepening of the sea, at a distance of 10 to 20 miles from the shore, along the Mekrán coast, has been supposed to represent a submerged cliff. More positive evidence of recent subsidence is to be found in the occurrence of a number of trees imbedded of mud, in the spot where they grew, at a depth of 12 feet below low-water mark on the east side of the island of Bombay,[1] and in the submerged forest at the western end of the Valimukam bay on the Tinnevelli coast described by Mr. Foote.[2]

Local alterations of level, accompanied by earthquakes, are known to have occurred on at least one occasion, namely the great earthquake of Cutch in 1819, when a considerable area in the Rann of Cutch was suddenly submerged.[3] A more doubtful instance is the elevation and subsidence which is said to have taken place on the Arakan coast in the middle of the last century, presumably during the great earthquake of 1762. A raised beach which is 9 feet above sea-level at Foul island and 22 on the

[1] *Records*, XI, 302, (1878).

[2] *Memoirs*, XX, 82, (1883).

[3] McMurdo, *Trans. Lit. Soc., Bombay*, III, 90, (1823); Lyell, "*Principles of Geology.*" See also A. B. Wynne, *Memoirs*, IX, 29, (1872).

north-west of Cheduba island, has been attributed to the effects of this earthquake which further north is said to have caused the permanent submergence of 60 square miles near Chittagong.[1]

Away from the sea coast, the Andaman and Nicobar islands have certainly at one time been connected with Arakan, and the intricate channels and long ramifying fjords which penetrate the great Andaman and adjoining islands indicate a considerable submergence. Along the coast there are, however, indications of minor oscillations of level, both upwards and downwards, within the recent period, the last movement being probably one of subsidence.[2] Off the west coast of India the coral archipelagoes of the Laccadive and Maldive islands probably mark the site of submerged land, though this is a matter still under dispute.

Besides the changes produced by rise and fall of the sea-level as compared with that of the land, there have been minor modifications of

Fig. 1.—Pagoda on the sea-shore at Tranquebar.

the shore line due to erosion and accretion of land. St. Thomé, a short distance south of Madras, is said to have formerly been situated 12 leagues inland and, 40 miles further south, the town of Mahábalipur is said to have been overwhelmed by the sea.[3] Still further south, erosion of the sea beach at Tranquebar is well attested by old records as well as the destruction of a large portion of an old pagoda, whose eastern gate tower had been partially destroyed in 1859[4] and has probably now been completely removed.

[1] *Phil Trans.*, VIII, 251, (1763); G. P. Halstead, *Jour. As. Soc., Bengal*, X, 433, (1841); F. R. Mallet, *Records*, XI, 190, (1878).

[2] S. Kurz, "Report on the Vegetation of the Andaman Islands," Calcutta, 1870; R. D. Oldham, *Records*, XVIII, 143, (1885).

[3] T. J. Newbold, *Jour. Roy. As. Soc.*, VIII, 250, (1846).

[4] W. King, *Memoirs*, IV, 362, (1864).

Evidence of the advance of land is to be found on the Tinnevelli coast, where the deserted town of Korkai, now five miles inland, has been identified with the "Kolkoi Emporium" of the classical geographers. About 600 B.C. this town was the capital of a kingdom and apparently an important sea-port. By the time that Marco Polo visited this coast in 1292 A.D., the advance of the land had necessitated the abandonment of the old port and the establishment of a new one at "Cail," a town which also has decayed and was forgotten till its site was discovered and re-

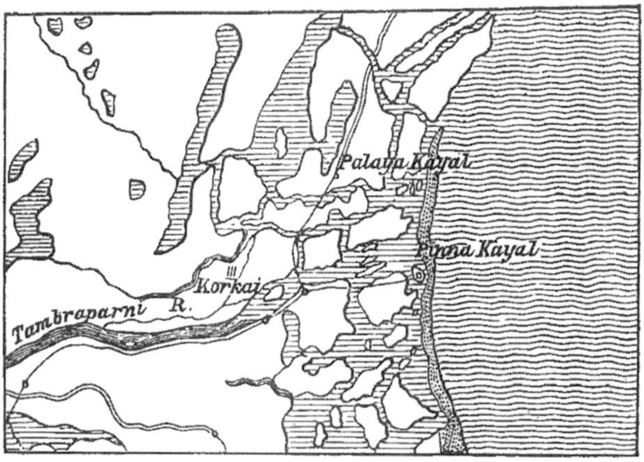

Fig. 2.—Part of Tinevelly coast shewing present position of old seaports.

cognised by Bishop Caldwell in the modern village of Kayal, and made public in Colonel Yule's edition of the travels of Marco Polo.[1]

A similar advance of the shore line is said to have taken place on the east coast of the Gulf of Cambay, and it is said that the Rann of Cutch was once a gulf of the sea with sea-ports on its shores, and that remains of ships have been found imbedded in the mud[2]. The Rann is now a sort of debateable land, being flooded during the south-west monsoon and a dry barren mud flat during the rest of the year; the change, which has indubitably taken place, if not so recently as has been supposed, was doubtless due to silting up, partly aided by a slight elevation of the land.

The evidences of alterations of level along the sea coast, which have been detailed above, point to a slight elevation of the land during the post-tertiary period, though too small to have any appreciable influence on the climate.

[1] 1st edition, Vol. II, p. 307, (1871).

[2] *Trans. Geog. Soc., Bombay*, XVIII, pp. lvi, lxix, lxxxv, (1868); A. B. Wynne, *Memoirs*, IX, 26, (1872).

This is important in its bearing on the evidence that there is of the cold of the glacial period having been felt in the peninsula of India There is no physical evidence, so far as is known, of a geologically recent cold epoch, and some geologists have doubted whether India was affected by the glacial period. In the Himálayas there is everywhere abundant evidence of the glaciers having extended to lower levels than they reach. Grooved and polished rock surfaces have been found now at as low a level as 7,500 feet in Pángi,[1] and in a higher latitude large boulders are found imbedded in the fine silt of the Potwar at an elevation of less than 2,000 feet above the sea.[2] Besides these there are many cases of large erratic blocks and supposed morraines which have been referred by some observers to glaciers and by others to river action. The positive and unmistakeable proofs of a period colder than the present are sufficient to enable us to discard all the more doubtful evidence, and more recent investigations have shown that it cannot be attributed, as was once suggested,[3] to a former greater elevation of the Himálayas than they now attain.

These indications do not point to a sufficient diminution of temperature of the Himálayas to make it probable, or even possible, that there should be any actual physical proofs of the glacial period having been felt in the Peninsula, and it is of importance to ascertain whether there is any collateral evidence of a cold period having affected India in later tertiary or post-tertiary times, it being remembered that a general refrigeration of the earth's surface, sufficient to produce an arctic climate in Europe, would not diminish the temperature of the Indian Peninsula beyond the average of the temperate zone at the present day.

The argument is, briefly, as follows. On several isolated hill ranges, such as the Nílgiri, Anamalai, Shevaroys and other isolated plateaux in Southern India, and on the mountains of Ceylon, there is found a temperate fauna and flora, which does not exist in the low plains of Southern India, but is closely allied to the temperate fauna and flora of the Himálayas, the Assam range (Gáro, Khási, and Nágá hills), the mountains of the Malay peninsula, and of Java. Even on isolated peaks, such as Párasnáth (4,500 feet high in Behar) and on Mount Abú in the Arávalli range, several Himálayan plants exist. It would take up too much space to enter into details; the occurrence of a Himálayan plant like *Rhododendron arboreum*, and of a Himálayan mammal like *Martes flavigula* on both the Nílgiris and Ceylon mountains, will serve as an example of a considerable number of less easily recognised species. In some cases there is a closer resemblance between the temperate forms found on the peninsular

[1] C. A. McMahon, *Records*, XIV, 310, (1881).
[2] W. Theobald, *Records*, X, 140, (1877).
[3] H. B. Medlicott, *Memoirs*, III, pt. ii, p. 155, (1864).

hills and those on the Assam range[1] than between the former and Himálayan species, but there are also connections between the Himálayan and peninsular temperate regions which do not extend to the eastern hills. The most remarkable of these is the occurrence on the Nílgiri and Anamalai ranges and on some hills further south, of a species of wild goat (*Capra hylocrius*), belonging to a sub-genus (*Hemitragus*), of which the only other known species, *C. jemlaica*, inhabits the temperate region of the Himálayas from Kashmir to Bhután. This case is remarkable, because the only other wild goat found completely outside the Palæarctic region is another isolated form on the mountains of Abyssinia.

The range in elevation of the temperate fauna and flora of the oriental region in general appears to depend more on humidity than temperature, many forms which are peculiar to the higher ranges in the Indian hills being found represented by allied species at lower elevations in the damp Malay peninsula and archipelago, and some of the hill forms are even found in the damp forests of the Malabar coast. The animals inhabiting the Peninsular and Singalese hills belong, for the most part, to species distinct from those found in the Himálaya and Assam ranges. In some cases even genera are peculiar to the hills of Ceylon and Southern India, and one family of snakes is unrepresented elsewhere. There are, however, numerous plants and a few animals inhabiting the hills of Southern India and Ceylon, which are identical with Himálayan and Assamese hill forms, but which are unknown throughout the plains of India.

That a great portion of the temperate fauna and flora of the Southern Indian hills has inhabited the country from a much more distant epoch than the glacial period may be considered as almost certain, there being so many peculiar forms. It is possible that the species common to Ceylon, the Nílgiris, and the Himálayas, may have migrated at a time when the country was damper without the temperature being lower, but it is difficult to understand how the plains of India can have enjoyed a damper climate without either depression, which would have caused a large portion of the country to be covered by sea, a diminished temperature which would check evaporation, or a change in the prevailing winds. The depression may have taken place, but the migration of animals and plants from the Himálayas to Ceylon would have been prevented, rather than aided, if the southern area had been isolated by sea, so that it may be safely inferred that the period of migration and the period of depression

[1] Only one species of plant, however, is mentioned by Hooker and Thomson ("Introductory Essay to the Flora Indica", p. 238) as being found both in the Khási hills and Nílgiris, but not in the Himálayas. One land-shell at least, *Bulimus nilagiricus*, has the same distribution, and the genus *Streptaxis* is found in Burma, the Khási hills, and the Southern Indian ranges, but not in the Himálaya west of Bhután. Several other instances might be quoted.

were not contemporaneous. A change in the prevailing winds is improbable so long as the present distribution of land and water exists, and the only remaining theory, to account for the existence of the same species of animals and plants on the Himálayas and the hills of southern India, is depression of temperature.[1]

The Indian empire can boast of one volcano, which is at present dormant, but has been in active eruption within the century. Barren island in the Bay of Bengal is not only a perfect model of a volcano, but is classical in the history of geological controversy. It has been repeatedly referred to and described by geological writers, but the earlier references were all more or less inaccurate, and it is only within the last few years that a careful description of it has been given by Mr. F. R. Mallet.[2]

The volcano of Barren island has an irregularly circular form of about 2 miles in diameter, composed of an outer rim, rising to a height of from 700 to 1,000 feet above the sea and surrounding a slightly elliptical amphitheatre, whose larger axis runs north-east and south-west with a length of about 9 furlongs. From the centre of this a cone of remarkably regular form rises to a height of 1,015 feet above the sea. The depression, where the slope of this cone and the inner slope of the amphitheatre meet, has a maximum elevation of 300 feet above the sea, and is almost entirely occupied by three distinct lava streams. It is evident that the island was once much higher than it is at present, and the outer rim marks the limit of the crater, produced by some great paroxysmal eruption which blew away all the upper portion of the old cone. The bottom of this old crater must have been much below the level of the sea and, for a time, before the new cone attained its present dimensions, the sea must, as is represented in the older descriptions of the island, have flowed round its base, between it and the foot of the internal slope of the amphitheatre, but there is no authentic record of any one ever having seen this stage. In 1789, when Blair saw the island, the sea did not penetrate into the amphitheatre, which had all been filled up to above sea-level. The volcano was then in active eruption, throwing out blocks and scoriæ, and it may be that it had been visited at some earlier

[1] The above is a meagre and condensed account of a very interesting subject, which requires further enquiry. One possible objection may be answered at once. It is true that many of the temperate damp-loving forms of the Nílgiris and Ceylon hills are forest forms, and it may be urged that they might have migrated when the plains of India were covered with forest. But, judging from what remains of the forest on the plains of the Carnatic, Deccan, Central Provinces, etc., the flora, even when the whole was forest, differed so widely from that of the hills, that it is improbable that any general diffusion of hill species could have taken place without a change of climate.

In a subsequent chapter reference will be made to the probable influence of the glacial epoch on the Siwálik mammalian fauna.

[2] *Memoirs*, XXI, 251, ff, (1885).

period, before the hollow was completely filled up, and that the only record left of this visit is to be found in the erroneous description which was at one time current in text-books.

It must be remembered that the portion of the volcano above sea-level, which is all that has been referred to in the description, is but an insignificant portion of its whole bulk. Soundings taken by Captain Carpenter, show that the cone rises from a depth of 800 fathoms below the sea, and that the total height is consequently some 6,000 feet at present, or was 8,000 feet before the upper part of the outer cone was blown away.[1]

At the time of Blair's visit there appears to have been no lava stream in the gap where the outer rim is breached, but in 1832 the lava was there, and still so hot that the water in contact with it was boiling. Since that period the flow has cooled down and the temperature of the water, which percolates beneath the lava and issues as a spring on the sea shore, has steadily diminished at each visit, till it was no more than 110° F. in 1886.[2] It seems certain, therefore, that this lava flow was poured out later than 1789, and probably within the present century.

Seventy-five miles north-north-east of Barren island lies the island of Narcondam, indubitably of volcanic origin like the former, but composed almost, if not quite, entirely of hornblende andesite lava with little or no volcanic ash. It is not certain whether this volcano ever had a crater, as it may have been of the so-called endogenous type, formed by the quiet extrusion of lavas unaccompanied by any crater-forming materials. The complete obliteration of the crater, if there ever was one, is in itself an indication of the period for which the volcano has been extinct, and in any case the deep ravines, with which its sides are scored, are an equally eloquent testimony of the time during which subaerial denudation has been uninterruptedly at work, so that this volcano has probably been longer extinct than either of the two that follow.

About 50 miles north-north-west of Yenangyoung and 25 to 30 miles east-south-east of Pagán, both large towns on the Irawadi, the extinct volcano of Puppa[3] rises to a height of about 3,000 feet above the undulating country composed of pliocene sands and gravels. The mountain has preserved its original form to some extent, but the crater has been greatly broken down by denudation, and the rim completely cut away at one point, where the drainage from the interior has made itself a means of exit. The peak consists of ash breccia, but lava flows, mostly trachytic, form the lower slopes and the surfaces around the base of the volcano. Among these flows are some of a very beautiful porphyry, with crystals of pyroxene.

The horizontal beds of gravels and sands around the base of the volcano

[1] *Records*, XX, 46, (1887).

[2] *Records*, XX, 48, (1887).

[3] W. T. Blanford, *Jour. As. Soc., Bengal*, XXXI, 215, (1862).

contain fossil wood and ferruginous concretions, and apparently belong to the pliocene fossil wood group. They are capped by the lava flows, contain pumice and volcanic fragments, and, in one place, a bed of ash breccia was found interstratified with them. It appears highly probable, therefore, that the volcano was active in pliocene times, but it may have continued to emit lava and scoriæ at a later period.

Far to the north the extinct volcano of Hawshuenshan near Momien in Yunnán has been described by Dr. Anderson[1]; and near Kanni, on the Chindwin, Dr. Noetling observed basalt breaking through the pliocene sandstones and forming a cone on their surface, but no detailed notice of this locality has been published.

In this connection mention may be made of a mass of trachyte which is found about four miles east by north of the village of Byangyi on the Bassein river and some 30 miles south of Bassein town. It is about six feet in diameter, there is no reason to suppose that it has been transported from a distance and no similar rock is known anywhere else in the province. No rock is seen in contact with the trachyte, but unaltered shales and sandstones of upper nummulitic age are seen not far off, dipping at low angles. Close to the block itself are some fragments that have flaked off, and among them a piece of shale, which had a somewhat baked appearance on one side, was found. This, taken in conjunction with the fact that it lies on the same general line as the volcanoes just described, suggested the idea that it forms part of an intrusive neck,[2] but the true relations of the mass are obscure.

It will be noticed that these old volcanoes lie along a line which, if continued to the south, would be continuous with the general direction of the great chain of volcanoes running through the islands of Java and Sumatra in the Malay peninsula, and this suggests that they form the northern termination of what is known as the Sunda chain of volcanoes. The observation is interesting and important in view of the fact that this chain has been supposed to find its final expression in the pseudo-volcanic phenomena on the Arakan coast which are described below.

But before passing on to this subject it will be well to notice some doubtful cases of volcanic action in the Indian Peninsula and on its shores. In 1756 a submarine eruption is said to have taken place off the coast of Pondicherri, which threw up large quantities of ashes and pumice and formed an island half a league long and of the same breadth. No exact details of locality are given, but the account is a very circumstantial one[3] and, unless a pure fiction, must refer to a true volcanic eruption. It may

[1] Report on the expedition to Western Yunnan, Calcutta, 1871, p. 87.

[2] *Memoirs*, X, 330, (1873).

[3] "Asiatic Annual Register," 1758, reprinted in *Jour. As. Soc., Bengal*, XVI, 500, (1847).

be noticed that the Admiralty chart of the Bay of Bengal marks a sounding of 5 fathoms in east longitnde 80° 42′, north latitude 12° 46′, with the remark 'Doubtful'; the position would agree sufficiently well with that indicated in the account, and the depth is that which would be produced by the action of the waves.

This also is the best place to notice a very curious crateriform lake, situated in the interior of the Indian Peninsula, near the village of Lonár, about 40 miles east by north of Jálna in the northern part of the Nizám's territory, and about half-way between Bombay and Nágpur. The surrounding country for hundreds of miles consists entirely of Deccan trap and in this rock there is a nearly circular hollow, about 300 to 400 feet deep and rather more than a mile in diameter, containing at the bottom a shallow lake of salt water without any outlet, whose water deposits crystals of sesquicarbonate of soda. The sides of the hollow to the north and north-east are absolutely level with the surrounding country, whilst in all other directions there is a raised rim, never exceeding 100 feet in height and frequently only 40 or 50, composed of blocks of basalt, irregularly piled, and precisely similar to the rock exposed on the sides of the hollow. The dip of the surrounding traps is away from the hollow, but very low.[1]

It is difficult to ascribe this hollow to any other cause than volcanic explosion, as no such excavation could be produced by any known form of aqueous denudation, and the raised rim of loose blocks around the edge appears to preclude the idea of a simple depression. It is true that there is no sign of any eruption having accompanied the formation of the crater, no dyke can be traced in the surrounding rocks, no lava or scoriæ of later age than the Deccan trap period can be found in the neighbourhood. The raised rim is very small, and cannot contain a thousandth part of the rock ejected from the crater, but it is impossible to say how much was reduced to fine powder and scattered to a distance, or removed by denudation.

Assuming that this extraordinary hollow is due to volcanic explosions, the date of its origin still remains to be determined. That this is long posterior to the epoch of the Deccan traps is manifest, for the hollow appears to have been made in the present surface of the country, carved out by ages of denudation from the old lava flows. To all appearance the Lonár lake crater is of comparatively recent origin, and if so it suggests that, in one isolated spot in India, a singularly violent explosive action must have taken place, unaccompanied by the eruption of

[1] Malcolmson, *Geol. Trans.*, 2nd series, V, 562, (1840); Newbold, *Jour., Roy. As. Soc.*, IX, 40, (1848) (with this paper there is a fairly executed view of the lake);—G. Smith, *Mad. Jour. Lit. Sci.*, XVII, 1, (1856). See also *Records*, I, 63, (1868), where other references are given.

melted rock. Nothing similar is known to occur elsewhere in the Indian Peninsula.

Associated with true volcanoes in name at least, even if, as is held by many geologists, in nothing else than name, are mud volcanoes, of which two principal groups are known, in Burma on the east, and Balúchistán on the west, of India, respectively.

Of the Burman ones the best known are those of Minbu on the Irawadi, and those of the islands of Ramrí and Cheduba on the Arakan coast.[1] A few others are reported, but they are small and isolated, and consist only of temporary outbursts.

The Ramrí mud volcanoes are more interesting than the others, since they alone, so far as is known, are subject to paroxysmal eruptions of great violence, and from them alone stones have been ejected and flames emitted. Some of the principal phenomena may be briefly described here. There are about a dozen or rather more vents in Ramrí island itself, more than half that number in Cheduba, and a few in the other neighbouring islands. Near Kyauk-pyú in Ramrí, six occur in a line, within a distance of about a mile and a half along the summit of a low broad ridge.

Many of the vents consist of truncated cones, built up of the dried mud ejected by outbursts of gas. The crater, filled with more or less liquid or viscid mud through which the gas escapes, occupies the top of the conical hillock. The majority, however, of the Ramrí mud volcanoes consist of mounds, composed on the surface of angular fragments of rock and having scattered over them a few small mud cones with craters at the top, varying in height from a few inches to eight or ten feet. When gas ceases to be omitted from a vent, the mud is rapidly washed away by rain and there remains a low mound, composed of angular fragments of rock which were ejected together with the mud, and the repetition of a similar process accounts for the formation of the mounds. The mounds in Ramrí are from 50 to 100 yards in diameter, with a height of from 15 to 30 feet, two, of exceptional size, near Pagoda hill in Cheduba, being 200 to 250 yards across. The cones in which the mud is viscid are very steep, being built up partly of small quantities of mud, spurted out by the evolution of gas so as to form a hard rim round the mud crater, partly of mud poured out from the crater down the slopes through broken portions of the rim.

Besides the gas and mud, a small quantity of petroleum is usually discharged from the vents. The gas consists mainly of marsh gas (light carburetted hydrogen), probably mixed with some of the more volatile hydrocarbons usually associated with petroleum. The mud is simply the

[1] For a description of the mud volcanoes of Minbu by Dr. Oldham, see Yule's *Narrative of the Mission to the Court of Ava in 1855*, appendix, p. 339. The Ramrí and Cheduba mud volcanoes are described, with full references to earlier accounts, by Mr. Mallet in *Records*, XI, 188, (1878). Sketches of the cones are given in both cases.

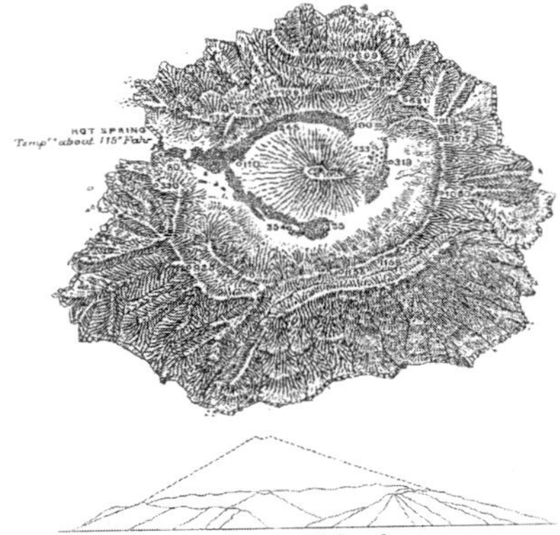

Barren Island

Nareondam Island

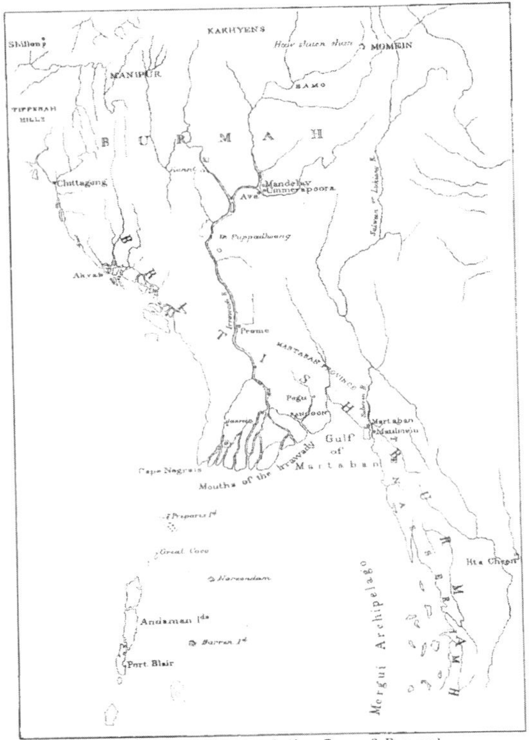

Volcanoes of Burma & the Bay of Bengal.

grey shale or clay of the tertiary rocks, mixed with water containing some salt in solution.

The association of petroleum with large quantities of marsh gas, and the frequent emission of, usually saline, water and of gas in abundance from borings for mineral oil are too well known to require the recapitulation of details. Both petroleum and gas are known to be found in many localities amongst the tertiary rocks of extra-peninsular India and Burma. Not unfrequently both gas and mineral oil issue with water in the form of a bubbling spring being, perhaps, forced to the surface by the pressure of the gas. Whether a spring of this kind forms a "mud volcano" or not depends evidently on the nature of the beds traversed on the way to the surface. As a rule the mud in these vents is either of the same temperature as the air or a little higher, but in the Ramrí craters a higher temperature has been recorded after the more violent eruptions.

These paroxysmal eruptions appear to occur at irregular intervals, are at times very violent and appear frequently to accompany earthquakes. Mud and stones are shot out with great force, accompanied by large quantities of inflammable gas, which in many cases takes fire and lights up the country for miles around. Some of these eruptions of ignited gas have taken place at sea off the coast of Ramrí, and in one case a small island was formed near False island, south of Ramrí and south-east of Cheduba, but it was soon washed away again. The ejected stones are in all cases fragments of the tertiary rocks, chiefly shale or sandstone, some of them being from half a cubic foot to a cubic foot in size and a few are larger, but the majority range from half an inch to five or six inches in diameter. The ignition of the gas is ascribed by Mr. Mallet to frictional electricity,[1] and not to the high temperature at which the various ejecta issue, the fact that the stones, and even fragments of lignite, thrown out during eruptions are, as a rule, entirely unchanged by heat, proving that the gas is not in a heated state previous to emission. A very few rare and exceptional fragments of burnt and reddened shale have been found, and these have probably been calcined by the flaming gas.

It should be noticed that even the fiery eruptions of the mud volcanoes of Arakan have nothing in common with the igneous outbursts of true volcanoes. In the former gas is emitted at, in all probability, a low temperature and is ignited, or perhaps occasionally exploded, when mixed with atmospheric air. In the latter red-hot lava and scoriæ are ejected, and the appearance of flame is due either to the high temperature of the substances

[1] *Records*, XI, 202. Mr. Mallet points out that the principle of the hydro-electric machine, in which the production of electricity of high tension is due to the issue of partially condensed steam through small orifices of such form as to produce great friction, is similar to that of violent evolution of gas from such vents as those of Ramrí. He also notices the well-known fact that lightning often accompanies volcanic eruptions.

projected into the air, or to the reflection of masses of glowing lava by condensed vapours or by clouds of volcanic dust.

One mud volcano is known in eastern Assam, but none have been found in the Himálayas or in the Peninsula of India. On the Balúchistán coast in Lus a number have been described, which do not appear to be subject to the same paroxysmal eruptions as those of Ramrí and attain a greater size, the largest being over 300 feet in height. How far this is due to the absence of paroxysmal eruption or to a greater constancy of the vents, and how far to the rainless nature of the climate, is not clear, but in all other respects they resemble those already described.

CHAPTER II.

METAMORPHIC AND CRYSTALLINE ROCKS.

Gneissose rocks of the Peninsula of two ages—The older gneiss—The newer gneiss—Bundelkhand gneiss—Newer gneiss of Singrauli—Contrast with gneiss of Bundelkhand—Chutiá Nágpur—Orissa—Central Provinces—Haiderábád—Nellore—Bellary—The southern Konkan—Malabar—Nilgiris—Trichinopoli and Arcot—Madura and Tinnevelli—Instrusive trap of Southern India—Gneissose rocks of the Arávallis—Extra-peninsular India—Afghánistán—Himálayas—Assam—Hills east of the Irawadi and Malay Peninsula.

The oldest rocks in India, and consequently the first to be considered in the arrangement adopted in this book, are those belonging to the great series of gneissose rocks, which covers a larger area in the Peninsula of India than all the newer formations put together, and forms the foundation on which they are built up. In spite of the great interest of the numerous problems presented by these rocks, they have as yet received but little attention from the Geological Survey, its energies having been mainly directed towards those districts where the most important results were to be expected in the shortest time, and we have consequently but little detailed information regarding the gneissose rocks, except in the immediate neighbourhood of the sedimentary formations.

How imperfect is our present knowledge of these rocks may be judged from the fact that it has been found impossible to distinguish, on the accompanying map, between true gneiss and intrusive granite. It is only within late years that the frequency with which granite has acquired gneissose characters has been recognised, and it is highly probable not only that some large exposures, of what has been described as granitoid or massive gneiss, are composed of gneissose granite, but that what háve in many cases been described as beds of gneiss, interbedded with non-felspathic schists, are in reality veins of granite which have been intruded along the foliation planes, and subsequently acquired a parallel arrangement of their constituent minerals. Under these circumstances it is impossible to attempt anything like a complete or connected account of the gneissose rocks, and the following pages must be taken as the merest sketch, whose main use will be to show how much has yet to be learnt.

In spite, however, of the imperfect state of our knowledge, certain leading features have come out from the investigation, so far as it has gone, and the most important of these is the recognition of the fact that the gneissose rocks do not belong to a single system, but are comprised in two or more distinct systems differing in age from each other.

The oldest of these, often described as the Bundelkhand gneiss from its having been first recognised in the country of that name, is characterised by its massive structure, with the foliation generally but obscurely developed, and the extreme rarity of accessory minerals. It has been recognised in Bundelkhand, in the gneissose inliers of Singbhúm and in the Bellary district, and it is possible that the massive granitoid gneisses of other parts of Southern India may belong to the same system. In the first and last named districts, and apparently also in the second, it formed the floor on which the oldest distinctly sedimentary beds of the Peninsula were deposited, showing that it must, even at that remote period, have been an ancient rock which had undergone a great amount of sub-aerial denudation. These facts indicate a greater unity of character than can be found in the very diverse characters of what are believed to be the newer gneisses, and seem to show that the Bundelkhand gneiss may be considered as a true rock system.

The second, or newer, type of gneiss exhibits a much better developed foliation as a rule, it is at times interbedded with schists and is distinguished from the first by the abundance and variety of the accessory minerals it contains. The rocks of this division are looked upon as newer than those already mentioned, firstly, because no unconformable contact of original deposition, between them and the sedimentary beds of the older transition systems, has been observed, secondly, because, where they are in contact with the latter, bands of gneiss have in several places been observed apparently interstratified with the distinctly sedimentary beds, and thirdly, because the general type of rock is, according to theories that are widely held, newer than that of the Bundelkhand gneiss. All three of these arguments are open to dispute. As regards the first it may be observed that the presence of a contact of original unconformity with the transition beds has of itself been held sufficient to show that the gneiss belongs to the older class, as regards the second, there is no evidence in the recorded observations to show that the supposed interbedded gneiss is not gneissose granite intrusive along the planes of bedding, or a foliated arkose, and as regards the third, the theories are by no means completely established. Still, though absolute proof is wanting, there is a distinct balance of evidence in favour of the more foliated gneisses being of newer origin than the massive forms, and this is in accordance with the type which the foliated gneisses and their accompanying beds of crystalline schist exhibit, intermediate between that of the granitoid gneiss on the one hand and the distinctly sedimentary, though metamorphosed, beds of the transition systems on the other.

In the only critical and detailed examination of the gneisses of the Peninsula that we have at present,[1] Lacroix has suggested the existence of groups of successive origin among the gneisses of the Salem district and Ceylon. His investigations were based entirely on specimens whose localities had been but imperfectly recorded, and the suggestion still lacks that confirmation from detailed observations in the field which alone could render it authoritative. It is, however, probable, as will appear in the sequel, that there are more than two successive series of gneisses represented in the Peninsula of India.

After these introductory remarks it will be best to proceed to a brief review of the principal points of our knowledge of the gneisses in different districts of the Peninsula, commencing with that of Bundelkhand.

This gneiss forms the basis of lower Bundelkhand, as distinguished from the higher portions of that district lying on the adjoining Vindhyan plateau. On the north-north-east border, for 200 miles, the gneiss is gradually covered by the superficial deposits, forming outlying and marginal portions of the Gangetic plains, at an elevation of 500 to 600 feet above the sea. Elsewhere the area is very sharply bounded by a scarp of overlying formations, whether Vindhyan or transition. Along the base of the scarp to the south-west the elevation varies between 1,000 and 1,200 feet above the sea, the scarp itself rising to 1,900. The gneiss sometimes forms hills, but the general features of the ground are flat, undulating uplands, sparsely cultivated, including shallow valleys and plains of alluvial land.

Gneiss is by far the commonest rock.[2] It may be described as consisting of some six minerals—red orthoclase felspar, a white plagioclastic felspar (probably oligoclase), quartz, hornblende, chlorite, and mica. In any one place the gneiss may contain only two of these, or it may include all of them. With regard to texture every variety is met with, from a homogeneous felstone, in which no individual mineral can be distinguished even with a lens, to a coarsely porphyritic rock, including felspar crystals more than two inches long. The orthoclase nearly always forms the main mass, and exceeds in quantity all the other minerals together. Its ordinary colour is a darkish red, but now and then it is much paler, and almost or quite white. In such cases it is not easy to distinguish the felspars on a fresh fracture, when both occur, but the difference becomes apparent on a weathered surface. The orthoclase has a vitreous, or somewhat pearly, lustre and translucent aspect, the other is

[1] *Bull. Soc. Franç. Minéral.*, XII, 83, (1889); *Records*, XXIV, 155, (1891).

[2] The characters of the rocks of this area are chiefly taken from the unpublished reports of Mr. F. R. Mallet.

dull and quite opaque, having been superficially altered into kaolin. The plagioclastic felspar is a comparatively uncommon mineral, when present it plays quite a subordinate part, and occurs in much smaller crystals than the orthoclase. As often as not, no free quartz can be detected in the gneiss, and it very rarely indeed occurs in large quantity. The rock is usually hornblendic, but sometimes the hornblende is partially or entirely replaced by chlorite and mica. The mica is of more than one species, both uniaxial and biaxial, and of varying colours, green, brown, black, and silvery-white, the last being rare and chiefly found in the southern part of the area.

Foliation is seldom well developed, the rock being very commonly to all appearance perfect granite, but no direct evidence has been obtained, beyond this, of the existence of true granite, and sometimes, by close examination, traces of foliation may be detected in rock which at first sight appears quite devoid of it. This foliation has generally an east-north-east direction, but varies to north-east and east-south-east. The planes are more or less vertical.

The subordinate varieties of rock, all combined, are of very insignificant extent in comparison to the gneiss. One of the most prominent is hornblende rock, which sometimes resembles trap very closely, parts of it being as fine-grained as the intrusive dykes of diorite, or the overflowing basaltic trap, and weathering into similarly rounded lumps, but this variety passes into a much coarser kind in which the felspar and quartz are well separated. The most trappean looking portions, moreover, contain thin strings and films of epidote, which have not been observed in the trap. Another well-marked variety contains about equal quantities of hornblende and white or greenish-white felspar, in crystals about an eighth of an inch long. This rock sometimes contains, in addition, an inconsiderable proportion of reddish felspar, quartz, and green mica, and very minute specks of iron pyrites.

Besides the hornblende-rock, talcose, hornblendic, chloritic, quartzose, and even argillaceous schist, and the combinations of these with each other, occur with the gneiss. Mica schist has not been observed. Schists are of very rare occurrence in the gneiss generally, but all the above varieties are to be found in some force in the southernmost part of the area, in the Maraura region. This peculiarity of distribution is so marked that it was thought that the schistose strata might here be separable from the gneiss, a suggestion which is much encouraged by the fact that the great quartz-reefs, elsewhere so prevalent in the gneiss, stop short of this ground, but it has not as yet been found possible to draw a line between the gneissic and the schistose subdivisions. Gneiss of the usual type is still a prevalent rock in the schistose area, and is the most southerly rock seen at Sháhgarh. It seems, too, to be truly associated with the schists. We may perhaps at least infer that in this region we have the top of the gneissic series of Bundelkhand.

It is a noteworthy fact that over the whole of this large area of gneiss not a single bed of limestone has been detected.

Pegmatite veins, from a few inches to a foot or two in breadth, are very common. If these were intrusive, it might be expected that they would be somewhat uniform in composition, irrespective of the nature of the surrounding rock, but it has been invariably observed that the felspar of the vein is the same as that of the rock adjacent, whether the latter is orthoclase or plagioclase, or includes both, the chief difference consisting in the larger crystallisation and in the usual absence of the third mineral (hornblende, etc.) in the veins. It is therefore presumable that the veins were formed by segregation at the time of the crystallisation of the gneiss.

The almost total absence of accessory minerals in these rocks is remarkable. Mr. Mallet only mentions epidote sparingly in the hornblende rock, schorl in some of the small quartz veins of the Maraura region, small grains of ilmenite in some of the pegmatite veins, and strings of altered kyanite in the quartzose rock of Dhánkua hill. Small pieces of galena have been sent from Jhánsi for analysis, but their locality is not known, and they may not have been procured from the gneiss. Iron ore has been extensively burrowed for at Dhawara, it is a decomposed earthy condition of the banded hæmatite and quartz. The absence of any trace or tradition of gold in connection with the quartz reefs is noteworthy.

All over lower Bundelkhand long narrow serrated ridges composed of quartz-reefs form a most striking feature of the landscape. They run in straight lines, generally with a north-east to south-west direction, sometimes attain an elevation of 600 feet over the surrounding country, and are exclusively confined to the gneissic series. They are pretty equally distributed over the ground, with the exception noticed in the southern region, which, moreover, lies right across the strike of the reefs to the north-east.

The prevalent strike of these quartz-reefs is about north-east by north, but varies from north-north-west to east north-east. The breadth of the veins varies from a few feet up to 100 yards, in the reef west of Bhagwa. Some of them are traceable in a direct line for more than 60 miles, the local interruptions which occur being sometimes due to removal by denudation, sometimes to strangulation of the vein itself. Other reefs, again, though of full thickness and very prominent at the surface, are short and end abruptly. There is a good case of this at Dehri, where a broad reef, 300 feet high, is only about a mile long, no trace of it occurring in the gneiss to the north or south. The narrow gaps by which the minor streams in many places cross the reefs give peculiar facility for the formation of lakes, as a very short dam is often sufficient to pond back a large surface of water, and many of the numerous artificial lakes in Bundelkhand are formed in this way.

The reefs are often affected by joint-planes, which sometimes give an appearance of horizontal bedding to the mass. When parallel with the direction of the reef itself they suggest, in a more puzzling manner, the impression of a bedded mass. Now and then the quartz is much shattered. At Deokalli and Bagpura samples might be taken for the Bijáwar hornstone-breccia, a description of which will be found on a subsequent page.

Foliation is often developed, owing to the metamorphic condition of the reefs themselves, and it usually extends to the contiguous gneiss, which is generally so amorphous. When the foliation in both rocks has the same direction as the reef itself, the quartz mass might be taken to be interbedded with the gneiss. Generally the foliation is oblique to the direction of the vein, while still the same as that of the gneiss.

Many of the reefs are of greyish-white quartz. Frequently they contain a large amount of impure serpentine, and occasionally they are formed almost entirely of this material. The more northerly of the two ridges at Dallipur is an example. At Rájápur a band of nearly black serpentine occurs, apparently a continuation of the quartz vein to the north. In many cases the gneiss is serpentinous for some distance on each side of a vein, there being no distinct separation between the two rocks, but serpentinous gneiss has not been observed except near a quartz vein. Steatite takes the place of serpentine in a few veins.

From his observations, as above sketched, Mr. Mallet concludes that the reefs were probably formed before the metamorphism of the gneis. was accomplished.

Many of the quartz-reefs, as well as the gneiss itself, are traversed by more recent, and much smaller, veins of pure white quartz, the thickest not much exceeding one foot in breadth. They are very frequently crystalline and drusy in the centre, and they are always sharply distinct from the rocks they traverse. Their direction is very irregular.

The gneiss of Bundelkhand is also remarkable for being traversed by extensive trappean intrusions, none of which penetrate any of the younger formations. These dykes, of true igneous rock, are more numerous than the quartz-reefs, and exhibit nearly as much regularity in their course, their prevailing direction being about north-west by north, so as to cut the reefs obliquely at an angle of about 70°. Some few run east of north, or due west. Many are of considerable size, a breadth of 100 feet being not unfrequent, while some are much wider. They are often persistent for great distances. The commonest type is an extremely hard and tough close-grained greenstone (diorite), in which the hornblende and the white felspar are sometimes clearly separated. The rock often weathers into large rounded blocks without any tendency to exfoliation. The small dykes are of a more earthy texture.

It is very rare to find any intersections of the dykes and reefs that can be taken as conclusive of their relative age. It is not so uncommon to find a dyke running close up to a reef on both sides without cutting it, but such an occurrence might easily happen, even though the reef were the older, as it may have offered a greater resistance to splitting. One good case of the converse carries much more weight. Mr. Mallet records an occurrence where a quartz vein, striking east-20°-north, abuts against a strong dyke running west-20°-north, traces of the quartz being found also on the other side. But he considers the general argument from the condition of the two rocks to be independently conclusive, the trap has certainly not undergone metamorphism, whereas the reefs as certainly have.

The history of this comparatively small area of gneiss would be very interesting. It has served as a shore or a bed for each of the great adjoining formations. The Bijáwars and the Gwáliors lie upon its margin north and south, but no detached outliers of either are found within its border, so that it may have been a well-elevated area at the period of their formation. The same may be said of the lower Vindhyan deposits. It is not so with the upper Vindhyans, of which the outliers are numerous, and lie at considerable distances from the scarp of the basin. In the east these form a portion of an ascending slope, the base of the capping Vindhyan sandstone being higher in the outliers than in the scarp. In the north-west it is curiously the reverse, the gneiss reaches high up all along the western scarp, but the outliers of Vindhyan sandstone to the eastward rest at the general level of the low country.

The next overlying formation is the Deccan trap, remnants of which are found on the low ground in the southernmost part of the area, and traces of the infra-trappean (Lametá) conglomerate occur more extensively in the same position. That this portion of the scarp-bounded area can have been so occupied almost necessarily implies that the whole of the gneissic ground must, at the period of the Deccan trap, have had a configuration very like what it has now, and, the source of the eruptive rock being presumably to the south or south-west, the lava must have poured from the plateau to the lowlands. Trap does, in fact, occur continuously from one level to the other in the Madanpur gorge, but its condition suggests no resemblance to a lava stream.

The newer type of gneiss is well developed south of the Vindhyan basin, in Rewá, Mírzápur, and Behar. It has been there examined by Mr. Mallet, from whom the account of the Bundelkhand area has been taken, and consequently we have not to allow for discrepancies of observation.[1] He gives the following tabular abstract of the constituents of the gneiss in Singrauli, a petty principality now absorbed in the Rewá state and adjoining districts :—

1.—Minerals occurring as constituents of the gneiss :—

Quartz, orthoclase, oligoclase, muscovite, biotite, hornblende, epidote.

2.—Occurring in beds in the gneiss :—

Limestone, dolomite, corundum, magnetite, quartz as quartzite and quartz-schist, hornblende as hornblende-rock, tremolite-rock and jade, mica as mica-schist, epidote.

3.—Occurring in veins in the gneiss :—

a.—Quartz in veins and reef-quartz, stilbite (?)

b.—In pegmatite veins (as constituents): orthoclase, oligoclase, quartz, mica.

c.—In epidotic veins : epidote, quartz.

4.—Accidental minerals in the gneiss :—

Magnetite, ilmenite, schorl, garnet.

5.—Accidental minerals occurring in the subordinate beds (2) of the gneiss :—

a.—In the limestone—magnetite, pyrite, hæmatite, serpentine, chrysotile, phlogopite (?), wollastonite.

b.—In corundum bed—schorl, euphyllite, diaspore.

c.—In jade bed (associated with corundum)—corundum, rutile (?), schorl, euphyllite.

6.—Accidental minerals occurring in the veins, etc., in the granite veins :—

a.—In the quartz-veins—micaceous iron, tremolite, augite, epidote, schorl, muscovite.

b.—In the quartz-reefs—galena, cerussite.

c.—In the pegmatite veins—schorl, garnet.

To this list may be added the minerals found by Mr. Mallet in the same zone further to the east, in Hazáríbágh: lepidolite, tourmaline, beryl, apatite, leucopyrite, and tinstone. Zircon is also said to occur.

The contrast between the minerals named in this table and the constituents of the Bundelkhand gneiss is very striking. The most marked differences are the abundance of the disseminated quartz, the comparative frequency of limestone, and dolomite, and of mica schist, and the general occurrence of accessory minerals in the Bengal gneiss.

The structural characters present another noteworthy point of contrast between these two gneissic series. In Bundelkhand the rock is generally homogeneous and amorphous, the foliation obscure and constantly in more or less vertical planes, as if due to the causes which produce cleavage. In the Bengal gneiss bordering the Bijáwar basin on the south, the foliation coincides with what appears to be the original lamination and bedding. It has a general east-north-east strike, corresponding with that of the main rock-boundaries, but the alternating strata frequently roll about at low angles of dip, or are crushed together confusedly, the foliation constantly agreeing with the lie of the beds. A third point of

[1] Mr. Mallet's work has been only partially published, *Records*, V, 18; VI, 42; VII, 32, (1872-74).

contrast is in the relation to the overlying transition formations, which rest nearly horizontal and undisturbed on the gneiss of Bundelkhand, and are uniformly disturbed, metamorphosed and subjected to granitic intrusions where they occur in contact with the gneiss of the Bengal area.

Quartz-reefs have been described in this gneiss also, but to a very subordinate extent, and their origin as veins is in many cases open to question. A common mode of occurrence of this quartz, or quartzite, is close to the boundary of the slate and gneiss series, but it does not coincide with their junction, and it is not in any sense a contact-formation separating sharply distinct types of rock. It occurs in the strike of the foliation and stratification, and may well be an altered quartzite.

There is a rock common in this northern area of the Bengal gneiss, perhaps occurring most typically within the zone mainly occupied by the transition series. It is known as dome-gneiss, from its weathering into great hemispherical or ellipsoidal masses of bare rock, the only divisional planes being concentric layers of exfoliation[1]. The domes are often several hundred feet high, and form a very peculiar object in a landscape. Foliation is always more or less traceable, and, in every respect of texture and composition, the rock is the same as that of the thin bands alternating with schists in the adjoining ground. Both are often porphyritic, the dome-gneiss generally so, containing large ill-formed (rounded) crystals of felspar. There can be no doubt that the peculiar form exhibited by this rock is due to the occurrence of large masses of more homogeneous composition than usual, but the question is how these conditions were produced, whether we must not suppose a partial degree of plasticity to have been attained, and whether the rock is not in a manner intrusive. At the Kálápahár and the Bhiaura hills on the northern fringe of the Hazáribágh plateau, and the Mandar hill of the Bhágalpur district in the same geological region, there are very typical examples of the dome-gneiss.

The comparative rareness of trap-dykes in the Bengal gneiss is another point of contrast with the Bundelkhand area. In some parts they are pretty frequent, perhaps most so in the vicinity of the basins of Gondwána rocks, where they are often continuous into such basins, thus fully establishing their comparatively recent date, but they are by no means generally distributed.

Pegmatite is not uncommon in the gneiss of Singrauli. Mr. Mallet does not consider this to be intrusive, its composition varying with the rock it traverses, as was explained in the case of the pegmatite of Bundelkhand. In northern Hazáribágh, however, he describes the extensive occurrence of intrusive pegmatitic granite, ramifying in the most intricate manner, in veins and dykes of from ½ inch to 50 yards wide, through both the gneiss and the transition schists, and maintaining its composition

[1] V. Ball, *Memoirs*, VI, 132, (1867); XVIII, 95, (1881).

irrespective of the enclosing rock. It is composed, in order of crystallisation, of tourmaline, mica, felspar, and quartz, all four being generally present, but their proportions vary greatly. Its texture is very uneven, the coarsest forms being often found in comparatively narrow dykes. It is in this rock that the mica-mines of Behar are worked. Not unfrequently the pegmatite assumes the curious form known as graphic granite.

The gneiss of the Chutiá Nágpur districts, up to the basin of transition rocks in south-west Bengal, is more or less freely interbedded with micaceous hornblendic and silicious schists, and occasional bands of the porphyritic granitoid variety. Patches also occur of less highly metamorphic schists.

The junction of the Chutiá Nágpur or Bengal gneiss with the transition rocks of Singhbhúm, is described by Mr. Ball[1] as a great fault. But within this basin of submetamorphic rocks there are extensive inliers of a gneiss which seems to be of an older date than that of Chutiá Nágpur, and is apparently equivalent to the Bundelkhand gneiss. It is very uniform and granitoid, there is a total absence of the thin-bedded gneiss, schists, etc., which abound in the main gneissic area to the north, and we again find a remarkable abundance of trap-dykes, forming two intersecting systems, having north-westerly and north-easterly courses, respectively. In contact with this Chutiá Nágpur gneiss the transition strata exhibit a minimum of alteration and disturbance. Mr. Ball describes them, at and near Cháibásá, as sandstones and mudstones resting immediately on the rough weathered surface of the granitic gneiss. There are local faults along the boundary, but it is certain that the original relation of the two series is like that between the Bijáwars and the Bundelkhand gneiss, as already described.

Further south, in the Tálcher country, the ordinary newer type of metamorphic rocks again prevails. The following rough classification of them is given by Mr. Blanford[2] :—

Gness, *a.*—Hard, coarse, and felspathic, becoming sometimes lithologically a perfect granite.
,, *b.*—Soft foliated, quartzose or micaceous.
,, *c.*—Compact, but sometimes soft, containing garnets, frequently decomposed.
Hornblendic gneiss or schist, soft and foliated.
Quartz-schist or schistose quartz, occurs frequently in bands separated by softer micaceous layers

The variations in composition coincide with the planes of foliation, the prevailing direction being west-north-west to east-south-east.

Higher up the Mahánadí valley in the neighbourhood of Sambalpur, Mr. Ball[3] observed syenitic and protogenic gneiss as common, hornblende-rock and schist as somewhat rare, strong quartzites forming the

[1] *Memoirs*, XVIII, 88, 130, (1881). [2] *Memoirs*, I, 39, (1856). [3] *Records*, X, 181, (1877).

most peculiar feature in the gneiss; mica-schist, quartz-schist, and shaly slate (and in one instance, near Kátikela north-east of Sambalpur, a conglomerate) were found associated with the gneiss. The strike in this region would seem to be very variable—east to west, north to south, north-west to south-east, and north-east to south-west, being all recorded.

On the same latitude, about Nágpur, Dr Blanford[1] has noticed the general resemblance of the gneissic rocks to those of Bengal. Here again there is much irregularity in the strike.

There is but little information available regarding the gneiss of the Nizam's dominions in Haiderábád. Westward of Kamamet the massive form is known to be largely developed in broad bands running about north-north-west, with the more foliated types intervening. From a short distance east of Kamamet the schistose forms are found,[2] and continue to the eastern limit of the gneiss area[3].

In the Nellore portion of the Carnatic, and the coastal regions of the Kistna and Godávari districts, Dr. W. King recognised four distinct varieties of gneiss, occupying separate areas and apparently indicating different periods of formation, which were distinguished as—[4]

4. Micaceous, talcose and hornblendic schists, with few quartz-schists or quartz-rock	Schistose gneisses.
3. Foliated gneisses with frequent quartz-schists or quartz-rock	
2. Grey gneiss, sometimes porphyritoid	Massive gneisses.
1. Red granitoid gneiss	

The red granitoid gneiss only occurs south-west of Venkatagiri and westwards into the upland of North Arcot and Cuddapah. The gneiss is generally a close-grained aggregate of quartz and felspar (orthoclase, oligoclase (?) and a little albite), hornblende being often scarcely discernible and mica even more rare. The quartz is of two forms, dull amorphous, and glassy, and the felspar usually of a pale flesh colour.

The grey granitoid gneiss forms a band immediately east of this, and extends northwards along the eastern edge of the main exposure of Cuddapah rocks. The actual passage from one type to the other is gradual, but on either side of a narrow interval the rocks differ decidedly in type. There is more variety in the rocks of this band than in the red gneiss, but the prevalent form is a rather rough gneiss of quartz, felspar, and hornblende. The rock is obscurely foliated, but the foliation can generally be detected.

The boundary between the granitoid and the schistose gneisses is said to be tolerably defined. The latter are distinguished by their distinct foliation, and consist of hornblendic, micaceous, talcose, and chloritic schists, and well foliated and more massive gneisses. No attempt was made to map the two subdivisions separately, but the relative age was regarded

[1] *Memoirs*, IX, 301, (1872).
[2] R. B. Foote, *Records*, XVIII, 28, (1885).
[3] W. King, *Memoirs*, XVIII, 201, (1881).
[4] *Memiors*, XVI, 126, (1880).

as that given in the tabular statement. The gneiss No. 4 is said to occur in a band separating the granitoid gneisses from No. 3, and to overlie both. It seems probable, however, that an outcrop of the Dhárwár system, which had not been separated when the survey was made, has been included with the metamorphics.

In the northern portion of this area a fifth form of gneiss is known, which has been described as the Bezwáda gneiss[1] from the town of that name, where it is typically developed. It forms a band along the edge of the alluvial area from the Kistna northwards to the Godávari, and, from its position as well as the less degree of metamorphism it has undergone, is believed to be newer than the other gneisses. With the exception of thin subordinate bands of quartz-schist and quartzose gneiss, the usual rock is of a dark brownish colour, composed principally of lustrous red murchisonite, which sometimes so predominates that there are seams, and even thick beds, of felspar rock, the murchisonite being then granular. Garnets are frequently scattered through the rock and often extremely abundant. Here and there graphite occurs in sufficient abundance to convert the rock into a graphitic schist.

These rocks appear to extend northwards along the coast as far as Vizagapatam[2], where they attain an extensive development, and, besides the types of rock seen further south, contain some bands of crystalline limestone.

In the southern part of this area important mica mines have lately been opened in some very coarse granite intrusions. At Inikurti the crystals are as much as 10 feet in diameter, sheets of 4 or 5 feet across

Fig. 3. Mica mines at Inikurti.

have been obtained, free from adventitious inclusions which would spoil

[1] W. King, *Memoirs*, XVI, 206, (1880). | [2] W. King, *Records*, XIX, 119, (1886).

their commercial value, but the size of the crystals can perhaps be best realised from the photograph reproduced in figure 3.

In the south Maráthá country, and in the Bellary district, the massive type of gneiss, resembling that of Bundelkhand, is largely developed. The outcrops form broad bands intervening between the strips of transition rocks, which rest on an eroded surface of the granitoid gneiss. There are two principal types of this,[1] first a fine or medium-grained reddish or grey rock, sometimes so homogeneous as to be hardly distinguishable from a felsite, whose most remarkable accessory mineral is pistacite, occurring in veinlets and films lining the joints, and also as grains scattered through the mass of the rock. The other type is coarse-grained, often highly porphyritic, and is principally developed in the west, while the finer-grained types are more common in the east of the district. The foliation in the coarse-grained porphyritic rock is very obscure, and the prevailing felspar a pink orthoclase.

Schistose-foliated gneiss is found in a band lying between the massive porphyritic rock of Bellary and the band of Dhárwárs to the south-west, the principal form of rock is quartzo-hornblendic, but no sections showing the contact of this rock with those on either side were seen.

Numerous veins of granite, composed of quartz, red or pink orthoclase, and pistacite, are found intersecting the granitoid gneiss near Gooty, and the schistose hornblendic gneiss further south.

The crystalline rocks in Bellary district are, as a rule, very unlike the mass of the gneisses in the east and south of the Peninsula, but bear a very strong petrographical likeness to the Bundelkhand gneiss. The resemblance is not only a petrographical one in hand specimens, but also a very striking one in the features of the landscapes of parts of these two widely remote regions—a likeness abundantly confirmed by comparison of good photographic views of the granitoids in both tracts.

The long, narrow, serrated edges of quartz reefs, which form such frequent and striking a feature in the Bundelkhand landscape, are nearly as common in Bellary district and other parts of the Ceded Districts, and, like those in Central India, they not unfrequently attain to heights of 500 to 600 feet and upwards above the general level. The granitoid gneisses in Bellary district and the adjacent districts of Anantápur, Karnúl, and Cuddapah, are also traversed by very numerous trap dykes, of great size and length, which often rise into bold crests and ridges, forming very conspicuous objects in the landscape. The relation of these to the gneisses and to the great quartz reefs is precisely the same as in the typical Bundelkhand area.

[1] The description in the text is based on R. B. Foote, *Memoirs*, XII, 37, (1876); *Records*, XIX, 100, (1886) and a forthcoming *Memoir* on the Bellary district.

A further noteworthy fact, in perfect agreement with the geological structure of the Bundelkhand gneiss, is the total absence, as far as our present knowledge goes, of limestones in the gneissic region of Bellary, the south Marátha country, the Raichur Doáb, and the districts of Anantápur, Karnúl, and Cuddapah. Yet another point in which the granitoid gneisses of Bundelkhand and Bellary show a strong similarity, is in the extraordinary rarity of accessory minerals—a point in which they differ much from the gneisses in the eastern Carnatic.[1]

The gneissic rocks of Sáwantwári and Ratnágiri in the Konkan, as the low country between the Sahyádri range and the sea is called, would seem, from Mr. Wilkinson's description, to be more varied than on the Deccan plateau above the ghâts. The distribution in separate bands of more massive and more schistose characters does not occur. The beds consist of true gneiss (*i.e.* a well foliated quartzo-felspathic rock), micaceous and hornblendic schists, quartzites, and altered micaceous sandstones, with some subordinate bands of granitic and syenitic gneiss, also occasional talcose, chloritic and actinolitic schists. The mass of porphyritic syenite forming Wajhiri hill, five miles from Vengurla, is considered to be intrusive.[2]

In south Malabar Mr. Lake[4] has described three principal types of gneissose rocks, which appear to be of different ages. They are, *firstly*, the quartzose gneiss, composed principally of quartz and hornblende, or sometimes mica, the other minerals are frequently absent, and thick bands of pure quartz occur; large masses and runs of very hornblendic rock are also found, and some beds consist principally of quartz and hæmatite, the latter apparently due to the decomposition of garnets. *Secondly*, the garnetiferous gneiss, usually a granular rock, with much quartz, a little felspar, and hornblende in varying, though seldom, large, proportion; garnets characterise the rock, and are occasionally very abundant, but usually subordinate to the quartz. *Thirdly*, the felspathic gneiss, in which felspar forms the principal constituent, quartz also occurs, but there is very little of any other mineral. There are also exposures of granitoid and hornblendic gneissose rocks which appear to be intrusive and do not cover any large area in the district.

The distribution of the three types of gneiss presents some features of interest. The quartzose type of rock is found in two areas, which are probably connected with each other in the intervening unsurveyed ground. The strike of the foliation is nearly north and south. The quartzose gneiss is succeeded to the west by the garnetiferous, and this again by the felspathic

[1] This and the two preceding paragraphs are extracted from Mr. R. B. Foote's unpublished *Memoir* on the Bellary district.

[2] *Records*, IV, 41, (1871). Some allowance must be made for discrepancies of nomenclature between different observers in these rocks; what some might call an altered micaceous sandstone others would name a quartz-schist.

[4] *Memoirs*, XXIV, 209, (1890).

gneiss, in which the foliation runs nearly east and west, but bends round the outcrops of the quartzose gneiss, in a manner suggesting very strongly that they are of different ages and that the latter is the oldest. The foliation is usually coincident with the banding in all three types, but where the east and west foliation bends round the north and south foliation, there are occasional discrepancies.

Our next note upon these rocks refers to the south-east Wainád, on the uplands of Mysore at the north-west base of the Nílgiris. In the little map published with Dr. King's report [1] on this ground the greater part of the area is shown to be within the region of the steady east-north-east strike which obtains in the Nílgiris and along the south-east edge of the Mysore plateau, but towards the north there is an area of troubled dips, centred round two masses of granitoid rock forming the Mani malai and Yedakal malai, north of which the foliation again passes into the normal north-north-west strike of the Sahyádri. Dr. King treats these granitic masses as doubtfully intrusive. This Nílgiri strike is noted as distinctly that of the lamination and bedding of the gneiss as well as of the foliation, the general dip here being southerly. Four belts of gneiss are recognised in the south Wainád, the quartzo-hornblendic gneiss of the northern face of the Nílgiris, and below (north) of it the Devala band of highly felspathic gneiss with two minor belts of chloritic gneiss. North of this is the quartzose and ferruginous band forming the Marpanmadi range, beyond which is a broad area of more varied gneiss. The auriferous quartz-reefs are perhaps most developed in the Devala band. Their lie is peculiar, the strike is north-north-west, corresponding with that of the gneiss in the country to the north, and at right angles to that of the rocks in which they occur, yet they generally have a low dip, from 10° to 30°, always easterly. One small trap dyke occurs in the Devala band, running east by north, nearly in the strike of the gneiss.

In the Nílgiris[2] massive (obscurely foliated) gneiss prevails, but it is of a very different type from the massive gneiss of the south Maráthá country, which is granitoid and copiously felspathic. It is in the very hornblendic variety of the gneiss, such as prevails over the northern portion of the Nílgiri plateau, that the foliation is least marked. The rock is described as hard, tough, and black, breaking with an even fracture and consisting of an intimate mixture of quartz and hornblende, with some garnets. It was taken by early observers for syenite and greenstone. A similar rock, but with a variable proportion of felspar, is very common in the central parts of the hills. There are also several strong courses of a quartzo-felspathic gneiss, which has been taken

[1] *Records*, VIII. 29, (1875).

[2] *Memoirs*, I, 218, (1858).

for graphic granite.[1] Locally this gneiss also contains garnets in great quantity.

A few thin dykes of trap have been observed in the Nílgiri hills, but no granitic veins. Small irregular veins of white quartz are common, but no reefs have been observed.

To the south, as well as to the north, of the Nílgiris the gneiss of the low ground becomes well foliated and schistose. South of Coimbatore a band of limestone has been observed in the metamorphic rocks. Granitic veins are also common in this neighbourhood, being especially conspicuous in the hill of Sankari Drug, but no intruded granite mass of large dimensions occurs. Mr. H. F. Blanford, from whom the notes on the Nílgiris are taken, describes[2] a band of granitic rock to the north of Trichinopoli, and points out that this band is possibly a continuation of the very similar rocks of Coimbatore. In Trichinopoli, as to the westward, there is no massive intrusion, but the whole band (about four to six miles wide) may be considered rather as a network of veins, running generally in the planes of foliation of a shattered band of highly foliated hornblendic gneiss, which is frequently twisted and contorted in every direction. The veins consist of a largely crystalline binary granite, mica occurring but rarely. The proportions of quartz and felspar vary greatly, and these ingredients sometimes affect the structure known as graphic granite. Mica is altogether a rare ingredient in the gneiss of this region of the Peninsula.

A considerable area of the gneissic rocks of Southern India, from the Cauvery northwards, has been mapped in some detail. The geology has been described by Messrs. King and Foote, and the leading features have been made out, or at least suggested.[3] The belt of granitic intrusion along the north bank of the Cauvery, already mentioned, is on an anticlinal axis. Beds of variable gneiss and schists, with some limestone, dip from it on both sides. To the north they pass under the great mass of rocks forming the several clusters of hills in the Salem district, where, as in the Nílgiris, a syenitoid hornblendic gneiss is very prominent. With it are associated the various magnesian schists from which the magnesite of the "chalk hills" is derived[4] and also the great beds of magnetite which have made Salem famous as an iron-producing district. These are not lodes, but regularly bedded masses of banded iron-ore and quartz, associated with the gneiss. With the aid of the very conspicuous outcrops formed by this rock, several great features of contortion have been made out, proving the strata to be frequently repeated at the surface.

[1] Recent unpublished observations by Mr. Holland indicate the probability that some of these are intrusive.

[2] *Memoirs*, IV, 30, (1863).

[3] *Memoirs*, IV, 269, (1864).

[4] Recent observations, as yet unpublished by Mr. Holland, show that the magnesite is probably derived from the decomposition of the ultrabasic intrusive rocks which he has detected in this district.

A collection of rock specimens from the Salem district, made by Leschenault de la Tour in 1819, has recently been made the subject of an important study by M. Lacroix,[1] who found among them an interesting series of acid and basic gneisses, scapolite bearing rocks, and mica schists. The greater part of his results are of a lithological and mineralogical nature, too isolated in their bearing on Indian geology to bear summarising here, and his suggestion, that the specimens belong to three separate stages, corresponding to those known as ζ^1, κ^2 and κ of the geological map of France, yet waits the confirmation of detailed work in the field. The paper is, however, of importance, apart from the valuable results it contains, as showing how much remains to be done in the study of the Indian gneissose rocks.

In South Arcot, to the east of the Salem hill-groups, a considerable area is occupied by rocks having a very granitic aspect, yet showing in many places an appearance of stratification and occurring in great continuous ridges, which apparently form anticlinal and synclinal folds. The rock is composed of quartz and white and pink felspar. It frequently contains blocks, both angular and rounded, of hornblende schist. Altogether, the nature of this rock and its position in the metamorphic series are still open questions.

North of Trichinopoli a change takes place in the direction of the strike of the metamorphic foliation analogous to that noticed in the Wainád. The east-north-east direction changes rapidly into north-north-east, parallel to the Coromandel coast. The regularity of the coast-line is no doubt connected with this fact, and it is interesting to note how the main structural features of the fundamental rocks thus determine the actual configuration of the Peninsula. All the fossiliferous deposits, and even the later azoic formations, are but patches on the weather-worn surface of this most ancient gneissic mass.

In the Madura and Tinnevelli districts the gneissose rocks are of two types,[2] one the ordinary granitoid gneiss, the other described as granular quartz rock, and Mr. Foote believed that he had recognised no less than six alternations of these in the northern portion of the Madura district. Such supposed stratigraphical successions, however, in rocks that have been so highly metamorphosed and disturbed as these, are open to grave elements of doubt. Both micaceous and hornblendic types of gneiss are found, granite intrusions are not abundant, quartz veins are rare, and trappean intrusions almost entirely absent. A few exposures of coarsely crystalline limestone in the gneiss are recorded.

[1] *Bull. Soc. Franç. Minéral.*, XII, 83, (1889). Translated by Mr. F. R. Mallet, *Records*, XXIV, 155-166, (1891).

[2] R. B. Foote, *Memoirs*, XX, 10, (1883).

A few words may here be said regarding the distribution of the trap dykes which traverse the gneisses of southern India. They are extremely rare in Tinnevelli and south Travancore, and very few are seen in Madura, Púdúkattái, south Trichinopoli, or in the band penetrated by granite veins north of the Cauvery. In north Trichinopoli, Salem, and South Arcot the number increases, but in Coimbatore they are not numerous On the Nílgiris they are few and small, and only one is known in the Wainád. In North Arcot they are very numerous, and often large and of great length, they continue to be numerous in south-west Cuddapah and Anantápur, becoming less so in Bellary and Karnúl. In Chengalpat, Nellore, and Kistna they are not common, and none very large.

In composition the trap is dioritic, usually of medium grain, though both very coarse and very fine-grained varieties are met with. The intrusions are of very ancient date, and probably connected with the volcanic outbursts of the Cuddapah system. The ancient volcanic neck of Wajra Karur, a notice of which will be found in a subsequent chapter, may be connected with them.

Turning northwards to the Arávalli range, we have very few observations recorded, regarding the gneissose rocks. In some exposures in the northern part of the range and on its eastern margin, beds belonging to one of the transition systems are said to rest unconformably on granitoid gneiss,[1] but in the central portion of the range the contact is said to be transitional and accompanied by an interstratification of the transition beds with the gneiss. It is not impossible that there is gneiss of at least two distinct ages in the range, but the apparent transition is probably due to the disturbance the beds have undergone, and the apparent interstratification to veins of gneissose granite intrusive along the planes of bedding. True granite is known to occur abundantly in the Arávallis.

The gneiss of Mount Abú and its neighbourhood is said to be highly felspathic, massive, and crystalline, but occasionally a few schistose beds occur.

At the southern end of the range, in the lower Narbadá valley, near Jobat, well foliated gneisses, with quartz and hornblende schists, are prominent, while mica schists and granitoid rocks are rare. Limestone and slates occur among the metamorphics, but whether the association is real, or only due to the disturbance they have undergone, is not clear.[2]

In extra-peninsular India the gneissose and granitoid rocks occupy a smaller proportion of the total area and, except in a small portion of the Himálayas, have been even less studied than in the Peninsula.

[1] C. A. Hackett, *Records*, XIV, 279, (1881). [2] P. N. Bose, *Memoirs*, XXI, 7, (1884).

Of the gneissose rocks of Afghánistán we have very little information. They are known to occur in the Safed Koh and in the Hindu Kush, and probably occupy the greater part of Káfiristán, up to the region of the Pámírs.

In the Himálayas the few observations that have been recorded are often of a misleading nature; owing to the difficulty there often is in distinguishing between true gneiss which has become so metamorphosed as to be passing into granite, and granite which has assumed the foliated structure characteristic of gneiss. The broad features of the distribution of the gneissose rocks in the north-western portion of the Himálayas are tolerably known, and it is found that the area occupied by them not only corresponds to the regions of special geological elevation, but is approximately coincident with the principal ranges of high peaks, the coincidence being due quite as much to the comparatively greater resistance offered to denudation by the crystalline rocks, as to the fact of these being areas of greater special upheaval.

In the region north-west of Kashmír, where the Himálayas and Hindu Kush meet, there is known to be a large extent of gneissose rocks, associated with some schistose slates and crystalline limestones, but no lithological details of importance are available.

In Kashmir territory the boundaries of the crystalline rocks have been approximately mapped, and the rocks themselves cursorily described by Mr. Lydekker,[1] but his description, being based on rapid traverses, contains few details of sufficient importance and certainty to be repeated here, and fails moreover in always distinguishing between the gneissose granite and granitoid gneiss. Mr. Lydekker believed that the gneissose rocks were of two ages,—for the one, the older, he retained Dr. Stoliczka's name of "central" gneiss, while the newer gneiss was regarded as conformable with, and in part formed by the metamorphism of, the sedimentary unfossiliferous slates, believed to be of silurian age, so largely developed in the Himálayas. It is more than probable that the Himálayan gneiss, like that of the Peninsula, belongs to more than one system, and is not all of one age; but the second conclusion, though in consonance with a somewhat discredited theory of metamorphism, is open to doubt. The apparent interstratification of gneiss with schistose slates and limestone appears, in at least some instances, to have been due to foliated granite, intruded along the bedding planes, and the apparent complete metamorphism of the slates into gneiss is probably due either to a complete obliteration of the sedimentary rocks by a gneissose granite, or to a misinterpretation of the scattered observations.

[1] *Memoirs*, XXII, 265, (1883). It must be remembered that the observations were made on widely separated traverses of the country which has not yet been examined in detail.

It is certain, however, that in Zanskar, Rúpshu, and the gneissose area which extends along the main axis of the Himálayas north of Kumáun, true gneiss is largely devoloped. In places this is well foliated, and exhibits a succession of parallel layers, of differing mineral constitution, which are strongly suggestive of an origin by some process of sedimentation.[1] But often the gneiss is extremely granitoid, with the bedding very obscurely exhibited, and it is at present uncertain whether this is merely the result of a more advanced metamorphism of the bedded gneiss or indicates a difference of age.

No careful and critical examination of these gneisses has been made as yet. The principal minerals are quartz, orthoclase, and mica, in varying proportions, but plagioclase felspars, schorl, garnet, and kyanite are not uncommon accessory constituents. In some of the beds porphyritic crystals of orthoclase are found, of lenticular shape with a well developed crystalline structure, exhibiting universally a twinning of the Carlsbad type, with the plane of twinning coincident with that of the foliation.

The gneiss of the Dárjíling district is of the foliated type, apparently a true gneiss[2] composed of translucent colourless quartz, opaque white felspar and dark brown and silvery mica, varying in texture from fine-grained to moderately coarse. The gneiss frequently passes into mica schist or felspathic mica schist, bands of quartzite are rare, while hornblendic rocks are extremely uncommon and in beds of insignificant thickness. Almost the only accessory minerals are kyanite, schorl, and garnet, the two last often forming large-sized crystals in the schists.

The well-bedded gneisses seen on the ascent from the Sutlej valley to the Babeh pass, were originally classed by Dr. Stoliczka[3] with his "central" gneiss, which he considered as forming the original central axis of the Himálayan chain, on either side of which the subsequent deposits were accumulated. The name has, after being current for many years, been abandoned of late, as it implies a theory which, to say the least, has not been proved, and because a granitoid porphyritic rock, which was included with these bedded gneisses and regarded as the typical member of the system, has since been shown to be a true granite in its mode of occurrence.

This rock consists of porphyritic crystals of orthoclase imbedded in a fine-grained matrix composed of quartz, orthoclase, some plagioclase felspar, and mica, both biotite and muscovite, as well as crypto-crystalline mica whose exact mineralogical species is indeterminable, with magnetite, garnet and schorl as accessory minerals.[4] The larger masses of this rock

[1] It is now well established that the apparent stratification of the crystalline gneisses is often deceptive. In the cases referred to in the text this cannot always be the case, for not only are acres of surface of a single bed occasionally exposed on the steep hill-sides, but the beds themselves occasionally show palpable indications of their detrital origin.

[2] F. R. Mallet, *Memoirs*, XI, 43, (1874).

[3] *Memoirs*, V, 15, (1865).

[4] C. A. McMahon, *Records*, X, 22; XV, 34 XVI, 129; XVII, 53, 168; (1877-84).

exhibit slight traces of foliation, generally recognisable in the field, but barely, if at all, in hand specimens. Besides the larger, slightly foliated, intrusions it is found in sheets, generally intruded along the planes of bedding, and traceable at times for distances of 20 miles with a fairly constant thickness of as many feet. Under these circumstances, it has a remarkable resemblance to a truly interbedded rock, a resemblance enhanced by the well-developed foliation, parallel to the bounding surfaces, which these thin sheets invariably exhibit. But there is one constant character which marks both the thin sheets and the larger, less foliated, masses as intrusive, and that is the presence of numerous porphyritic crystals of orthoclase, one and all twinned on the Carlsbad type, varying in size from half an inch to as much as six inches across, and lying with their axes pointing in every direction.

The small amount of disturbance with which the intrusion was accompanied, is in many cases remarkable. It is especially conspicuous in the case of the Chor mountain, which rises to a height of nearly 12,000 feet, 25 miles south-east of Simla. All round this mountain the stratified rocks are comparatively little disturbed, dipping inwards on all sides, and in the centre of this quaquaversal dip the granite has risen, as if it had simply dissolved and absorbed the rocks whose place it occupies. There is some direct evidence that such really was the case, for in the south-eastern portion of the intrusion, where it has replaced volcanic beds of the carbonaceous system, the fine-grained matrix is coloured dark green from the amount of hornblende it contains, a mineral which is usually conspicuous by its absence. Even the thin sheets appear to have been intruded to some degree in the same manner, for the further they are traced from their parent mass the more micaceous, that is to say the more impure, do they become, till the mica in some cases becomes so abundant that the rock splits easily into large flags. This contamination of the rock only extends to the matrix, and the porphyritic orthoclase crystals are unaffected, showing that they had already solidified, and crystallised out from the still fluid or pasty magma, when the rock was intruded into its present position.

A similar porphyritic granite is found in Hazára, where it has been graphically described by Mr. Wynne,[1] it occurs in the core of a highly compressed anticlinal fold in the Pír Panjál and Dháola Dhár ranges, and extends eastwards along the ranges bordering the Sutlej valley. It is also found in the gneissic area to the north of this line, associated with the true gneisses of Zanskar and Rúpshu, and, to the south of the main axis of intrusion, is found in the Chor mountain already referred to, and in the Dudatoli mountain in Garhwál. Intrusions also occur in Kumáun which have not yet been mapped in detail.

[1] *Records*, XII, 118, (1879).

The date of these intrusions is still undetermined, and the evidence is contradictory. On the one hand, we have the recorded occurrence of blocks of it in the so called Panjál conglomerate of Kashmír,[1] but on the other, the intrusive masses of the Pir Panjál and Dháola Dhár ranges must have been in a plastic condition subsequent to the period of deposition of the beds in which these boulders were imbedded. The probable explanation is that the boulders were derived from an ancient land area, composed of a rock very similar to the porphyritic granite of the Dháola Dhár, and that in the great compression and disturbance which caused the elevation of the Himálayas this rock was once more fused and intruded into the position where it is now found.

Another very characteristic intrusive rock of the Himálayas is a white granite, occurring in veins of various sizes and degrees of coarseness of texture. The granite consists of quartz, white felspar, which in at least one instance is oligoclase,[2] and muscovite. Schorl is a very common accessory constituent, and beryl, fluor spar, and garnet are found. The rock is a very common and conspicuous one along the principal axis of the Himálayas, occurring in intrusive masses and innumerable veins, ramifying through the gneiss and schists, and even penetrating the slates. Its intrusion appears to have been, generally, of later date than the porphyritic granite.

Syenite is largely developed in the range north of the Indus and west of Leh, where it is unconformably overlaid by the eocene beds of the Indus valley. Another outcrop of syenite is known, east of Chakráta, in Jaunsar, but this is intrusive in the slates, and is very probably of later origin than the Ladákh rock.

From a geographical point of view, Assam and the Shillong plateau could not be affiliated to the Peninsula, but, geologically, this would appear to be their proper connection, since the prevailing rocks closely resemble the gneissic and transition formations of Bengal, and differ widely from the rocks of the adjoining mountains to the north and east. The structural characters bear the same relation; on the edges of the Shillong plateau secondary and tertiary strata lie quite horizontally, while much younger deposits have undergone intense disturbance in the contiguous Himálayan and Burmese regions. The plateau thus forms a wedge-like mass of neutral ground, occupying an acute angle between two regions of contortion.

The ground to which these remarks apply is not known to extend north-eastwards, beyond the Dhaneswari (Dhansiri) river, though it is likely that the gneissic rocks stretch for some distance under the alluvium of

[1] *Memoirs*, XXII, 280, (1883).

[2] *Memoirs*, V, 14, (1865). The felspar was at first incorrectly described as albite. See F. R. Mallet, *Records*, XIV, 238, (1881).

upper Assam. The principal area is the continuous hill-mass, 250 miles long, between the Dhaneswari and Brahmaputra. The whole of the lower Assam valley may be included in the same geological region, for the numerous hills protruding through the alluvium, north of the Brahmaputra, consist of the same gneiss, and not of the Himálayan type of metamorphic rocks.

The most interesting of these outcrops in the low ground of the Brahmaputra valley is one observed by Mr. Mallet[1] within 200 yards of the tertiary sandstone at the base of the Himálayas, on the left bank of the Raidak river, in the western Bhután Dwárs. It is really within the sub-Himálayan zone, being up a river-valley, inside the mean outer boundary of the sandstones. The rock is thick-bedded hornblende-schist, a common type of rock in the Bengal gneiss, but one that is rare in the Dárjíling gneiss of the adjoining mountains. This is the only instance of close proximity of the azoic rocks of the Peninsula to the Himálayas.

The only observations hitherto made on this Assam gneiss prove little more than that it has a likeness to the Bengal rock, and that the general strike is the same. Some granitic intrusions occur in the transition rocks of the Shillong area, in connection with which they will be noticed.

Gneissose rocks are largely developed in the hill ranges which run southwards from the eastern termination of the Himálayas, to the east of Burma and Tenasserim, but very little is known of their constitution. More or less granitoid gneiss, hornblendic gneiss, crystalline limestone, quartzite and schists of various kinds are found. In many places the gneiss becomes a true granite, and much of the area is occupied by a rock which has been described by various observers as an eruptive granite. Some of the granitoid portions of the rock weather into remarkable rounded masses resembling perched blocks, isolated from each other by the disintegration of the intervening rock.[2] The hornblendic gneiss appears to be less common than in the Peninsula of India, while crystalline limestone is not uncommon.

These gneissic formations are known to be metalliferous in several places. Tin occurs in abundance in Tenasserim and the Malay Peninsula, lead and silver mines—one of them at least, the famous Bau-duen-gyee, of very large dimensions and highly productive—exist in the Shan States north-east of Ava, while the valuable and productive ruby mines of Mogouk and the less important ones nearer the capital, are situated in the same series of metamorphic rocks.

From the foregoing brief account of the present state of our knowledge of the gneissose rocks of India, some idea of its imperfection can be

[1] *Memoirs* XI, 44, (1874).

[2] Sketches of these are given by C. Parish, *Jour. As. Soc. Beng.*, XXXIV, pt. 2, plates vi, vii, viii, (1865).

gathered. As was stated in the opening paragraph of this chapter, the attention of the Geological Survey has, with few exceptions, been only incidentally devoted to the gneissose rocks, the exceptions being almost exclusively surveys made at an early date, when the true nature and origin of these rocks was far less understood than at present. The analogy of other countries makes it impossible to doubt that results of great importance and value will be obtained when the great gneissose area of India comes to be examined in detail.

In the meanwhile there seems to be only one thing certain, that what has been described as the older or Bundelkhand gneiss had already solidified, acquired most if not all of the characteristics it now has, and been exposed to extensive denudation, at a period anterior to that of the oldest of the distinctly sedimentary rocks of the Peninsula. How far the banded gneisses are newer than this, or how far they are the result of subsequent changes of the older gneiss, it is at present impossible to say. The researches that have been made in other countries renders it almost certain that, in many cases, what has been described as stratification in the gneiss was not produced by any process analogous to sedimentation, but it is a result of the deformation and other changes set up in the rock by pressure, heat, and the intrusion of foreign igneous rocks. The recognition of this will necessitate a profound modification of many of the passages in this chapter, a modification whose extent and limit it is not yet possible to indicate.

CHAPTER III.

TRANSITION SYSTEMS.

Transition systems—Difficulty of classification—Definition of term—The Dhárwár system—The Bijáwar system—A possibly older system—Transition rocks of Behar—of the Assam hills—of south-west Bengal—The Gwalior system—The Arávalli system—The Delhi system—Kirána and Chiniot hills—Champáner beds—Maláni rocks—Relative ages of the Transition systems—Transition rocks in extra-peninsular India—The Vaikrita system—The Daling series.

Resting upon the crystalline gneisses, and intervening between them and the oldest fossiliferous beds of the Peninsula, there are a number of rock systems whose age it is impossible to determine with certainty, whether relative to the European sequence or, in many cases, relative to each other. This difficulty is due to the complete absence of any recognisable fossil, an absence which in many cases is easily explicable by the disturbance and partial metamorphism the beds have undergone, but equally often they have undergone little or no disturbance, and are apparently admirably fitted for the preservation of the remains of such animals as may have lived when they were deposited.

Failing fossil evidence, we are compelled to fall back on lithological characters and the degree of disturbance or metamorphism the beds have undergone, and on these grounds they can be divided into a number of separate systems, representing different periods, or at any rate different areas, of deposition, and these systems can again be divided into two groups,—an older which, as a rule, has undergone considerable disturbance and some degree of metamorphism, and a newer which, as a rule, shows only gentle dips and a much less degree of alteration than the other. The characters are by no means constant; the Gwalior rocks, which we will have to include among the older group of systems, are almost undisturbed, while the Cuddapahs, which will fall among the newer, are in places highly compressed and contorted. On the whole, however, the distinction appears to be valid, as it certainly is convenient for the purposes of description, though it is impossible to say how far the eras represented by the two may not have overlapped each other.

The newer of these groups of rock systems we can, with some degree of propriety, class as older palæozoic, and for the older the name 'transition' has become customary. This word is not intended to imply any theory as to the mode of origin of the deposits, but merely to indicate their position, intermediate between the older crystalline gneisses and the newer sedimentary deposits, which have undergone a less degree of disturbance and metamorphism.

The most recently recognised of the transition rock systems, though probably the oldest in point of age, is that which has been described by Mr. Foote under the name of Dhárwár system.[1] As at present seen, it occupies a series of long bands and elongated outliers of highly disturbed beds, folded and faulted into the gneiss, which have a general north-north-west and south-south-east strike, and extend from the southern limit of the Deccan trap to the Cauvery valley, south of which only a few small outliers are known. These outcrops exhibit two types of structure. They are either a sharply folded synclinal or series of synclinals, or else, showing on one side, usually the west, a natural boundary of original contact, they are cut off on the other by a fault of great throw, by which the softer beds of the upper Dhárwárs have been dropped down, and so preserved from entire removal.

The most westerly of these bands commences in the north as a series of inliers in the Kaládgi basin north-east of Belgáum. From the southern edge of this basin it runs down, with a width of 10 to 16 miles, past Dhárwár to the Tungabhadra river. Here it spreads out and covers a large area in north-western Mysore, sending one offshoot down to within 40 miles of the city of Mysore, while another may possibly run southwards from Shimoga down into the low country of south Kánara. The next of these bands starts at the southern boundary of the Kaládgi rocks and runs by Dambal and Chitaldrúg to Chiknáyakanhalli. South of this the band bifurcates,—one branch extends to Seringapatam, the other, running somewhat east of south, crosses the Bangalore-Mysore railway. A third band runs along the north-east boundary of the Kaládgi rocks, east of Bellary, and on to the Penner river. Between this and the last-mentioned band there is a tract of Dhárwár rocks forming the Sandúr hills and copper mountains west of Bellary. A fourth band runs southwards through the Shorápur district past Maski to near the Tungabhadra.

Besides these larger bands there are a number of outliers, the most important of which are those which form the gold-fields of Kolár, and the three bands, further east, which run out from under the Cuddapahs.

The rocks of the Dhárwár system are hornblendic and chloritic schists, phyllites and conglomerates, associated with contemporaneous trap, banded jasper, and hæmatitic quartzites. The degree of metamorphism

[1] The principal published descriptions of the Dhárwár system are in *Records*, XXI, 40-56, (1888), and XXII, 17-39, (1889). The notes in the text are principally based on the unpublished *Memoir* on the Bellary district by Mr. R. B. Foote.

they have undergone appears to vary considerably and to be connected with the varying and often extreme degree of compression they have undergone. Generally speaking, the beds are distinctly schistose and frequently well-characterised schists, but in places they are described as argillites, which readily weather into soft shale. The dioritic traps, which are usually found conspicuously developed in the lower part of the series, are replaced by hornblende schists on some of the highly disturbed sections, and the pebbles in the conglomerates have occasionally been deformed into long rod-like forms.

Conglomerates are of frequent occurrence at and near the base of the system, some being of the ordinary type of true conglomerate, others of that type, consisting of boulders scattered through a fine-grained matrix, for which the name of 'boulder beds' has been suggested. The included pebbles and boulders consist of various varieties of schist, quartz, quartzite, grit, banded hornstone, and gneiss.

The hæmatitic quartzites are composed of alternating layers of quartzite and hæmatite in proportions varying from a rich, pure, hæmatite iron ore on the one hand, to a banded hornstone containing little or no iron on the other. These hæmatite beds are everywhere found in the lower portion of the system, and owing to their hardness, they stand up as sharp ridges from the softer schists, which have been denuded away from their sides, thus acquiring a conspicuousness quite out of proportion to their real importance. They are specially abundant in the Sandúr state, where there are vast supplies of a rich, pure, hæmatite iron ore, formerly worked to a considerable extent, by the natives of the country. The industry is now almost extinct, as much in consequence of the reckless destruction of forests as the competition of imported iron.

The Dhárwár system is economically important as it carries all the paying gold reefs that have yet been discovered in Southern India, all the known gold fields being on the Dhárwár outcrops, and all trials outside them having so far led to nothing but disappointment. The gold appears to be most abundant in the outcrops situated in Mysore territory, where gold-mining has, within late years, passed through the phase of fierce speculation into a well-established and remunerative industry. The reefs have been mined in prehistoric times by miners, whose workings, in spite of their primitive appliances, penetrated to depths of over 200 feet in places and, besides the abundant mines, old dressing floors can still be found, with the mortars in which the quartz was crushed, generally small hollows in which the quartz was pounded, but occasionally large saucer shaped depressions in which huge blocks of granitoid gneiss of a ton and more in weight were rolled round and round.

The relation of the Dhárwár system to the granitoid gneiss is one of most unequivocal unconformity. Wherever a section showing the original contact is found, the bottom beds of the Dhárwárs are found to rest on an uneven eroded surface of the granitoid gneiss. Yet gneissoid beds are occasionally found in the Dhárwárs and a section, east of Memkal in Bellary, is recorded, where several alternations of micaceous gneiss with quartzites and hornblende schists are seen. The section is, however, exceptional, the gneiss is described as differing more from the typical granitoid than from the Dhárwár schists with which it is stratigraphically connected, and it is possible that the interbedded gneiss may be metamorphosed arkose. There is of course the alternative interpretation that the Dhárwárs are here locally in contact with a gneiss newer than that to which they are so distinctly unconformable.

In a northerly direction the Dhárwár beds of the central and western bands run under the great spread of Deccan trap, but not before their upturned and denuded edges have been seen to be unconformably covered by the Cuddapah beds of the Kaládgi area, and to the south-east a similar relation subsists between the Dhárwárs and the typical Cuddapahs of the Cuddapah area. The Dhárwár system is thus completely isolated, both geologically and geographically. The unconformable breaks above and below are so great that they indicate nothing more than the necessity for an utter separation of the system from any other occurring in its neighbourhood, while the distance which separates it from the north of the peninsular area is too great to allow of its identification on mere lithological grounds with any of the transition systems there seen. Yet these are the only ones on which we can venture even a guess at its correlation and, so far as they go, the resemblance is greatest in the case of the Gwalior system; contemporaneous dioritic traps occur in both, while the hæmatitic quartzites of the Dhárwárs resemble the hæmatite beds of the Gwaliors and not those of the Bijáwars in their structure. On the other hand, the small amount of metamorphism or disturbance which the Gwaliors have undergone sharply distinguishes them, and it is probable that both they and the Bijáwars are newer than the Dhárwárs of Southern India.

There can be little doubt that, when the gneissic area of the Madras presidency is more fully surveyed, other outcrops of Dhárwár rocks will be discovered. At present one, at least, can be indicated in the Nellore Carnatic, where a distinct band of more eminently schistose rock[1] is said to occupy the western edge of the field, the schists being talcose, micaceous and chloritic, with frequent intercalations of hornblendic bands. Interbedded quartzites are common, and a laminated hæmatitic quartz schist occurs south of the Swarnamukhi. Associated with these more foliated rocks there is a development of trap, both as dykes and as a large irregular mass of diorite and greenstone, which was regarded as intrusive on the whole, but in certain cases[2] distinctly stated to be interbedded. This

[1] *Memoirs*, XVI, 133, (1880).

[2] *Memoirs*, XVI, 150, 168, (1880).

association of rocks is strongly suggestive of the presence of an outcrop of the Dhárwár system, which had not been separated from the crystallines at the time the survey was made.

A good deal of doubt attaches to the mapping of this area owing to the occurrence of what appear to be outliers of true Cuddapah quartzites, and the difficulty of distinguishing between the less altered and disturbed Dhárwár quartzites and those of the Cuddapah system, where they have undergone much disturbance and alteration. Between the Penner and the Swarnamukhi, the narrow strips of quartzite appear all to be associated with contemporaneous traps and schists, and are probably Dhárwárs. North of the Penner there seems good reason for supposing that they are Cuddapahs.

Turning now to the northern part of the Peninsula, we find a great system of transition rocks, which has been distinguished under the name of Bijáwar, from the town of that name in Bundelkhand. By far the greater part of the area, over which these rocks originally extended, is now covered up by the newer Vindhyans and Deccan trap and they are only exposed in a series of outcrops, of varying size, which extend from Bundelkhand to south of the Narbadá, a distance of about 100 miles from north-north-west to south-south-east, and from Jobat to the upper Son valley, some 500 miles from west-south-west to east-north-east.

The commonest bottom-rock of the Bijáwar system in Bundelkhand, is a quartzite[1] that might locally be called a sandstone. It is generally fine-grained, but sometimes, at the base, coarse and conglomeratic from containing pebbles of white quartz. It rests quite horizontally, or with a slight dip, upon a denuded surface of the gneiss, even in that most western part of the area, where the uppermost portion of the Bundelkhand gneissic series is supposed to be found.[2]

With this quartzite a hornstone-breccia and a limestone are intimately associated. They sometimes replace the quartzite as the bottom rock, or else are interstratified with or overlie it. The hornstone is compact quartz, more or less transparent or opaque, of yellow, brown, and red tints, the angular fragments included in it being generally of white quartz, and always paler than the matrix. In some cases, if not in all, they are clearly not the result of fracture caused by contortion, for the breccia mostly lies quite flat upon a firm support. Occasionally the former continuity of the detached pieces is evident, the mass looking as if thin bands of quartz had been shattered by concussion or shrinkage, then re-cemented in place, and the interstices filled by a more jaspideous form of quartz. The limestone, too, is highly silicious, the quartz appearing both as thin layers and as shapeless, irregular segregations of chert.

[1] *Memoirs*, II, 11, (1860). | [2] *Supra*, p 27.

These bottom rocks of the Bijáwar system in Bijáwar are very irregular in their distribution. In some sections there is no quartzite, in others no hornstone-breccia or limestone. The total thickness nowhere exceeds 200 feet. This unevenness of the basement-bed tends to suggest the unconformity of the succeeding deposits, but no confirmation has been found of this suggestion, on the contrary, sub-schistose shales, like those of the upper part of the system, are sparingly intercalated with the limestone and quartzite.

More or less earthy ferruginous sandstone, locally somewhat conglomeratic, is the prevailing upper rock, and is associated with incipiently schistose rusty shales. The iron in these rocks is locally concentrated into a rich hæmatite which has been extensively worked. Several thick, but discontinuous, beds of dioritic trap occur in the bottom part of the group.

The whole Bijáwar system in the typical Bijáwar area is probably not more than 800 feet thick. The strata generally either have a very low south-easterly dip, or are quite horizontal, but in a few places to the south, before they become covered up, they are seen to have undergone a considerable amount of crushing, which has not in the least affected the lower Vindhyan rocks immediately overlying. The general immunity from disturbance in this small area may be due to the original shallowness of the deposits here, where they thinned out over the mass of gneiss, which afforded an unyielding support against compression. It is probable that the transition basin deepens rapidly to the southward, beneath the Vindhyan rocks, and that the complete unconformity between the Bijáwars and the lower Vindhyans, as observed in the Son valley, rapidly replaces the general parallelism of stratification that obtains in the Bijáwar area. East of the Ken the Bijáwar rocks soon disappear, being totally cut out by the Vindhyans overlapping on to the old gneiss. From a little west of Allahábád all the lower azoic rocks are concealed by the Gangetic alluvium, which stretches up to the base of the Vindhyan scarp.

At the Ken the character of the bottom Bijáwar rocks changes rapidly, the strong quartzite thins out suddenly, and a prominent rock, on the continuation of its strike, is a peculiar, sharply cellular quartzite, much quarried for quernstones, but the beds associated with this quartzite are sandstones and shales like those of the upper part of the series. In the river, and certainly below the horizon of the bottom quartzite of the Bijáwars west of the Ken, there are two or more steady outcrops of pebbly sandstones, having the same low south-easterly dip as the adjoining Bijáwar strata, but occurring in the midst of thick pseudo-crystalline gneissic rocks. It is important to notice these observations with a view to their verification or correction, for these sandstones seem to have escaped the notice of the later observers, and they are important as fixing the affinities of the associated gneissic strata with the transition series rather than with the normal gneiss of Bundelkhand. Very similar rocks are found far to the east in an analogous position at the base of the transition series in Behar, and again extensively in the Arávalli region, and the question arises as to whether we must not recognise in the great gneissic series some rocks that are not metamorphic in the full sense of the word, but are merely reconsolidated granitic or gneissic detritus.

Proceeding from Bijáwar in a south-west direction obliquely across the plateau, where the Vindhyans are for the most part covered by the Deccan trap, we should strike the Narbadá about Handiá, at the west end of the wide alluvial plain, 200 miles long, which, in India, is specially designated as the Narbadá valley. West of Handiá there is a considerable area occupied by transition and gneissic rocks. They abut on the west against the Vindhyan rocks of the Dhár forest area, but appear again in the north of this area, and west of it about Bárwai. These transition strata have been fully recognised by Mr. Mallet,[1] as bottom Bijáwars, consisting of quartzite hornstone-breccia and chert-banded limestone, identical with those of Bundelkhand. No associated trap rock was observed.

These rocks are more disturbed here than in Bijáwar, but Mr. Mallet describes their relation to the gneiss as the same, the quartzite being often found quite flat and surrounded by vertical strata of the metamorphics. It is only possible to question this view by supposing that what we take to be stratification in the metamorphics is a result of molecular forces acting on lines of cleavage. This possibility has been forcibly argued with reference to this very area, and is connected with the suggestion that the two series may be very closely allied, the gneiss being more or less a metamorphic condition of the Bijáwars.

Upon the settlement of this question as to the relations between the metamorphic and transition series, it will depend whether the gneiss of the Dhár forest should be affiliated to that of Bundelkhand, or to the supposed younger gneiss of Bengal. The composition of the Dhár forest gneiss is in favour of the former, as well as the relation stated to subsist between it and the Bijáwars.

Here, as so often elsewhere, a doubt occurs as to the intrusive character of the more granitoid varieties of the gneiss. Some hornblendic and earthy schists of this area, as above Mortaka where the Indore railway crosses the Narbadá, have been included with the gneiss, but it may be questioned if they do not belong to a transition group older than the Bijáwars.

In the lower Narbadá valley the Bijáwar formation has been recognised[2]

[1] *Memoirs*, VI, 199, (1869). [2] *Memoirs*, VI, 200, (1869); XXI, 14, (1884).

specifically identical with the beds in the Dhár forest, near Bágh and Jobat, the two localities being separated by 80 miles of Deccan trap.

All the most characteristic rocks of the formation are well represented at Bágh,[1]—quartzite, hornstone, breccia, and cherty limestone,—and here again interbedded trap occurs, though not found in the Dhár forest area. Clay-slate, too, is more prominent, and sometimes becomes conglomeratic through the presence of pebbles, which are more or less drawn out in the direction of the cleavage planes. The town of Bágh stands near the south boundary of the small triangular area of Bijáwars, where it is covered by cretaceous rocks, the other two boundaries, with the gneissic rocks, being faulted. The area only extends 7 miles to the north-north-west, and 5 miles to the east of Bágh. The rocks are highly disturbed and cleaved, but the metamorphism is local and moderate.

Jobat is about 15 miles west-north west of Bágh, and stands at the southern point of another small patch of transition strata. The conditions are peculiar and puzzling. The only recognisable Bijáwar rock is a very typical one, a locally brecciated, ferruginous jasper, with veins of hornstone. It lies almost horizontally, forming a low scarped plateau. Along the north-east border, south of Anthi, black and grey schistose slates appear between the jasper and the metamorphics, the foliation and apparent bedding in the schistose slates and gneiss being parallel, with a high dip to the south-west. Both rocks are highly charged with vein-quartz, suggesting local crushing or faulting.

These disturbed rocks were originally regarded as Bijáwars and classed with the horizontal jasper beds. Mr. Bose has, however, described patches of slates, quartzites, and limestones, which are said to pass insensibly into the metamorphics, but are quite distinct from the horizontal jasper rock.[2] In this case the latter alone can be regarded as Bijáwar, and this appears to be the more probable interpretation.

Proceeding eastwards up the Narbadá valley from Handiá, no rocks are exposed under the Vindhyan scarp, on the northern side for a distance of 120 miles to where the Bijáwars form low hills in the Narsinghpur district. The cherty limestone and breccia are the only beds seen here, but this may be because the lower rocks are covered by alluvium. The gneiss does not appear again on this side of the valley.

Along the south side of the river, on the edge of the Gondwána formations of the Sátpura hills, there are more frequent outcrops of the transition rocks. The most westerly are near the Moran river, about 30 miles east of Hardá, where some narrow ribs of the cherty limestone protrude through the Deccan trap, which covers all the rocks to the west of this point. On this south side of the valley, the cherty limestone, generally

[1] *Memoirs*, VI, 303, (1869).

[2] *Memoirs*, XXI, 13 (1884).

much contorted and brecciated, is also the rock most frequently seen, but other beds do occur, as in the Bári hill 15 miles east of Soh́ágpur, where a considerable thickness of trappoid and earthy rocks is exposed, the latter being so little altered as to be easily mistaken for the Talchir shales of the contiguous Gondwána area. In many places on this south side of the valley gneissic rocks of doubtful character occur close to the Bijáwars, and the relation between the two series is certainly not simple superposition, both being found at the same level in closely adjoining positions.

At the head of the Narbadá valley in the north of the Jabalpur district, there is a continuous exposure of Bijáwar rocks between the Vindhyan and Gondwána areas. All the leading characters of the system already noticed are represented here, with a greater development of the argillaceous element. Fine earthy slates of reddish tints are the lowest strata seen. Their upper beds are associated with the quartzite, which underlies the limestone and is intercalated with it, and the limestone itself is not so constantly cherty as has been described elsewhere. Ribboned jasper beds, passing locally into bluish quartzite, among which rich hæmatite beds are well developed above the limestone, and both jasper and quartzite are frequently brecciated. Locally conglomeratic, earthy schists are also freely associated with this band.

Above the iron-band there is again a considerable thickness of earthy schists. Bedded trap occurs throughout the series.

These rocks are not on the whole greatly disturbed. Low undulating dips prevail, although locally there is much contortion. The highly inclined planes, so general in the schists, are of cleavage, not stratification. The thickness of the whole series exposed cannot be great, probably it is under 1,200 feet, and there is scarcely any presumption that the conformable slates beneath the limestone attain any great thickness.

Notwithstanding these conditions the rocks are in an advanced state of metamorphism. The limestone is generally crystalline. The schists are often highly micaceous, hornblendic, and garnetiferous, and the iron-ore is mostly the micaceous form of hæmatite. The section in the Narbadá, at the well known marble rocks, 10 miles south-west of Jabalpur, exhibits the high degree of alteration and local disturbance to which the Bijáwars have been subjected in this region.

Immediately to the east of the flat watershed between the Son and the Narbadá the band of transition rocks is entirely concealed by an extensive spread of laterite and alluvium, and beyond this we get into the region of the lower Vindhyans, which stretch to the south of the scarp of the upper Vindhyan plateau until they nearly come into contact with the Gondwána deposits. After crossing the Son, however, the band of transition rocks again expands gradually to a width of 25 miles in the south of the

Mirzápur district, and it is here we encounter the question whether one or two formations occur within this basin of transition rocks.

The northern half (about 10 miles wide) of the transition band, at a little west of the Rer river, is formed of regular Bijáwar rocks, such as we have hitherto seen them—quartzites, hornstones, banded jasper and hæmatite, limestones and slates or schists, with an abundance of intercalated trap. The whole band strikes against, and under, the lower Vindhyan strata, where the Son takes a southerly bend opposite Agori. The southern half of the transition band, 15 miles wide, as exposed in the Rer, is entirely composed of fine slates, with intrusive trap only, the dykes being mostly transverse to the bedding. Both groups are so intensely crushed together that no decisive section of the junction has been found in the low jungle-covered hills. Mr. Mallet mentions an instance at Ubra, at the north end of the section in the Rer, where a quartzite of the northern set seems to cap a ridge of the slates, but the case is not clear and the question of the relation is quite open, except that it certainly is not one of horizontal transition, as the two contrasting deposits are in full force and character in close juxtaposition to each other.

The western extension of the section into the Rewá country has been but imperfectly examined. The slates have already disappeared at the Gopat, and the northern band of true Bijáwars is in contact with the gneiss. In this region, where the Son takes a bend into the area of the transition rocks, there is a good instance of local metamorphism, the transition rocks along the lower Vindhyan boundary, distinctly recognisable as Bijáwars throughout the whole length between the Gopat and the Son at Marai, being in a gneissose condition and intrusive granitic rocks occur in them. The character of the contact of these beds with the main gneiss to the south is, however, of the kind described by Mr. Hacket north of Jabalpur, abrupt rather than transitional, but it is certain that they themselves are locally gneissic, and have been effected by granitic intrusions.

If it were certain that this character of the contact of the Bijáwars with the southern gneiss is constant and not due to faulting, and also that the gneiss of the Rer and the Gopat are the same, we could at once affirm the distinctness of two transition systems in this ground, for the junction of the slates of the southern band with the main gneiss is perfectly transitional—a gradual alternating passage from the strong gneiss, through gneissose and other crystalline schists, into the fine clay-slate, as is well seen in the section in the Rer. But while doubts exist upon these two conditions, it must remain possible that these slates of the Rer are only a bottom member of the Bijáwar system.

East of the Rer and the Kanhar several large inliers of gneiss and of granitoid rocks, of more or less intrusive character, occur within the slate area, and gneiss is the only rock seen below the Vindhyans at the Koel.

This encroachment of the crystallines upon the zone of the transition rocks is extended in Behar, where, for some miles north of the Grand Trunk Road west of Gayá, gneiss reaches quite across the strike of the slates. Several hills isolated on the alluvial plains in this neighbourhood are of thorough granite.

Immediately east of Gayá, transition rocks appear again on the prolongation of those in the Son valley, and having the same strike. They form several groups of hills in east Behar, known as the Maher, Rájágriha (Rájgir), Shaikhpúra, Kharakpur, and Gídhaur hills, which stand clear of the main gneissic area and more or less isolated in the alluvial plains, and those of Mahábar and Bhiaura on the northern margin of the gneissic upland. The aspect of all these hills at once shows that they must be formed of very different rocks from the Bijáwars of the west, and suggests also that all these Behar rocks belong to one system. They generally present scarped faces formed of massive quartzites on every side, the associated schists or slates appearing obscurely in the valleys. All the peculiar Bijáwar rocks are wanting. There is no limestone, hornstone, jaspideous ironstone, or bedded trap. The only similar rocks in the west are the slates of the Rer section, and there the quartzites, which form such a prominent part of the transition series of Behar, are absent.

We have a somewhat detailed description of the Mahábar and Bhiaura hills by Mr. Mallet,[1] and the relation of the gneiss and transition rock-series is shown to be very peculiar. The transition series here consists of three divisions, an upper, composed exclusively of strong quartzites as seen in Mahábar hill, a thick middle band, in which fine mica-schists largely predominate and a basal member, in which quartzites again occur, sometimes in great force, as in the Bhiaura ridge, though they may be altogether wanting at no great distance. The frequent presence of these quartzites here is of great service, by removing the doubts that so often arise, as to whether planes of lamination in schistose rocks, of uniform composition, are due to bedding or to cleavage.

It would be difficult to draw a more irregularly intricate line than the transition and gneiss boundary on Mr. Mallet's map. Near the Bhiaura and Mahábar ridges there is some approach to an average east and west strike of the boundary, but, as the plane of junction between the two series rises to the south, its line of outcrop meanders about in the most devious manner. This is not due, as might easily occur, to the irregular denudation of two deposits in flat parallel superposition. Here, as a rule, the lower (older) rock forms the prominences, between which the schists are deeply buried, yet the bedding in both rocks is found to follow

[1] An abstract of it is published in *Records*, VII, 36, (1874).

the intricate twistings thus produced, the actual junction being generally inclined at a high angle.

The boundary can always be fixed with precision, even in the absence of the bottom or Bhiaura quartzites, on account of the strong contrast between the fine mica-schists and the coarse gneiss, yet the transition rock seems to have fully partaken in the metamorphic action, for it is a thoroughly crystalline, garnetiferous mica-schist up to the base of the Mahábar quartzites. Variations are found in the gneissic series at the contact. On the north side of the Bhiaura ridge the bottom quartzites lie steeply against the dome-gneiss, elsewhere schistose gneiss occurs at the boundary. The dykes and massive outbursts of pegmatitic granite of this region are principally exhibited in the transition series.

A very close connection is thus established in this position, by conformity of stratification and by a common metamorphism, between the transition rocks of Behar and the gneiss in contact with them, and it is probable that a large part of the gneiss of Bengal is of the same age as that at the boundary of the transition series. There is, for instance, a very distinct outlier of the Mahábar schists and Bhiaura quartzites on the plateau just north of the Grand Trunk Road at Barhí, 30 miles to the south of the boundary in Behar.

There can scarcely be a doubt that the rocks of the Rájágriha and other detached hills of Behar are of the same formation as those of Mahábar, and so the contrast of their mineral condition is interesting. The latter have undergone general crystalline metamorphism, the former have only very locally suffered this change, being for the most part still in slaty condition. Yet it would seem that they too are closely surrounded by crystalline rocks, for whenever rock is exposed, through the alluvium near these hills, it proves to be granite. At one spot near Ghansura, on the north side of the Rájágriha range, there is a contact showing distinct intrusion of granite into the soft earthy schists. It is an ordinary ternary granite, not like the pegmatitic granite of the Mahábar region. In the immediate neighbourhood of Gayá many forms of special metamorphism and of contact-action are well exhibited.

The amount of disturbance is rather greater in the detached hills, where the rocks are less metamorphic than in the Mahábar region, and the very peculiar confused form of contortion, noticed in the surface of junction where the transition rocks rise against the gneiss to the south, is well exhibited throughout the formation, but in larger proportions in the top beds of the series. The Mahábar ridge itself is a typical instance of this structure. It is a long, narrow, synclinal ellipse, the quartzites dipping at a high angle all round, towards the centre, and curving continuously at each end of the axis. The Rájágriha range contains a pair of such ellipses compressed together, the quartzites being for the most part quite vertical along the sides. The Kharakpur hills, which form the largest of these groups, are a congeries of these discontinuous flexures, little or no regularity being observed in the direction of the axes of contortion.

We have still to notice the rock underlying the quartzite in the small ridge of Shaikhpúra, and in the little hills a few miles to the east at Luckeeserai, the junction station for the chord and loop lines of the East Indian Railway. There can be little doubt that the quartzite of these localities is the bottom rock of the Behar transition series, the Bhiaura quartzite. In the Shaikhpúra ridge it rests steeply against a rock having the texture of a thoroughly crystallised coarse granite, but completely decomposed. The relative position of the two rocks is precisely that of the Bhiaura quartzite and the dome-gneiss. Along a steady outcrop of some two miles long no feature of special intrusion was observed, and there is no extra metamorphism at this junction. The only contact-action that occurs is of secondary origin, in the formation of layers and vein-like strings of a sharply cellular quartz-rock much used for making hand-mills.

This section is noticed in connection with a more decided one at Luckeeserai, only a few miles to the east on the same strike, where the quartzite again rests against an amorphous mass of pseudo-crystalline granitoid rock, of much less sharply defined texture than at Shaikhpúra, in which strings of pebbles can be detected. This is underlaid by strong beds of coarse conglomerate, having the same dip as the overlying quartzite. The pebbles and boulders in this conglomerate are mostly sub-angular, and are exclusively of varieties of quartzite like those of the overlying formation, none being of crystalline rocks. They often appear elongated in the direction of the foliation, and adhere firmly to the matrix, which is a quartzose, sub-gneissose schist. Just east of Dharárah station some masses of this rock protrude through the alluvium close to the base of the Kharakpur hills. Another outcrop of conglomeratic schist was observed under the east end of the Gídhaur range and dipping towards it.

These Lukeeserai beds remind one forcibly of the pseudo-gneiss observed conformably underlying the Bijáwars in the section of the Ken river in Bundelkhand, and the suggestion revives, however slightly, the question of the possible correspondence of the transition groups in the two areas.

There is another rock frequently found with the undulating gneissic rocks of Behar, and elsewhere in this zone or protruding from the alluvium near the hills, that suggests the same connection. It is a jaspideous quartzite, often brecciated, and not unlike the bottom Bijáwar rock of Bundelkhand and the Dhár forest. It commonly has the same moderate dip as the rocks with which it occurs, but, when vertical or crushed, it is readily mistaken for fault-rock or vein-stone.

Suggestions of an opposite tendency can, however, be pointed out from observations recorded in preceding paragraphs. It was stated that the contrasting groups of transition rocks in the northern and southern portions of the section in south Mirzápur cannot be, in any degree, representative of each other by horizontal transition, and the presumption would be strongly in favour of the southern beds—the slates of the Rer—being the older of the two. If the Behar rocks had to be affiliated to either of these exclusively, it would certainly be to the latter group.

It has been already explained that the gneissic formations in lower Assam and the hills to the south are more closely allied to those of the peninsular region of India than to the metamorphic formations of the Himálaya. This relation holds also for the transition rocks, which are largely developed on the south side of the hills, where the sub-metamorphic beds are for the most part covered by the horizontal cretaceous rocks of the plateau, but are exposed in the deep ravines that penetrate to the very axis of the range. The lateral extension of these transition rocks has not been ascertained. On the central cross-section in the Khási country they stretch for 30 miles from near the south margin of the plateau to beyond the watershed north of Shillong, the culminating ridge with its summit 6,450 feet high being formed of the quartzites of the transition series, which have hence been described as the Shillong series.[1]

These Shillong beds have a general resemblance to those in Behar. They consist of a strong band of quartzites overlying a mass of earthy schists. Great masses of granite and of basic trap rock also occur intrusively. The former may well represent the similar rock seen to be intrusive into the slaty schists of Rájágriha, and the latter resembles certain trappoid rocks in Behar. Thus altogether the affinity is sufficiently marked to introduce the notice of the Shillong area in sequence with that of Behar. In the lofty and deeply eroded ground of the Assam hills the sections are much more favourable for study than on the alluvium-smothered plains, and some very puzzling observations have been recorded regarding the relations of the hypogene rocks to the Shillong series.

The lithology of these Shillong rocks varies much, according to local conditions of metamorphism. In places the quartzites, generally very firm and more or less schistose, are quite friable and might be called sandstones, but this state is probably due to decomposition, for the texture always reveals the effects of chemical change. It is coarser grained than is common in the Behar quartzites, and at the base, immediately over the slates or schists, there usually occurs a conglomerate, often of considerable thickness, made up chiefly of quartz pebbles, but with some rounded fragments of coloured quartzites. Still, so far as has been made out, the quartzite is conformable to the schists, but in troubled ground it is difficult to make sure of such a point. The schistose beds also exhibit

[1] H. B. Medlicott, *Memoirs*, VII, 197, (1869).

much variety of texture, from ordinary clay slate to well foliated schists and gneiss. These changes are simultaneous in both quartzites and schists, and it is noteworthy that the increase of metamorphism is towards the south, away from the area of the old gneiss.

The relation of the transition rocks to this gneiss has not been made out. On the only section of which we have critical observations, nearly due north and south through Shillong, the boundary with the gneiss occurs in the low jungle-covered hills, where observation is almost impossible. The dividing line between the two series crosses the high range to the west of our section, and it is there that the junction should be examined. The observation already noted, that the metamorphism increases to the south, would suggest that the junction of the schists with the main gneiss to the north may be lithologically abrupt. At the southern boundary there is a steep plane of contact between the highly altered transition rocks and the great accumulation of bedded eruptive rock, known as the Sylhet trap, supposed to correspond with that of the Rájmahál hills and, therefore, to be of jurassic age. The cretaceous sandstones lie evenly and unconformably on both formations.

In the midst of the transition area there is an extensive exhibition of eruptive rock, of very different character from the Sylhet trap. It is a dense, massive, basic diorite or greenstone. The high road between Surarim and Mauphlong crosses this rock continuously for five miles in the gorges of the Kálapáni and Bogapáni rivers. The direction of the road is oblique to the strike of the rocks, but at right angles to its outcrop the greenstone is fully a mile wide. It nowhere betrays any bedded structure, and its intrusive character is not so marked as might be expected with so extensive a display of igneous rock. There is, however, sufficient evidence of intrusion for this greenstone, as a well defined dyke passes from the main mass into the quartzite of the ridge, about half a mile south of Mauphlong. Elsewhere one may walk for miles along the junction of the two rocks without finding any signs of penetration of one by the other.

The relation of the granite, or at least of the larger masses of the crystalline rock, to the transition rocks, is also very puzzling. Two such masses adjoin the high road across the Khási hills. One is the Myllím (Molím) area just south of the Shillong ridge, and close to the road between Mauphlong and Shillong. The other area is much more difficult of approach, the granite being only exposed in the deep gorges under the sandstones of the plateau, as on both sides of Surarim.

The rock is a thorough granite. It commonly affects a spheroidal structure, and it is composed of pale pink orthoclase, often in large crystals, a small proportion of very pale greenish oligoclase, a little dark-green or brown mica, and an abundance of disseminated hyaline quartz. There can be no question that these great granitic masses are of later origin than the transition series, for the total want of symmetry in the arrangement of the surrounding sedimentary rocks forbids the supposition that they could have been deposited round the granite, yet the absence of any apparent connection between the form of the intrusive masses and the disturbance of the transition rocks is very difficult to understand. The quartzites (the upper member of the transition series) are generally found at the boundary, but their dip and strike are quite independent of the granite, as if their contortions had been fully established before the granite was introduced, and remained quite unaffected by it. The facts seem totally to preclude the notions of fracture and compression commonly associated with the word intrusion. The supposition of the mass being faulted into position also lacks any corroborative evidence, the boundary lines are all rounded and show no symptoms of fissuring. It is as if a great hole had been burned out of the old stratified rocks and the crystalline mass let in, or as if the transition rocks had been converted into granite up to a certain boundary, without affecting the area beyond that line. Yet the junction is quite sharp, the quartzites not being more altered at the very contact with the granite than away from it. In keeping with all these negative characters is the fact that no dykes or veins of granite have been observed issuing from the great mass of Myllim, nor even in its neighbourhood. This is the more remarkable, because dykes and veins of similar granite are not uncommon in the southern part of the area, where the general metamorphism of the transition series is so much greater as to suggest that the focus of hypogene activity lay in that direction, beyond the present southern limit of these formations. It is also in agreement with the facts and suggestions recorded to note that the granite is younger than the old dioritic Khási trap; several small dykes of granite are seen ramifying through the diorite in the bed of the torrent east of Surarim.

The gneissic uplands of Hazáribágh and Chutiá Nágpur, about 120 miles wide, separate the transition rocks of Behar from those which occupy parts of Mánbhúm and Singbhúm in south-west Bengal and stretch far to the west, the whole transition area being about 150 miles long from east to west, and 80 miles wide[1].

Although the total thickness of this series must be great, no distinctive zones are marked in it. From top to base it seems to be an indiscriminate alternation of quartzite, quartzite sandstone, slate and shales, hornblendic, micaceous, talcose, and chloritic schists, the latter passing into potstone, and some bedded trap. Well-preserved ripple marks are found in the slates and shales, and some of the latter are so little metamorphosed that they might be mistaken for Talchirs, but for the quartz veins that penetrate them.

[1] *Memoirs*, XVIII, 73, 124, (1881).

Some large inliers of gneiss occur within this basin of transition rocks. Around some of these inliers the boundary is in its original condition, as at Cháibásá, where shales and sandstones rest flatly and quite unchanged upon the coarse gneiss of the principal inlier, and the unconformity of the two series is further proved in this case by the fact that the underlying gneiss is profusely traversed by trap dykes, which do not penetrate the overlying deposits. The boundary between the transition rocks and the main gneiss of Bengal on the north is said to be a fault, on account of the more or less continuous presence along it of a rib of veinstone. This boundary occurs, however, at the base of a long descending section of the transition rocks and the beds along the line of junction are such as elsewhere appear as bottom-beds of the transition series. There are besides outliers of the slate series beyond the supposed faulted boundary to the east, about Súpur, and an inlier of gneiss a short distance inside it at Borobhum. We can at least conclude that the junction here, whether faulted or not, is abrupt, that is to say, without any gradation of stratigraphical or mineralogical characters. In this part of the basin the maximum of disturbance and of metamorphism seems to occur away from the boundaries. Further to the west, however, the junction of the slate and gneissic series is described as transitional, and granite veins penetrate the slates without much affecting them.

The most striking feature of this area is a mass of dioritic trap running continuously, but with varying width, nearly east and west, through the centre of the transition basin. Dalmá hill, 3,050 feet high, is formed of this rock, and here the outcrop is nearly 3 miles wide. The trap is described as a great dyke, but its composition is described as complex and obscurely bedded. A section north of Rámgarh is given as—[1]

1. Indurated chloritic schist.
2. Porphyritic trap.
3. Indurated chloritic schist.
4. Compact and amygdaloidal trap.
5. Indurated chloritic schist.
6. Brecciated trap.
7. Indurated chloritic schist.
8. Brecciated trap.

Several other cases of similar variation were observed. The supposed dyke is found along the axis of a greatly compressed synclinal fold, and has evidently been subject to much crushing, the description is not such as one would expect in the case of a truly intrusive rock, and it is at least possible that it is composed of contemporaneous volcanic rocks whose structure has been obscured by disturbance.

As in Southern India, these transition rocks carry metalliferous lodes of gold, silver, copper, lead, etc., but so far none of these have proved remunerative—except to promoters of joint stock companies and a limited number of speculators in mining scrip.

In Chutiá Nágpur a few exposures of quartzites and schist have been

[1] *Memoirs*, XVIII, 80, (1881).

separated from the gneiss, and it is probable that more will be found when the gneissic area, intervening between the patches of Gondwána rocks, is more closely examined. Some of these patches are indicated on the map, but they do not exhibit sufficient peculiarity to require detailed notice.

West of Ráipur, in the Central Provinces, is another stretch of transition rocks, known as the Chilpi ghát or Sáletekri beds,[1] which have been but little examined. They consist of quartzites, dark green and buff slates, and shaly beds, coarse conglomerates and numerous beds of trap. They are for the most part much disturbed, but sometimes lie at easy dips. The succession of the beds has not been worked out, and little more is known of them than their approximate boundary.

To the south-east Dr. Ball found, on the eastern side of the plateau south of Tarnot,[2] some much disturbed shales, with subsidiary quartzites underlying the horizontal quartzites of the plateau. These very probably belong to the same series as the Chilpi ghát beds, and on the accompanying map have been coloured as such.

Turning to the north-west corner of the Peninsula the beds of the Gwalior system[3] are found, about 120 miles from the Bundelkhand outcrop of the Bijáwar system, resting upon the gneiss in precisely the same mechanical relation, horizontally or with a gentle slope. The denuded outcrops of the quartz reefs traversing the gneiss are in both cases covered by the bottom deposits of the overlying transition groups, but a slight difference is noticeable at the actual contact. The bottom layer of the Bijáwars is commonly more or less adherent to the gneiss, the result of the partial metamorphism that the Bijáwars, even in Bundelkhand, have undergone, whereas in the Gwalior rocks the bottom contact-layer is still unaffected in contact with the gneiss.

A general list of the rocks of the Gwalior formation would not suggest any separation from the Bijáwars. Each contains sandstone or quartzite, limestone, jasper, iron bands and bedded traps. The arrangement of these different strata is, however, markedly different in the two cases, and the general facies of these two series does not suggest to the observer that they are representative. Still, the Gwaliors are more nearly allied by their mineral characters to the Bijáwars than to the lower Vindhyans, and on this account, on account of their relations to the slaty series of the Arávallis, and of the great unconformity which subsists between them and the upper Vindhyans, they are here classed with the transition rather than the older palæozoic systems.

[1] The only published description will be found in *Records*, XVIII, 187, (1885).

[2] *Records*, X, 175, (1877).

[3] There is a short notice of the Gwalior series by Mr. C. A. Hacket in *Records*, III, 33, (1870); it is also mentioned in Mr. Hacket's paper on the north-east Arávalli region, *Records*, X, 84, (1877).

The area occupied by the Gwalior system is only 50 miles long, from east to west, and about 15 miles wide. It takes its name from the city of Gwalior, which stands upon it, surrounding the famous fort built upon a scarped outlier of Vindhyan sandstone. The composition of the Gwalior formation is very mixed, and admits of only a twofold and very unequal subdivision. There is constantly at the base a sandstone, or semi-quartzite, a fine-grained stone, pale grey in colour, regularly and thinly bedded, often conglomeratic at the base, called the Pár sandstone, from a town 12 miles south-west of Gwalior. It varies from 20 to 200 feet in thickness, and is overlaid by about 2,000 feet of strata, consisting mainly of thin, flaggy, silicious, ferruginous shales, copiously interbanded with hornstone and jasper, frequently of a brilliant red colour. Limestone, more or less cherty, occurs on two principal horizons in these shales, but not continuously, and there are two principal zones of a dense basic dioritic trap. All these upper beds amounting to about 2,000 feet in thickness, and constituting the bulk and the characteristic portion of the Gwalior formation, have been distinguished from the Pár sandstone as the Morár group, the name being taken from the military station close to Gwalior.

With the exception of some very local slips and crushing, the Gwalior rocks are undisturbed, having a steady, low, northerly inclination of only three to five degrees. The features of the area correspond with this arrangement of the rocks. There is a continuous broad plateau-range on the south, from 300 to 500 feet high, formed largely of the Pár sandstone. On the west it is connected at right angles with the Vindhyan scarp, which lies at a slightly lower level, and it stretches thence eastwards to the Sind river, forming a steep scarp overlooking the gneissic area of lower Bundelkhand to the south. There are two other ranges on the north, parallel to the Pár scarp, but they are much broken by cross-drainage, the two longitudinal valleys between the three ridges being due to greater decomposition and erosion along the two outcrops of bedded trap. It is only at the west end, near the Vindhyan plateau, that these trappean bands are well exposed.

The general easterly direction of the Pár scarp is very steady up to the Sind river, but the line is much serrated by bays and headlands, in which the nature of the junction with the gneiss is well exhibited. At the most southerly points of the range the gneiss reaches to within a few feet of the summit, and is capped by only a few feet of sandstone, but the surface of the gneiss gradually slopes downwards in the valleys cut back to the north, and the thickness of the overlying Gwalior beds increases. This slope of the junction is largely due to the original form of the basin in which the beds of the Gwalior series were deposited, for close to the edge of the scarp the thickness of the Pár sandstone is small, near Pár only about 20 feet, while on the north side of the range, wherever sections are exposed, as at Badhano, ten miles south-east of Morár, the thickness is not less than 200 feet. No trace of an unconformity between the Pár and Morár groups having been detected, this thickening must be due to the form of the basin of deposition, and would seem to show that the present southern limit of the Gwalior series represents very closely the original limit of deposition of the Pár sandstone.

On the top of the Pár sandstone there occurs locally a compact calcareo-silicious bed that is worth noticing, because the peculiar coralloid forms it exposes by weathering were thought by Dr. Stoliczka to be of organic origin.[1] This rock is best seen just south of Bára, 25 miles east by south of Morár.

The lower zone of bedded trap is about 400 feet from the base of the Morár group. There are two or more flows with intervening shales well seen on both sides of the Indore road, at from 6 to 10 miles south-west of Gwalior. The thickness of these flows is very various. From 70 feet they thin out to nothing, but they are probably nowhere absent on this horizon, obscure outcrops of them having been observed at several places in the valley formed along their strike to the east. At some spots there is an appearance of the trap having burst up through the underlying shales. Thus, in the stream near the Trunk Road north-west of Bela, there is a low section showing the shales and trap in vertical contact, but otherwise the interstratification is unbroken.

In connection with this lower zone of trap there occurs a rock that will again come under notice in the Cuddapah system, and also in the lower Vindhyans. It is a compact porcellanic rock, as sharply and regularly bedded as the associated jaspideous shales. Occasionally it is obscurely porphyritic, having small indeterminate crystals scattered through it. An analysis of a specimen from the Gwalior beds gave a composition approaching to that of orthoclase felspar.[2] But there is no presumption that this porcellanic rock, or hornstone, which has more than once been described as volcanic, has any connection with volcanic activity, and its association here with trappean beds of highly basic composition is probably quite fortuitous.

The upper zone of trap is on a much larger scale. The whole plain of Morár is underlaid by it, at least on the north side, and, if allowance is made for the small dip, the flow can hardly be less than 500 feet thick. It is admirably exposed in the undercliff of the Vindhyan scarp in the fort hill and the promontories to the westward. In a small plateau about three miles to north-north-east of the fort it is overlaid horizontally by typical, rusty, jaspideous shales of the Gwalior series. Several detached hills in the plain lying east by north of Morár are formed entirely of this massive trap.

Limestone occurs principally on two horizons, in and above the lower trappean zone and, in the northern hills, above the great trap flow. In

[1] *Records*, III, 35, (1870). | [2] *Records*, III, 37, (1870).

both positions it is very uncertain and discontinuous. Within a space of 100 yards a mass of limestone, more than 50 feet thick, is found to be totally replaced by ochreous shales.

The iron ore, which is largely mined in the Gwalior system, is quite different from that found in the Bijáwars. The latter is a massive concretionary hæmatite irregularly concentrated in ferruginous earthy sandstones. The Gwalior ore is a fine wafer-like shale, composed of thin flakes of hæmatite with still thinner films of clay. It is a decomposed condition of the jaspideous shales, from which the amorphous silica has been dissolved out, leaving the iron ingredient in a very favourable state for mining and smelting. The conditions for this change seem only to have obtained near the base of the group, as the mines are in the shales a few feet over the Pár sandstone.

To the east and north the Gwalior system is covered by the great alluvial plains. On the west it passes under the upper Vindhyans, and two inliers, exposed by the removal of these covering rocks, are crossed by the Trunk Road. The only specific identification of the Gwalior beds beyond this area is at the nearest point on the opposite side of this northern extension of the Vindhyan basin, 70 miles to the north-west of Gwalior. At Hindaun there is a narrow ridge of banded jasper and ferruginous shales, which Mr. Hacket considers to be indubitable Gwaliors[1]. The Gwaliors at Hindaun are more or less vertical, and in contact with them, but not conformably, are some quartzite sandstone and red and black slaty shales, with irregular bands of limestone, which will be again referred to when dealing with the Delhi system.

The antiquity of these rocks is shown not only by this section at Hindaun, but by the very extensive denudation they had undergone pre-

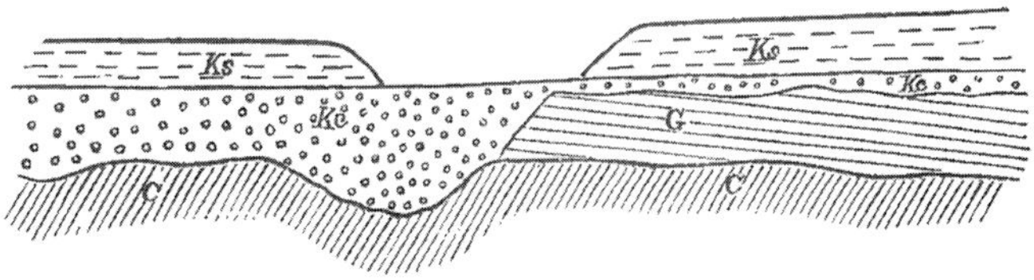

Fig. 4. Section showing the relation between the Gwalior and Vindhyan systems at the junction of the Pár and Vindhyan scarps. Ks. Káimur sandstone. Kc. Káimur conglomerate. G. Gwalior beds. C. Bundelkhand gneiss.

vious to the deposition of the upper Vindhyan sandstones. At the western end of the Pár scarp, the Káimur sandstones and conglomerate are deposited against a scarped face of Pár sandstone and rest on the gneiss at a lower level than the base of the Gwaliors close by[2].

The Arávalli system was formerly taken to comprise all the transition

[1] *Records*, III, 40, (1870); X, 90, (1877). | [2] *Memoirs*, VII, 58, (1869).

beds of the Arávalli range, including those belonging to what we will now distinguish as the Delhi system[1]. Little can be said of the petrography or of the relations of the Arávalli system, as limited by the exclusion of these newer beds, and even the validity of the separation is in doubt. Mr. Hacket was inclined to believe[2] that the schists of the Arávallis were merely the metamorphosed equivalents of the lower portion of his Delhi system, but the disproportion, in thickness as well as in metamorphism, between the beds of the great schist area of the Arávallis and the much smaller thickness of slates and limestone, which alone can be included with certainty in the Delhi system, renders this doubtful. Moreover he has recorded some observations, noticed below, which point to an unconformable break between the two systems.

Accepting the validity of a distinct Arávalli system of transition age, it may be described as consisting of quartzites and limestones, often containing coccolite, hornblendic and mica schists, abounding in crystals of andalusite, staurotide and garnet, and felspathic schists and gneisses.

The contact of the schists and gneiss shows a gradual transition[3], both in the centre of the range, and where they are in contact with the granitoid gneiss of Dhariawad and near Chitor, which has been regarded as belonging to the Bundelkhand or older gneiss series. This transition is in part apparent, and due to the true gneiss not having everywhere been distinguished from gneissose forms of intrusive granite, but is not improbably to some extent real.

The relation of the Arávalli schists to the Delhi system is somewhat doubtful; where the lower beds of the latter have undergone metamorphism they are difficult to distinguish, and it is probable that in those sections which appear to show a passage between the two, the break occurs between beds which it is difficult to distinguish from each other. This is rendered probable by the unmistakable unconformity which is shown by some sections, such as the one near Nithahar, where the Alwar quartzites rest upon the edges of nearly vertical argillaceous and quartz schists, and in the hills south of the Basi railway station, where a thick band of coarse conglomerate occurs at the base of the Alwar quartzites immediately above what were regarded as beds belonging to the Raialo group.[4]

East and south-east of Udaipur, in the heart of the range, conglomerates, containing numerous boulders and pebbles of quartzite in a schistose quartzite matrix, occur close to ridges of quartzite, which were regarded as of Alwar age[5]. The position of these conglomerates is not very well established. The nature of the boulders they contain would lead one to suppose that they were of later date than the quartzites of the ridge close by, but their position would indicate that they came between the quartzites and the adjoining older schists.

[1] *Manual*, 1st ed., p. 49.

[2] C. A. Hacket, MS. Report, 1886.

[3] C. A. Hacket, *Records*, XIV, 282, (1881). MSS. Reports, *passim*.

[4] Records X, 86, (1877); XII, 4, (1879).

[5] C. A. Hacket, MS. Report, 1885.

Thesesections appear to leave but little room for doubting that an unconformable break exists, and for this reason it appears advisable to recognise the existence of a distinct system of schists and slaty beds, underlying and older than the Delhi system, though it is impossible to attempt any subdivision or detailed description of them.

Beds of compact silicious rock and jasper, slightly resembling those of the Gwalior system, are recorded as occurring in the Raialo group in Shaikhâwatí, near Chenpura, north-east of the Basi railway station, and near Muhammadpur, south of Kherly railway station. At the time the observations were made the term Raialo covered all the beds below the Alwar quartzite, except the gneiss, and it is consequently uncertain whether these beds belong to the Delhi system or not, but the occurrence of jasper pebbles in the conglomerates of the Alwar group makes it probable that they should be referred to the Arávalli system, as here restricted.

In the central part of the Arávalli range the Arávalli schists are profusely penetrated by granite veins, and have in consequence undergone great metamorphism. But west of Udaipur there is an area where granite is wanting, and the beds are almost as unaltered as the slates and limestones below the Alwar quartzite, west and south of Nímach. This was considered to be an indication that the last-mentioned beds were represented in a more metamorphosed condition by the Arávalli schists, but it is equally possible, and on the whole more probable, that they are an outlier of the newer beds or, belonging to the older system, have locally escaped metamorphism.

The rocks of the Delhi system extend, in a number of isolated outcrops of varying size, from Delhi to beyond Nímach, a distance of about 340 miles from north-north-east to south-south-west, and for a width of about 150 miles in a direction transverse to this. The name applied to the system by Mr. Hacket in 1881[1] has proved an unfortunate one, as it is but ill exposed near Delhi, and we must look to the hills near Biána, and those of Mandsaur and the neighbourhood of Nímach, for the typical sections. It is, however, the name which has been used on the maps and in the publications of the Survey, and a change would lead to greater inconvenience than its retention.

The beds comprised in the system consist of a lower group of slates and limestones, and an upper, very much thicker, group of quartzites, known as the Alwar quartzites. The lower group was, in the first instance, named the Raialo, but as this name has subsequently been applied to all the beds below the Alwar quartzites, including those which there seems good reason for separating as an independent system, its use will be abandoned here.

[1] *Records*, XIV, 281, (1881).

The Alwar quartzites may be described generally as well-bedded quartzites, of light grey colour and fine grain, in which ripple markings and sun-cracks on the surface of the beds are common. Coarse grained beds are of frequent occurrence, and slaty bands are met with, arkose is often found near the base of the group where it rests on gneiss or granite, and the earlier part of the period during which it was formed appears to have witnessed an outburst of volcanic activity.

In the Biána hills the Alwar quartzites have been divided into five groups as follows:—[1]

5. Wer quartzites and conglomerates.
4. Damdama quartzites and conglomerates.
3. Biána white quartzite and conglomerate.
2. Badalgarh quartzite and shale.
1. Nithahar quartzites and bedded traps.

These groups are all separated by slight unconformities of denudation and overlap, but the distinctions appear to be quite local, for, even in the Biána hills, they are distinguishable on some sections, while in other outcrops it has been impossible to recognise them. All the groups vary very much in thickness, and are completely overlapped near Nithahar by the Wer quartzites, which rest directly on the schists.

West and south of Nímach[2] a very similar succession of beds is seen, consisting at the base of a conglomeratic sandstone, overlaid by about 200 feet of shales and limestone, and capped by a varying thickness of quartzite. The beds are very little disturbed and the exposures are completely isolated, but, northwards of Sadri, there are exposures of highly disturbed quartzite which, forming a series of ridges running northwards, can be correlated with the beds of Mandsaur and Sadri on the south, and those of the Biána and Alwar hills and Delhi on the north. The identification depends partly on the observed relation of the quartzites to the older rocks, partly on the similarity of lithological character, a perfectly justifiable method over such short intervals as we are dealing with, and is helped out by the frequent occurrence of beds of contemporaneous trap in the lower part of the series, though these have not as yet been subjected to a critical examination.

The relation of this system to the gneiss west of Nímach, and to the schists of the Arávalli system, as now restricted, is one of complete unconformity, there being usually a conglomerate at or near the base of the section, in which pebbles of the underlying gneiss are stated to occur near Daulapáni. In the ridges north of Sadri the same unconformity has been observed, and a similar unconformable contact, accompanied by a basal conglomerate, has been observed near Nithahar where the quartzites rest on vertical schists, near Talra south of Alwar where they rest on granitoid gneiss and contain gneiss pebbles, and at Marot, north of the Sámbhar salt lake, where the bottom beds of the quartzite, conglomeratic with rolled

[1] *Records*, X, 86, (1877).

[2] *Records*, XIV, 293, (1881).

and subangular fragments of quartz and felspar, rest on micaschists penetrated with granite intrusions[1].

In the Nímach area the lower slates and limestones of this system rest unconformably on the gneiss, but no case of unconformity with the Arávalli beds appears to have been recorded. The complete overlap of the slates by the quartzites which are conformable to them is, however, in itself an ample proof of the unconformity of the Delhi system to the underlying rocks.

The central sections of the Arávalli exhibit an apparent passage of the quartzites into the gneiss, one section in particular being mentioned, where alternating mica schists and gneissose beds are overlaid by schists and quartzites, then a two-foot band of gneiss, capped by the Alwar quartzites[2]. In these cases it is probable that a gneissose granite, intrusive along the bedding planes, has been described as a gneiss, or the so-called gneiss may be merely a metamorphosed arkose, in neither of which cases would there be any proof of a transition between the true gneiss and the Alwar quartzites.

In the hills near Hindaun quartzite sandstone, associated with red and black slaty shales and irregular bands of limestone, occurs in close proximity to the jasper beds which are considered to belong to the Gwalior system.[3] No actual contact is seen, but the distribution of the two types of rock leaves no room for doubting that the quartzites and slaty shales are the newer of the two. In spite of their likeness to the beds of the Delhi system, and the absence of anything at all resembling them among the Vindhyans, they were referred to the latter system[4]. This reference appears to have been due to the supposed difficulty of finding time for the deposition of all the transition beds of the Arávalli range, and their subsequent disturbance, between the close of the Gwalior period and the commencement of the Vindhyan, the beds of both these systems being almost undisturbed. The difficulty regarding the time required for the deposition of the transition beds vanishes if we recognise two distinct systems in the Arávallis, the older of which may be contemporaneous with, or older than, the Gwaliors, while the newer is younger; and as regards the disturbance, we will find when we come to deal with the extra-peninsular mountain ranges, that an intense and extensive disturbance of the strata has taken place during the tertiary period, and even within the latter half of it. The time required for the disturbance of the Arávallis may consequently be reduced to a very short period, geologically speaking, and if the suggestion,[5] that the Vindhyans bear much the same relation to the Arávalli range that the deposits of the Indo-Gangetic plain do to the Himálayas, is correct, the

[1] *Records*, XIV, 294, 296, 298, (1881).
[2] *Records*, XIV, 297, (1881).
[3] *Records*, X, 90, (1877); *Supra*, p. 67.
[4] *Manual*, 1st edition, p. 52.
[5] *Infra*, p. 104.

disturbance would not be anterior to, but contemporaneous with, the deposition of the Vindhyan beds.

Under these circumstances the correlation subsequently adopted,[1] which classed the quartzites of Hindaun with those of the Alwar group, seems the most probable one, and we have the Delhi system established as newer than the Gwaliors. This conclusion is supported by the occurrence of pebbles of jasper, closely resembling that of the Gwalior system, in the lower beds of the Alwar quartzites, north of Dhaulapáni and in the Biána hills.

Before leaving this system we must notice a peculiar form which the quartzite locally assumes at Kaliána, near Dádri, in Jind. The rock is here extensively quarried for millstones, and in some of the quarries it has become locally converted into what is known as itacolumite, or flexible sandstone. The quartzite in its natural form is glassy, and the individual grains of sand have become coated with an outgrowth of secondary quartz, giving them an irregular outline when seen in section. Generally the rock appears to withstand weathering extremely well, and is as hard and glassy a few inches from the surface, as in the depths of the quarries; locally, however, decomposition has been able to penetrate into the rock, and it has weathered into a mass of very irregular-shaped aggregates of quartz grains, held together by the interlocking of their irregularities, but capable of a certain amount of freedom of movement over each other. There is nothing to show why this peculiar form of weathering should have taken place in some places, and not in others. It is not confined to particular beds, nor is it continuous for many feet along the strike in the same bed[2].

Far to the north west of the termination of the Arávallis, after a wide interval of plains traversed by the Sutlej and the Rávi, some hills occur on the sides of the Chenáb at Chiniot and Kirána. These hills are only 40 miles distant from the Salt-range, but the rocks are totally different from any that occur there, and correspond well with those seen in the Arávalli range. They consist of strong quartzites with associated clay slates, forming steep ridges with a north-east to south-west strike. The highest summit is stated by Dr. Fleming to be 957 feet above the plain. The rocks seem, from the uncertain observations given of them, to be in a less metamorphic state than those nearest them to the south east, a fact which agrees with their remoteness from what is presumably the centre of disturbance of the region. The oldest rocks of the Salt-range are probably, from their contrasting petrological conditions, very much younger than the strata of Kirána, and, as the former are at least Cambrian, we thus obtain a small hint of the age of these transition deposits.

[1] C. A. Hacket, *Records*, XIV, 288, (1881). H. B. Medlicott, *Records*, VII, 30, (1874);

[2] C. A. Hacket, *Records*, XIV, 285, (1881); R. D. Oldham, *Records*, XXII, 53, (1889).

East of Baroda, at the south-west extremity of the Arávalli region, there is an outcrop of rocks which must be referred to one of the transition formations. It extends some twenty miles east from the Páwagarh hill, for eight miles south from Champáner, and to a considerable but unknown distance to the north. The beds of this exposure, while resembling those of the Bijáwar system in general character and state of metamorphism, do not contain any of its characteristic rocks, while the most remarkable rock of the exposure is wanting in the Bijáwars. For this reason it is not possible to refer them to the latter system. So far as can be judged from the description, they are more like the rocks of the Delhi system, but it is impossible to definitely refer them to it, owing to the long stretch of unexplored ground that separates the two. Under these circumstances it will be best to treat them under the name of Champáner, from the capital of the old Mahomedan kingdom of Gujarát, which stands upon their margin.

The principal constituent of the Champáner beds is a quartzite or quartzite sandstone, the other beds being conglomerates, slates, and limestones, with occasional ferruginous bands. The conglomerates are the most distinctive beds of the Champáner area; the matrix is a coarse, gritty sandstone, containing pebbles and boulders, often a foot in diameter, and occasionally ranging to three feet, consisting of granite, quartzite, talcose slate, and crystalline limestone, but none of typically Bijáwar rocks. Cleavage, which is well developed in all the beds which are susceptible of it, is occasionally seen in the pebbles of the conglomerate, but is rarely distinguishable in the matrix.

The passage from the Champáner beds to the gneiss appears to be gradual, so much so that it is frequently almost impossible to determine where the boundary should be drawn. Within the tract occupied by the metamorphics, quartzites are found, and a true conglomerate, containing rolled fragments of quartzite and very similar to that of the Champáner beds, is found among the gneiss west of Jámbughora. This area has not as yet been subjected to a close examination, and it is impossible to say whether the apparent transition is a real one, or the result of the intense disturbance which both the metamorphics and the Champáner beds have undergone.[1]

In the south and south-eastern portion of the country west of the Arávalli range there is a series of very ancient eruptive rocks, named after the Maláni district of the Jodhpur state. They consist principally of very silicious felsites, so hard that they are not scratched by quartz, and have

[1] The description in the text is based on that of Dr. W. T. Blanford, *Memoirs*, VI, 202, (1869);— Mr. R. B. Foote, in a letter received as this work is going through the press considers that the Champáner beds, by their mineral character and degree of metamorphism should be referred to the Cuddapah rather than the transition systems.

frequently the appearance and texture of jasper. They vary greatly in colour, from black or dark-brown to pink, blue, or white, the dark-coloured rock being always hard and undecomposed, whilst the light-coloured varieties are softer and appear to be altered. The most constant character is the presence of small crystals of felspar, usually of a pink or red colour, in addition to which small grains of transparent silica are frequently disseminated throughout the rock[1]. Their extremely silicious nature may be due to alteration, but their porphyritic character, and the occasional occurrence of ash beds, sufficiently attest their volcanic origin.

In places diorite was found associated with these rocks, and in some of the hills west of Bálmer coarsely crystalline granitoid syenite and pegmatite are intercalated in large masses with the porphyritic felsites. True granite may occur, but in the few hills examined mica was absent, although the character of the rock was distinctly granitic. The presence of similar granitoid rocks elsewhere is rendered probable by the occurrence of pebbles and boulders in some of the later formations.

The Maláni rocks must be very ancient, but no idea can be formed of their geological position, as they are nowhere associated with rocks of known age, except where underlying beds of comparatively recent date, and nothing resembling them appears hitherto to have been detected elsewhere in India.

They have been regarded[2] as of lower Vindhyan age, since they occur undisturbed in close proximity to the highly disturbed slates and schists of the Arávallis. They have not as yet been found in actual contact with the older rocks, but small hills of both are found standing up from the recent alluvium and irregularly interspersed with each other. On the other hand, nothing at all resembling the Maláni felsites has as yet been found in the lower Vindhyans east of the Arávallis. The general type is that of a much more ancient rock, and felsites, closely resembling those of Jodhpur except that they are not porphyritic, are found in the Toshám hill[3], dipping at high angles with the older beds of the Arávalli range. Besides this, the unconformity between them and the overlying sandstones, which are regarded as upper Vindhyans, is most marked, contrasting with the very much less pronounced unconformity between the lower and upper Vindhyans of the typical area. The correlation of the sandstones with the upper Vindhyans is, however, conjectural, and if they belong to the uppermost members of the system, this unconformity would not be inconsistent with a lowermost Vindhyan age for the Malánis. The age of these last must remain doubtful for the present, but they appear to belong to the transition rather than the Vindhyan rocks.

[1] W. T. Blanford, *Records*, X, 17, (1877).
[2] *Records*, XIV, 303, (1881).
[3] C. A. McMahon, *Records*, XIX, 164, (1886).

Reference has just been made as to the doubtful propriety of classing the Maláni beds with the transition systems and a similar doubt may almost be expressed with regard to the Gwalior and Delhi systems. The former of these finds its nearest analogue, as regards both mineral composition and degree of induration, in the Cuddapah system of Southern India, and Mr. Foote's suggestion that the equivalents of the Champáner beds must be looked for among the Cuddapah rather than the Dhárwár deposits of Southern India has been referred to.[1] As will be noticed in the next chapter, there are grounds for questioning whether the Cuddapah system should not be classed with these, among the newer of the transition systems, rather than with the Vindhyans. However this may be, there seem good reasons for accepting the Delhi and Gwalior systems as the newest of those described above. Next after them would come the Bijáwar and Behar systems, the latter being the older of the two, and finally the transition rocks of south-west Bengal and the Dhárwárs of Southern India, the last of these being marked out as the oldest by the greater degree of disturbance and metamorphism it has undergone, as well as by the manner in which the eroded edges of its upturned and metamorphosed strata are covered by the nearly horizontal basement beds of the Cuddapah system.

There can be little doubt that rocks corresponding to the transition systems will be found extensively developed in the extra-peninsular mountain ranges, but as yet these have not been sufficiently explored to allow of their separation as distinct rock series, except in a few isolated localities.

In Hundes and Spiti Mr. Griesbach has separated, under the name of Vaikrita[2], a series of beds which overlie the granitic gneiss. It is described as of great thickness, varying much in lithological composition, composed principally of micaceous schists, talcose rocks, phyllites and gneiss. The beds are now found occupying the cores of highly compressed synclinal folds, the crests of the intervening anticlinals having been denuded away till there is now an apparently continuous succession of strata across the folds.

Somewhat similar schistose beds occupy large areas in the central part of the range, and appear to extend far towards its southern margin in Nepál[3].

In the Dárjíling district Mr. Mallet has described a series of beds, said to be transitional with the gneiss, under the name of the Daling series.

[1] *Supra*, p. 73. foot-note.

[2] Said to be the Sanskrit for metamorphosed, *Memoirs*, XXIII, 41, (1891).

[3] *Records*, VIII, 93, (1875).

They consist of light green, slightly greasy, slates, sometimes interbanded with a dark greenish grey kind, passing insensibly into ordinary clay slates and more or less earthy or silvery according to the degree of alteration they have undergone. There are also bands of quartzite and quartz flags, occasionally some hornblende schist, sometimes slightly calcareous and passing into an impure dolomite containing crystals of actinolite. This is, however, a rare and exceptional rock, the most prominent lithological distinctions between these and the succeeding Baxa series being the almost complete absence of lime and the rarity of the brilliantly-coloured alternations of slates.[1]

The distribution of these beds is peculiar, and led to an erroneous idea of their position being formed in the first instance. They occur along the outer (southern) edge of the gneissic masses of Dárjíling and Daling, in the valley of the combined Tístá and Ranjit rivers. They separate these two areas of gneiss and extend on the northern side up each of the valleys, dipping inwards towards the gneiss on all sides, and the junction is described as transitional, except for a portion of the boundary north of Dárjíling, which is faulted. On the south the apparent passage is somewhat rapid, but on the inner sections the Dalings are more metamorphosed, and the distinction between them and the gneiss more difficult to draw. The form of the outcrop and the direction of the dips combine to convey the impression that the Dárjíling gneiss lies in the centre of a synclinal and is newer than the Daling series.

At the time the description was written a belief in the possibility of regional metamorphism, that is to say, of ordinary sedimentary rocks being converted, within a moderate distance, into true schists and gneisses, was still held by many geologists, but the whole tendency of recent investigations has been adverse to this opinion, and the opinion now prevalent is that of two contiguous series of beds the one which exhibits the greatest degree of metamorphism is *primá facie* the older. Added to this, the apparent dip of newer beds under older is a common feature of Himálayan sections, and when we find that the apparent relation of the gneiss to the Dalings is the same as of these to the Damudas, and again of the Damudas to the tertiaries, it is impossible to escape the belief that the true sequence is the reverse of the apparent one.

[1] *Memoirs*, XI, 40, (1874).

CHAPTER IV.

OLDER PALÆOZOIC (CUDDAPAH AND VINDHYAN) SYSTEMS OF THE PENINSULA.

Older palæozoic rocks—Reason for adopting the name—SOUTHERN INDIA—Cuddapah system—Cuddapah area—Kaládgi area—Karnúl series—Cuddapah area—Bhímá area—Godávari valley—Pakhal series—Pengangá beds—Chhatísgarh—Sullavai series—CENTRAL INDIA—Lower Vindhyans—Upper Vindhyans—Relation of upper Vindhyans to the Arávalli range—Vindhyans west of the Arávalli—Source of the diamond—Relative ages of the rock systems described.

In dealing with the newer group of systems, intervening between the gneiss and the lowermost fossiliferous beds of the Peninsula, we are met by the same difficulty as with the transition systems,—the absence of any fossil evidence by which we can judge of the true position of the beds. In this case the absence of fossils is the more extraordinary as many of the strata appear well adapted for the preservation of organic remains and have undergone no disturbance which could account for their subsequent obliteration.

The selection of a general name for the beds described in this chapter is a difficult task. Omitting purely local names, they have been classed as upper transition, azoic, or Vindhyan, but none of these are completely satisfactory and the best course to pursue will be to take into consideration the strongly marked unconformity that exists between the newest of them and overlying beds, of upper palæozoic age, together with their general lithological character, and class them as older palæozoic. This much we know, that they must be considerably older than permian, but it is as impossible to decide whether some of the oldest may not be precambrian, as to determine whether they may not to some extent be contemporaneous with part of those classed with the transition systems.

The older palæozoic strata, as defined here, are principally developed in two separate areas, one in the Madras presidency, the other in central India. There are besides a number of exposures in the Godávari and Mahánadí valleys in which the beds are not so well exposed and have been less studied than in the two principal areas.

There can be no doubt that the oldest rocks of this group of systems are those that have been described as the Cuddapah system in Madras, and they will consequently stand first for notice here.

The rocks of the Cuddapah system occupy a large area about the middle of the east side of the Peninsula, where the coast line bends from a northerly to a north-easterly direction. This feature is probably connected with the form of the Cuddapah basin, which is of a roughly crescent shape, convex to the west. The north-east horn of the crescent is known as the Palnád, and reaches to Jaggayyapet, a few miles north of the Kistna river; the southern termination at Tirupati (Tripetty) hill is 30 miles north-west of Madràs, or only 18 if measured to the outlier at Nágari Nose. The town of Cuddapah stands in a south-central position near the Penner river. Karnúl is on the northern edge and, further south, Gooty is just outside the western border, at its centre. The length of the basin is about 210 miles and its width 95, the area being nearly 13,500 square miles.

The eastern edge of the basin constitutes a well-defined segment of that vaguely expressed general feature known as the Eastern Gháts. The actual face of the highlands is locally known as the Yellakonda ridge. It is a flanking member of the Nallamalai range, which is formed by a belt of contortion of the Cuddapah rocks along this side of their basin. Between the hills and the sea there is a zone of low country, formed of metamorphic rocks and alluvium, about 50 miles wide, constituting the plains of the Carnatic, or Páyan Ghát (country below the Gháts), in the Guntúr, Nellore, and North Arcot districts. The elevation of this ground at the base of the hills is under 200 feet, the crest of the Yellakonda rising to about 1,000, and the summits of the Nallamalai to 3,500. The centre of the Cuddapah basin is occupied by the broad valley of the Kundair, the rocks rising again to form a steep range along the western margin of the basin, 2,000 feet above the sea and overlooking the gneissic upland of Mysore and Bellary, the elevation of which near the range varies from 800 to 1,800 feet. The Madras railway enters the basin at Gooty and leaves it at the southern point of the crescent, while the Kistna river adopts a very similar course in the northern limb. The watershed of the basin lies far to the north, and the Penner receives most of the drainage.

More than a third of the area, within the boundaries indicated, is taken up by the overlying Karnúl series, which occupies all the low ground of the Kundair valley in the middle of the basin and another large space in the Palnád.

The Cuddapah formation has been divided into the following groups[1] :—

Kistna group, 2,000 feet	Quartzites (Srishalam). Slates (Kolamnala). Quartzites (Irlakonda).
Nallamalai group, 3,400 feet . . .	Slates (Cumbum). Quartzites (Bairenkonda).

[1] *Memoirs*, VIII, 126, (1872).

Cheyair group, 10,500 feet	Slates (Pullampet). Quartzites (Nagari).
Pápaghni group, 4,500 feet . . .	Slates (Vémpalli). Quartzites (Gulcheru).

The groups are all more or less unconformable to each other, and all in turn overlap the others and rest directly on the gneiss, but there is so marked a unity of character running through all that it is necessary to regard them as a single system.

The distribution of these groups relieves us in some measure of the enormous aggregate thickness of 20,000 feet given in the list. Although the succession may be taken strictly in order of time, it is scarcely to be supposed that there was ever at one spot a continuous superposition of these strata to the extent of their aggregate thickness. Even within the present rock-basin, which must be taken as only a part of the area of deposition, the groups are local and discontinuous, each in turn overlapping the one below it and resting on the gneiss. In each case, however, there is more or less of denudation-unconformity, as well as overlap, so that the groups are much more than mere horizons of variation in deposition.

The original characters of deposition, and the induced characters of disturbance, are closely related to the actual boundaries of the field. All round the western boundary the junction is natural, and the deposits rest as originally laid down upon the gneiss, the strata having undergone comparatively little disturbance. On the east side of the basin, on the contrary, there has been much contortion of the strata, the boundary is represented as faulted and the beds often inverted. The lower groups are found to the south-west, and are gradually overlapped to the north and east.

In each of the groups of the Cuddapah series sandstones or quartzites prevail at the base and earthy deposits forming shales or slates above, limestones often occurring with the latter. The Pápaghni group is only found between the Tungabhadra and the Cheyair, being overlapped in both directions by the Cheyair beds. It takes its name from the river, in the gorge of which the best sections are seen. Its bottom member, the Gulcheru quartzite, rests upon an uneven surface of the gneiss, and rises up to the west to form steep cliffs, over an undercliff of the crystalline rock. Although the contact is quite sharp the two rocks are often connected together into an adhering mass. A considerable thickness at the base is coarsely conglomeratic, the pebbles consisting of the brecciated veinstones and banded jasper-rocks, which form prominent outcrops in the adjoining metamorphic area, but no pebbles of gneiss or granite were found except at one spot.[1] These bottom beds are described as shore deposits.

In the Vempalli subdivision of the Pápaghni group limestone is

[1] *Memoirs*, VIII, 158, (1872).

largely associated with the shales, and intrusive sheets of trap are also of frequent occurrence. In contact with, or near, the trap the limestone often contains bands of serpentine and steatite, as may be seen close to Karnúl, where the Vempalli band has overlapped the bottom sandstones, and rests directly on the gneiss.

The Cheyair group is well exposed on the Cheyair river. It is divided into two areas by the Karnúl formation stretching southwards, west of Cuddapah, into contact with the Pápaghni rocks. The constitution and relation of the Cheyair group in the two positions are somewhat different. In the north-west area, traversed by the Penner, the bottom band of sandstones and conglomerates is comparatively unimportant. It is there described as the Púlivendala (Pulavaindla) subdivision, from a town 40 miles west by south of Cuddapah. North of the Kistna it overlaps the Vempallis, and rests upon their denuded surface in the Penner ground, the conglomerates and breccias being largely made up of the characteristic chert-bands of the Vempalli limestone. Here, too, intrusive sheets of trap occur in the Púlivendala band. The corresponding beds in the southern area are described as the Nágari quartzites, from the well-known hill near Madras. They form for the most part the bottom-rock of the Cuddapahs resting on the gneiss in this region. The conglomerates are here made up of pebbles of quartz and quartzites (which are themselves sometimes conglomeratic), and occasionally of red-banded jasper, being thus more like the Gulcheru beds of the Penner area.

The upper dand of the Cheyair group in the Penner area is described as the Tádputri (Todapurti) beds, named from a principal village of the district. They comprise a great series, in which slaty shales predominate, with limestones, eruptive rocks both intrusive and contemporaneous, ferruginous chert, and jasper beds. Although not greatly disturbed, the shales are to some extent affected by cleavage and are hence qualified as slaty. Limestone occurs in two principal bands. It is a finely crystalline grey rock, with much segregated chert, which often assumes very fantastic shapes, especially in the upper part of the beds and near trap-flows. Of these eruptive rocks there are many strong outcrops, in two principal bands, a main one near the base of the group, and another two-thirds up. The only rocks that can be certainly classed as eruptive are coarse-grained, dark, basic diomrites, someties compact and of grey or pale-green colours. They are shown to be contemporaneous by their outcrop being continuous for long distances between well-marked bands of aqueous deposits, but the intervening deposits frequently cease, and the flows locally coalesce; moreover, they are distinctly confluent with intrusive dykes, as is well seen in the small bay below the southern flanks of the Opalpád plateau, 20 miles east of Gooty Perhaps the strongest argument for the contemporaneity of the bedded traps on this horizon is the fact that no intrusive igneous rock is known to occur higher in the formation, or in the Karnúls, and this could hardly be the case if the massive bands in the Tádputri zone were intruded after the completion of the sedimentary series.

In this group there are, associated with the traps, porcellanic beds resembling those of the Gwalior system. They have been regarded as of volcanic origin, but there is a great difficulty in supposing so highly silicious an ash could be produced by the same series of eruptions as gave birth to the unmistakeable igneous rocks in the section. They do not in any way resemble any known product of volcanic activity and their associations with the lava flows is probably fortuitous.

In the Cheyair area the Pullampet slates and limestones represent the Tádputri beds of the Penner. The traps and porcellanic beds are absent. The limestones are again silicious, and sometimes they are brecciated in a very unaccountable manner, without any disturbance of the strata. Some beds present a rugged humpy surface, suggestive of a coralline formation, but no organic structure has been detected.

The Nallamalai occupies a larger area than the other groups, principally on the east side of the basin, and takes its name from the range. The Bairenkonda summit, 3,500 feet above the sea, gives its name to the bottom band of quartzites. In the Pálkonda range, east of Cuddapah, these quartzites rest with slight unconformity upon the Cheyair group. In the Penner area the strong quartzites of the Gondicotta hills, overlying the Tádputri shales, are on the same horizon. Here the beds have a gentle north-easterly slope and pass under the Karnúl formation, but when they rise again to the east, in the Nallamalai, contortion is the rule, often to so extreme a degree as to produce folded flexures and inversion. In the synclinal troughs of these contortions the upper member of the group, called the Cumbum slates, is found, the underlying quartzites rising up to form the ridges.

The Cumbum slates are by far the thickest member of the group, and cover the greater part of the area. They are not very uniform in composition. There are several subordinate bands of quartzite, which it is not easy in broken ground to distinguish from the underlying Bairenkonda rock, and the slates themselves present many varieties, from fine, silvery, talcose beds to coarse, earthy clay slates, of many shades of colour. Occasionally they are foliated and schistose, and not easily distinguished from the schistose beds of the adjoining gneissic area, when the two happen to come in contact. As a rule, however, quartzites are found at the junction. Strong bands of limestone are frequent in the Cumbum slates. It is generally compact or finely crystalline, micaceous or talcose, of a slate grey colour, with purple tinges. The old lead mines near Nandiálampet, 16 miles north of Cuddapah, occur in a dark silicious variety of this rock.

At the north end of the Nallamalai, just south of the Karnúl and Guntúr road, there is a great dome-shaped mountain known as Eshwarakupam. It is composed of lower Cuddapah rocks dipping away from the hill on all sides, and surrounded by Nallamalai beds. A great thickness of strata is exposed, but it is not easy to identify them specifically with the groups already described.

The plateau through which the Kistna has cut its gorge, known as the Kistna Nallamalai, is formed of beds higher than the Nallamalai group and unconformable to it. These beds are therefore distinguished as the Kistna group. They comprise three well-marked divisions; the Irlakonda quartzites, forming the plateau of that name on the west, where they are 1,200 feet thick; the Shrishalam quartzites, forming a higher plateau to the north and east, called after a well-known shrine on the Kistna; and the intermediate shales, which are called Kolamnala, after a stream that traverses them. To the north the group spreads out over a flat surface of gneiss, and to the east it passes under the Karnúl beds of the Palnád, in which region the rocks are, again, intensely disturbed on the east. This group is supposed to be also represented further south in the Nallamalai, but the evidence is not decisive.

In the south Maráthá country, on the southern border of the great area occupied by the Deccan trap, and in great part separating the trap-region from the gneissic area of Mysore, there is a basin of somewhat similar rocks named after the town of Kaládgi[1], which lies near its eastern end. Its peculiar position is in a manner accidental, for it is certain that the whole of this basin was once overspread by the trap, which still stretches continuously along the crest of the Sahyádri for some distance to the south, and elsewhere outliers of trap are found resting on the gneiss. The strata rest with total unconformity on the crystallines, quite unaffected by metamorphism, and are considered to belong to the Cuddapah system on the strength of a general resemblance in lithological character, although the particular sub-divisions of the Cuddapah area cannot be recognised.

From the Kistna, below its confluence with the Gatparba, the Kaládgi rocks stretch continuously westward for more than 100 miles and then disappear under the trap forming the crest of the Sahyádri. In this direction several inliers are exposed by the local removal of the basaltic covering, the largest of which, at the foot of the Phonda Ghát, in the Konkan, is probably continuous with the main basin. On the north there is a large inlier at Jamkhandi. In all of these inliers, however, only the lower beds occur, so it is probable that the formation does not extend far beneath the trap. On the south borders of the basin there are numerous outliers of the bottom quartzites resting on the gneiss, both on the uplands of the

[1] R. B. Foote, *Memoirs*, XII, 70, (1876).

Deccan and in the Konkan. The Vengurla rocks and other small islands off the coast all consist of the very hard rocks belonging to the quartzite series. The former continuity of all these patches of rock cannot by any means be asserted, for it is evident that the deposits took place upon a very uneven surface of the crystallines, of which there are extensive inliers within the main basin, as at Gokák.

The series is divisible as follows[1]:—

Upper Kaládgi.

	Thickness.
6. Shales, limestones and hæmatite-schists . . .	2,000 feet
5. Quartzites, local conglomerates, and breccias .	1,200—1,800

Lower Kaládgi.

4. Limestones, clays, and shales	5,000—6,000
3. Sandstones and shales	3,000—5,000
2. Silicious limestones, hornstones, or cherty breccias	
1. Quartzites, conglomerates, and sandstones . .	

The bottom conglomeratic rocks are made up of the debris of the adjoining crystallines, and vary with the composition of the latter. They generally slope up towards the boundary of the area and form a scarp over a basement of gneiss. The cherty breccias form the most peculiar and conspicuous member of this part of the series. Mr. Foote suggests, with much probability, that they are formed by the decomposition and crushing of the highly silicious limestones that occur on the same horizon. A large proportion of the total area, forming a continuous margin to the basin, very wide on the south, and including all the outliers, is formed of the lower members (Nos. 1, 2, 3) of the series, and in this position the rocks are very little disturbed, and scarcely at all altered.

The limestones and shales forming the fourth division of the Kaládgi series are only found in a special basin of depression and contortion on the north-east side of the area. They generally occupy low ground and are much concealed, but may be fairly seen about the town of Kaládgi, exhibiting much disturbance. Several varieties of the rock are very homogeneous in texture and variously tinted, making pretty marble.

The only remnants of the upper Kaládgi group are found in the axes of synclinal flexures within this special basin, their preservation being evidently due to their being thus let in and encased by the folding of the whole series, so that the maximum of disturbance and of metamorphism is exhibited in these remains of the topmost beds of the formation. The principal of these elliptical synclinal areas of the upper groups are all within a short distance of Kaládgi. The direction of the axes of disturbance is very steady between west by north and west-north-west. This is also the direction of the major axis of the basin itself, in which all the special

[1] *Memoirs*, XII, 71, (1876).

contortion seems to have been concentrated on the north side, along what is now the lower valley of the Gatparba.

Only four cases of intrusive rock have been observed in the Kaládgi area, and all in the region of disturbance, in the highest beds; three in the Lokapur basin, and one in the Arakere synclinal valley. They are of compact, green diorite, unlike the older diorites of the gneissic area.

The rocks of the Karnúl series lie almost entirely within the basin of the Cuddapah system, where they are found in two separate areas, the larger of which occupies the whole of the Kundair valley and stretches to beyond the Kistna, while the other lies in the district known as the Palnád. The series has here a total thickness of only 1,200 feet, less than that of the smallest group of the Cuddapahs, and might be regarded as a member of that system, a view which has been urged[1] on the ground that the unconformity between the two is not much greater than those between the different sub-divisions of the older system, and that on the east side of the basin the Karnúl series has felt the full effects of the disturbance which they have undergone.

Such was not the opinion of the actual observers, who described the unconformity between the Cuddapah system and the Karnúl beds as sufficient to

N by W.

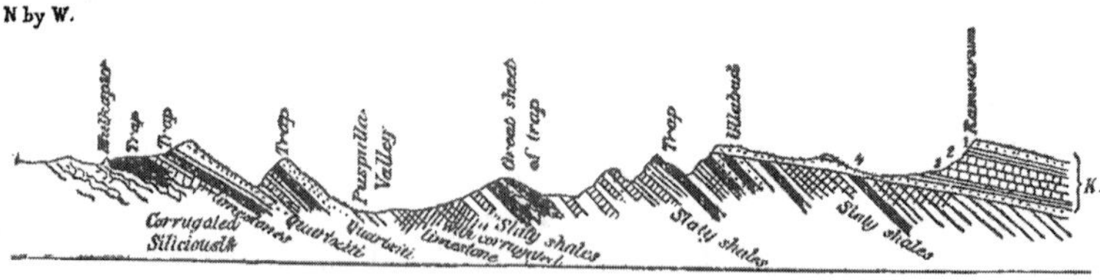

Fig. 5. Sketch section illustrating the relation of Cuddapah and Karnúl rocks, after King.

justify the separation of the latter.[2] This claim is strengthened by the occurrence of an outcrop of beds, referred to the Karnúl series, under the edge of the Deccan trap plateau in the valley of the Bhímá river. Although the westernmost point of this basin is separated by only 8 miles from the boundary of the Kaládgi area of Cuddapah rocks, and though each basin is over 100 miles in length, yet no representative of any of the rocks found in the one has been recognised in the other. If the reference of the rocks of one area to the Cuddapah system and the other to the Karnúl series, on the strength of general petrographical resemblances, is to be trusted, this indicates a change in the areas of deposition of the two periods sufficient to justify the separation of the rocks formed in each.

[1] *Manual*, 1st ed., p. 70.

[2] The nature of this unconformity is exhibited by the sketch section, fig. 5, which, at the least, shows the interpretation put by the actual observer on the observations made by him.—W. King, *Memoirs*, VIII, 125, (1872).

The Karnúl series, which is mainly a limestone formation with subordinate bands of sandstone and shale, has been divided into the following groups[1]:—

Kundair group	Shales (Nandiál). Limestones (Koil Kuntla).
Pániam group	Pinnacled quartzites. Plateau quartzites.
Jamalamadúgú group	Shales (Auk). Limestones (Narji).
Banaganpalli group	Sandstones.

The Banaganpalli group consists of sandstones, generally coarse, often earthy, occasionally felspathic or ferruginous, and usually of dark shades of red, grey, and brown colours. Pebble beds are frequent, the pebbles being small and numerous, composed of quartzite and various coloured cherts, jaspers and hardened shales, evidently derived from the cherty shales of the Cheyair group, on which the Banaganpalli beds rest.

The Banaganpalli beds are of interest as being the principal source of the diamond in the Cuddapah area.[2] There are many places on or near the Karnúl group where diamonds have been worked for in surface gravels, but at Banaganpalli these workings are carried on in the solid rock. Shallow pits, not more than 15 feet deep, are sunk in the rock, which is hard and quartzitic at the surface, but turns soft and easily worked underground, where short galleries are driven in the diamond layer, at, or close to, the base of the group. The diamonds occur in some of the more clayey and pebbly layers. Dr. King has recorded the opinion that they are innate in this rock, an opinion based principally on the perfection of many of the crystals. In the case of so hard a mineral the argument is not conclusive, and the nature of the rock in which they are described as occurring certainly suggests that they are of detrital origin. It is rather mysterious why the rock-workings should be so crowded as they are over certain spots, whilst large adjoining areas of apparently the very same deposits are left quite untouched. If this irregular distribution of the mines be only due to a delusion of the diamond-seekers there is still a very large field awaiting exploration.

The Jamalamadúgú group takes its name from a large village on the west side of the Kundair valley. It is composed at the top of buff, white, and purplish non-calcareous shales, well seen near the village of Auk (Owk). They have a maximum thickness of 50 feet, and pass down gradually into a finely crystalline or compact limestone, generally blue-grey, sometimes nearly black, and occasionally of pale buff and fawn colours. A very inferior lithographic stone used to be obtained from these

[1] *Memoirs*, VIII, 30, (1872).

[2] *Memoirs*, VIII, 96, (1872.

beds, and the rock is now much used for building, large quarries having been opened near the railway at the village of Narji, by which name the stone is known. West of Banaganpalli the Narji limestone is about 400 feet thick, but thins out both to the south and north. In the Ráichúr Doáb, about Karnúl, it rests on the metamorphic rocks, where it becomes cherty and brecciated in a peculiar manner and is described as a shore deposit.

Between the open Kundair valley and the western ranges, or Yellamala, there are in the Karnúl district some low flat hills, such as the plateaux of Upalpád and Undutla. These low plateaux are composed of a sandstone or quartzite, locally intercalated in the Karnúl limestones and known as the Pániam group, after the town of that name. The greatest thickness of the quartzites is only 100 feet, and the group disappears altogether to the north and south, nor has any sign of it been observed on the eastern edge of the basin. An upper portion, formed of firm white sandstone, has been distinguished as the 'pinnacled quartzites' from its mode of weathering, the lower beds, or 'plateau quartzites,' are coarser, more earthy and ferruginous, of various rusty tints.

In a basin of slightly disturbed strata the uppermost group must cover the largest area, and so the Kundair beds occupy the whole valley of the Kundair. There is a thickness of 500 to 600 feet. The upper two-thirds of purple calcareous shales and earthy limestones, distinguished as the Nandiál shales after a large village of that name, pass insensibly down into purer, compact and crypto-crystalline, flaggy limestones, known as the Koil Kuntla band, from a town 10 miles south-east of Banaganpalli, in which small papillæ resembling casts of *Cypris*, and numerous discoidal markings of half an inch to two inches in diameter are found.[1] The town of Cuddapah and all the large villages in the centre of the valley are on the Nandiál shales. In this position the rock is soft and crumbling, but to the south and east, on the margin of the mountain region, these uppermost beds of the whole sedimentary basin are quite slaty, being cleaved and contorted proportionally with the underlying formations. The lithological character of this group, as in some of the Cuddapah groups, changes to the north-west, and in the proximity of the metamorphics the Koil Kuntla beds are described as shore-deposits, which never extended much beyond their present boundary.

In the Palnád there is a large exposure of limestones which are believed to be of Karnúl age, and even the sub divisions have been in a manner specifically recognised in the south-west part of the ground.[2] The limestone is underlaid by a diamond-bearing sandstone, which has consequently been supposed to represent the Banaganpalli rock. In the Palnád country, however, there is great difficulty in distinguishing this rock from a closely

[1] *Memoirs*, VIII, 46, (1872).

[2] *Memoirs*, VIII, 107 (1872).

associated sandstone, clearly belonging to the Cuddapahs, but of the Kistna group at the very top of the Cuddapah series and several thousand feet higher, stratigraphically, than are the beds of the Cheyair group underlying the diamond sandstone of Banaganpalli; such at least is the position made out for the bottom sandstone on the south-west of the Palnád, towards the expanding rock-basin. On this side, too, some slight unconformity has been pointed out between the Palnád limestone and successive masses of the sandstone, and it has been remarked that the diamond workings here are confined to the rock close under the limestone, so as to suggest the limitation of diamonds to the horizon of the Banaganpalli group. All round the north-east corner of the basin, however, this sandstone, there known as the Jaggayyapet quartzite, rests directly upon the gneiss.

The leading structural character of the Cuddapah basin is maintained in the Palnád. On the west side the strata are comparatively undisturbed, while on the east border they are cleaved, foliated, and contorted, and appear to be overlaid by a natural ascending sequence of shales, limestones and quartzites, above what have been described as the Palnád limestones, and so these upper rocks would be newer members of the Karnúl formation. According to another, more probable, view this sequence is deceptive, being due to total inversion of the strata, the top quartzite being really a Cuddapah rock.

On the north-western border of the Cuddapah basin the Karnúl deposits are described as overlapping the formations upon which, for the most part, they rest, and lying upon the gneiss for a short distance up the Kistna valley. Seventy-five miles further in this north-westerly direction there is another area of rocks, having a strong likeness to the Karnúl deposits, and resting throughout their entire south-east border, for a distance of more than 100 miles, immediately upon the gneiss, while along their entire north-western border they are covered by the Deccan trap. The width of the basin thus exposed is exceedingly variable, both boundaries being very irregular in outline. It is greatest, about 25 miles, where the Bhímá river crosses the outcrop nearly at its middle, and from this circumstance the name of the river has been taken for the local designation of the rock basin.[1]

The Bhímá series is mainly a limestone formation which has been divided as follows in the central portion of the basin :—[2]

Upper.

	Thickness
(*g*) Red calcareous shales	30 feet.
(*f*) Flaggy limestone beds	
(*e*) Buff shales	18 „

[1] *Memoirs*, XII, 139, (1876). [2] *Memoirs*, XII, 160, (1876).

Upper—contd.

		Thickness.
(*d*)	Quartzites (sandstone)	200 feet.
(c^2)	Blue thick-bedded splintery limestone, brecciated in part	200 „
(c^1)	Thin-bedded limestone, with chert . . .	20 „
(*c*)	Blue and grey splintery limestone, occasionally brecciated	200 „

Lower.

(*b*)	Purple, red, drab, and dark-green shales, with calcareous flags at top	100 „
(*a*)	Quartzites (sandstones) and conglomerates . . .	60 „

It is principally in the south-western part of the area that the bottom sandy beds are developed to any extent. The pebbles of the conglomeratic bands are derived from the adjacent metamorphics, upon a very uneven surface, of which the Bhímá deposits were laid down, as is shown by the very winding outline of the boundary and by the occurrence of gneissic inliers, some of which are found near the trap of the north-western edge of the area. There is thus no presumption that the sedimentary basin extends far beneath the eruptive rock.

At Bachimali, the extreme easterly point of the southern expansion of the Bhímá basin, there is a basement pebble-bed much resembling the diamond layers of the lower Kistna valley. It is much broken up by small pits, as if at one time it had been searched for diamonds, but there seems to be no local tradition of any having been found.

The upper quartzite is quite a local intercalation, so that in some sections the series is almost exclusively made up of limestone. This is a very fine-grained rock for the most part, with a texture approaching that of lithographic stone. The colours are very various. Grey prevails, but drab and pink tints are common. The rock generally occurs in flaggy beds, and is much used for building, the pale cream coloured variety being preferred, although the grey stone is the more durable.

The formation has undergone very little disturbance, and the inclination of the strata very rarely exceeds from 2° to 5°. At a few places near the boundary some crushing and faulting has taken place, as at Gogi, where the lowest beds are vertical.

There are some patches of a singular limestone-breccia resting on the gneiss within the confines of the Bhímá basin, as west and north of the village of Yeddihali in the Agani valley. The brecciation has clearly been caused *in situ*, and Mr. Foote conjectures that these patches may be remnants of a former spread of the Kaládgi rocks.

With the exception of a doubtful fragment of silicified wood (or bone) found by Mr. Foote in the basement conglomerate close to the village of Kasakanahal, just within the Agani valley, no traces of organic remains were obtained from any of the Bhímá rocks. Mr. Foote speaks of the limestones as a pelagic formation, and remarks that there is a good deal to suggest that they were once continuous with the like rocks of the Karnúl area, and that they have been separated only by denudation.

North of the Kistna some outliers of highly disturbed quartzites and interbedded limestones have been recognised as belonging to the Cuddapah system[1], and north of these a large spread of Cuddapahs, originally described[2] as the Pakhal series, extends up the Godávari valley.

In the hills east of the Godávari at Albaka the series is described as consisting of two well-marked members, namely, a more or less slaty division, with many strong bands of altered arenaceous beds, and at least two bands of limestone, best exposed near Pakhal; and an upper division which is more generally arenaceous in its composition. The thicknesses of the divisions, where best developed, may be reckoned as—

2. Albaka division	2,500 feet.
1. Pakhal division	5,000 „

Near Pakhal the lower division can be divided into sub-groups as follows :—

5. A slaty band with thick seams of quartzite sandstone .	3,600 feet.
4. Grey and fawn-coloured silicious limestones . .	300 „
3. Clay slates and quartzites	500 „
2. Silicious limestone	150 „
1. Quartzites, with a few slates	700 „

These sub-groups are not constant and overlap each other. The lowest beds are frequently conglomeratic and at times pass into strong conglomerates, which rest unconformably on the gneiss. The upper division, more fully represented east of the Godávari in the hills which run parallel to the river and touch it at Albaka, is described as essentially a sandstone and quartzite formation.

The Pakhal outcrop extends in a north-west direction to the Maner river, a short way beyond which it is abruptly cut off by a fault. The rocks become gradually less indurated in this direction, the shaly beds finer in grain, and the limestones less prominent. In the Maner valley the shales, fine in texture, green and purple in colour, and much banded with thin calcareous seams, closely resemble those of the Cheyair group.

On the east side of the Godávari there is another large area extending from about 10 miles south-east of Albaka for over 100 miles to the north-west. In this exposure the upper arenaceous beds are largely developed, but there appears to be a slight unconformity in the cliff sections

[1] *Records*, XVIII, 20, (1885). | [2] W. King, *Memoirs*, XVIII, 209, (1880).

on the west of the range of hills, and it is not impossible that part of what has been here classed, and coloured on the map, with the Pakhal series belongs in reality to the Sullavai service which succeeds it.

In the upper Pránhíta valley, west of the great Wardhá valley coalfield, a series of limestones and shales of ancient date, known as the Pengangá beds, have been recognised by Dr. King as identical with the Pakhal beds of the Maner valley. They fall naturally into a lower limestone and an upper shaly group.

The limestone group consists of pale or dark grey or buff-coloured, seldom red limestone, well-bedded, with occasional layers of ribbon jasper. It is overlaid by the shale group, composed of fine-grained earthy shales, usually some shade of red in colour, with occasional beds of flaggy limestone. The shales have often a nodular structure and weather into small thin discoidal fragments like the Talchir shales, for which an isolated outcrop might easily be mistaken. It is extraordinary that no fossils have been found in these beds, whose texture is eminently fitted for the preservation of organic remains, while there has been no subsequent disturbance or metamorphism to account for their obliteration.

The Pengangá shales and limestones are usually found resting directly on the gneiss without any representative of the lowest quartzose conglomeratic zone of the Pakhals. North-west of Edlabad, however, there are said to be sandstones which appear to dip under the limestone, and a quartzite is recorded as occupying a similar position, in the hills north of Aksapur.[2]

These Pengangá beds were regarded by the earlier observers, and have always been referred to in the Survey publications, as of Vindhyan age. Further on we will return to the discussion of the validity of this correlation, but in the meanwhile it may be noticed that the evidence in favour of identifying the Pakhal and Pengangá beds with the Cuddapahs is as strong as it can be in the case of unfossiliferous rocks, where there is an absence of absolute continuity of outcrop. They were unhesitatingly identified by Dr. King, who examined both areas. The general lithological resemblance is described as very close, and, though the particular subdivisions cannot be recognised in the different areas, the Pakhal beds were regarded as answering to the Kistna and Nallamalai groups.[3] They exhibit much the same degree of induration and a similarity in their relation to the gneiss, and to the rocks of the transition period, and the case is much strengthened by the occurrence of a series of small outliers, in the space intervening between the northern limit of the main Cuddapah area and the southern extremity of the Pakhal outcrop.

[1] W. T. Blanford. MS. Report, 1866.
[2] *Memoirs*, XVIII, 224, (1881).
[3] *Memoirs*, XVIII, 212, (1881).

In the degree and nature of the disturbance they have been subjected to, the Pakhal beds resemble the Cuddapahs. Lying at low and gentle dips throughout the western portion of the area they occupy, they are compressed and folded at high angles about their eastern limit in the neighbourhood of the Singareni coal-field.

The Cuddapah beds have been recognised in Bastár territory, where the Indrawatí river, at Chitarkot, falls over quartzites that rest horizontally on the gneiss. They are overlaid by limestones and red shales, which overlap on to the gneiss and are overlaid by a yet higher series of quartzite sandstones. The country has not been examined in detail, and it is not yet certain whether the latter belong to the Cuddapahs or to the overlying unconformable Sullavai series.[1]

Further north Dr. Ball recognised the same beds on the plateau south of Tarnot,[2] and they spread out and occupy a large area of the Mahánadí valley in Chhatísgarh, where they have been regarded by observers working from the north as Vindhyan. They are described as consisting of a lower group, composed principally of quartzitic sandstones at times pebbly or even conglomeratic at the base, and an upper group of limestones and shales.[3] The shales are nearly always of a red purple colour, very rarely green or dirty grey, the limestones are fawn-coloured, grey or even black, sometimes pink or pale reddish purple, thick-bedded, compact, splintery or shaly in composition, graduating into shale, often seamed with chert bands. These beds, which were recognised in 1866 by Dr. Blanford as very similar to the Pengangá beds, occupy the centre of the basin and lie with easy rolling dips, the underlying sandstone band forming a margin round the outcrop, where it is often turned up at high angles.

The evidence for classing these with the Cuddapahs is not at present as good as in the case of the Pakhals, seeing that the intervening ground has been only cursorily visited, but the general resemblance is close, and certainly much closer than to the rocks of the overlying Sullavai series, which must now be noticed.

Besides the beds of Cuddapah age Dr. King recognised, in the Godávari valley, a series, unconformable to the Pakhals, which he called the Sullavai series.[4] They consist typically of a massive quartzite sandstone and conglomerate with a few slaty beds, overlaid by generally salmon or chocolate coloured sandstones, and capped by a strong series of thin and thick-bedded, very pebbly and gravelly, quartzites or indurated sandstones, which weather in the peculiar and picturesque manner characteristic of the pinnacled quartzites of the Karnúl series in its typical area.

[1] *Memoirs*, XVIII, 224, (1881).
[2] *Records*, X, 174, (1877).
[3] *Records*, XVIII, 173, (1885).
[4] *Memoirs*, XVIII, 227, (1881).

The thickness of the series where best developed in the Dewalmari hills and in the western outcrop near Kápra are stated to be—[1]

	Kápra.	Dewalmari hills.
3. Kápra quartzites and conglomerates .	100 feet.	700 feet.
2. Venkatpur salmon and chocolate beds .	200 „	300 „
1. Encharám quartzites	900 „	600 „

The unconformity of the Sullavais on the Pakhals is indicated by both overstep and overlap, yet on the whole there is a remarkable parallelism of dip between the two near Sullavai itself. Some sections are, however, recorded where the Sullavai sandstones rest almost horizontally on the eroded edges of the nearly vertical Pakhal slates and quartzites.

These Sullavai beds were recognised in the hills near Dewalmari and the observation is important, as the sandstone of these hills had already[2] been identified with that of the exposures, regarded as of Vindhyan age, east of the Wardhá valley coal-field. According to Dr Blanford these are white and purplish quartzite sandstones, breaking with a distinct conchoidal fracture, and, in the great exposure extending from Chimúr to Mul, they are associated with a more or less felspathic coarse grit, which decomposes into a very soft rock, easily mistaken for Damuda sandstone.

There is a much wider and more distinct barrier between the great northern Vindhyan basin and the Chhatísgarh, or upper Mahánadí area, than between the latter and any of the affiliated rocks to the south. The ridge of gneiss which, to the west, forms the well-raised base of the basaltic plateau throughout the districts of Mandlá, Seoní, Chhindwárá, and Betúl, and to the north-east forms the highlands of Chutiá Nágpur, is interrupted at this point and the Gondwána deposits stretch across from the Son to the Mahánadí valley. The watershed between the Son and the Mahánadí drainage is pretty high, and is occupied by Talchir rocks, probably of no great thickness, so that the gneiss most probably forms a rock-barrier from east to west, though of course it is open to question when this was produced. It may well be of post-Gondwána age. To the north of this barrier of gneiss the Cuddapahs and Karnúls of the south appear to be represented by a great series, principally composed of sandstones, long known to geologists under the name of Vindhyan.

The name Vindhyan, one of the oldest introduced by the Geological Survey, was used to designate the great sandstone formation of Bundelkhand and Málwá, and was adopted from the name currently applied by

[1] *Memoirs*, XVIII, 231, (1881).

[2] W. T. Blanford: MS. report, 1866.

Anglo-Indian geographers to the scarped range along the north side of the Narbadá valley. The Vindhyan system ranks third in superficial extent within the rock-area of the Peninsula, occupying in a single basin a larger surface than the combined areas of any other formation except the gneiss and the Deccan trap. The form of the basin is peculiar. There is a great area, 250 miles long, between Chitor on the west and Ságar on the east, and 225 miles broad from Indargarh on the north to Bárwai (or Mortaka) on the south, all presumably occupied by upper Vindhyans, although a very large part of it is covered by the trap of the Málwá plateau. From Ságar a long arm, with a maximum width of 50 miles, stretches eastwards for 340 miles to Sásserám in Behar. Another broader tract extends northwards from Ságar, and passes under the Gangetic alluvium between Agra and Gwalior. The gneissic mass of Bundelkhand lies between these prolongations. The exposed surface of the Vindhyan deposits is about 40,000 square miles and, with the area beneath the trap, the basin would occupy about 65,000.

Throughout the greater part of their border the Vindhyan sandstones are unconformably related to transition or gneissic rocks, but in the eastern branch of the area in Bundelkhand and the Son valley, and in the neighbourhood of Chitor and Jhalra Pátan, they rest, with little or no unconformity, upon deposits of very different character. These lower beds were at first noticed under local names in the several areas, but the convenience and fitness of having a common name for deposits so nearly related was soon felt, and the term lower Vindhyan has been used in this sense in spite of the very disproportionate importance of the two divisions so established, and a doubt as to whether they are really members of the same conformable system.

On the map prepared for this Manual it has been found necessary to make one colour serve for the lower Vindhyans and the Cuddapahs, but as the lower Vindhyans, in the sense here used, are confined to the margins of the Vindhyan basin, this is not likely to lead to confusion. From Sásserám, at the extreme east end of the area, the lower Vindhyans are continuous at the base of the Káimur scarp for 240 miles, disappearing at the Son-Narbadá watershed, where the upper Vindhyans sweep across into contact with the transition rocks. The greatest width of the lower Vindhyans across their outcrop in this their typical area is 16 miles, just where the Son enters its main valley from the south. At some points on the lower reaches of the river their outcrop is less than two miles wide. Some small inliers, appearing through the alluvium in Behar at a short distance east and north of the termination of the Vindhyan plateau, are most, if not all, of them of lower Vindhyan rocks, which also crop out from beneath the upper Vindhyans in some of the valleys on the north side of the plateau west of Sásserám. In this direction, however, the lower Vindhyans soon disappear and at the lowest level, where the Ganges washes the base of the plateau at Chanár (Chunar), only upper Vindhyans are exposed. The concealment of the lower groups is probably only due to depression in the main axis of the basin, for the very same rocks appear again beneath the Káimur sandstone as it rises towards the gneissic mass of Bundelkhand.

It may be considered certain that the Semri rocks, under the Káimur scarp in south-eastern Bundelkhand, are the same as the lower Vindhyans of Son valley, but their appearance on the north is much more irregular in every way, a circumstance which is easily accounted for. From Chebu, close to the Jumna, they are seen at intervals below the Vindhyan scarp for 160 miles to beyond the Dhasán. The principal exposures are for 20 miles east of the Dhasán, and for 12 west of the Ken (Cane). East of the latter river the beds are totally concealed for long distances, where the upper Vindhyans pass over them on to the gneiss, and the lower formation is only visible in the gorges of the principal streams. About Karwí again, where the main scarp begins to trend eastwards, oblique to the general strike of the basin, the lower Vindhyans are freely exposed, but at Bhita, where the Jumna first touches the rocks of the plateau a few miles above Allahábád, the upper Vindhyans are at the water level, the position being more to the dip of the basin.

The third of the principal exposures of lower Vindhyans lies in the extreme south-west corner of the main outcrop of the system and extends from about 10 miles north of Chitor in a southerly direction to the edge of the Deccan trap, whence it turns eastwards and occupies a narrow strip, with irregular boundaries, extending to about 25 miles south-east of Jhalra Pátan. Besides these three principal exposures there are some small outliers of what are believed to be lower Vindhyans resting on the disturbed Alwar quartzites on the eastern margin of the Arávalli region, south-west of Karauli.

The classification of the lower Vindhyan beds wants the definiteness that is attainable in the upper Vindhyans. There are no well-marked zones of sub-division, and all the members of the group are not to be found on every section, the irregularity being partly due to thinning out and partly to a lateral change of mineral character. The want of constancy is more conspicuous in the lower members than in the upper, a direct result of the mode of deposition. The first beds formed were deposited on an uneven floor of the older rocks, and as the irregularities of this became smoothed off, and the area of deposition enlarged by gradual subsidence, the conditions of sedimentation became more uniform and gave rise to more uniform and constant stratification of the succeeding beds.

The following sub-divisions have been recognised in the Son valley:

11. Limestone. } Rohtás group.	5. Porcellanic shales.
10. Shales. } Rohtás group.	4. Trappoid beds.
9. Limestone. } Rohtás group.	3. Porcellanic shales.
8. Shales and sandstone.	2. Limestone.
7. Limestone.	1. Conglomeratic and calcareous sandstone.
6. Shaly sandstone.	

These lithological characters by no means indicate well-defined or constant horizons in the series. They are all variable and pass into each other, both vertically and horizontally, by interstratification and thinning out, or a horizontal replacement of one form of sediment by another.

The lowest two groups, which, strange to say, appear to be in some degree equivalent to each other, are only found in the Son valley, and doubtfully in some outliers near Sásserám. They exhibit great and capricious variations of thickness, which can only be explained by their having been deposited on an uneven surface. The conglomeratic beds vary in type from a coarse thick-bedded conglomerate, composed of slightly rolled fragments of the underlying older rocks, 6 to 8 inches in diameter, to a coarse sandstone, with pebbles of quartz. The limestone No. 2 is also capricious in its distribution. It is in part a tolerably pure limestone, but is for the most part hard and silicious.

The porcellanic and trappoid beds are almost sufficiently described by their name. The porcellanic beds, mostly grey in colour, are very much like the beds described under the same name in the Gwalior and Cuddapah systems. The other beds interstratified with them were called trappoid from their resemblance, in mineral constitution and mode of weathering, to traps, but they are in fact composed of the debris of crystalline rocks which has undergone a subsequent induration. Their distribution is in accordance with their origin, as they are conspicuous where there is an abundance of crystalline rock close to the boundary, and absent or very slightly developed where slates are the chief rock exposed.

The divisions Nos. 6, 7, 8 form a sub-group of limestones, shales, and sandstones with a band of limestone, thicker and more prominent than the others, about its centre. Some of the sandstone is described as lithologically similar to that of the Káimur group, and owing to the dark, often black, colour of the shales, they were once mined into in the hope of obtaining coal.

The three uppermost members, Nos. 9, 10, 11, form another group for which the name Rohtás has been suggested[2], derived from the ancient fort of Rohtásgarh. Taken together, they are by far the most constant of any of the groups of lower Vindhyans in the Son valley.

[1] F. R. Mallet, *Memoirs*, VII, 28, (1871).

[2] *Manual*, 1st ed., p. 78.

The beds of No. 6 are described as being almost universally ripple-marked in the Mahánadí tributary of the Son, besides which sun-cracks and the marks of rain-drops are very common.

In Bundelkhand the Semri, or lower Vindhyan, beds were classified by Mr. Medlicott as follows[1]:

5. Tirohán limestone.	3. Dulchipur sandstones.
4. Pulkoa schists.	2. Semri shales and limestone.

1. Semri sandstone.

These are not all represented on every section. In the easternmost exposures only the Tirohán (Tirhowan) limestone, and what was believed to be a representative of the Semri sandstone, are seen. The Semri shales die out near Sháhgarh and the Semri sandstone, thinning out about the same place, can only be traced to the Dhasán river. Coincident with the decline in thickness of these two groups, the Dulchipur sandstones, whose most easterly limit is near Chopra, increase in thickness and importance and come into direct contact with the Semri sandstone by the overlap, or thinning out, of the Semri shales. At the western end of the exposure of the lower Vindhyans the Dulchipur sandstone is the only member of the group represented.

Though none of the sub-divisions of the lower Vindhyans in Bundelkhand can be identified with those in the Son valley, the general resemblance in lithological character, and more especially in their relations to the upper Vindhyans, is such as to make their identity certain.

The two uppermost members very circumstantially represent the Rohtás group of the Son—the thin, sharply bedded, fine grained limestone of very variable composition, both in chemical and mechanical ingredients, and the flaky silicious shales between which, even more capriciously than in the Son area, occur the most complete vertical and horizontal transitions. When the Bundelkhand ground was first described the equivalence of these different rocks was not detected, and consequently it was supposed that the shales had suffered denudation before the deposition of the limestone, and the limestone again before the deposition of the Káimur sandstone, which is found resting directly on both. In one form or the other, as shale or limestone, this group is found from end to end of the outcrop, being, like the Rohtás group, the only constant member of the series.

There is one character connected with the limestone in Bundelkhand that does not occur in the Son region. It is almost constantly overlaid by a silicious breccia, not detrital, but apparently composed of thin layers of agate, chert, and jasper, shattered in places either by concussion or desiccation, and re-cemented by sintery or hyaline silica, free

[1] *Memoirs* II, 6, 1860).

from sand or other detrital matter. This breccia, which is adherent to the limestone and also fills cracks in its upper surface, is connected rather with the Tirohán limestone than with the overlying Káimur sandstone, which often has at its base a breccia conglomerate very different in character from the Tirohán breccia. This bed is sometimes 40 feet thick, as on Panwári hill, south-east of Tirohán.

In the middle area, at and west of the Ken, the Semri sandstone and the overlying shale and limestone band are well developed. The latter is also fairly seen in the gorges of the Ranj and the Boghin, east of Panner, but in the eastern area, about Karwí, the Tirohán (Rohtás) limestone, very free from its familiar shales, is with one exception the only member of the series. The exception consists of a very peculiar bottom rock covering the granitoid gneiss. Where found under the limestone this rock might readily be referred to the Tirohán group, for it often has layers of dense, fine limestone just like that rock, and is otherwise cherty, as is often the case with the limestone, but it is largely a detrital rock composed of quartz-sand, felspar-grains, and (characteristically) glauconite. Cherty segregation in many forms,—spongy, pisolitic, amygdaloidal or disseminated,—gives a most peculiar aspect to the bed. This rock is traceable in the hills south-west of Karwí, the most north-westerly of which about Akbarpur are altogether of metamorphic rock, and have a pointed or rounded outline, the next have a thin cap of Káimur sandstone, but the sedimentary beds thicken steadily to the south-east, and at the sacred hill of Chhattarkot the gneiss is only seen at the base on the north-west side. At the high elevation of the junction there is only a remnant of the cherty contact rock coating the gneiss under the Káimur sandstone, but in the Chhattarkot hill the contact rock occurs under the limestone, holding its position as a true bottom-rock. At a few places in the eastern area the flaggy sandstones of this band are well marked, as in the gullies to the south-east of Chhattarkot hill, and they become more developed to the west or north of Panna, on Bisrámganj Ghát, where they are 50 feet thick. In this way they are traceable into relation with the Semri sandstone, in which also glauconite grains are of common occurrence.

This peculiar contact-rock of the east has been more specially noticed because of a conjecture that it may possibly be an original nidus of the diamond. A common form of it is a semi-vitreous sandstone, or pseudo quartzite, of a greenish tinge, the result of the local solidification of sandstone by diffused silica. Large pebbles of this rock are very abundant in the conglomeratic diamond bed of the Rewá shales at the Panna mines, and it is said they are broken up in the search for diamonds.[1] The diamond-bearing beds of the upper Vindhyans are now at a much higher level than any existing outcrop of the Semri beds, but it is very probable

[1] *Memoirs*, II, 71, (1860).

that this peculiar rock once extended over the then elevated surface of the gneissic area.

Both in the Son valley and in Bundelkhand there are indications that the present limit of the lower Vindhyans is not very far from that of their original extension. The irregularity of the lower groups in the Son valley, together with the coarseness of texture of the lowest member of the group, show that they were deposited on an uneven floor of deposition. The thinning out of the subdivisions, except those forming the Rohtás group, east of Bardhi, and the complete absence of Nos. 6, 7, and the sandstone of No. 8, are evidently due to this area having escaped the sedimentation, which went on elsewhere. Another observation of importance is that in the outliers to the south the lowest conglomerate is much thicker than in the main exposure, indicating an approach to the limits of deposition.

In Bundelkhand the original limitation of the lower Vindhyans is most unmistakeably exhibited by the overlap of the Káimur group on to the gneiss and Bijáwars. This is clearly enough seen on a large-scale geological map, where, in all the northern prominences and outliers, the upper Vindhyans are in direct contact with the older rocks, while in the deep cut valleys draining from the south the lower Vindhyans intervene. It is confirmed by the record of sections, where the lower Vindhyans are seen to be banked against a sloping surface of Bijáwars.

The lower Vindhyans of Chitor and Jhalra Pátan have not been so fully described as those of the other two areas. They consist of shales, limestone, and sandstone, the latter often conglomeratic near the base of the series and sometimes containing boulders that range up to three feet across, but do not appear to contain any of the volcanic or pseudo-volcanic beds found in the Son valley. No unconformity with the upper Vindhyans, or trace of one, is mentioned as occurring in this area, and the beds do not exhibit that degree of compression which is seen further east. At their western limit they rest unconformably on the gneiss and transition formations, but along the southern margin the boundary is formed by the overlying Deccan trap. The lower Vindhyans are here exposed in an anticlinal, whose southern half is for the most part concealed, but at Jhalra Pátan and south of Rámpurá there are outliers of the upper Vindhyan sandstone, intervening between the lower Vindhyans and the edge of the trap.[1]

The justice of classifying these beds with those next to be described in a single system is open to question. There is most certainly an unconformity between them and the so called upper Vindhyans. This might be inferred from the complete overlap of the lower Vindhyans by the Káimur group But there is better evidence than this in the very different facies of the two,

[1] C. A. Hacket, *Records*, XIV, 291, (1881).

the lower being on the whole argillaceous and calcareous, and the upper arenaceous and argillaceous, and more especially in the sudden and wide spread change from the fine grained deposits of the Rohtás to the coarse sandstone of the Káimur group. Apart from this, the occurrence of debris of lower Vindhyan beds 100 feet above the base of the Káimur group shows that the lower Vindhyans must have undergone some disturbance and been then exposed to denudation.

Two deceptive features have, however, given grounds for exaggerating the importance of the break between the upper and lower Vindhyans. Before the equivalence of the upper shale and limestone of the Rohtás group had been recognised, it had to be assumed that one or the other had been very extensively denuded before the deposition of the Káimur beds, an assumption which involved much irregular superposition, although none could be detected in actual sections. The other deception more important, because it involves the introduction of disturbance unconformity, is the apparent contortion of the lower Vindhyans before the Káimur period. This view rested upon the fact that the lower Vindhyans are often found sharply twisted in close proximity to the perfectly undisturbed Káimurs in the Son valley. The upper Vindhyans themselves have, no doubt, undergone considerable flexure in this zone, as may be seen in the Son area, on the west at Bilheri and on the east at the Ghaggar. But these broad undulations were not at first thought sufficient to include the frequent flexures seen in the lower rocks[1] which, though sharp, never seem to carry the beds much out of an average horizon. This opinion had, however, to give way to the fact of invariable complete parallelism of the layers of the two formations whenever a contact could be observed, even in proximity to those contortions. It is important to dwell upon this observation, because some unconformities of this class, reported and insisted on elsewhere, rest upon no other evidence than that found to be fallacious in this case. It may even be suggested that such appearances might possibly be produced independently of any general disturbance of associated thick and thin, or hard or soft, deposits merely by pressure from an adjoining elevated mass upon yielding underlying beds, as occurs in the familiar case of the 'creep' in coal mines.

The classification of the strata composing the upper or true Vindhyans is as follows:—[2]

BHANDER (BUNDAIR)	Upper . .	13. Sandstone.
	Lower . .	12. Shales (Sirbu).
		11. Sandstone.
		10. Limestone.
		9. Shales (Ganurgarh).

[1] F. R. Mallet, *Memoirs*, VII, 50, (1871).

[2] *Memoirs*, II, 56, (1860); VII, 27, (1871).

REWÁ . . .	Upper . .	8. Sandstone.
	Lower . .	7. Shales (Jhiri).
		6. Sandstone.
		5. Shales (Panna).
KÁIMUR (KYMORE) .	Upper . .	4 Sandstone.
		3 Conglomerate.
	Lower . .	2. Shales (Bijaigarh).
		1. Sandstone.

The general composition of the Vindhyan rocks is as uniform as their general arrangement. Although chiefly made up of sandstones, which are the coarser type of detrital deposits, the fineness of the rock throughout the formation is remarkable. With the exception of the Káimur conglomerate, which is constantly present as a bottom bed all round the boundary in Bundelkhand, pebble beds are of rare occurrence. The Káimur conglomerate is everywhere conspicuous through the prominence in it of bright red jasper pebbles, presumably derived from the jasper bands so abundant in the Gwalior formation. Where the Vindhyans rest upon the Gwalior beds, the rock is rather a breccia than a conglomerate, the included fragments being quite angular. The amount of this debris throughout such a length of outcrop, to such a distance from the nearest known area of Gwalior deposits, suggests the extensive removal of these peculiar rocks from the position now occupied by the gneiss.

There are general characteristics peculiar to each of the great sandstones. The Káimur rock is fine grained, greyish, yellowish or reddish white, sometimes speckled brown. False bedding is frequent and massive beds are abundant, but on the whole the bedding is of moderate thickness, sometimes flaggy and shaly. The Rewá sandstone is somewhat coarser, and generally presents a mixture of massive strata and false-bedded flags. The Bhander sandstone is softer than that of the lower bands, very fine grained and generally distinguishable as of deep red with white specks, or of pale tints with or without red streaks. The beds are generally thinner, and not more than 6 to 18 inches in thickness, but massive beds also occur, as is exemplified by the great monoliths cut from the quarries at Rúpbás near Bhartpur. Ripple marking is common throughout the greater part of the Vindhyans, and occurs in great profusion and variety in the upper Bhanders.

The different shale bands of the upper Vindhyans do not present any constant distinctive characters. Thin, sharply bedded, flaggy, silicious or sandy, sometimes micaceous shales, of greenish and rusty tints, form the prevailing type throughout. Purely argillaceous shales are rare.

In the main Vindhyan basin diamonds are only known to occur in the upper Vindhyans. Here, as elsewhere, the great majority of the diggings are alluvial, but the principal workings, upon which most labour is spent, are in a bed at the very base of the Rewá shales. Notwithstanding the immense range of this group, it is only known to be productive within a small area of the Panna state, on the borders of the Bundelkhand gneiss, and the surface diggings are confined to the same neighbourhood. Here again, as already noticed of the Banaganpalli mines in Southern India, the diamond layer is conglomeratic and the inference would seem to be that the diamond occurs as a pebble with the others. The observation recorded[1] that a particular kind of those pebbles at the Panna mines is broken up and searched for diamonds, and that these particular pebbles are derived from a peculiar bottom bed of the lower Vindhyan series, would of course point to this latter rock as the original nidus of the gem. But the observation in question needs confirmation.

The search for diamonds in Panna is not, however, confined to positions in which the gems could be derived from any existing outcrop of the Rewá shales. There are numerous pits (all apparently surface diggings) in the gorges and on the slope of the upper Rewá sandstone south of Panna, and at a much higher elevation than any present outcrop of the bottom shales or of the lower Vindhyans.

The Bhander limestone is the most variable rock of the series. Sometimes there is a considerable thickness, as much as 260 feet, of firm stone; elsewhere there is very much less, the carbonate of lime being apparently disseminated amongst the calcareous shales associated with the limestone and partly taking its place. The limestone is generally earthy and compact, of grey, yellow or reddish tints, sometimes purer and either compact or crystalline. It was in this rock, at Nagode, that fossils were thought to have been found long ago by Captain Franklin; they were supposed to be *Gryphæa*, and the rock was on this account assigned to the lias. It is not known what became of the specimens, and repeated search at the same locality has failed to verify the discovery. It is highly probable the objects discovered were not organic at all, and quite certain that the specific determination of them was fanciful.

The mutual relation of these sandstones, shales, and limestones is most intimate throughout the upper Vindhyan series. The passage upward, from shale into limestone, or into the great bands of sandstone, is always more or less gradual, by interstratification, while the change into shale at the top of the great sandstone beds is as generally abrupt. Both the chief and minor subdivisions are wonderfully persistent over the whole of the great basin, all being found in both the eastern and northern areas into which the main area is divided by the Deccan trap. The lower Bhander and lower Rewá sandstones are very attenuated in certain directions, but there is an equivalent increase in the thickness of the enclosing shales. In certain positions also the great bands of shales thin

[1] *Supra*, p. 97; *Memoirs*, II, 71, (1860).

out altogether, and the main sandstones coalesce. These reciprocal variations in the distribution of the coarser and finer deposits have distinct relation to position with reference to the border of the area, the shales being in force towards the middle of the basin, and being replaced by sandstones near the margin, showing that this border is approximately an original limit, and that the actual basin corresponds pretty closely with the basin of deposition. There are local exceptions to this condition, and it is in the direction in which these exceptions occur, on the Arávalli side, that the only recognisable distant outliers of the upper Vindhyans have been observed.

A formation so constituted, and for the most part but little affected by disturbance, can result in but one form of surface. Accordingly the upper Vindhyan area presents a three-fold plateau, each step formed of one of the main groups, with minor plateaux, terraces or ledges corresponding to the various subdivisions. The thick sandstones form vertical scarps over a talus of the underlying shales. There is, moreover, a basin shaped lie of the beds, apparently to a great extent original, whereby the surfaces are rendered more or less concave, and the edges of the successive scarps of sandstones scarcely higher than the outer one, composed of the Káimur rock. From this arrangement it follows that the upper group occupies by far the larger part of the area, even the middle step of the plateau, the edge of which is determined by the Rewá sandstone, being chiefly occupied by the lower Bhander shales.

Over almost the whole of the area occupied by the upper Vindhyan beds they lie little disturbed and almost horizontal, and any violent effects of disturbance are restricted to the south-south-east and the north-west margins of the basin. Two local exceptions to this rule may be noticed. In the Panwari ridge, south of Tirohán, the Rohtás limestone is capped quite horizontally by Káimur sandstone. The hill is more or less detached from the main plateau, and in the broken ground intervening, the sandstone is found dislocated and dipping in the most irregular fashion, quite inexplicable by any ordinary mode of disturbance. The displacement is probably due to the underground solution and removal of the Rohtás limestone and the consequent subsidence of the sandstone.

The other special instance of disturbance is not local in the same sense as the last, as it is probably only a symptom of much more that is concealed. It has been said that over the wide expanse of Vindhyan rocks between Gwalior and Nímach, the Bhander and Rewá beds lie quite flatly, and it has been presumed that to a considerable extent they stretch in this manner under the trap of Málwá. Close to Jhalra Pátan, however, at the northern edge of the basaltic plateau, a sharp axis disturbance passes from the south-east, beneath the trap, to the north-west, throwing up the Vindhyan strata in an anticlinal flexure, with dips of 70° on each side. Along this steep outcrop the standstone weathers into long narrow ridges. This feature gradually dies out to the north-west. It is a hint that the disturbance, which so violently affects the Vindhyans of the Dhár forest, extends far to the north under the traps of Málwá.

The disturbance of the strata along the south-south-east border of the Vindhyan basin, to as far west as Hoshangábád, is plainly a recurrence, on the same lines, of the compression which had produced the contortion and cleavage in the adjoining transition and gneissic rocks. It seems to have taken different forms in different parts of the ground. Along the whole Sen valley, there is little or no faulting in the zone of disturbance, but at the Son-Narbadá watershed one or more faults occur at and close to the boundary, the east-north-east strike being remarkably steady throughout. Down the Narbadá valley towards Hoshangábád, the dips in the Vindhyans become unsteady. At Hoshangábád, and again in the Dhár forest, there is a decided predominance of a north-west, south-easterly strike, and as the east-north-east strike remains constant in the contiguous transition and metamorphic rocks, it may be inferred that the former strike is the later of the two. It is that to which the features of the Vindhyans south of Nímach and at Jhalra Pátan conform.

The north-western boundary of the Vindhyans is in the main a fault of great throw, along which the almost horizontal Bhander sandstone is brought into contact with the highly disturbed Arávalli beds. Beyond this fault there are a few small, but important outliers, composed of the lower members of the system. The largest of these occur south-west of Karauli, where a narrow ridge of Alwar quartzites is faulted against undisturbed sandstones of upper Vindhyan age on the south-east, and on the north-west is overlaid by two alternations of sandstone and limestone, the lower being regarded as lower Vindhyan (Rohtás group) and the upper as Káimur.[1] The beds have been compressed and are exposed in two narrow synclinals, about 20 miles in length, but there are small outliers of the lower beds to the south-west of Naráoli, as far as the parallel of Ranthambhor.

From just north of Búndi, extending almost to Indargarh, a narrow strip of Káimur sandstone rests with little disturbance on the slates immediately west of the great boundary fault.

The throw of this fault must be at least 5,000 feet, and there is naturally some difficulty in accounting for a single fault of so great a throw having been formed subsequent to the deposition of the Vindhyans, and among beds which have undergone so little subsequent disturbance as they have. But we will find when treating of the Himálayas that the nature of the boundary between the Vindhyans and the disturbed Arávalli

[1] *Records*, XIV, 288, (1881).

beds is very similar to what we may infer is the contact between the undisturbed deposits of the Indo-Gangetic plain and the disturbed beds of the Himálayas. Moreover, along the foot of the Himálayas, there is a strip of upper tertiary beds which have been disturbed, but to a less degree than the older beds of the range, while the equivalents of these beds are believed to occur under the alluvial plain, in perfectly conformable sequence with the most recent alluvium. Now, if these suppositions are correct, as is almost certainly the case, we can imagine that, after ages of denudation, the upper tertiary rocks of the Siwálik zone will be almost removed, and the northern boundary, of what is now known as the Indo-Gangetic alluvium, will then exhibit very much the same features as the boundary of the Vindhyans towards the Arávalli range now does. The upper beds will be in contact with highly disturbed rocks of much more ancient date along a great line of fault. Beyond this will be a few outliers composed of the lower beds of the series, the Siwáliks of the present classification, and to the north of these there will be a broad exposure of the wreck of a mountain range.

In the case of the Himálayas the fault has been gradually formed *pari passu* with the deposition of the Indo-Gangetic alluvium, which is contemporaneous in its origin with the principal elevation of the Himálayas and formed of the debris of that range.

It is natural to suppose that the similar structure in the case of the Arávallis indicates a similarity of origin, and that the great Vindhyan spread of Central India is formed of deposits which bore the same relation to that range as the Indo-Gangetic alluvium does to the Himálayas.

The suggestion is an important one since it would fix the period of the formation of the Arávalli range, or at any rate of its principal importance, as contemporaneous with the deposition of the upper Vindhyan rocks that were formed of its debris. It would account for the greater prevalence of sandy beds near this margin of the deposit and would place the original limit of deposition not very far beyond the present limit of the outcrops.

Allusion has been made[1] to some small outliers, believed to be of Vindhyan age, which occur on the north and west Bundelkhand gneiss. They differ much in character, and their peculiarities of composition may help to explain their apparently anomalous position. Although the gneiss reaches high up under the scarp of Káimur conglomerate all round the western border of the area which is described as a local edge of deposition, these small outliers occur at the level of the low country. If they agreed in composition with the rocks of the main area, which are so strikingly constant in this respect within that area, the fact might be at once explained by a subsequent change of level, but such is not the case.

[1] *Supra*, p. 29.

The most curious of these outliers form a very broken chain running to south-south-east from close under the Pár scarp at Ládera (7 miles east of Antri) to Uchár on the Sind river. Most of the exposures are quite level with the plain, or only to be seen in the beds of streams. In a few cases, as at Ládera and Pichor, they form narrow ridges up to 300 feet in height. The rock is sandstone of the upper Vindhyan type, and at the north end, close to the Pár scarp, it contains large angular pieces of the banded Morár shales. Elsewhere it is quite free from coarse debris of any kind. From many clear sections it is quite evident that these ribs of sandstone once filled a more or less continuous run of cracks, fissures, chasms, or small valleys in the gneiss. On both sides of the Pichor ridge the gneiss reaches well up on the sides of the sandstone mass, with vertical surfaces of contact. In the low ground, at the point of the ridges, and in the small outliers, thin vein-like runs, of 3 feet wide and upwards, of the sandstone are well seen, completely let into the gneiss, as it might be filling an emptied trap-dyke, the rootlets of the wider chasm above. Even in the larger masses no bedding is visible but sometimes, at the edge of the mass, planes of pseudo-lamination and even ripple marked surfaces occur parallel to the vertical wall of gneiss. The lines of ripple were horizontal, and the steep face of the ripple turned downwards in every case observed. Some of these features seem to necessitate the supposition that the sandstone was let into this position by disturbance, but all the other circumstances have suggested the explanation given.

At Mahárájpur, 10 miles south of Antri and 14 miles east of the Vindhyan scarp, there is a small group of hills, about three square miles in extent, formed of fine sandstone overlying about fifty feet of flaggy shales, both of Vindhyan type. The strata are greatly disturbed, but most irregularly, as if compressed from every side. Although so much broken, the rock is quite free from vein quartz, which is also a general character of the Vindhyans as compared with the Gwalior strata.

The small hills of Sonár, 10 miles south-east of Narwar, and of Mohár 16 miles farther in the same direction, present the same characters of composition and disturbance as at Mahárájpur. At Mohár a trace of the Káimur conglomerate occurs in the sandstone above the shales, which cover a considerable area round the base of the hill, and may be looked upon as lower Káimur.

A consideration of all the peculiar circumstances of these outliers would seem to suggest that they may represent small local basins of the upper Vindhyans. It seems that the process of denudation all round the Vindhyan area has been to decompose, and remove the chemically-constituted metamorphic rocks which once formed high land around the sedimentary basins, whether this relation were original or due to subsequent warping of the surface, leaving the softer but undecomposable detrital rock to project where once had been depressions of the surface.

To the west of the Arávalli range there are numerous exposures of horizontal sandstones, mostly hard and usually more or less red in colour, with occasional strings of pebbles.[1] They are frequently false bedded and show ripple marks on the surfaces of the slabs. With these there occur exposures of a compact grey cherty limestone which is believed to overlie the sandstone, though no actual contact sections have been observed.

Wherever these sandstones rest on older rocks, whether of the Maláni series or the tilted beds of the Arávallis, the relation is one of complete unconformity, usually accompanied by a band of strong conglomerate at the junction. On the east of the Arávallis a similar unconformity is to be observed, but the western beds are even more markedly superficial with respect to the older rocks, which must already have been disturbed and elevated into a mountain range at the time of their formation.

The outcrops of these beds are scattered over a large area in western Rájputána, but they are for the most part individually small and surrounded by sandhills and alluvium. It is consequently very difficult to make out the true thickness, or even the superposition, of the beds in the different exposures, and it is possible that, in the neighbourhood of Pokaran, beds which have been coloured as Vindhyan are really much newer. The sandstones are darker in colour, somewhat softer, and contain more pebbles than near Jodhpur and are distinctly underlaid by a boulder bed consisting of a fine matrix through which numerous large blocks, many of which show distinct signs of glaciation, are scattered. There is no record of such a formation having been observed at the base of the Vindhyans further east, but a similar glacial boulder bed near Báp was regarded as Talchir. If the two are identical the Pokaran sandstones would belong to the Gondwána system. They differ, however, in type, from any of the sandstones of western Rájputána which can be referred to that period, and the glacial beds differ from those of Báp in being interbedded with dioritic trap and ash beds, and in the absence of boulders of the supposed Vindhyan limestone, which are extremely abundant in the boulder bed of Báp.[2]

In degree of induration, lithological character, and relation to the tilted beds of the Arávallis, these sandstones and limestones closely resemble the Vindhyans to the east, and they are regarded, with a strong show of probability, as of that age But, on the hypothesis of the relation between the Arávalli range and the upper Vindhyan sandstones which has been proposed above, there can never have been any continuity between the outcrops on either side, and it is incorrect to speak of the Jodhpur beds as outliers of the Vindhyan basin. They must have been formed contempora-

[1] *Records*, XIV, 299, (1881). [2] *Records*, XXI, 32, (1888).

neously in an independent basin of deposition and the connection between the two would be similar to that which subsists between the recent deposits north of the Himálayan range and the Indo-Gangetic alluvium.

The rock systems which have been described in this chapter are remarkable as being the source to which all the diamonds found in India—if we except the reputed occurrence of diamonds near Simla—can be traced, though the actual workings are more commonly in recent stream gravels.[1] In Southern India the principal source of the diamond is the Banaganpalli group of the Karnúl series near the town of that name, but they are also obtained at other places within, or just outside, the boundary of the Cuddapah basin. In the Mahánadí valley they are found near Sambalpur, just outside the boundary of the Cuddapah area, but have not been found *in situ* as yet. The third diamond district of India is near Panna, where the original home of the diamond is in the lower part of the upper Vindhyan series, though the possibility of its derivation from lower Vindhyan beds has been indicated.

This repeated occurrence of the diamond at, or about, the same geological horizon might be held to be corroborative of the general resemblance, which has led the rocks in which it is found to be classed together. But the evidence is of small value, for the diamond is in every case derivative, and its original source has not yet been found.

About six years ago it was supposed that an original source of the diamond had been discovered in a volcanic neck near Wajra Karur, filled with a substance which closely resembled the blue clay of the Kimberley diamond fields. A more detailed examination has shown that the Wajra Karur rock is the product of decomposition of a basic volcanic rock, not a peridotite, like the Kimberley blue clay.[2] Previous to this the rock had been thoroughly prospected for diamonds, but none were found in the matrix though diamonds have certainly been found on the surface after rain, and, as there are no outliers of Cuddapah or Karnúl conglomerates within the drainage area, it was supposed that they must have been washed out of the decomposed volcanic debris filling the neck.[3]

The conclusion is not a necessary one, for the rock filling the neck has been more largely removed, being more easily weathered than the surrounding gneiss, and the neck now forms a depression in the general surface of the country into which the diamonds might have been washed. A different origin for the diamond has been suggested by M. Chaper,[4] who believed that he had obtained diamonds, sapphires, and rubies from the

[1] Details of the known diamond localities will be found in Vol. III of this Manual, pp. 1—5, and *Records*, XXII, 39-49 (1889).

[2] P. Lake, *Records*, XXIII, 72 (1890).

[3] *Records*, XIX, 110 (1886).

[4] Sur une pegmatite diamantifère de l'Hindostan, *Bull. Soc. Géol. de France*, 3rd series, XIV, 330, (1886); *Records*, XXII, 39, (1889).

debris of a pegmatite vein in the gneiss. He did not, however, find any of these gems in their matrix, nor did he apparently himself procure them from the debris. The evidence produced is of the slightest and has not materially helped forward the solution of the problem of the original source of the diamond.

It remains now to consider briefly the relations of the rock systems described in this chapter to each other. In the first edition of this manual the Cuddapah system was classed, along with the Gwaliors, among the upper transition deposits, but subsequently the rocks of the Godávari valley, which had previously been regarded as Vindhyan were identified with the Cuddapahs of Madras, and Mr. Medlicott, abandoning his previous opinion, accepted the Cuddapahs as the equivalent of the lower Vindhyans.[1] Adopting this conclusion we have in the southern area two unconformable series or systems, of which the uppermost is quite unimportant compared with the lower. In the central area much the same is the case, but in the northern area the two series are said to be nearly conformable and the lower is certainly much less in thickness than the upper. The first, and most obvious impulse would be to class the lower Vindhyans, the Pakhals, and the Cuddapahs together, as an older system, and the upper Vindhyans, the Sullavais, and the Karnúls, as a newer. The classification would be to some extent in accordance with the lithological resemblances of the rock series, and may be correct in the main, but the truth must be less simple than this.

It is impossible to suppose that the lower Vindhyans of Central India can, in any proper sense of the word, be the equivalents of the great Cuddapah system, and if the account of the relation between the upper and lower Vindhyans given above is correct, it indicates a difference from that which subsists, between the Pakhal and Sullavai beds of the Godávari valley, or between the Cuddapahs and Karnúls of Madras. If the Karnúl series represents any portion of the Vindhyan system, whether upper or lower, it is difficult to escape the conclusion that the commencement of the Cuddapah epoch must date further back than the oldest of the typical lower Vindhyan deposits, and may be in part contemporaneous with the newest of those described among the transition systems, a supposition strengthened by the resemblances which may be observed between the banded jaspers, hornstones, and porcellanic beds of the lower Cuddapahs, and the similar rocks of the Gwalior series. The resemblances are, however, not such as amount to identity, and it is certainly more convenient for purposes of description to class these beds with these newer rather than those older than them.

[1] *Records*, XV, 2, (1882).

CHAPTER V.

OLDER PALÆOZOIC SYSTEMS OF THE EXTRA-PENINSULAR AREA.

The Salt-range—Central Himálayas—Unfossiliferous slates of the outer Himálayas—The Jaunsar system—The Deoban limestone—The Baxa series—Eastern Tibet and Burma.

In the last chapter an older palæozoic age was accepted for the less disturbed unfossiliferous rocks of the Peninsula, on the ground of the great discordance that exists between them and the next succeeding beds, which are known to be of upper palæozoic age. In the extra-peninsular area we have no need to content ourselves with such indirect evidence as this, for the presence of older palæozoic rocks has been proved, in three distinct areas, by the discovery of cambrian and silurian fossils.

The most important, because most fully studied, of these is the Salt-range, where an extensive series of conformable strata, nearly 3,000 feet in thickness, is divided into the following groups:—

Salt pseudomorph zone	450 feet.
Magnesian sandstone	250 "
Neobolus beds[1]	100 "
Purple sandstone	450 "
Salt marl[2]	1,500 "

The complete sequence is only seen in the eastern part of the range, for the series is unconformably overlaid by the succeeding one, and the groups are successively thinned out to the west by denudation, as one after the other becomes the uppermost to be seen.

The salt marl consists typically of a fine grained rock, varying from dull purple to bright red in colour, composed of very fine grained clayey

[1] This differs from that given by Wynne, (*Memoirs*, XIV, 69) as the thicker sections of his include part of the overlying carboniferous beds.

[2] Dr. Waagen (*Pal. Ind.*, series xiii, IV, 44) has added, below this, a grey gypsum group and a lower purple sandstone. They are only seen at one place in the Khisor hills, and there does not seem to be sufficient evidence for this addition to the series, especially in view of the doubts cast on the sedimentary origin of the gypsum (*infra*, p. 111).

matter, mixed with some disseminated gypsum and carbonates of both lime and magnesia.[1] The marl itself never exhibits the slightest signs of stratification,[2] nor are any coarser grains of sediment found mixed with it, but it acquires an appearance of stratification from containing beds of gypsum and rock salt, a few layers of dolomite, some beds of sandy dolomite in the lower part of the series, and near Kheura a six inch band of bituminous shale and some irregular patches of an obscure, dioritic-looking, dark purple trap, said to be associated with paler purple volcanic ash.

The most interesting and important of these, both as regards thickness and economic value, are the salt and gypsum deposits. The former appears to be somewhat irregularly developed and, except where mining operations are carried on, difficult to observe, but it is described as occurring in regular beds which exhibit distinct lamination. The greatest development is in the Mayo mines at Kheura, where there are over 550 feet of salt. Half of this thickness, or 275 feet, is made up of five beds of nearly pure salt, which is mined and placed upon the market without being refined, the other half, known as *kalar*, is too earthy and impure to be used in its natural state, and has, consequently, no marketable value.

The impurities of the salt are principally sulphate of lime and chlorides of magnesium and calcium. In the Mayo mine a lenticular band, with a maximum thickness of 6 feet, has been found, composed of a mixture of ylvine (chloride of potassium) and kieserite (sulphate of magnesium, with one equivalent of water), which also prevailed through a thickness of about seven feet below the sylvine band. Glauberite (anhydrous sulphate of soda and lime) has also been found. Epsom salts (sulphate of magnesia, with seven equivalents of water) commonly crystallises in the passages of the mines on the surface of the salt marl and the *kalar*, and, as this salt would result from the absorption of water by kieserite, it appears to indicate that the magnesian salt is of common occurrence in the rock.

Besides being disseminated through the red marl, the gypsum occurs abundantly in beds and irregular masses, overlying the salt as a rule, but also occurring more doubtfully in lower situations. It is found everywhere accompanying the red marl and, at Mari and Kálabágh on the Indus as well as in the Khisor range, contains more or less numerous bipyramidal crystals of quartz.[3] Sometimes layers of hard flaggy dolomite are found in thick masses of the gypsum and at one or two places numerous large and perfect casts of hopper shaped crystals of salt are found in the dolomite layers.

[1] *Memoirs*, XIV, 70, (1878).

[2] C. S. Middlemiss, *Records*, XXIV, 28, 1891).

[3] A. B. Wynne, *Memoirs*, XIV, 73, (1878); XVII, 239, (1880); T. H. Holland, *Records*, XXIV, 231, (1891).

Till quite recently there would have been little hesitation in regarding this group of beds as sedimentary in their origin, and older than the sandstones which overlie them on most sections. It is true there were always difficulties in the way of this interpretation, among which were the absence of stratification in the marl, the enormous masses of sea water whose evaporation would be involved in the formation of the great beds of salt, the great age of the beds containing so soluble a mineral and, more especially, the fact that much of the gypsum was known to be not fully hydrated, and Mr. Middlemiss[1] has recently declared himself in favour of the theory, broached by more than one of the earlier observers, that the marl, with its gypsum and salt, is hypogene, and has come into its present position by a process analogous to that of igneous intrusion.

The arguments in favour of this conclusion are two-fold and derived, firstly, from the abnormal stratigraphical position in which the marl is sometimes found and, secondly, from the mineralogical character of the marl and its peculiarities of structure and included minerals.

As regards the position of the salt marl, it is ordinarily below the purple sandstone and the form of the boundary, imbaying up the deep valleys, is such as to suggest that it is in normal infraposition, but, according to Mr. Middlemiss, there is no transition between the two beds, and in the sections quoted by him there are numerous fragments of the sandstone scattered through the upper layer of the marl. Besides this, the salt marl is found at higher horizons intruded along the cores of flexures and along fold faults. None of these observations are, however, of vital importance, as the features might easily be the result of pressure acting on a soft material like the marl.

The second argument is derived from the condition of certain included fragments of dolomite. These are described as pitted, corroded and honeycombed, and showing, in one section at Kavhad, an ultimate passage into the red marl. In connexion with this, another section, two miles north of Burikhel, must be noticed, where the salt marl is immediately overlaid by the glacial boulder bed forming the base of the upper palæozoic system of the Salt range. The marl here contains numerous large aggregates of a hard compact dolomite, whose peripheral portions are pitted or honeycombed and the cavities filled by gypsum, only the central portions being unaltered. In the boulder bed no trace of the marl could be found, whether as included fragments or by a colouration of the matrix, and none of the numerous dolomite pebbles, resembling those of the central portions of the lumps in the subjacent marl, exhibit any pitting or corrosion. From this it was concluded that the corrosion of the dolomite in the marl was of later date than the boulder beds. Here the

[1] *Records*, XXIV, 26, (1891).

doubt arises whether the dolomite fragments in the boulder bed might not have been derived from the magnesian sandstone group, rather than from the dolomite in the marl, unless indeed it is claimed that the marl has here replaced the dolomites of the magnesian sandstone group.

It will be seen that the evidence is not conclusive, but taken as a whole it throws a considerable doubt on the hypothesis that the marl in its present condition was formed superficially, and this doubt is strengthened by the manner in which inliers occur among the upper tertiary sandstones, in situations where its presence is difficult to account for, if it is of cambrian age.

In Kohát there is another great development of salt which has been regarded as of tertiary age, and it would be natural to regard this, with its associated gypsum and marl, as the equivalent of that in the Salt range. It is said, however, that the Kohát salt exhibits a difference of colour so constant and characteristic that it is always possible to say with certainty from which area the salt came. Too much weight must not, consequently, be attached to the other resemblances, but it would certainly be strange if two separate developments of salt, each on so large a scale, had taken place in closely adjoining areas at such widely different periods as the lower cambrian and tertiary.

The discussion of the origin of the salt marl has been helped forward greatly by a recent investigation by Mr. Holland,[1] who has shown that the associated gypsum was certainly formed by the union of water with anhydrite. Not only are the inclusions in the quartz crystals all anhydrite, but the matrix proves to be a mixture of anhydrite and gypsum in varying proportions according to the degrees of hydration it has undergone. Anhydrite could not have been formed at the surface or at a lower temperature than 125° C. (257° F)[2], nor could the quartz crystals in it have been exposed to a red heat, as at that temperature they would have been attacked by the sulphate of lime. He suggests that the anhydrite is due to the action of acid vapours on a pre-existing limestone or dolomite, and in this we probably see the true explanation of the marl. It is not a hypogene rock intrusive in its present position, nor is it a sedimentary rock formed superficially as such with its associated gypsum and salt, but is due to the alteration of pre-existing sediments, whose exact composition is unknown, by the subterraneous action of acid vapours or solutions.[3]

The group next succeeding the salt marl is composed of even-grained sandstones, of a dull purple colour, containing carbonates of lime and magnesia. The lower 50 feet to 100 feet are more argillaceous and may indicate a transition to the beds below, but with this exception bands of

[1] *Records*, XXIV, 231, (1891); XXV, 54, (1892).

[2] Hoppe-Seyler *Poggendorf. Annalen*, CXXVII, 161, (1866).

[3] *Records*, XXIV, 243, (1891).

clay are rare or absent throughout the group. At the upper limit the colour of the sandstones becomes paler and some buff bands are seen. The sandstones show ripple marks but rarely, and the only trace of life they contain are a few, almost equally rare, obscure fucoid markings.

The purple sandstone group is succeeded abruptly, but conformably, by dark or blackish shales, associated with sandy and calcareous beds only some 20 to 100 feet in thickness, but extremely important on account of the fossils they have yielded. These have come from two horizons, one near the base of the group from a shale band containing *Neobolus*, the brachiopod which has given its name to the group, and the other close to the top, immediately below the magnesian sandstone.

The fossils from the lower horizon comprise the following species[1] :—

Discinolepis granulata.
Schizopholis rugosa.
Neobolus warthi.
" *wynnei.*
Lakhmina linguloides.
" *squama.*
Lingula kiurensis.
, *warthi.*
Fenestella, sp.

The fossils collected by Mr. Middlemiss from the higher horizon have not yet been examined, but some fossils collected by Dr. Warth, which probably came from the same horizon, include[2]—

Conocephalites warthi.
Olenus indicus.
Hyolithes wynnei.
" *hussakensis.*
Orthis warthi.

Besides which a trilobite belonging to the genus *Olenellus* has been found by Mr. Middlemiss.

None of the species from this group have been found in other parts of the world as yet, but *Conocephalites warthi* is said to be very close to *Solenopleura cristata* from the Paradoxides beds of Sweden. The general facies of the fauna, however, leaves no room for doubt that the beds are of cambrian age, and, consequently, the oldest in India whose age can be determined with any approach to certainty.

The fossiliferous beds are succeeded by the magnesian sandstone group, which forms conspicuous scarps along the eastern part of the range. It consists of from 10 to 300 feet of hard, light cream coloured or white, dolomitic sandstone or sandy dolomite, associated with which are beds of light coloured sandstones, and sometimes oolitic or flaggy bands, the latter occasionally covered with fucoid markings and separated by greenish or dark coloured shales.

The topmost member of the system, as developed in the Salt range, is composed of thin bedded and flaggy sandstones, with interbedded shales or

[1] *Pal. Indica*, series xiii, IV, 92, (1891).

[2] *Pal. Indica*, series xiii, IV, 104, (1891).

clays, characterised by their bright red colour and by the numerous pseudomorphic casts of salt crystals which cover the bedding surfaces. These casts are found indenting the lower surfaces of the separate flags, and there can be no doubt that they are due to salt having crystallised out, by the evaporation of salt water, and having then been covered by fresh sediment, which, adapting itself to the angularities of the crystals, preserved their form in the casts now seen. As might be expected, these beds are unfossiliferous, the only organic remains found being obscure fucoid impressions and worm tracks.

In the Central Himálayas of Hundes and Spiti a system of conformable strata, of older palæozoic age, was described by the late Dr. Stoliczka[1] as the Babeh and Muth series. Mr. Griesbach, as the result of a more extended survey, has recently rearranged these beds as follows[2] :—

Carboniferous . .	8. White quartzite, with limestone. 7. Red crinoid limestone.
Devonian (?) . .	6. Dark coral limestone.
Siulrian . . .	5. Flesh-coloured and brown quartzites and shales. 4. Coral limestone.
Haimanta system .	3. Red quartz shales and slates. 2. Shales and silky phyllites, with a great thickness of quartzites. 1. Quartzite, generally purple and a great thickness of conglomerate.

The whole of this sequence of rock groups is described as perfectly conformable throughout and separated by a slight unconformity from the beds above; it will therefore be treated as a single system for descriptive purposes.

The three lowest groups are united by Mr. Griesbach under the name of the Haimanta (snow clad), a convenient designation for the great thickness of beds, intermediate between the silurian and the gneiss and schists which are either entirely unfossiliferous or have only yielded organic remains too badly preserved to be identifiable.

The Haimanta series is divided into three groups, the lowest of which is characterised by a great development of conglomerates, composed of rolled and subangular fragments, among which quartz and gneiss predominate. The matrix is a hard, often deep purple coloured, sometimes partially schistose quartz rock. The junction with the underlying schists is said to be transitional and often not determinable with accuracy, but the prevalence of coarse conglomerates points to an unconformity and the apparent transition may be due to the compression and metamorphism the beds have undergone.

[1] *Memoirs*, V, 17-24, (1865).

[2] *Memoirs*, XXIII, 49-64, (1891).

The conglomerate group is overlaid by a great thickness of greenish grey phyllites, slates and thicker bedded quartzites, traversed by quartz veins, towards the upper part of which reddish brown and pinkish quartz shales are intercalated. They are said to resemble, lithologically, the Simla slates and contain some very obscure fossil remains which have been referred to *Bellerophon*, with some *Crinoid* stems and casts of bivalves.

The uppermost member of the Haimantas is described as very constant and conspicuous and, consequently, very useful horizon in unravelling the structure of the hills. It is described as consisting of bright red and pink quartz shales passing upwards into greenish grey shales and quartzites with pink and shaly partings. Some thin bands of deep red limestone are occasionally found in the lower part of the group.

The total thickness of the Haimantas is about 3,000 to 4,000 feet, of which from 250 to 500 belong to the upper group.

The passage from the uppermost Haimanta be ls to the lowest, classed as silurian, is gradual, and the two types of rock are interbedded at the junction. The lower silurian beds, according to Mr. Griesbach, consist at the base of dirty coloured greyish pink quartzite, with calcareous partings, passing upwards into grey shaly quartzites, alternating with dark blue to black coral limestone,—limestone being the prevailing rock of the group.

The lower beds of silurian age are comparatively thin, being only about 300 feet thick, and are succeeded by from 1,000 to 1,200 feet in the Niti sections, and much more in Spiti, of dirty pink to flesh coloured quartzites, with greyish green intercalated shales, and some limestone bands in the lower part. In Spiti Dr. Stoliczka observed some contemporaneous trap in these beds,[1] but none was seen in the Hundes sections.[2]

The upper Silurian quartzites pass gradually into dark grey or black limestone, generally concretionary, showing sections of *Corals* and *Brachiopoda* in large numbers, on the weathered surface. These again pass gradually into bluish grey and brownish red limestones, containing fragments of *Crinoids*.

The uppermost member of the system is composed of white quartzite, ranging from 350 feet to 800 feet in thickness It was classed by Dr. Stoliczka with his Muth series of doubtfully silurian age,[3] but according to Mr. Griesbach, it is overlaid in Spiti with partial interstratification, by limestones of carboniferous age containing *Athyris royssii*, *Productus*, etc.[4]

[1] *Memoirs*, V, 20, (1865).

[2] A large number of silurian fossils from General Strachey's collections were described by Messrs. Salter and Blanford in 1865, and the collections made by Griesbach, which are still in course of description, will doubtless add to the number. Until this fauna has been worked out and its relations fully determined, there does not appear to be any benefit in printing a nominal list of the species that have been described. None of the fossils found in Spiti by Dr. Stoliczka were specifically determinable.

[3] *Memoirs*, V, 23, (1865).

[4] *Memoirs*, XXIII, 63, (1891).

On the accompanying map a very large area, in Kashmír and the north-west Himálayas south-west of the snowy range, has been coloured as silurian, although a very considerable, if not the greater, portion of this area is probably occupied by post-silurian beds. This course has been adopted owing to the impossibility of distinguishing generally between the older and younger slates, where all have been equally disturbed and are, as a rule, equally unfossiliferous. In Kashmír the whole area mapped by Mr. Lydekker as belonging to his Panjál system has been coloured as silurian, though it has since been shown[1] that a considerable portion of this system is probably of upper palæozoic age, and the same colour has been used in the Hazára district for that series of slates with limestones and quartzites which are known as the Attock slates.

The Attock slates, which derive their name from being particularly well seen in the hills along the Indus south of Attock, consist of dark coloured slates and limestones, some sandstones and trap, both intrusive and interbedded. Nothing is known of their age except that, in the Sirban mountain near Abbottábád, they underlie, with a strongly marked unconformity, beds which are older than the trias.[2] They must consequently be, in part at least, of older palæozoic age. On the other hand, it is certain that beds of secondary age are folded up amongst them, for limestone containing *Dicerocardium* has been found east of the road between Khánpur and Haripur, and *Ammonites* east-south-east of Haveliyan[3]; quite lately, too, some fossils of decidedly cretaceous appearance were found by Mr. Griesbach in the area mapped as Attock slates. It becomes obvious from this that the area coloured as silurian contains both upper palæozoic and mezozoic rocks, but, as it is impossible to separate these at present, it has been thought best to give one uniform colour on the map to the area occupied by the generally unfossiliferous, slaty series, except where sufficient information is available to justify the separation of a distinct area of post-silurian rocks.

In Kashmír the Panjál system of Mr. Lydekker includes, besides slates and associated beds of transition and older palæozoic age, a great series of beds which there is good reason to believe are carboniferous and permian. These form the principal portion of the typical sections of the Panjál system, and as they will be dealt with further on,[4] there remains nothing definite to be said of the lower beds in this place. The Panjál system was regarded as perfectly conformable to the underlying gneisses, but it is probable that unconformable breaks exist, which were not recognised in the course of a rapid exploration.

Further to the south-east a system of strata has been distinguished,

[1] *Records*, XXI, 139, (1888); *infra*, p. 136.
[2] *Memoirs*, IX, 335, (1872).
[3] *Records*, XII, 121, (1878).
[4] *Infra*, p. 134.

under the name of Jaunsar, which exhibits considerable resemblance to the Haimantas and silurian of the central Himálayas. The lowest beds, known to belong to this system, are characterised by the prevalence of purple quartzites with intercalated red slaty beds and, low down in them, a boulder bed, composed of large fragments of quartzite, dispersed through a fine grained matrix.

In the upper Pábar valley the quartzites are found lying unconformably on the gneiss. In eastern Sirmur they appear to be underlaid by a great thickness of grey slate with a band about 300 feet thick of grey limestone, but the true position of these beds has not been established with certainty. In the Bangál valley of eastern Sirmur the purple quartzites are overlaid by about 200 feet of a dark grey felsitic trap, covered by as much more of mixed trap and ashes, which are of subaerial origin. In northern Jaunsar these are replaced by a great thickness of slaty beds, mixed with bedded lava flows and impure volcanic ashes, and a band of limestone, some 300 feet thick. Above the volcanic beds comes a great thickness of sub-schistose slates.[1]

The purple quartzite beds exhibit considerable lithological resemblance to those of the Babeh series (Haimanta) and the volcanic beds may well be contemporaneous with those of silurian age mentioned by Dr. Stoliczka.

South of the snowy range the Jaunsar system has not been recognised with certainty away from Jaunsar and eastern Sirmur, but, low down in the Simla slates, as the beds which underlie the Blaini group in the Simla district have been called, there are some purple quartzites which may belong to the system, while there can be little doubt that the purple quartzites and volcanic breccias described by Mr. Middlemiss in western Garhwál[2] are of the same age.

The Jaunsar system is unconformably overlaid by a great limestone series, which forms, and derives its name from, the peak of Deoban, north of Chakráta. It consists of a pale to dark grey limestone, often more or less dolomitic, with interbedded grey shales. The limestone is frequently mephitic, in places contains numerous cherty concretions, and is occasionally oolitic.[3]

The age of this limestone series is unknown, it appears to be unconformable to the Jaunsar system and is certainly overlaid, with marked unconformity, by beds which there is reason to believe are of carboniferous or permian age. Outside Jaunsar it has not been identified with certainty, but is probably represented by the great limestone formations seen south-east of the Chor, in the Shálí mountain, in Kúlu, and in Garhwál and Kumáun. The correlation, which is based on general lithological resem-

[1] *Records*, XXI, 131, (1888). [2] *Records*, XX, 34, (1887).
[3] *Records*, XXI, 133, (1888).

blance and relative position in the series, derives great support from the discovery, in the limestone of Kúlu[1], Jaunsar[2], and Kumáun[3], of curious structures, exhibiting considerable resemblance to each other, which may be merely concretionary but are quite possibly of organic origin.

In the eastern Himálayas Mr. Mallet distinguished a series of beds named after the hill fort of Baxa,[4] composed of variegated slates, quartzites and dolomite, the latter being the most prominent member of the series and having a thickness of 1,500 feet. The dolomite is described[3] as generally massive with obscure bedding, but frequently shaly and passing at times into a dark grey slate. It is saccharoid, light grey, rarely white in colour, with nests of more coarsely crystallised calcite and drusy cavities, lined with the same mineral, scattered through it. There is also a strong band of quartzites and quartz schists, and carbonaceous slates have been observed. This series was regarded as younger than the Damudas, but, as has already been explained,[5] it is probable that the section was ministerpreted, and it should be looked on as older, probably considerably older than the Damudas, and consequently of lower palæozoic horizon.

Further to the east devonian fossils were found by the Abbé Des Mazures near Gouchou in eastern Tibet,[6] but no other older palæozoic rocks are known along the eastern frontier till near Mandalay, where Dr. Noetling has recently faund a shaly limestone containing *Crinoids*, *Orthoceras* and a species of *Echinosphærites*.[7] The latter of these alone would be sufficient to stamp the beds as silurian. These older palæozoic beds are no doubt largely developed in the country east of the Irawadi valley, but no details have as yet been published.

[1] *Geol. Mag.*, 3rd dec., V, 257, (1888).
[2] *Records*, XXI, 133, (1888).
[3] *Jour. As. Soc., Beng.*, III, 628, (1834).
[4] *Memoirs*, XI, 33, (1874).
[5] *Supra*, p. 75.
[6] *Comptes Rendus*, LVIII, 878, (1864).
[7] *Records*, XXIII, 79, (1890).

CHAPTER VI.

CARBONIFEROUS AND TRIASSIC ROCKS OF EXTRA-PENINSULAR INDIA.

The Salt range—Central Himálayas—Carbonaceous system of the outer Himálayas—Kashmír—Hazára—Afghánistán—Tenasserim—Unfossiliferous slaty series in southern Afghánistán—Arakan—Manipur and Nágá hills.

The upper palæozoic rock groups, of peninsular and extra-peninsular India alike, bring forcibly before us the impracticability of a rigid application of the European divisions of the geological scale. Almost everywhere the palæozoic rocks pass upwards, without an unconformable break of any importance, into beds of mesozoic age, and it will be found convenient to class the upper palæozoic and trias together in the description, as it has been found necessary to do on the map. But with this exception the carboniferous and triassic rocks of the two areas present so strong a contrast that it is necessary to treat them separately. In the Peninsula they are represented by a great system of subaerially formed river deposits known as the Gondwána system ; in the extra-peninsular area the rocks of the same age are of marine origin, and as their age can be more satisfactorily established, and they form an important link in the chain of argument by which the age of the Gondwána system is determined, they will stand first for description.

The Salt range—but now in its western half—is again our typical area, where the series is best exposed, most fossiliferous, and has been most completely studied. It has there received a considerable amount of attention from the Geological Survey, as well as from independent observers, and the classification now adopted is more complete than the simple division, into speckled sandstone and carboniferous limestone, originally adopted by Mr. Wynne.[1] The first of these names may, however, remain to distinguish one of the principal divisions, and for the other the name Productus limestone, proposed by Dr. Waagen, will be used, as the term carboniferous is misleading.

[1] *Memoirs*, XIV, 69, (1878). The description of these rocks is based partly on Mr. Wynne's memoir, and partly on Dr. Waagen's account in *Pal. Indica*, series xiii, IV, (1890-91). The palæontological data are all taken from Dr. Waagen's account, *op. cit.*, Vol I.

The oldest member of this system was distinguished by Mr. Wynne as the speckled sandstone. It rests unconformably on the older palæozoic rocks, and is usually, if not invariably, characterised at its base by a boulder clay, formed of a fine grained matrix of shale or sandy shale, usually olive-green in colour and weathering into acicular fragments, through which are scattered blocks of hard rock, ranging to several cubic feet in size, almost invariably subangular and frequently showing faces that have been smoothed, polished and striated in the manner characteristic of glacier action. These fragments are principally composed of slates, quartzites, and crystalline rocks, whose nearest analogues are to be found 750 miles to the south, in the syenites and porphyritic felsite of the Maláni series, but mixed with them are numerous fragments of the older palæozoic beds of the Salt range, more especially of the magnesian sandstone. The glacial origin of these beds is so obvious, and the resemblance to a boulder clay is so close, that it is difficult now to understand how there should ever have been any doubt regarding its mode of origin, were it not for the difficulty of accounting for the presence of glaciers and floating ice at sea level, in so low a latitude, at a period when, on *a priori* grounds, a milder climate might have been expected to prevail than at the present day. The difficulty is increased by the apparent derivation of the included fragments from the south, but it is part of a much larger problem which cannot be fully dealt with as yet, and we must be satisfied with accepting the fact of the existence of glacial conditions at sea level.

Many of the glaciated pebbles of this bed show a peculiarity which is rare in other glacial boulder clays. Instead of being smoothed and striated on one or two surfaces only, they often have a number of flat surfaces meeting at obtuse angles and each showing a different direction of striation.[1] It is in fact evident that, after one face had been ground and smoothed, they had been slightly shifted so as to offer a fresh surface and, by a repetition of this process, many of the pebbles have acquired a facetted appearance, as if they had been ground by a lapidary.

In the trans-Indus extension of the Salt range the boulder beds are underlaid by grey shales, in which three species, *Hyolithes orientalis*, *H. sp.* and *Cardiomorpha indica*, have been found,[2] but east of the Indus a more extensive and most interesting fauna has been obtained from the beds which immediately overlie the

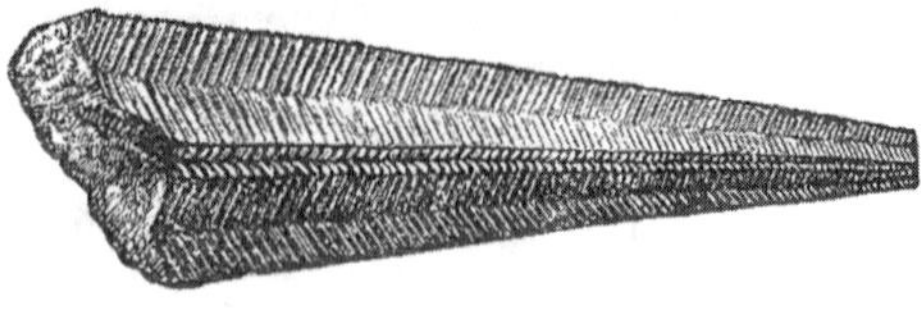

Fig. 6.—*Conularia warthi*, Waag., nat. size.

[1] A. B. Wynne, *Geol. Mag.*, 3rd dec., III, 131, (1886); H. Warth, *Records*, XXI, 34, (1888).

[2] *Pal. Indica*, series xiii, IV, 114, (1891).

boulder clay. The first indication of this was the discovery by Dr. Warth in 1885, of *Conularia* near Pid, but as his fossils were obtained from transported pebbles, their importance in fixing the age of the beds they were derived from was disputed. Subsequent research has shown that, though the nodules in which they are preserved are certainly transported pebbles, they must have been of almost contemporaneous origin with the bed in which they are now formed. Moreover, some fossils from a sandstone band above the conglomerate indicate a precisely similar homotaxis to those in the pebbles, and in discussing the relations of the fauna we may take it as a whole.

The fossils obtained from the conglomerate band are given in the following list, where the letter *A* indicates that the species is also found in the marine beds below the coal measures of New South Wales[1] :—

A. *Pleurotomaria nuda.*
Bucania warthi.
A. *Conularia lævigata.*
A. *Conularia tenuistriata.*
Conularia warthi.
A. *Sanguinolites* cf. *mitchelli.*
„ *tenisoni.*
Nucula, sp.
Pseudomonotis subradialis.
A. *Aviculopecten* cf. *limæformis.*
A. *Martiniopsis darwini.*
A. *Chonetes cracowensis.*
Discina, sp.
Discinisa warthi.
Serpulites undulatus.
„ *warthi.*
„ *tuba.*

From the associated sandstones the following four species were obtained :[2]—

A. *Eurydesma globosum.*
A. „ *ellipticum.*
A. *Eurydesma cordatum.*
A. *Mæonia gracilis.*

The most noticeable point about this fauna is the very large proportion of species, thirteen out of twenty-two, which are identical with forms found in the marine carboniferous beds of New South Wales, a proportion which not only shows that the two are approximately contemporaneous in origin, but that they must have been deposited in the same great marine area, with free communication between the two localities. The Australian beds were formerly regarded as equivalent to lower carboniferous of the European sequence, but are now considered as upper carboniferous, if not homotaxial with the permo-carboniferous of Europe.

The number of identical species would of itself be sufficient to place the Salt range boulder beds on the same horizon as the marine carboniferous beds of Australia, and it is a remarkable fact, whose importance will appear hereafter, that these same marine carboniferous beds of New South Wales have been found to contain large boulders of foreign rock, exhibiting distinctly glacial smoothings and striations, imbedded in fine grained silt, along with delicate *Fenestellæ* and bivalves whose valves are still united in the position in which they lived and died.[3]

[1] *Pal. Indica*, series xiii, IV, 60, 145, (1890-91). | [2] *Pal. Indica*, series xiii, IV, 147, (1891).
[3] *Records*, XIX, 41, (1886).

Tabular Statement of the distribution of Fossils in the groups of the Productus series of the Salt range.

Dr. Waagens classification.			Pisces.	Crustacea.	Cephalopoda.	Gasteropoda.	Lamellibranchiata.	Brachiopoda.	Bryozoa.	Vermes.	Echinodermata.	Coelenterata.	Protozoa.	Total species.	Ranging from groups below	Ranging to groups above.	Species peculiar to the group.	Species ranging both ways.	Classification adopted in this work.	
Productus Limestone.	Upper	Topmost beds (Chidru)	3	..	2	18	31	6	1	...	1	...	1	63	17	...	46	...	Chidru group	Productus Beds.
		Cephalopoda beds (Jabi)	1	...	4	5	10	53	6	...	1	10	...	90	56	7	30	3	Upper	
		Lower and Middle divisions (Khundghát) .	4	1	13	22	18	46	3	1	2	12	4	126	52	55	47	28		
		Upper division (Kálabágh) . . .	1	...	4	7	5	58	4	...	1	7	...	87	53	52	15	34	Middle	
		Middle division (Virgal)	3	...	1	4	7	74	14	1	6	25	...	135	21	69	62	17		
		Lower division (Katta)	...	...	1	3	...	34	...	...	4	2	...	44	10	16	23	5	Lower	
	Lower or Upper speckled sandston		1	...	...	6	4	41	2	1	...	3	5	63	...	18	45	...		

Australia is not the only place where glacial boulder beds of upper palæozoic age have been found, and we shall have frequent occasion to refer to évidences of glacial action and will use such evidence for purposes of correlation, when palæontological evidence is wanting. But as the real importance of this horizon is in establishing the homotaxis of the Gondwána system, the fuller consideration of the recorded observations in other countries, and the discussion of the validity of this method of correlation will be deferred till that rock system is dealt with.[1]

This group of glacial and fossiliferous beds is overlaid by about 400 feet of mostly light coloured, reddish or purplish sandstones, unfossiliferous except for some obscure fucoid markings and plant impressions, known as the speckled sandstone group.[2] The sandstones are interbedded with some red shaly bands, and lavender coloured and purple argillaceous and gypseous bands, which are especially prominent near the top of the group. Ripple marks and oblique lamination are frequent, and the weathered surfaces of the sandstone is frequently studded with rounded knobs due to a local concentration of the calcareous cement.

Above the speckled sandstone comes a great series of beds which have long been known for the wealth of fossil remains they contain, and for the presence in them of Ammonites associated with Brachiopods of palæozoic type. The detailed classification of these beds is a matter of some difficulty; the first column of the accompanying tabular statement shows that adopted by Dr. Waagen, but if the distribution of the fossil remains, as shown in the tabular statement, is examined, it will be seen that this grouping ignores the two most prominent palæontological breaks, that between the Katta and Virgal beds and the still more striking one between the Jabi and Chidru beds, nor is it, so far as can be judged from the published descriptions, what would be adopted on purely lithological grounds. The most important modification required appears to be the separation of the topmost, or Chidru, beds from the rest of the series, a separation demanded not only by some slight indication of a physical break and by the small proportion of species which are found in lower groups, but more especially by the complete change in the type of the fauna from one marked by the prevalence of Brachiopoda to one in which their place has been taken by the Lamellibranchiata. Excluding these beds, the other groups fall naturally into three divisions of two groups each, as indicated by the brackets on the right hand side, a grouping that will be adopted in the following description.

The two groups of the lower Productus beds are more distinctly separated than in the overlying divisions. The lower group consists mainly of

[1] *Infra*, chap. VIII.

[2] *Memoirs*, XIV, 90, (1878).

soft sandstones, coaly near their base, full of fossils, contrasting in this point strongly with the lithologically not very dissimilar beds of the speckled sandstone, in which no fossils have been found. The group becomes more calcareous to the west, and in the trans-Indus extension of the Salt range is said to consist of limestone. The upper group is not very sharply defined from the lower, but is more calcareous; where normally developed it contains many beds of light coloured yellowish or whitish limestones, intercalated with marly and sandy beds.

The first point to notice about the fauna of this group,[1] whose general facies is decidedly upper carboniferous, is the entire absence of a single species also found in the fossiliferous beds at the base of the speckled sandstone, the only species presenting any degree of alliance being *Bucania kattaensis*, which might be a modified descendant of *B. warthi* from the lower beds. The change is complete, and with it disappears all connection with the Australian carboniferous fauna, to be replaced by a relationship with upper carboniferous and permian faunas of Europe. Not counting allied species, the lower group contains no less than sixteen species identical with forms found also in the carboniferous and permian of Europe, or in that series of strata, intermediate between the two, which the Russian geologists have distinguished under the name of permo-carboniferous. The names and distribution of these species will be easiest explained by the tabular statement below:—

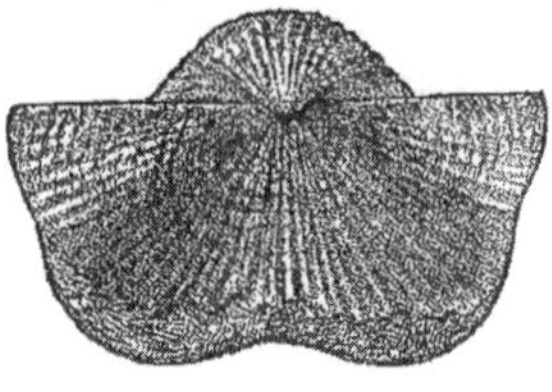

Fig. 7.—*Productus semi-reticulatus*, Mart., lower productus beds.

	Lower carboniferous.	Upper carboniferous.	Permo-carboniferous.	Permian.
Dielasma elongatum	...	...	*	*
Athyris royssii	*	*	...	*
Spiriferina cristata	...	*	*	*
Spirifer striatus	*	...	...	...
„ *marcoui*	...	*	*	...
„ *moosakhailensis*	...	...	*	...
„ *alatus*	...	*	*	*
Martinia cf. *glabra*	*	...	...	...
Orthis indica	...	*	*	...
„ *pecosii*,	...	*	...	...
Reticularia lineata	*	*	*	...
Streptorhynchus pelargonatus	...	*	*	*
Productus lineatus	...	*	*	...
„ *cora*.	...	*	*	...
„ *semireticulatus*:	*	*	*	...
„ *spiralis*	...	*	*	...

[1] Complete lists of the fossils will be found in *Pal. Ind.*, series xiii, IV, 160, 182, (1890-91).

From this it will be seen that four of the five species which are found in the mountain limestone range up into higher beds, and all of the five found as high as the permian are also found in lower beds, while there are fourteen species found in upper carboniferous and permo-carboniferous beds, seven of which have never as yet been found in newer or older beds. This fixes the homotaxis of these beds as upper carboniferous, or intermediate between that and permian, a conclusion in accordance with the general aspect of the fauna.

The upper group of the lower division only contains four species that are found in the permo-carboniferous of the Ural mountains, of which only one (*Strophalosia horrescens*) is not found in the lower group. It also contains two species, *Dielasma itaitubense*, and *Spirifer marcoui*, which have also been found in South and North America respectively.

We have here evidence of a great change in the distribution of land and water from the time when the lower beds of the speckled sandstone were being deposited and the sea stretched continuously from the Salt range to New South Wales. The sea no longer flowed over eastern Australia, where fresh water sandstones and shales with beds of coal were being formed, and the barrier which, at an earlier period, shut out the European forms of life had been submerged, allowing the western fauna to invade the Salt range permo-carboniferous sea.

The same sea appears to have extended eastwards into China, for, from thin beds of limestone above the coal of Lo Ping, a fauna of upper carboniferous type has been described, which contains eleven species also found in the Salt range. Of these *Reticularia lineata*, *Martinia glabra*, *Orthis pecosii*, *Productus semireticulatus*, *Strophalosia horrescens*, and *Richthofenia sinensis*, are found in the lower group of the lower Productus beds, while another species, *Productus compressus*, is found in the upper group of the same division. The other five species are found only in the middle or upper division.[1]

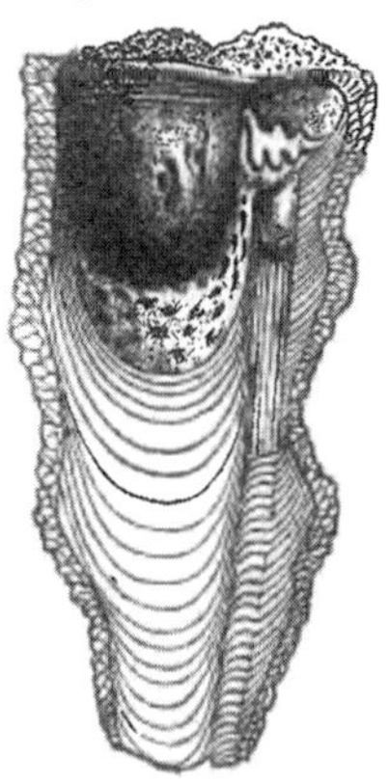

Fig. 8.—Section of *Richthofenia laurenciana*, lower and middle Productus beds.

The middle division of the Productus beds is the most important in thickness, and the most conspicuous owing to its forming great precipices in the western portion of the range. It consists principally of compact grey limestone, frequently dolomitic and full of fossils which are usually difficult of extraction, the very numerous specimens obtained having been mostly derived from marly beds interstratified

[1] *Pal. Indica*, series xiii, IV, 163, (1891).

with the more compact limestone. Corals are common and, in places, they are accumulated in great reefs in which no bedding can be traced. The lower group of the middle division, which has a thickness of 200 to 300 feet, so far as can be gathered from the descriptions, is overlaid by from 10 to 50 feet of marly beds intercalated with thin limestone, which form the upper group. Silica, which is always present in varying quantity in the lower group, and sometimes forms large flinty concretions, is much more prevalent in the upper group, and the fossils are beautifully silicified, especially at Musa Khel, and are generally very numerous.

The fauna of this division is a very extensive one, comprising 169 distinct species, and might easily be extended by systematic collecting.[1] Of these twenty-two have been found in other parts of Asia and Europe whose distribution in time is shown by the following tabular statement :—

	Carboniferous.	Permo-carboniferous.	Permian.
Macrochcilus avellanoides	...	...	*
Pseudomonotis gosforthensis	...	..	*
Camerophoria humbletonensis	...	...	*
" *superstes*	*	...	...
Spirigerella derbyi	*	...	...
Athyris royssii	*	*	*
Spiriferina cristata	...	*	*
Spirifer moosakhailensis	*	*	..
Orthis cf. indica	...	*	...
Productus lineatus	*	*	...
" *cora*	*	*	...
" *humboldti*	*	...	*
" *abichi*	...	...	*
" *tumidus*	...	...	*
Marginifera typica	...	*	...
Fenestella perelegans	*	...	*
Synocladia virgulacea	...	...	*
Thamniscus dubius	...	...	*
Acanthocladia anceps	...	..	*
Spirorbis helix	...	...	*
Geinitzella columnaris	...	...	*
Stenopora ovata	*	...	...

It will be seen from the tabular statement that the fauna is distinctly newer in type than that of the lower division, and contains a larger proportion of exclusively permian forms. It is also marked by the appearance of certain species with decidedly mesozoic affinities, namely *Nautilus peregrinus*, and *Oxytoma atavum*, which are allied to jurassic species, *Hemiptychina inflata* has its nearest ally in the trias of Italy. *Pecten* is

[1] A full list of fossils will be found in *Pal. Indica*, series xiii, IV, 60, 186, 198, (1891).

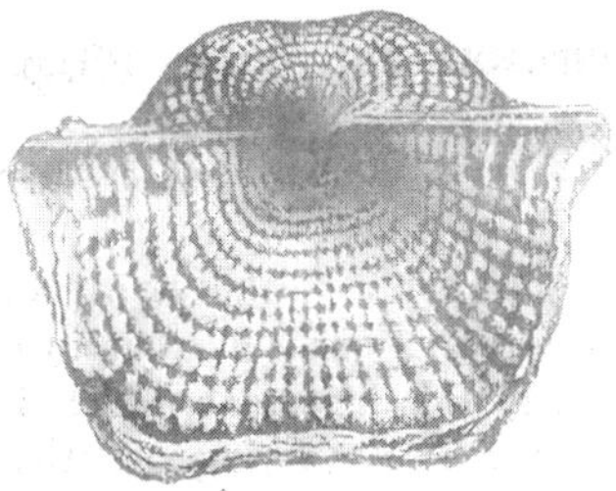

Productus subcostatus, Waag.

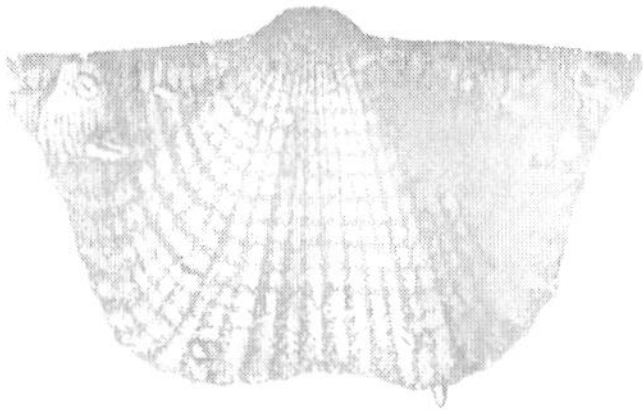

Productus indicus, Waag.

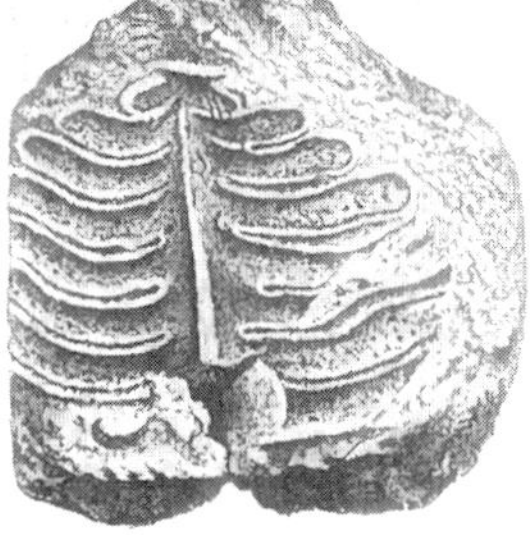

Lyttonia tenuis, Waag.

Pseudomonotis [illegible], King.

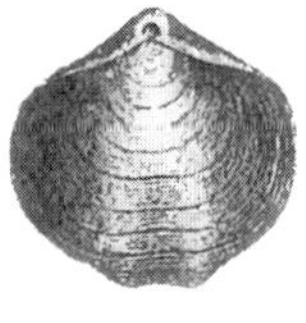

Athyris royssii, Lev.

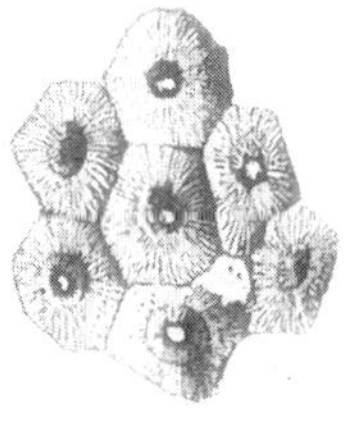

Lonsdaleia salinaria, Waag & Wentzel.

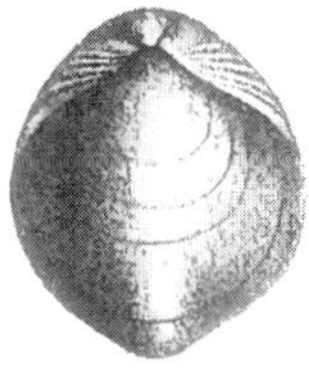

Spiriferella derbyi, Waag

Camerophoria purdoni, Dav.

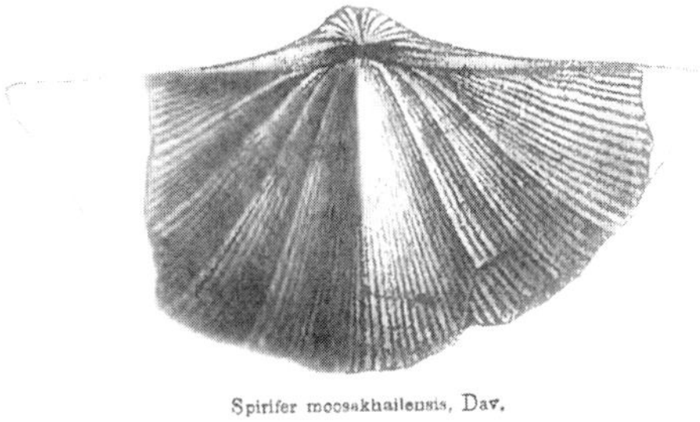

Spirifer moosakhailensis, Dav.

PERMO CARBONIFEROUS (Middle Productus Limestone) FOSSILS.

Calcutta Phototype Co.

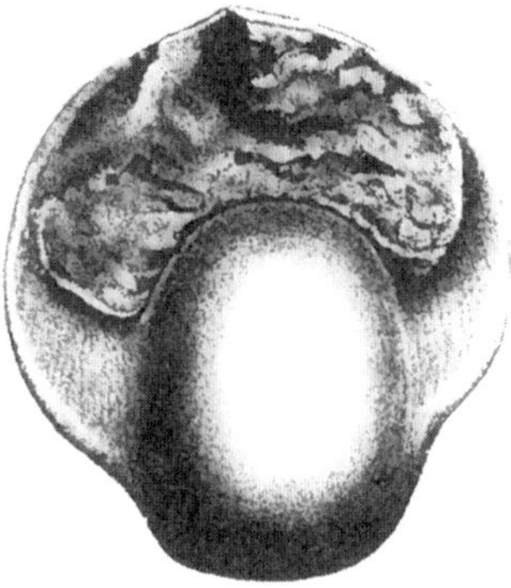

Bellerophon jonesianus, Kon.

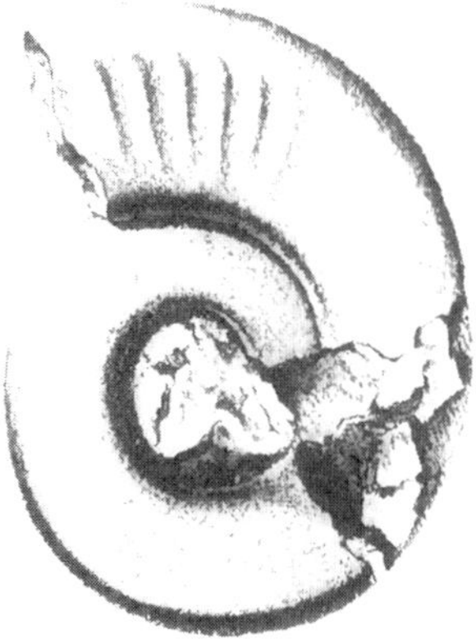

Xenodiscus carbonarius, Waag.

Synocladia virgulacea, Waag.

Bellerophon jonesianus, Kon.

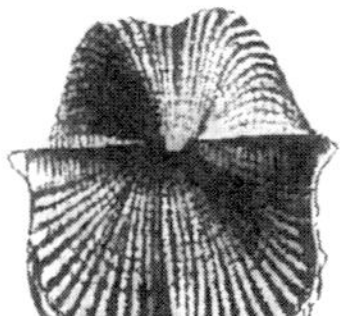

Productus gratiosus, Waag.

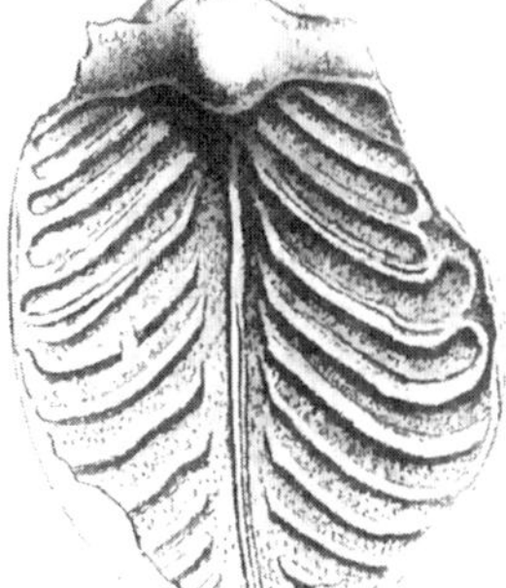

Oldhamina decipiens, Kon.

Entalis berculea, Kon.

PERMIAN (Upper Productus Limestone) FOSSILS.

seldom found below the trias, and the genus *Oldhamina* has its nearest relation in the rhætic *Pterophloios.* These mesozoic forms preclude us from assigning the group to an older date than the permian, but the relationships are not close or extensive enough to justify assuming a more recent date than is indicated by the rest of the fauna.

The upper Productus beds, whose whole thickness does not exceed 100 feet, consist of light yellowish sandy dolomite impregnated with silica. The fossils are abundant and always silicified, but usually only on the inner and outer surfaces of the shell. Rusty coloured ferruginous dolomites, which are occasionally found in the lower division, are not entirely absent from the upper one, but they are very rare.

There is said to be a somewhat abrupt lithological change from the divisions immediately underlying this, but there is no corresponding change of the fauna, over one-third of the total number of species being also found at a lower horizon. The fauna is a very extensive one, no less than 175 species having been recorded, and Dr. Waagen has expressed an opinion that this number might be doubled by systematic search.[1]

The general facies of the fauna is permian, though only eighteen species have been found elsewhere in beds that are believed to be of permian age, half in the permian of Europe and half in Armenia. But, mixed with these palæozoic fossils, there are numerous forms which show more distinctly mesozoic affinities than those found in the groups below.[2] Among these the most remarkable are the *Ammonitidæ* represented by *Cyclolobus oldhami, Arcestes antiquus, A. priscus, Xenodiscus carbonarius, X. plicatus, Sagoceras haueriarum.* When these were discovered, some twenty years ago,[3] the finding of well characterised ammonites in beds containing numerous brachiopoda of palæozoic types, and believed to be of upper carboniferous age or intermediate between that and the permian, was regarded as so extraordinary that doubt was cast on the accuracy of the observation. Subsequent researches have shown that, even if we exclude all the species having mesozoic affinities, there is nothing in the fauna to warrant us in assigning it an older date than the zechstein, or upper permian, of Europe, while the presence of the mesozoic forms, and their superposition on distinctly permian strata, leads us to regard the beds as even newer than this. The presence of true ammonites in these beds has consequently become less surprising than it was at first, owing on the one hand, to the establishment of the true age of the Salt range *Ammonitidæ* as uppermost permian, or even newer, and on the other, to the subsequent discovery of *Cephalopoda* closely allied to the *Ammonitidæ* in upper palæozoic rocks.

[1] Full lists will be found in *Pal. Indica,* series xiii, IV, 63, 210, 221, (1890-91).

[2] *Pal. Indica,* series xiii, IV, 213, 223. (1891).

[3] *Memoirs,* IX, 351, (1872).

The topmost productus beds of Dr. Waagen, here separated as the Chidru group, are only about 15 feet in thickness, composed of soft light yellow sandstones, with coaly bands at the base in some of the sections. No actual unconformity between these beds and those below has been detected, but a considerable interval of time probably intervened between the deposition of the two, for only seventeen of the sixty-three species are also found in the beds below, and there is a complete change in the type of the fauna from one in which the Brachiopoda comprise more than half the total number of species and the Lamellibranchiata less than one-tenth, to one in which the proportions are almost exactly reversed. The fossils are all described as more or less rare, except *Margaritina schwageri*, which is said to be very common.[1]

The general type of the fauna is distinctly less palæozoic than any of the preceding ones. Only four species *Schizodus rotundatus*, *Nucula trivialis*, *Pseudomonotis radialis*, *Athyris sub-expansa*, are identical with permian forms of other countries, while of the species peculiar to this group twenty-four have palæozoic affinities, and no less than twenty-two are allied to mesozoic forms. So far as the palæontological evidence goes we are already well on the way into the secondary era, even if the beds cannot be regarded as lowermost trias in age.[2]

The Chidru group closes the conformable sequence of beds containing palæozoic fossils. No unconformity between them and the next overlying beds has been established as yet, but one is suggested by a section, recorded by Dr. Waagen, in which the mesozoic beds followed immediately on the upper Productus beds, with a basal conglomerate but without the intervention of the Chidru group.[3] However this may be there must certainly have been a considerable time interval between the two, for not a single species bridges the interval and is found both above and below the separation of the Chidru group and the ceratite beds which overlie it.

The general aspect of the triassic ceratite beds is such that they might easily be classed with the Productus beds, the succession varies much as to details, but consists generally of a thin limestone with *Ceratites* at the base, succeeded by a thick marly zone which yields readily to weathering, and turns a light greenish colour. It is overlaid by grey sandstone, and flaggy limestone with many ceratites, passing upwards into grey nodular marls capped by hard limestones and calcareous sandstones. Some of the bands of limestone contain glauconite, and beds of conglomerate occasionally occur.[4]

Owing to these beds having been at first confounded with the under-

[1] A list of species is given in *Pal. Indica*, series xiii, IV, 60, 228, (1890-91).

[2] *Pal. Indica*, series xiii, IV, 230, (1891).

[3] *Pal. Indica*, series xiii, IV, 227, (1891).

[4] *Memoirs*, XIV, 96, (1878).

lying Productus beds, and to the later collections not having as yet been described, it is impossible to give a list of fossils. Ceratites abound and, most, probably, all of the species described by de Köninck[1] are from triassic beds. Besides these, which are the characteristic fossils of the group species of *Orthoceras*, *Anoplophora*, *Cardinia*, *Gervilia*, and *Rhynconella* are found, among which the bivalves predominate. The most remarkable form, however, is a species of *Bellerophon*, a genus not known to occur in rocks of later than palæozoic age in Europe.

In the central Himálayas there is no marked unconformable break between the lower and upper palæozoic rock systems, such as is found in the Salt range, and the carboniferous follows with perfect conformity on the underlying beds.

The oldest rock group which can be regarded as carboniferous is a crinoidal limestone, usually red in colour.[2] Mr. Griesbach's collections have not yet been examined, but some fossils brought by Mr. Hughes from a white crinoid limestone in the Milam pass were ound by Dr. Waagen to include:—[3]

Hemiptychina himaloyensis.	*Spirifer glaber.*
Notothyris subresicularis.	*Productus semireticulatus.*
Athyris royssii.	*Lyttonia, sp.*

The horizon of this fauna is regarded by Dr. Waagen as about that of the upper portion of the lower Productus beds. The crinoid limestone is overlaid by a series of fine grained, hard, white, quartzites, in thick beds with a few shaly partings, which were originally included by Dr. Stoliczka in his Muth series. In Spiti they are, according to Mr. Griesbach overlaid by, and partially interstratified with, flaggy dark grey to blue limestones, which contain *Athyris royssii*, and *Productus*, sp., marking them as carboniferous in age

In the Spiti valley General McMahon has recorded the occurrence, in two places, of beds of fine grained slate, through which small rounded quartz pebbles are scattered. The similarity of this bed to that of the Blaini group, which will be described further on, as well as its structure, are suggestive of the action of floating ice. The exact horizon of the bed has not been determined, and it is not certain whether it should be classed with the group just described, or that which overlies it.[4]

The white quartzites, with their overlying limestones, are abruptly overlaid by a group of shales. The junction is said to be unconformable, the

[1] *Quart. Jour. Geol. Soc.*, XIX, 11, (1863).
[2] *Memoirs*, XXIII, 59, (1891). *Supra* p. 114.
[3] *Pal. Indica*, series xiii, IV, 167, (1891).
[4] *Records*, XII, 63, (1879). See also *Records*, XXI, 151, (1888).

unconformity being accepted on the strength of the sudden lithological change, and of the shales resting in different sections on different horizons of the carboniferous strata. These shales, apparently together with part of the underlying quartzites, were distinguished by Dr. Stoliczka[1] as the Kuling series and regarded by him as carboniferous. Mr. Griesbach has divided them into a lower portion, composed of dark, generally black, somewhat micaceous shales, often carbonaceous, with coaly traces here and there, of permian age; and an upper portion of very similar shales, difficult to distinguish lithologically, though they are somewhat less earthy and micaceous, which he regarded as lower triassic in age.

Of the fossils from the Kuling series, described by Dr. Stoliczka, which were probably for the most part derived from the lower part of the shales, *Productus semireticulatus* and *P. purdoni* are found in the Productus beds of the Salt range, the former in the lower, the latter in the middle division, but are both rare. *Spirifer moosakhailensis* is found in all three divisions, but is abundant only in the middle one.

The shaly beds pass conformably upwards into a great series of limestones, the Lilang and Pára of Dr. Stoliczka, which have been subdivided by Mr. Griesbach as follows[2]:

TRIAS	Upper	10. Liver coloured limestone with greenish grey shales in Niti, strong limestones elsewhere. *Corbis*, sp.		Keuper.
		9. Friable shales and earthy beds in Niti, limestones and shales in Spiti and Nilang. *Spirifer lilangenis.*		
		8. Limestone. *Tropites*, sp.		
		7. Earthy limestone and shales		
		6. Black limestone, flags and dolomites. *Daonella*, sp.		
		5. Black limestone flags and partings of splintery black shales.		
	Lower	4. Hard grey concretionary limestone in thick beds. *Ptychites gerardi.*		Muschelkalk.
		3. Earthy limestone. *Rhynconella semiplecta.*		
		2. Limestones alternating with shales.	*Posidonomya*, sp.	Buntsandstein.
		1. Dark shales and limestone partings.	*Otoceras woodwardi.*	

Speaking generally, the lower part of the central Himálayan trias may be described as a series of very dark, almost black, hard limestones with partings of shales; the upper part varies more, being represented by a great thickness of friable shales in the Niti and Milam sections, but in Spiti and to the east it is a limestone formation. The total thickness of the trias is about 4,000 feet in Niti, but probably exceeds that to the east; in Spiti

[1] *Memoirs*, V, 24, (1865).

[2] *Memoirs*, XXIII, 69, (1891).

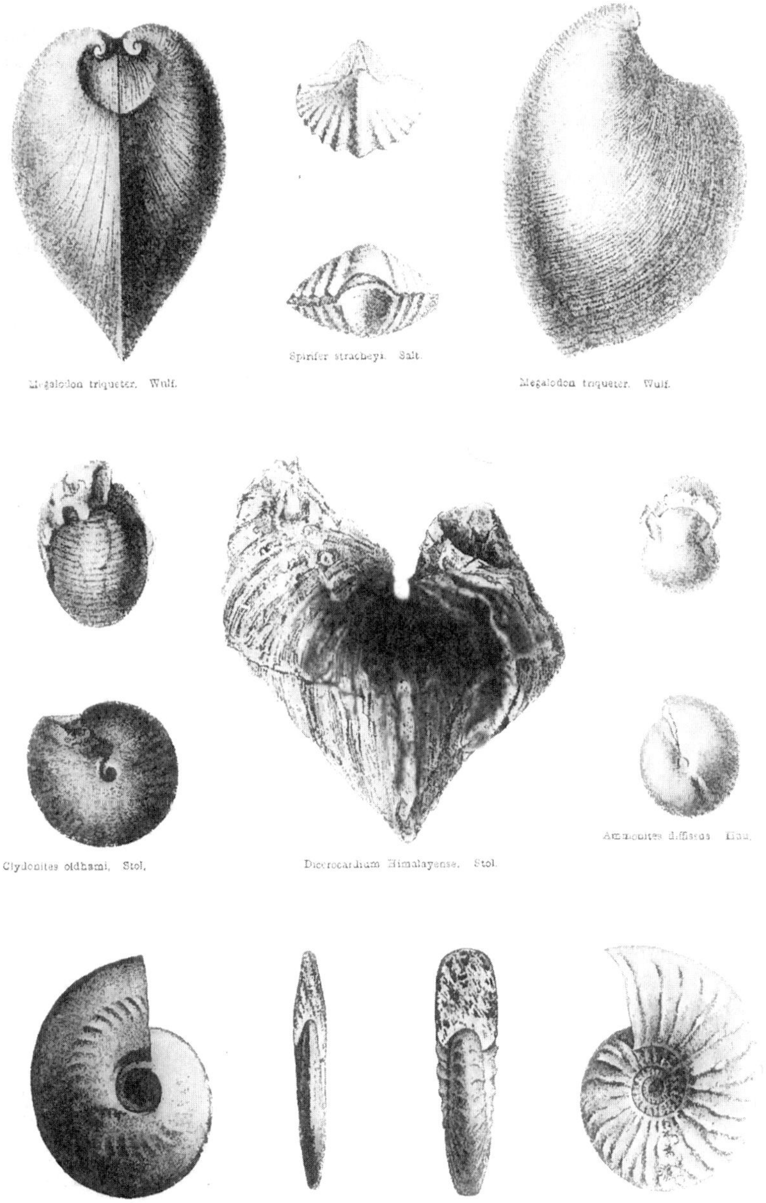

TRIASSIC FOSSILS.

The Calcutta Phototype Co.

Dr Stoliczka estimated the thickness at 1,000 to 2,000 feet. The underlying Productus shales are quite insignificant in comparison to this, being only 150 feet thick. The most noteworthy feature of the Himálayan trias is the abundant cephalopod fauna of the lowest beds, a fauna extremely abundant in specimens, though not so extensive as regards species, and remarkable for its transitional character between a palæozoic and mesozoic facies.

In the central Himálayas the trias is succeeded with perfect conformity by a thick series of limestones, formerly regarded as rhætic and liassic, above which comes an abrupt lithological change, not known to be accompanied by unconformity, at the base of the jurassic series.

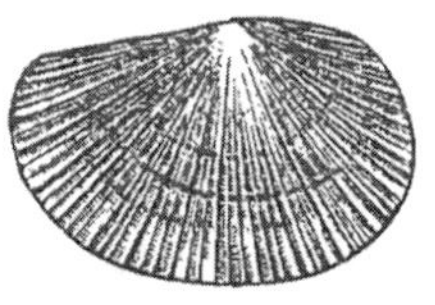

Fig. 9.—*Daonella* (*Halobia*) *lommeli*, Wissm. (Trias).

According to Mr. Griesbach's description the combined rhætic and lias have a thickness of 2,500 to 3,000 feet, of which the lias only comprises 100, and are divided as follows : [1]

5. Black shales and dark earthy limestones with oolitic structure; lower lias fossils.
4. Grey crinoid limestones in irregular thin beds; *Terebratula horrida*, *Gervilia inflata*.
3. *Lithodendron* limestone in thick beds.
2. A great thickness of limestones and dolomities, *Megalodon*, sp.
1. A great thickness of dolomites and flaggy dark limestones with thick-bedded dolomites, which pass downwards into the upper trias.

In his original description of the rocks of Spiti, Dr. Stoliczka adopted a classification different from the more recent one of Mr. Griesbach for the beds above the carboniferous. It is given here in abstract,[2]—

Upper tagling (*lias*).—Dark earthy or dolomitic bituminous limestone; thickness nearly 1,000 feet.

Lower tagling (*lower lias or rhætic*).—Dark grey, brown or black limestone, often earthy or bituminous, weathering into a rusty brown; more than 1,000 feet thick.

Pára Limestones (*rhætic or upper trias*).—Black, dolomitic, strongly bituminous limestones; 700 feet thick.

Lilang Series (*upper or middle trias*).—Dark limestones, calcareous shales and slates; the limestone compact or oolitic and quasi-concretionary in some of the lower beds; 1,000 to 2,000 feet thick.

In the preceding pages a brief outline of the published descriptions of the newer palæozoic and older mesozoic rocks of the central Himálayas has been given. No details have been entered into, nor has any attempt been made to clear up the discrepancies which are apparent, as the fossils that have been collected in this area are at present being examined and described, and it has already become evident that this examination will result in a considerable modification of the correlations adopted up to now. The

[1] *Memoirs*, XXIII, 73, (1891).

[2] *Memoirs*, V, pt. i, (1865).

most important results obtained so far, according to the distinguished palæontologists, whose names it would be unjust to couple with conclusions so provisional in their character, are, that there is a very strongly marked palæontological break between the silurian and the carboniferous, that in the carboniferous beds two successive faunas can be recognised, one allied to that of Australia, the other to the Productus fauna of the Salt range, and that the beds formerly regarded as rhætic and lias should be classed with the trias, leaving a distinct palæontological break between the trias and jura of the central Himálayas. The publication of these results, which it is hoped will not be long delayed, will be a most important addition to our knowledge of the stratigraphy of the central Himálayas, and will doubtless clear up many of the inconsistencies and uncertainties which now obscure it.[1]

The beds which have been just described are found in two great basins of disturbance, one of which stretches along the north of Kumáun and Garhwál, the other occupies the valleys of the Spiti and Zanskar rivers and stretches beyond them to the neighbourhood of Kartse. They are found again in the Kashmír valley, but before proceeding to the description of the outcrops it will be best to turn to the Simla district.

The older rock systems of this area have already been described and we now come to that which has been named the carbonaceous system.[2] The beds of this system present certain marked peculiarities by which they have been recognised with a greater or less degree of certainty over a large area south of the snowy range, from the western borders of Nepál to the confines of Kashmír, but it is only in the neighbourhood of Simla and the protected hill states that they have been studied in any detail.

The lowest member of the system appears to be part of what has already been referred to under the name of Simla slates.[3] This name was applied to a great series of slates, gritty slates and quartzites, in which no break has yet been detected, though it is highly probable that one or more will be established by detailed survey. Whether any of these beds should be classed with the carbonaceous system is open to doubt, but the uppermost ones appear to be perfectly conformable with a group of beds, which have so marked an individuality that they are of the greatest importance in unravelling the complicated structure of the hills, and in establishing the homotaxis of this system. The characteristic

[1] As these pages were going through the press a preliminary note on the Cephalopoda of the Himalayan trias by Dr. Mojsisovics has been published. The whole range of the European trias has been definitely recognised, and attention is drawn to the remarkable abundance of cephalopoda in the lowest beds, where they are poorly represented in Europe *Sitz. K. Akad. Wiss.*, Wien, CI, pt. i, (1892); *Records*, XXV, 186, (1892).

[2] *Records*, XXI, 133, (1889).

[3] *Supra*, p. 117.

member of this group, which Mr. Medlicott named Blaini[1] after the stream of that name, flowing westwards from Solan, is a conglomeratic slate, composed of rounded pebbles of quartz, ranging up to the size of a hen's egg, or, in other cases, angular and subangular fragments of slate and quartzite of all sizes up to some feet across, which are scattered at intervals through a fine grained matrix. It contains numerous fragments of the volcanic beds of the Jaunsar system, where exposed in the neighbourhood of their outcrop in the Naira valley in eastern Sirmur, and here even the matrix has much the appearance of a volcanic ash,[2] though as a rule there is not the slightest trace of volcanic material either in the matrix or in the included fragments. This remarkable rock has been observed from east of Mussooree at intervals to beyond Simla, and is generally, though not invariably, accompanied by a band, 20 or 30 feet thick, of thin bedded, usually pink, dolomitic limestone, which lies on top of the zone in which the beds of conglomeratic slates occur. The agency by which blocks of stone were dropped over so large an area into a tranquil sea, in which alone the matrix could have been deposited, must have been no local one, and the only one that appears at all adequate is that of floating ice. No smoothed and striated fragments have been found as yet, though one is occasionally met with showing striation resembling those produced by glaciers, but the rock has invariably undergone much compression and disturbance, at times accompanied by a distortion of the shape of the included fragments, which might account either for the obliteration of distinctly glaciated surfaces, or for the production of those scratches which have been observed. No certain conclusion can, consequently, be drawn from the occasional presence, or general absence, of striation, but the only alternative hypothesis, that the rock is in fact the volcanic breccia which at times it resembles, appears to be excluded by the infrequence of volcanic material in it, and the absence of any other associated volcanic beds, while the included fragments are too numerous to be accounted for by the action of floating drift wood.

The boulder beds are overlaid by a series of shales or slates, characterised by the greater or less prevalence of carbonaceous matter, which were originally described as infra-Krol[3], from the fact that they underlie the limestone of the Krol mountain. The name has since been so commonly used in the publications of the Geological Survey that, in spite of a certain awkwardness, it cannot well be abandoned now.

The carbonaceous impregnation of these shales is very irregularly distributed, being often extremely conspicuous, especially where the rock

[1] *Memoirs*, III, pt. ii, p. 30, (1864). The name was originally spelt Blini, being the spelling adopted in the Atlas of India. The spelling in the text was subsequently introduced as more correct. *Records*, X, 204, (1877).

[2] *Records*, XX, 156, (1887).

[3] *Memoirs*, III, pt. ii, 29, (1864).

has undergone crushing, but at other times wanting, at any rate near the surface. Not infrequently the blackest and most carbonaceous beds weather almost white by the removal of the carbonaceous element. Above these beds there is usually a series of quartzites of very variable thickness, varying from about 20 feet in the sections south of the Krol mountain to some thousand feet in western Garhwál. They are very noticeable at Simla, forming the whole of the Boileauganj hill and the lower part of Jutogh, where they have been called Boileauganj quartzites.

In western Garhwál, between the Tons and Pábar rivers these quartzites contain fragments of undecomposed felspar, usually subangular but in some of the beds large and angular, mixed with quartz, mica and fragments of the accesory minerals of the Himálayan gneissose granite and gneisses, the whole forming a rock which, having become foliated by a subsequent slight metamorphism, is sometimes difficult to distinguish at first glance from the true gneissose granite.[1]

The uppermost member of the system is another group of carbonaceous slates, associated with carbonaceous or graphitic limestones[2], which pass upwards, in western Garhwál, into blue limestones. In the Krol mountain the uppermost beds are blue limestones with associated shaly bands, mostly grey in colour, though there is one distinct zone of red shales, but as no carbonaceous beds are associated with them, and as the underlying quartzite exhibits remarkable variations in thickness, it is uncertain whether these limestones of the Krol group are the equivalents of those just referred to or belong to a later unconformable system.

The beds of the carbonaceous system contain, on most of the sections, interbedded basaltic lava flows and more or less impure volcanic ashes, either recognisable as such, or represented by hornblende schists, where the rocks have become schistose. The range of the volcanic beds varies on different sections. Their usual position is in the upper band of carbonaceous shales, but they are also found among the quartzites and in the upper part of the infra-Krol, though they never, so far as is known, extend down as far as the Blaini group.

In Kashmír fossiliferous beds of upper palæozoic age are underlaid by a great slaty series, the whole of which was grouped by Mr. Lydekker in his Panjál system[3] and regarded as silurian and cambrian, partly on account of its underlying supposed carboniferous beds, partly on account of the opinion regarding the age of the Blaini group which was prevalent when he wrote. A large part of these slates are doubtless of older palæo-

[1] *Records*, XX, 160, (1888). [2] *Records*, XX, 148, (1888).
[3] *Memoirs*, XXII, 209, (1883).

zoic age, but part at least appear to be more properly classed with the upper palæozoics.

The oldest of the beds with which we are at present concerned is a conglomeratic slate, composed of subangular fragments and rounded pebbles of slates and quartzites, imbedded in a matrix of fine grained slate. The rock is in every way similar to the Blaini group of the Simla area, and the same arguments in favour of a glacial origin are applicable in both cases. Besides this, the Kashmír boulder bed occupies the same position relative to fossiliferous beds of carboniferous age as the glacial boulder clay of the Salt range does relative to the Productus beds, and as it is reasonable to suppose that the extreme cold which affected the one area must have extended to the other, we may take it as certain that the so called Panjál conglomerate is also of glacial origin.

The boulder slate is overlaid by a series of quartzites and black carbonaceous slates, in the upper portion of which there is an abundance of contemporaneous volcanic rock, and above these there are thin bedded, light blue and white fossiliferous limestones from which a number of fossils have been collected by different observers. The following list, quoted on the authority of Mr. Lydekker, includes all those known up to now:—[1]

CEPHALOPODA—
Orthoceras, sp.

LAMELLIBRANCHIATA—
Avicula, sp.
Aviculopecten, sp.
Solenopsis, sp.

BRACHIOPODA—
E P. *Athyris subtilita.*
A. E. P. „ *royssii.*
Camerophoria, sp.
Chonetes (?) *austeniana.*
A. E. „ *hardrensis* var. *tibetensis.*
„ *lævis.*
Discina kashmiriensis.
Orthis, sp.
E. A. P. *Productus cora.*
E. P. „ *costatus.*
E. P. „ *humboldti.*
„ *lævis.*
E. A P. „ *longispinus.*
E. A. „ *scabriculus.*
E. A. P. „ *semireticulatus.*
E. „ (?) *spinulosus.*
E. P. „ *striatus.*
Retzia, sp.
Rhynconella barusiensis.
„ *kashmiriensis.*
E. P. „ *pleurodon* var. *davideuxiana.*
Spirifer barusiensis.
„ *kashmiriensis.*
S. „ *keilhavii*, (*S. Raja*).
A. P. „ *moosakhailensis.*
A. P. „ *striatus.*
„ *vercheri.*
„ *vihiana.*
E. A. P. *Spiriferina octoplicata.*
E. A. P. *Streptorhynchus crenistria.*
E. A. *Strophomena romboidalis* var. *analoga.*
Terebratula austeniana.
E. „ *sacculus.*

POLYZOA—
P. *Fenestella* (?) *lepida.*
P. „ *sykesi.*
P. „ *megastoma.*
A. *Protoretepora ampla.*
E. *Vincularia multangularis.*

CRUSTACEA—
E. *Phillipsia* (?) *seminifera.*

[1] *Memoirs*, XXII, 158, (1883).

To these may be added *Lyttonia*, which, according to Mr. Lydekker, is seen on the weathered surface of the limestones.[1]

In the foregoing list the letter E denotes that the species is also found in the mountain limestone of Europe, A in the marine carboniferous of New South Wales, P in the Productus beds of the Saltrange, S in the Kuling series of Spiti. The determinations are all on the authority of Dr. Feistmantel, as quoted by Mr. Lydekker. Dr. Waagen, who has adopted a narrower definition of a species, finds only two, *viz. Athyris subtilita*, Hall. = *Spirigerella derbyi*, Waag., and *Spirifer moosakhailensis*, Dav., identical with Salt range forms[2] *Discina kashmiriensis*, Dav., is said to be almost identical with *D. warthi*, Waag., from the beds at the base of the speckled sandstone, and Dr. Waagen is inclined to place the Kashmír beds at a horizon intermediate between these and the Productus beds, representing in fact part of the unfossiliferous speckled sandstone. The number of species identical with, or closely allied to, Australian forms is discounted by the fact that three-quarters of them are also found in Europe, but so far as it goes is more in accordance with Dr. Waagen's correlation than with a later date, and we may conclude that the carboniferous fossiliferous beds of Kashmír are somewhat older than the lower Productus beds of the Salt range.

The carboniferous of Kashmír is overlaid by a series of limestones, exposed in synclinal folds of various sizes, at either end and north of the valley. They are the equivalents of the triassic limestones of Spiti. They are sparingly fossiliferous, and though a considerable number of forms have been obtained from them, the only ones specifically identifiable are *Ammonites* (*Ptychites*) *gerardi*, *Megalodon gryphoides*, and *Spirifer stracheyi*. Besides these, stems of *Crinoids*, *Orthoceras*, doubtful *Ceratites* and *Goniatites*, and several genera of Gasteropods and Corals, all more or less doubtful, have been obtained.[3]

It will be seen from the descriptions that there is a great similarity between the sections in Kashmír and the Simla area. In both, boulder-bearing slates of presumably glacial origin are overlaid by a series of slates and quartzites, characterised by a carbonaceous impregnation and by the presence of contemporaneous volcanic beds, and in both the uppermost member is a limestone. The resemblances are not mere lithological ones, between rocks such as have always been in process of formation at every period of the earth's history. They are exhibited by rocks which owe their origin to wide reaching causes which have only occasionally acted, and it is difficult to resist the conclusion that they are evidence of the contemporaneous origin of the two rock series, and not merely accidental.

[1] *Records*, XVII, 37, (1884).
[2] *Pal. Indica*, series xiii, IV, 166, (1891).
[3] *Memoirs*, XXII, 158, (1883).

The country intermediate between the Simla and Kashmír areas has not as yet been examined in any detail, but we know from Colonel McMahon's descriptions that similar carbonaceous beds with associated volcanic basaltic traps, underlaid by a conglomeratic slate resembling the Blaini of the Simla area, are found in Chamba and near Dalhousie.[1]

These observations serve to link the Kashmír and Simla sections and strengthen the conclusion that was based on lithological resemblances. It has not as yet been corroborated by the discovery of fossils in the south-eastern area, though the rocks are in many places perfectly adapted for the preservation of organic remains. Even in Kashmír territory, once the drainage area of the valley is left, fossils become rare. As we go south-eastwards they become more and more sporadic in their occurrence, and, except in one small area in western Garhwál,[2] not a single fossil of older date than tertiary has yet been found south of the first snowy range.

The correlation of the beds of the carbonaceous system has long been a stumbling block in the way of our knowledge of Himálayan geology. When Dr. Stoliczka visited Spiti in 1864, the rock systems below the Blaini had not been identified, and the section along the road through Simla was believed to represent pretty fully the sequence of stratified rocks in the lower Himálayas. He attempted accordingly to distribute the beds of the Simla section over the section seen in Spiti, and suggested that the Blaini 'conglomerate' was the equivalent of certain conglomerates in the Muth series—Haimantas according to the classification adopted in this work—and consequently at least as old as silurian.[3] It does not appear that Dr. Stoliczka offered this correlation as more than a guess, to which he probably himself attached small value. It seems certain that he did not recognise the peculiar character of the Blaini rock, but regarded it as an ordinary conglomerate, and he never saw the rock in the Spiti river, which is an exact equivalent of the Blaini conglomeratic slate. Yet the glamour of his genius has shed an importance over this guess which it was never intended to possess, and time after time the Blaini group has been unquestioningly referred to as silurian. The more probably correct correlation was pointed out in 1888[4] and the evidence, then was practically as strong as it now is. It comes as near certainty as is possible in the absence of palæontological evidence, while there is really no evidence worthy the name in favour of the older view. Yet such is the vitality of error that the older palæozoic age of the Blaini has been accepted without question in one of the latest publications on Himálayan geology, and the very writer who first drew attention to the probability

[1] *Records*, XIV, 305, (1881); XV, 34, (1882); XVII, 34, (1884).

[2] See *infra*, p. 229.

[3] *Memoirs*, V, 141, (1865).

[4] *Records*, XXI, 142, (1888).

of its being carboniferous or permian, is quoted as supporting the view which he combated.[1]

The other correlations of Dr. Stoliczka, of the quartzites of Boileauganj with the Kuling, and of the Krol with the Lilang limestone of Spiti, are probably correct, and curiously enough an apparent confirmation was published, about the same time as his memoir, in Professor Gümbel's description[2] of a specimen from the Schlagintweit collection, said to have been obtained at Dharampur near Solan in the Simla district, containing three fossils, *Lima lineata* and *Natica gaillardoti*, found also in the Muschelkalk of Europe, and a new species, *N. simlaensis*. Dharampur in the neighbourhood indicated is, however, a well known locality on tertiary rocks, and specimen in question must have come from a totally distinct ground, probably in Tibet.

In the Kágán and Kishengangá valleys, north-west of Kashmír, there are a number of small outcrops of carbonaceous slates, overlaid by white or buff crystalline limestone, folded into the gneiss and schists.[3] They are probably representatives of the upper palæozoic and triassic rocks of Kashmír. They have not been closely examined or surveyed, and are mentioned here merely as indications of the former extension of these rocks, and as occupying a geographical position intermediate between the carbon-trias of Kashmír and of the Hazára district of the Punjab.

At the extreme north-western extremity of the Himálayas, fossiliferous rocks are found south of the snowy ranges in the district of Hazára. No fossiliferous beds of carboniferous age have yet been identified in this corner of the Punjab, and the only indication of their presence, west of the Jehlam and north of the Salt range, is the discovery, by Mr. Lydekker, of *Productus humboldti* in a loose block of limestone near Hasan Abdál.[4] The age of the rocks underlying the triassic group of the Sirban mountain near Abbottábád is uncertain, but they are quite unconformable to the underlying Attock slates, and may be carboniferous. They comprise two divisions; the lower consists of sandstones, shales, and silicious limestones, all red in colour, with an argillaceous breccia, full of fragments derived from the underlying rocks at its base, but the published descriptions are insufficient to determine whether this may or may not represent the glacial boulder bed at the base of the carboniferous in the Salt range and in Kashmír. The upper division is composed of dolomites only, lighter in colour than the lower beds, often highly silicious and of considerable thickness. These dolomites are over-

[1] *Memoirs*, XXIII, 54, (1891).

[2] *Sitzungsber, K. Bair. Akad Wiss*, München, 1865, Bd. II, p. 364.

[3] *Memoirs*, XXII, 205, (1883).

[4] *Manual*, 1st edition, p. 501.

laid by a group of hæmatitic rocks, quartz breccias, sandstones, and shales which may belong to the trias.[1]

In western Hazára there is a great series of much contorted rocks to which Mr. Wynne has given the name of Tanáwal (Tanol),[2] from the name of the district they are found in. They comprise an enormous thickness of grey and drab quartzites and quartzose beds, in rapid alternation with dark earthy beds, flaggy, shaly or slightly schistose, associated with conglomeratic slates containing pebbles of quartz and quartzite, ranging up to the size of a goose's egg. In the synclinal folds are thick zones of various coloured pseudo-brecciated, silicious, cherty or compact, grey, black and buff dolomitic limestones, with which are occasionally associated intensely black graphitic and sulphurous shales, or else purple and red sandstones and slaty bands.

These rocks, whose general description accords fairly well with that of the carbonaceous system of the Simla region, are regarded as the equivalent of the beds below the trias of the Sirban mountain, chiefly on account of their superposition and probable unconformity to the Attock slates, and partly because of the occurrence of red and purple slates and quartzites at the base of each.

Triassic rocks attain a great development in Hazára, being 1,500 to 2,000 feet thick in the Sirban mountain, and some 3,000 to 4,000 feet near Khánpur. Owing to the disturbance they have undergone they occupy a number of small exposures, too small to be shown on the map, in the areas coloured as silurian and nummulitic, respectively. In the Sirban mountain they consist chiefly of black or dark grey, distinctly bedded, limestone, with thick zones of massive dolomite, some of which contain numerous opaque laminæ of quartz. Near Abbottábád, where the series is complete, dolomites form the lowest beds, and are followed by thin bedded, fossiliferous limestones, containing *Megalodon*, *Dicerocardium*, *Chemnitzia*, and *Gervilia*. The dolomites are not always present, and the base of the series may be formed by the limestones, which are succeeded by quartzites and dolomites of considerable thickness, again overlaid by thin bedded limestones and slaty shales containing *Nerinea*, *Neritopsis*, *Astarte*, *Opis*, *Nucula*, *Leda*, and *Ostrea*.[3]

The other exposures of triassic rocks in Hazára, while exhibiting some variations, do not differ essentially from those of the Sirban mountain, and it is not certain how far the differences which have been observed may be only due to the obscuring effect of the intense disturbance they have undergone.[4]

Carboniferous and triassic rocks can be traced along the southern slopes

[1] *Memoirs*, X, 335, (1872).
[2] *Records*, XII, 122; (1879).
[3] *Memoirs*, IX, 336, (1872).
[4] *Records*, XII, 124, (1879).

of the Pír Panjál and Dháola Dhár ranges, but no fossils have so far been found, except some obscure gasteropods in the Jehlam valley. They are of the ordinary Kashmír type of quartzites and carbonaceous slates, underlaid by the boulder-bearing slates and overlaid by limestones. There are some inliers of massive grey limestone in the tertiary area, which are faulted up on their south-west side. No fossils have been found in the limestone, which is bedded, compact, dark grey to black in colour, resembling the limestones of the Himálayas, and is probably of triassic age or older. They have been coloured brown on the map, as that represents their most probable age, but it must not be left out of sight that they may well belong to the older, precarboniferous series of limestones of the Himálayas or to a later post-triassic age.[1]

To the north of Kashmír a series of limestones, slates, and quartzites are found north of Iskardo (Skardo) in Baltistán, which are probably triassic and carboniferous in age. Further eastward beds belonging to these periods are known to occur in the Chang-cheng-mo valley and the Kara-koram range, only isolated details are, however, known, and it is impossible to give a connected account of them, but they cannot be passed over without a notice of that remarkable group of fossils, allied to the *Foraminifera*, known as *Syringosphæridæ*, which are found in dark shales, below a limestone taken to be of triassic age, on the northern side of the Kara-koram pass. They are small, rounded or oval bodies of about an inch in diameter, and had long been known as Karakoram stones. Almost the last work of the late Dr. Stoliczka was the collection of a number of specimens which were described by the late Professor Duncan[2] under the generic names of *Syringosphæra* and *Stoliczkaria*, the former including seven, and the latter one species.

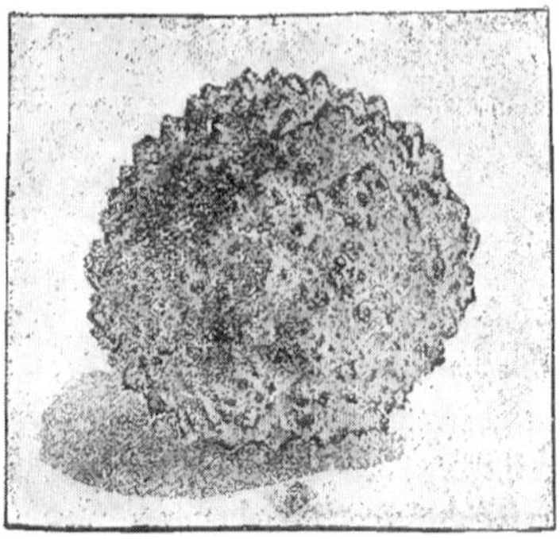

Fig. 10.—*Syringosphæra verrucosa*, Duncan.

Marine carboniferous rocks are known to occur in Afghánistán, in the Herát province, in the Hindu Kush,[3] and north of the Safed Koh,

[1] *Records*, IX, 53, (1876); *Memoirs*, XXII, 202, (1883).

[2] Scientific Results of the Second Yarkand Mission, Syringosphæridæ, Calcutta, 1879. *Records*, XXIII, 80—88, (1890).

[3] C. L. Griesbach, *Records*, XVIII, 62, (1885); XIX, 49, 240, (1886); XX, 17, (1887).

where they have undergone considerable metamorphism and are penetrated by granite veins.[1]

According to Dr. Fleming boulders of Productus limestone were found by him in the streams which flow eastwards from the Suláimán range,[2] but subsequent observers have not been able to detect carboniferous rocks in this range.

The marine carboniferous rocks are overlaid, in Herát and Turkestán, by a series of plant-bearing sandstones with seams of coal, which appears also to be represented south of the Safed Koh.[3] These are lithologically and stratigraphically the equivalents of the Gondwána system of the Indian Peninsula, and as such their description will be deferred to a subsequent chapter.[4]

In Tenasserim the only other region where marine deposits of carboniferous age are known to exist, there is a great accumulation of pseudo-porphyritic sedimentary beds known as the Mergui group,[5] whose principal feature is derived from imbedded fragments of felspar. The rock in its normal form is earthy, but highly indurated, passing into slaty masses without the conspicuous felspar fragments on the one hand, and on the other into grits and conglomerates. Resting on these grits, are dark coloured earthy beds, finely laminated, with hard quartzose grits. These rocks cannot be less than 9,000 feet in thickness, and in places they must be 11,000 or 12,000 feet. They have only been noticed hitherto near Mergui, and nothing is known of their relations.

The beds of the last group in the Tenasserim valley are succeeded by the Maulmain groups of hard sandstones, often in thin and massive layers, with thin earthy partings, sometimes in fine laminæ, the prevailing colour

[1] C. L. Griesbach, *Records*, XXV, 71, (1892). The statement that "lower silurian fossils from the Khyber hills were found by Dr. Falconer in the gravel of the Cabul river" was made by Colonel (then Captain) H. H. Godwin-Austen in 1866, in a paper which appears to have been drawn up from field notes without means of access to published information, and no reference is supplied to any original authority; *Quart. Jour. Geol. Soc.* XXII, 29, (1866). No notice of the discovery of such fossils can be found in Falconer's published writings, and the only original published statement is in a footnote to a paper by Captain Vicary, *Quart. Jour. Geol. Soc.* VII, 45, (1851). Vicary himself obtained "a small *Spirifer*, *Orthis* in abundance, a *Terebratula* and some *Polyparia*" from limestone boulders in the watercourses near Peshawar. In a footnote he adds, "Dr. Falconer obtained specimens of *Spirifer*, *Orthis*, and other palæozoic forms from these mountains several years ago." Also, in a note by Sir R. Murchison, prefixed to Vicary's paper, the discovery of palæozoic fossils is mentioned. Now, it is quite possible that the fossils collected by Falconer and Vicary have been examined and their age determined, but as this is not stated, some doubt remains whether the fossils may not have been carboniferous, as they were said to be by Verchere [*Jour. As. Soc. Beng.*, XXXVI, pt. ii, 21, (1867)], the *Orthis* being perhaps *Orthisina* or *Streptorhynchus crenistria*, formerly included in the genus *Orthis*.

[2] *Quart. Jour. Geol. Soc.* IX, 348, (1853).

[3] C. L. Griesbach, *Records*, XXV, 79, (1892).

[4] *Infra*, p. 196.

[5] T. Oldham, *Sel. Rec. Govt. India*, X, 32, (1856).

is a reddish tint, and some of the layers are calcareous. Some of the more soft and earthy beds contain marine fossils. Over these sandstones occur grey shaly beds, also sometimes calcareous and fossiliferous, with occasional beds of dark sandstone, then come 150 to 200 feet of fine soft sandstone, thinly bedded, with grey and pinkish shaly layers intercalated, and upon these again, hard thick limestone. The fossils found appeared to be of carboniferous age, *Spirifer* and *Productus* being the commonest forms, but the species have not been determined, and it is rare to obtain specimens in a state suitable for identification.

The thickness of these beds is estimated at about 5,000 feet, exclusive of the limestone, which is itself 1,100 feet thick near Maulmain.

Near Maulmain the limestone is extremely conspicuous, and forms large hills and ranges, extending far to the south-south-east up the valley of the Attaran and Zamí. The same rock occurs east of the Salwin, but does not extend far into Martaban, and is wanting in the Sittaung valley. Farther up the Salwín, however, in Karen-ni, and elsewhere beyond the British frontier, large tracts of limestone occur, probably belonging to the carboniferous series. Limestone is said to abound in the Mergui archipelago, and may very probably be, in parts at least, identical with that found near Maulmain.[1]

Besides the rocks already described, which can be ascribed to a carboniferous or triassic age with more or less certainty, there are some rocks coloured brown on the accompanying map, in south-east Afghánistán and in the hills east of India proper, whose true age is doubtful.

The first of these areas to be dealt with is in southern Afghánistán, where some unfossiliferous slates have been coloured on the map as carbon-trias. They form the Khwája Amrán range, the hills north of the Pishín valley, and on either side of the upper Zhob valley. They are slates and quartzites, whose similarity to the slates of the Simla area has more than once been noticed. In the first published description they were regarded as a flysch type of the nummulitic shales and limestones to the east of Quetta.[2] Subsequently the same observer considered that they were more probably lower cretaceous,[3] but in the absence of fossil evidence there is no more ground for this than almost any other correlation. One thing seems certain, that they are not altered nummulitics, for near Spira Raga, on the frontier road to Pishín, the same beds are

[1] Dr. Noetling's discovery of silurian fossils in the similar limestone of the Shan hills makes it possible that this limestone is silurian not carboniferous.

[2] *Memoirs*, XVIII, 32, (1881).

[3] *Records*, XVIII, 59, (1885).

found within a few miles of typical nummulitics which show all the groups characteristic of that series in Balúchistan.[1]

In this neighbourhood and in the hills bounding the upper Zhob valley about Hindubágh, the beds are penetrated by intrusions of serpentinous rock, porphyritic with crystals of diallage, precisely resembling the intrusive serpentine of Burma and Manipur. Among the slates a bed containing subangular fragments, ranging to six feet in diameter, was observed. The bed very much resembles the Blaini boulder bed of the Simla area, but no great weight can be attached to a single isolated observation like this. It has, however, taken in conjunction with the general lithological facies, and the resemblance of the serpentine intrusions to those of the Arakan hills, been allowed to influence the choice of the colour to be adopted in the map, and these rocks have been coloured as carbon-trias, with a warning note that the age is unknown and may very likely be younger than that indicated by the colour.

There is some independent evidence of the possibility of triassic rocks being found in Balúchistán, for Dr. Cook has recorded the finding of *Orthoceras* near Khelát. The other fossils found for the most part indicate a cretaceous horizon and the section as described fits in very well with the known cretaceous and tertiary groups of Balúchistán; there can then be no doubt of its correctness. On this ground it might be natural question the identification, but the brief description Dr. Carter gives is incompatible with the idea that it was a *Baculites* or the phragmocone of a *Belemnite*.[2] It is more probable that there is an undetected unconformity, and that both the cretaceous and the trias are represented in the section.

Between the Irawadi valley and the Arakan coast a tract of country has been coloured as carbon-trias on the map, with a note against it that newer rocks are known to occur and probably form a large portion of the area. The beds so mapped compose the Arakan Yoma, a forest clad range only traversed by a single road and by a few difficult paths at wide intervals, and, in general, absolutely inaccessible, except along the tortuous beds of streams. Anything like satisfactory geological surveying becomes almost impossible in such a region, unless some well marked and prominent beds occur to afford a clue to the stratigraphy, or fossiliferous belts are numerous. In the Arakan range neither is the case, the rocks of the main range consist of rather hard sandstones and shales, greatly squeezed, contorted, and broken, traversed by numerous small veins of quartz, often slaty, and sometimes schistose, but there is a marked, deficiency of any conspicuous strata. The few bands of limestone which occur are thin, isolated, and as a rule unfossiliferous. The rocks on

[1] R. D. Oldham, MS. Report, 1891. It is not established that these rocks are the same as those of the Khwája Amrán range, though they appear to be continuous with them.

[2] *Jour. Bombay. Br. Roy. As. Soc.* VI, 190, (1892).

the western, or Arakan, side of the range seem, on the whole, less altered than those on the eastern, or Pegu, slope, and unaltered nummulitic rocks appear, on both sides throughout a great part of the area, although not continuously on the outer spurs.

The crushed, hardened, and somewhat altered rocks of the Arakan Yoma were originally separated by Mr. Theobald from the newer-looking nummulitics under the name of axials, and considered as comprising the oldest tertiary beds and their immediate predecessors in the series. Although there is a well marked difference between the nummulitic beds and the axials, there is no distinct break between them. The two present an appearance of conformity, and it is far from clear that some of the axials are not merely nummulitic strata, greatly crushed and contorted. But subsequent to the preliminary examination of the area, a cretaceous ammonite was found in Arakan, and amongst some rather obscure fossils discovered near the former frontier of British and Native Burma, west of Thayetmyo, were a few specimens referred by Dr. Stoliczka to the typically upper triassic *Halobia lommeli*. It became, therefore, necessary to distinguish both triassic and cretaceous beds amongst the axial rocks of the Arakan range.

To the former has been referred a series of hard sandstones and shales, with grits and conglomerates, and a few bands of impure limestone, which form the crest of the Arakan range at the old frontier of Lower Burma, and extend southward, nearly to the parallel of Prome. The only characteristic beds are some white speckled grits, interbedded with shales and sandstones, and attaining a thickness of 1,300 feet, in the Hlwa (Lohwa) stream, 35 miles west of Thayetmyo; a band of dark blue shale, part of which is calcareous, 33 feet thick below the grits with conglomerate; and some thick bedded shales, passing into massive sandy shales with hard nodules interspersed, attaining a thickness of 110 feet, and containing a *Cardita* and some undetermined *Gasteropoda*. The calcareous conglomerate passes into a rubbly limestone, and appears identical with the beds containing the supposed *Halobia lommeli*. To the northward a band of limestone, much thicker and purer than that of the Hlwa stream, has been traced in several places. The speckled grits and conglomerates are, however, more conspicuous and more characteristic, and it is mainly by means of them that the area of supposed triassic beds was mapped.

The whole thickness of the group appears to be rather less than 6,000 feet, the characteristic beds just noticed being near, but not at the base of the group. To the eastward these beds are in contact with nummulitic strata; to the westward it is believed that cretaceous beds come in, but the country is difficult of access, and has not been surveyed. The area occupied within the limits of Lower Burma is elongately triangular, broadest at the frontier, where it extends for fifteen miles from east to west, and terminating in a point to the southward, west-by-north of Prome.[1]

The remainder of the rocks forming the Arakan Yoma, excluding those of cretaceous age, are either unfossiliferous or the few organisms which have been detected, mostly the indistinct remains of plants and mollusca, are insufficient to afford any trustworthy indication of age. They have been classed by Mr. Theobald as Negrais rocks, the name being derived from Cape Negrais, the south-western point of Pegu, and the extreme southern termination of the Arakan Yoma.

The Negrais rocks differ in no important particulars from the beds already noticed. They consist principally of hardened and contorted sandstones and shales, intersected throughout by numerous small veins of quartz and carbonate of lime. Limestone is not of common occurrence. Where seen, it does not generally appear in regular strata, but in huge detached blocks imbedded in the shales and sandstones, as if the latter had yielded without fracture to the pressure which dislocated the limestone. Conglomerates also occur, sometimes passing into breccias.

The alteration of these beds is most capricious and irregular. Frequently for a long distance they are apparently unchanged, except in being somewhat hardened; then they become cherty, slaty, or sub-schistose, and cut up by quartz veins. One not uncommon form of alteration is exhibited by the rocks affecting a greenish hue, due to the presence of chlorite, such rocks being generally much cut up by quartz veins. In a few instances, apart from the serpentine intrusions to be mentioned presently, irregular dykelike masses of either serpentine or a decomposed steatitic rock are found, but this is far from being of frequent occurrence. A more common form of alteration, seen along the coast north of Cape Negrais, is apparently due to the infiltration of silica in large quantities, and is shown by the intense, and often abrupt, alteration of beds of sandstone into cherty masses.

No satisfactory classification of these the main rocks of the Arakan Yoma has been practicable. They must be of great thickness, but the stratification is too confused for a clear idea as to the succession of different strata to be formed, in the absence of any well defined horizon. Some of them appear to be a continuation of the Ma-i, or cretaceous group, but on the other hand it is impossible to draw any definite line of boundary between the hill rocks and the nummulitics of Pegu. In

[1] It must be remembered that the specimen ascribed to *Halobia lommeli* was a mutilated and ill preserved one. Recent investigations, conducted while this work was passing through the press, and as yet incomplete, have shown that the supposed triassic rocks contain nummulites, and make it probable that most of the rocks of the Arakan Yoma are lower tertiary. As the results have not yet been fully worked out, the text is allowed to stand substantially as originally written.

Pegu, away from the base of the hills, comparatively soft, unaltered, fossiliferous beds, belonging to the older tertiary period, are found, which appear to rest upon the hill beds, for, away from the axis of the range, both have an eastwardly dip. The two rocks contrast strongly, the nummulitics being soft and unchanged, the hill beds hardened, crushed, and in places almost schistose, but it is impossible to fix a precise limit to either. The two are never seen in contact, there is no evidence that they are faulted against each other, and there appears to be a belt, often two or three miles wide, of rock in an intermediate condition. It appears possible that the rocks of the Arakan Yoma comprise representatives, slightly altered, of both cretaceous and nummulitic rocks, but there is no clear proof that these Arakan Yoma beds are identical with the Pegu nummulitics, and it appears best to distinguish the hill rocks by a separate name, though it has hitherto proved impossible to draw a line between the two.

From the foregoing description it will be seen that fossils have only been found at two places near the northern limit of the coloured area, and that there is no certainty as to the extent of the older rocks they indicate. The Negrais group was originally regarded as very possibly nummulitic, the lithological difference and greater induration, as compared with the undoubted nummulitics of Pegu, being attributed to the disturbance it had undergone. It is very doubtful, however, whether the explanation is sufficient, and in view of the probability of their distinctness, and of the fact that they are described as exhibiting a greater degree of induration than the beds which were supposed to be cretaceous, it has been considered advisable to adopt the course pursued in the preparation of the accompanying map, and colour them the same as the rocks known to be of carboniferous and triassic age, appending a warning note that their true age is unknown.

The intrusive serpentine which has already been noticed generally occurs as irregular shaped bosses of varying dimensions,[1] but dykes also occur, especially north-west of Prome. The rock is a characteristic dark coloured serpentine. It frequently becomes a gabbro, contains porphyritic crystals of bronzite, and is intersected by veins of gold coloured chrysotile, or, sometimes, of carbonate of magnesia. Occasionally it appears to be replaced by a form of greenstone which may possibly be distinct, although the two rocks occur in the same neighbourhood. The hills formed of serpentine may be distinguished at a distance by their barrenness. They appear to support little except grass and a few bushes, while the greenstone hills are covered with luxuriant forest. In all

[1] None are sufficiently large to be marked on the map issued herewith.

probability the serpentine and greenstone outbursts were originally the same or nearly the same, and the former rock has undergone a chemical change.

In the neighbourhood of some of the larger masses of serpentine the sandstones and shales are converted into greenstone and chloritic schist, but the effect varies, and in some instances the neighbouring rocks appear almost unaltered. It is, however, worthy of notice that, except far to the northwards, all the outbursts of serpentine appear confined to the Pegu, or eastern, side of the range, and that, as has already been stated, the rocks on this exhibit, as a rule, more alteration than those on the western slopes of Arakan. To the northward, near the northern frontier of Pegu, serpentine occurs on the highest hills of the Yoma, and, in one instance at least, on the western side, but elsewhere all the outbursts detected are not only east of the main range, but near the eastern limit of the hill rocks. Not a single intrusion has been detected in the unaltered nummulitic rocks.

It is unnecessary to describe the distribution of the serpentine masses in any detail. They are principally collected in three groups, the most northern of which consists of the largest mass known, a horseshoe shaped intrusion, some five miles in length, forming the Bidoung hill, nearly due west of Thayetmyo. Several masses occur north-north-west of Prome, and one of these, forming a long dykelike mass for about five miles along the boundary between the nummulitics and the supposed trias, appears to alter the triassic rocks, but not the nummulitic beds, although the latter are greatly crushed. Probably the difference is owing to the eastern boundary being a fault. The third group is west of Henzada, where twenty-one distinct and isolated intrusions occur, scattered over a length of twenty-six miles from north to south, close to the edge of the unaltered nummulitic area. The largest of these masses is about three miles long by perhaps half a mile broad, but the majority are less than a mile in diameter. Besides the principal groups a few small and unimportant outbursts are found isolated here and there, but none are found south of the area west of Henzada.

Further north a series of slates and indurated sandstones, which very much resemble the axial beds of Burma, is found in Manipur. They occupy the hills surrounding the valley of Manipur, and are penetrated by intrusive serpentine of the same type as in Burma. The intrusions, moreover, are confined to he neighbourhood of the eastern limit of the hill rocks. Our only information regarding the geology of Manipur[1] is derived from

[1] *Memoirs*, XIX, 217, (1883).

rapid traverses, where the movements of the geologist were determined by political considerations, and there is consequently no detailed information available. It was believed, however, that two unconformable, pretertiary, rock series were observed, the upper one being composed of red slaty shales overlaid by limestones, lithologically identical with those of the Ma-í group in Arakan. Some beds of volcanic ash, observed on the slopes of the Kachao mountain, were believed to be attributable to this group, but the correlation is questionable.

The rocks seen in Manipur, show some resemblance to those of the carbonaceous system of the Simla area, though the disturbance they have undergone is less intense. A bed of conglomeratic slate, containing rounded boulders of quartzite imbedded in a fine grained matrix, resembling in structure the conglomeratic slates of the Blaini group, was seen on the road between Manipur and Kohima associated with black carbonaceous slates.

Further to the north, in upper Assam, Mr. Mallet distinguished, under the name of Disang, a group of shales overlaid by sandstones, which are separated by a faulted boundary from the coal measures and overlying tertiary rocks lying between them and the alluvial plain of the Brahmaputra valley.[1] So far as their lithology goes, they agree fairly well with the older rocks of Manipur and the Nágá hills, and have been coloured the same on the map, though their true age is very uncertain.

[1] *Memoirs*, XII, 286, (1876).

CHAPTER VII.

THE GONDWÁNA SYSTEM.

GONDWÁNA SYSTEM—Probably of fluviatile origin—Relation to present river valleys—Division into groups—LOWER GONDWÁNAS—Talchir group—Karharbári group—Damuda series—Barakar group—Ironstone shales—Ránfganj group—Motur and Bíjori groups—Kámthí group—Pánchet group—Almod group—UPPER GONDWÁNAS—Mahádeva beds—Rájmahál series—in the Rájmahál hills—and on the east coast—Kota-Maléri group—Chikiála group—Jabalpur group—Plantbearing beds of Cutch and Káthiáwár.

The upper palæozoic and older and middle mesozoic formations of other countries are represented in the Indian Peninsula by a great system of beds, chiefly composed of sandstones and shales, which, except for some exposures along the east coast, appear to have been entirely deposited in fresh water, and probably by rivers. Remains of animals are very rare in these rocks, and the few which have hitherto been found belong to the lower vertebrate classes of reptiles, amphibians, and fishes. Plant remains are more common, and evidence of several successive floras has been detected. The subdivisions of this great plant bearing series have been described under a number of local names, of which the oldest, and best known, are Talchir, Damuda[1], Mahádeva, and Rájmahál, but the term Gondwána has now been adopted by the Geological Survey for the whole system. This term is derived from the old name for the countries south of the Narbadá valley, which were formerly Gond[2] kingdoms, and now form the Jabalpur, Nágpur, and Chhatísgarh divisions of the Central Provinces. In this region of Gondwána the most complete sequence of the formations constituting the present rock system is to be found.

Taken as a whole, the Gondwána system has a wide extension in the Indian Peninsula, but in extra-peninsular India, its representatives have hitherto only been detected in north-western Afghánistán and along the base of the eastern Himálayas in Sikkim, Bhután, and the Aká and

[1] More correctly Tálcher and Dámodar, but the spelling in the text has been so universally used that it is retained when the names are used in their acquired geological sense. The more modern and correct spelling is adopted when they are used geographically.

[2] For the information of non-Indian readers it may be well to add that the Gond is one of the principal Dravidian, or so called aboriginal, tribes, who are believed to have inhabited the country before the advent of the Aryan Hindu race.

Daphlá hills.[1] Representatives of the highest Gondwána groups are found in Cutch, resting upon marine jurassic rocks and capped by neocomian beds; some rocks containing plant remains, which underlie jurassic limestones in the desert between the Indus and the Arávallis, closely resemble portions of the Gondwána series in lithological characters. while representatives of beds high in the Gondwána series, in this case frequently containing marine fossils, extend down the east coast. But, with these exceptions, no representatives of the system are found in the Peninsula north of the valleys of the Narbadá[2] and Son, nor south-west of a line drawn from the sea at Masuli patam through Kamamet and Warangal, north-east of Haidarábád, till it enters the trap area near Nirmal. The main areas of Gondwána rocks are in the Rájmahál hills and Dámodar valley in Bengal, the Tributary Mahals of Orissa, Chhatísgarh, Chutiá Nágpur, the upper Son valley, the Sátpura range south of the Narbadá valley, and the Godávari basin.

It has already been mentioned that, with the few exceptions noted, the whole of the Gondwána series is believed to consist of strata deposited in fresh water, and the only question which arises is whether the beds are lacustrine or fluviatile. The coarseness of the rocks in general, the prevalence of sandstones, and the frequent occurrence of bands of conglomerate, render it improbable that these strata are of lacustrine origin, while the absence of mollusca almost throughout is, on the whole, rather more consistent with river than lake deposits, although it is difficult to account for on either hypothesis. The few fish and reptiles which occur might have inhabited either lakes or rivers, and the *Estheriæ*, which are common in several subdivisions of the series, might either have lived in lakes or in the great pools and marshes which often occupy so large an area in broad river valleys. The plants might have been preserved amongst either lacustrine or fluviatile deposits, except that it is difficult to conceive the formation of beds of coal at the bottom of lakes. It is more probable that the coal originated in marshy forests, such as frequently occur in the valley plains of rivers. The physical characters of the strata, the frequent alternation of coarse and fine beds, the frequency of current marking on the finer shales and of oblique lamination, due to deposition by a current, in the coarser sandstones, and the circumstance of the upper portions of a bed, such as a coal seam, being locally worn and denuded when a coarse

[1] Mallet, *Memoirs*, Vol. XI, 14, (1874); Godwin-Austen, *Jour. As. Soc. Beng.*, XLIV, pt. ii, 37, (1875); La Touche, *Records*, XVIII, 121, (1885). Perhaps the occurrence of a representative of the Rájmahál stratified traps on the flanks of the Khási hills might be quoted as another instance, but though the identification of the two sets of beds is highly probable, it has not been confirmed by the discovery of fossils.

[2] Outcrops have been found north of the river Narbadá westward of Hoshangábád, but far south of the watershed. The Narbadá, above the neighbourhood of Jabalpur, runs south of the general line of division, and Gondwána rocks occur north of the river.

sandstone is deposited upon it, a phenomenon of frequent occurrence, are quite consistent with the theory of deposition in a river valley, but opposed to the conception of lacustrine origin. A river constantly changes its course, and deposits coarse sediment near its channel and finer materials from the overflow of its flood waters, the area within which each form of sediment is deposited varying frequently. In a lake the coarse deposits must be limited to the margin, and finer sediment accumulates away from the shore, where there is no current to sweep away the surface of a recently deposited coal or shale bed, and to throw down coarse sand in its place. On the whole, the evidence is decidedly in favour of a fluviatile origin for the Gondwána rocks, and it is probable that they were deposited in a great river valley, or series of river valleys, not unlike those which form the Indo-Gangetic plains at the present day. There is a possible exception in the lowest beds of the series, the fine silts which form the basement beds of the Talchir group. These may be of lacustrine origin, but there is no clear proof that they are, and their remarkably persistent character throughout an immense tract of country is rather opposed to the idea of their having been formed in a lake or a series of lakes.

Concerning the relations of this great series to the older and newer formations in India but little can be said. No older fossiliferous deposits are known in the area to which the Gondwána rocks are restricted, and wherever these rest upon any older formation, there is complete unconformity between the two. The areas in which the upper Vindhyan and Gondwána systems are exposed being distinct from each other, the latter have nowhere been found in contact with the former, which are the next series in descending order, but pebbles of upper Vindhyan rocks are occasionally found in Gondwána rocks.[1] The Talchir and Damuda formations in the country south of Nágpur, on the Godávari below Sironchá, and in the Mahánadí valley near Sambalpur, occasionally rest unconformably upon strata believed to belong to the Cuddapah or lower Vindhyan series,[2] but in general the Gondwána beds are found to have been deposited upon metamorphic rocks.

On the other hand, the rocks of the Gondwána series are but rarely covered at all by a higher formation, except where the Deccan traps and their associated infratrappean formation, the Lametá group, rest unconformably upon the various subdivisions of the Gondwána series, from the lowest to the highest, in the Narbadá valley and the Nágpur country. There are, however, localities in India in which sedimentary formations of cretaceous age rest upon upper Gondwána beds. The first of these is in

[1] *Manual*, 1st ed., 205; *Memoirs*, IX, 304, 1872.)

[2] *Supra*, p. 91.

Cutch, where the Umia group, containing some fossil plants, found also in the uppermost Gondwána beds in the Narbadá valley, underlies a stratum containing *Cephalopoda* of upper neocomian (Aptian) age. The second is in the Narbadá valley near Bárwai, where Bágh beds (upper greensand or cenomanian) rest unconformably on representatives of the upper Gondwána series. The remaining two localities are near the east coast. One is in southern India, at Utatúr, north of Trichinopoli, where the plant beds containing Rájmahál fossils underlie the Utatúr (cenomanian) group, unconformably in places but elsewhere with apparent conformity. Lastly near Ellore, where the upper Gondwána beds contain Rájmahál plants, and marine fossils of upper jurassic age occur in the higher layers, the age of the strata resting unconformably upon the Gondwána strata is not equally well defined. The overlying beds consist of two fossiliferous bands, one underlying a flow of basalt believed to belong to the Deccan trap series, the other interstratified between the lower basaltic flow and a higher one. The igneous beds, like the Deccan traps elsewhere, are believed to be of uppermost cretaceous or lowest tertiary age, but the fossils in the upper, or intertrappean, bed differ from those in the lower, or infratrappean, and it has not hitherto been practicable to refer either to a definite horizon. Neither bed, however, can be older than upper cretaceous.

The manner in which the areas of Gondwána rocks are distributed throughout the country is peculiar, and there is still some difference of opinion concerning the interpretation to be placed on their mode of occurrence. As a general rule, these rocks are found occupying basin shaped depressions in the older formations, and such depressions sometimes, though not always, nor even generally, correspond to the existing river valleys. Occasionally the basins of Gondwána beds are scattered over the surface of the country, as in Bírbhúm, and in this case there can be no doubt of their representing the undenuded remains of strata which were once continuous over a much larger area. Whether the basins now remaining owe their preservation to disturbance of their originally horizontal position, and to their having been preserved from denudation through having sunk to a lower level than neighbouring portions of the same bed, or whether they were originally deposited in hollows in the older beds, is a point on which opinions differ. There can be no question that the former is the explanation of these basins having been preserved in some instances, but cases may also be cited in favour of the latter view, and it is certain that the Gondwána beds were originally deposited on an uneven surface.

A few instances will suffice to show the phenomena presented in the Dámodar valley in western Bengal, where some of the most important and best known Gondwána coal-fields occur. A number of detached basins are found, all in low ground on the banks of the river, and all presenting the very remarkable peculiarities that the lowest groups appear on the northern side of the basin, that there is a general dip from north to south, and that all are cut off abruptly on the southern edge, which in most cases is a straight or nearly straight line. Similar geological relations exist in many other areas, although the beds are not always, as in the Dámodar area, confined to the valley of a single river. Thus, in the great basin of south Rewá and Sargúja, again in the Sátpura area, and especially in the Tálcher field in Orissa, the rocks dip from one side of the basin, and are cut off on the other, but in all these cases the general dip is north, not south, and the beds are abruptly cut off along the northern border. The exact directions of the abrupt east and west boundaries vary, but they are always the same, or nearly the same, throughout each tract of country, that is to say, the boundaries of different fields are parallel to each other, and they are also, as a rule, identical in direction with the foliation of the underlying gneiss. In some cases, and especially in the northern part of the great area which occupies so large a portion of the Godávari valley, both boundaries, which run nearly north-west to south-east in the last named case, are straight, nearly parallel, and abrupt.

These abrupt boundaries are almost invariably accompanied by considerable disturbance of the beds in their neighbourhood. In some cases there is strong evidence that such boundaries are great faults, one of the best proofs being that the fault occasionally divides, as along the northern edge of the Tálcher field, and beds belonging to the lowest group are exposed between the different subdivisions of the main dislocation, the lowest Gondwána group (the Talchir in the instance mentioned) being faulted against Kámthí beds, much higher in the Gondwána system, on one side, and against metamorphics on the other. In some cases, as along the boundary of the Tálcher field and also on the eastern portion of the northern boundary in the Sohágpur field, the line of fault is marked by a breccia, containing fragments of the Gondwána sandstones. It is generally considered that all the fields which are bounded by an abrupt line cutting them off on one or both sides (and these, as will be seen, comprise a very large majority of the basins known) occupy areas of depression, produced subsequently to the deposition of the beds by a fault along the abrupt boundary, the connection of existing river valleys with these Gondwána areas being dependent on the fact that, the Gondwána rocks being much softer than the Vindhyan, transition, or metamorphic beds upon which they rest, the rivers have worn their way through the easiest channel,—in short, that the existing drainage, so far as it coincides with the distribution of the Gondwána rocks, has been determined by the disposition of those rocks, produced by disturbance and denudation, and has no necessary connection with their original areas of deposition.

A different view is held by others. They consider that, with a few exceptions, there is no sufficient evidence of faulting, that the appearance of straightness in the boundaries is partly fallacious and due to the rocks being ill seen at the surface, that the abrupt boundaries are caused by the deposition of the Gondwána rocks against cliffs forming the original sides of river valleys, and that the present disposition of the beds is a close approximation to that of the original areas in which they were deposited. They consider further that the vertical development of the different groups varies so much within small distances that there is no reason to believe that any great thickness of beds abuts against the abrupt cliff like boundaries, and that there is evidence in some cases that the different groups thin out towards the margins of the existing basins. They conclude that the present river valleys differ but little from those which existed in mesozoic times.

It is possible that there may be some truth in both views. It should be remembered that the conflict of opinion in this case is between observers who have chiefly been engaged in mapping widely separated regions. The view that the present basins closely correspond to ancient areas of deposition being supported chiefly by observations made in the Son and Narbadá valleys, and the opposite opinion, that the present Gondwána basins are chiefly due to faulting, being held by geologists who have especially studied the Gondwána rocks of Bengal, Orissa, and the Godávari valley. The strongest arguments against the existence of faults along the abrupt boundaries of the various Gondwána fields is founded on the fact that, in the Sátpura field to the south of the Narbadá valley, certain of the uppermost Gondwána beds overlap the boundary, but this may be due to the circumstance that the supposed line of fault, which cuts off the field on the northward throughout the greater portion of its extent, is more ancient than the topmost groups of the Gondwána series. A difficulty in the way of admitting that the abrupt boundaries of the Dámodar fields are due to deposition against inland cliffs is to be found in the improbability that all such precipices should be found on one side of a river valley, while there are some important observations in favour of the limits of the basins in the Dámodar valley being due to disturbance. Talchir and Damuda beds are found on the Hazáribágh table-land, immediately north of the Dámodar valley, at a height of about 1,000 feet above the surface of the same rocks in the valley itself, and the presence of fragments, apparently derived from lower Gondwána beds, in a conglomerate at a similar or higher elevation on the Chutiá Nágpur highland to the southward points to the former existence of the parent rock at a still greater elevation. In either case there is evidence of disturbance for the low level exposures must have been depressed, or the high level ones elevated; in other words, the Gondwánas must have undergone disturbance since they were deposited, and this disturbance cannot have been without effect on the present limitation of the outcrops.

The tracts of country occupied by rocks of the Gondwána series are, as a rule, covered with a poor sandy soil and ill suited for cultivation. The result is that, in many parts of India, they form wild uninhabited forests. Such tracts are always the last to be surveyed topographically, and, as a rule, minor details are omitted on the maps prepared. Moreover the upper Gondwána rocks are principally sandstones and decompose readily into loose sand, which covers the whole surface of the country and greatly conceals the rocks. These two circumstances—deficiency of maps and concealment of the surface—have combined to delay the geological survey of the upper Gondwána formations, and to render the examination of the beds exceptionally tedious and difficult.

The groups of which the Gondwána system is composed vary greatly, both in number and mineral character, in the several isolated areas in which they are found, the variation being much greater amongst the middle and upper than amongst the lower members of the series. The two lowest Gondwána groups, the Talchir and Barakar, which consist largely of shales, whilst the uppermost formations are chiefly composed of coarse sandstone, grit, and conglomerate, preserve their mineral character almost unchanged throughout the area in which the lower Gondwána beds are known to occur.

The system may be divided into an upper and a lower series, the distinction having been first established in western Bengal, where it is of a most trenchant nature, characterised by a marked stratigraphical discordance, by an almost complete absence of any species common to the two divisions, and an utter change in the type of the flora, equisetaceous plants prevailing in the lower subdivision, and cycads and conifers in the upper,[1] ferns being found commonly in both. Some *Equisetaceæ* occur, however, in the upper Gondwánas, and several species of cycads and conifers in the lower, but the genera are in most cases distinct in the two subdivisions. As the examination of the Gondwána system in the Sátpura ranges and in south Rewá has progressed, it has been found that the stratigraphical break there is not nearly so marked, and it is possible that a number of distinct floras will ultimately be found, bridging over the gap in western Bengal.

[1] The Mahádeva series has, however, hitherto proved almost unfossiliferous.

Table showing the probable representation of the Gondwána groups by each other.

	GENERAL SEQUENCE.	I.—Rájmahál hills.	II.—Bírbhum, Deogarh and Karharbári.	III.—Dámodar valley.	IV.—Son, Mahánadí, and Bráhmani valleys.	V.—Sátpura region.	VI.—Godávari valley.	VII.—East coast region.	VIII.—Cutch.
Upper Gondwána.	Umia and Jabalpur.	...	...	...	...	...	...	...	Umia.
		...	...	...	Jabalpur.	Jabalpur. Bágra	Chikiála.	Tripetty, Pávulur and Sattavédu ?	
						Denwa.	Kota-Maléri	Ragavapuram, Sripermatúr and Utatúr.	
	Rájmahál and Mahádeva.	...	...	...	Mahádeva				
			...	...					
		Rájmahál.	...	Mahádeva.		Pachmarhí	...	Athgarh, Golapilli and Budaváda.	
		Dubrájpur.							
Lower Gondwána.	Pánchet.	...	...	Pánchet.	...	Almod ?	Kámthí (including Mángli).		
		...	...	Ráníganj.	Kámthí (Hingir).	Bijori.			
	Damuda.	...	...	Ironstone shales.	...	Motúr.			
		Barakar.	Barakar.	Barakar.	Barakar.	Barakar.	Barakar.		
	Talchir.	...	Karharbári.	Karharbári.	Karharbári.	Karharbári.			
		Talchir.	Talchir.	Talchir.	Talchir.	Talchir.	Talchir.		

I.—The Lower Gondwána Series.

The lowest member of the Gondwána system is known as the Talchir group, thus named from its having been first clearly distinguished in the small district of Tálcher[1] one of the tributary mahảls of Orissa. When present—and it is rarely absent over a large area—this group forms the base of the Gondwána series, and consists in general of fine silty shales and fine soft sandstone. The shales are usually of a greenish grey or olive colour, sometimes slaty. They are of exceedingly fine texture, traversed by innumerable joints, and break up into minute, thin, angular fragments, sometimes elongate or acicular, which cover the surface of the ground in places. Occasionally the shales have a dull Indian red colour, but this is not common. They are frequently mentioned in the Survey reports under the name of mudstones and needleshales. Not unfrequently they are somewhat calcareous, and in some places large concretionary masses of impure carbonate of lime have been found amongst them.

The most characteristic sandstones are soft, fine, and homogeneous in texture, composed chiefly of quartz and *undecomposed* pink felspar, and in colour pale greenish grey, buff, or pale pinkish, almost of a flesh tint. They are frequently rather massive, though distinctly stratified, but they are also commonly interstratified in thin layers with the shales. In many places they break up, where exposed on the surface, into polygonal fragments, three or four inches across, whence they have been called tesselated sandstones.

These beds pass into coarser sandstones of less marked character, which vary in colour, and are sometimes, though rarely, conglomeratic. It is an almost invariable rule, contrary to what is found to be the case in most rocks, that in the Talchir group the beds of finest texture, the shales, are found at the base, and that the sandstones are higher in position, the coarser sandstones, moreover, overlying those of finer texture. A thin coal seam has been found amongst the Talchir beds in the Jhilmilli field, in Sargúja[2], but this formation is, as a rule, distinguished by the absence of coal seams, and even of carbonaceous shale.

There are three peculiarities of the Talchir group which still require notice, as all of them are of considerable importance.

The first is the frequent occurrence, amongst the shales and fine sandstones, generally towards the base of the group, but very frequently some hundreds of feet above the bottom, of pebbles and boulders, always rolled and usually well rounded, varying in size from small fragments quarter of an inch or an inch across to huge blocks fifteen feet in diameter and thirty tons in weight, fragments from six inches to three feet in diameter being common.

[1] *Memoirs*, I, 46, (1856).

[2] *Manual*, 1st ed., 205.

The distribution of the boulders is most irregular, in some parts of the area occupied by Talchir beds none are to be found over many square miles of country, but generally some are met with at intervals, and occasionally large numbers occur within a limited tract.

In very many instances there is every probability that the boulders have been transported from a distance, no rocks of similar character being found in the neighbourhood. If only one or two such cases had been observed, it might be supposed that the rock, from which the blocks were derived, had formerly existed in the immediate vicinity and been removed by denudation, but the cases in which there is reason to believe that the rounded blocks have been transported from afar are so numerous that this theory cannot be accepted. The boulders, it should be remembered, are frequently found imbedded in the finest silt. It is evident that deposition from water in rapid motion is here out of the question, as any stream which could have moved and rounded the boulders would have swept away the silty matrix in which they are deposited, and the only suggestion, as to the cause of their occurrence, which appears to account satisfactorily for their presence, is to suppose that they were originally rounded by torrents and then transported to their final position by ice. This theory has received strong confirmation from the discovery of smoothed and scratched surfaces on some of the large boulders found on the banks of the Pengangá river, about ten miles west-south-west of Chándá, Central Provinces.[1] The surface of the limestone rock underlying the Talchirs was also in this case found to be polished, scratched, and grooved.

The second peculiarity is the remarkable resemblance to a volcanic rock occasionally presented by the more compact forms of shale, and by a variety of the sandstone. So great is the similarity between the shale and a consolidated volcanic ash that two experienced surveyors have, at different times, marked the beds as trappean, whilst the sandstone occasionally simulates a decomposed basalt in colour and mode of weathering.

The third noteworthy feature of the Talchir beds is their power of resisting disintegration, and the entire barrenness, provided they are not covered by alluvial deposits derived from other rocks, of the ground where they appear at the surface, a natural consequence of their not decomposing to form soil. In many places along the edges of the coal-fields, where the Talchir beds occupy the ground, it is possible to walk for miles through very thin jungles, free from grass, over a surface composed entirely of the finely comminuted greenish grey shales.

South of the Pengangá river a peculiar rock was found by Mr. Fedden in the Talchirs near Chárli, and again in the Khairgaon nala west of

[1] T. Oldham, *Memoirs*, IX, 324, (1872); Fedden, *Records*, VIII, 16, (1875).

Neuropteris valida, Fstm.

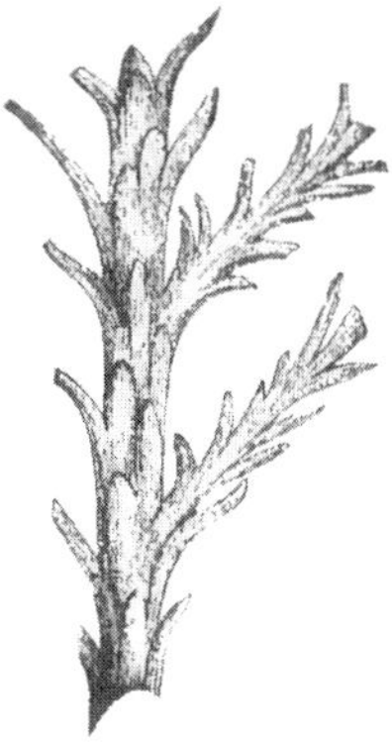

Voltzia heterophylla, Bgt.

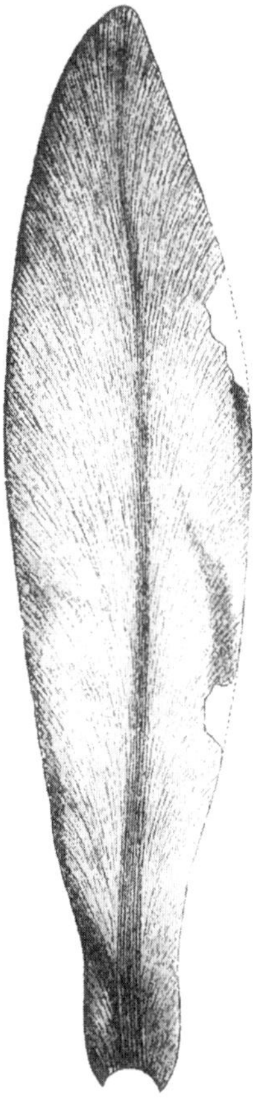

Gangamopteris cyclopteroides, Fstm,

Gangamopteris major, Fstm,

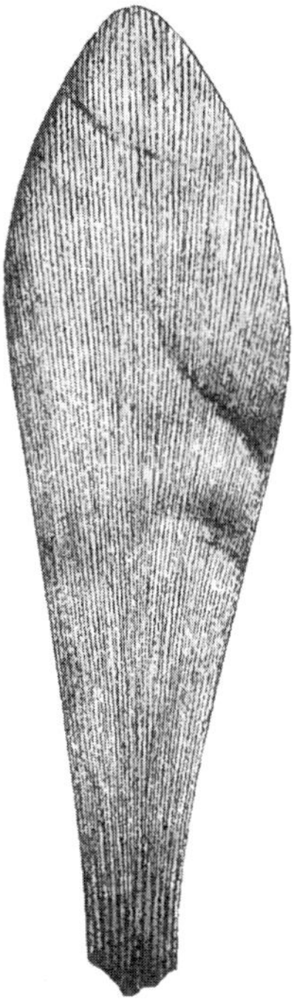

Nœggerathiopsis hislop, Bunb,

Calcutta Phototype Co,

TALCHIR KARHARBARI PLANTS.

Nándgaon. It is a calcareous sandstone, whose calcareous cement has assumed the form of ophitic crystalline masses quarter of an inch across. These crystals have been irregularly attacked in the weathering and the rock split up into a number of irregular aggregates of quartz grains, separate from each other, but held together by the interlocking of their irregularities. The rock, as a whole, has consequently a certain amount of flexibility analagous to that of what is known as itacolumite.[1]

The Talchirs preserve all their peculiarities throughout the area in which they occur,—an enormous tract of country, extending from the flanks of the Rájmahál hills to the Godávari, and from the Ráníganj field on the borders of the alluvium of Lower Bengal to the neighbourhood of Hoshangábád, Nágpur, and Chándá.

The thickness of the Talchirs nowhere appears to exceed about 800 feet, their extreme measurement where fully developed in part of the Ráníganj coal-field.

The fossils[2] hitherto discovered in the Talchir rocks are very few in number. Of animal remains only the wing of a neuropterous insect and some annelid tracks have been discovered, whilst the plant remains consist of *Noeggerathiopsis hislopi* and three ferns, *Gangamopteris cyclopteroides*, *G. angustifolia*, and *Glossopteris communis*. The only evidence of vegetable life hitherto found has been in the higher beds of the group, and there is a remarkable absence of plants in the lower shales, which are admirably suited for preserving vegetable impressions. Even in the upper beds of the group fossils are of singularly rare occurrence.

Reference has already been made to the possibility of a lacustrine origin for the Talchir beds, or at least for the lower portion. The chief reason for suggesting that these beds may have been deposited in lakes is the great thickness of very fine sediment accumulated at the base of the group, and the very frequent occurrence of much finer beds below than above. The latter, on the hypothesis of a lacustrine origin, may be explained by the gradual silting up of a lake basin, in which fine sediment would be deposited at a distance from the margin, whilst coarser beds would be thrown down by rivers as their deltas advanced into the lake and filled it up. This evidence, however, is quite insufficient by itself to prove that the Talchirs are a lacustrine deposit, and it is at least equally probable that they were formed in a river valley, like the overlying members of the Gondwána system.

At the same time the large size of the boulders and their generally

[1] *Memoirs*, XIII, 16, (1877). Compare the Káliána flexible sandstone (*supra*, p. 72); see also, for a more detailed description, *Records*, XXII, 54, (1889).

[2] The determinations of fossil plants belonging to this and other groups of the Gondwána system are taken from Dr. Feismantel's descriptions in the *Pal. Ind.*, series ii, xi, xii, (1877-86).

rounded aspect suggest that they were produced by rapid flowing streams, whose beds had a steep gradient. The great unconformity between the Talchirs and the underlying rocks points to a long continuance of dry land conditions, unfavourable to the accumulation of sediment, and one of the first effects of those land movements which caused its accumulation might well be to split up the river valleys into large lake basins, and steep stream valleys leading into them. Were this the case, all the conditions essential to the explanation of the features ordinarily exhibited by the Talchir boulder bed would be existent, if winter ice were superadded.[1]

This explanation is not, however, sufficient to account for the smoothed, polished and striated surfaces of the fragments included in the Talchir conglomerate, and of the underlying Vindhyan limestones in the Pengangá valley, which appear to be due to the action of a true glacier. The boulder beds largely developed near Báp, in western Rájputána, which can hardly be other than of Talchir age, being unconformably superimposed on the Vindhyan limestones and older than the upper Gondwána beds of this district, contain numerous well glaciated fragments, and a similar boulder bed near Pokaran is seen to rest on a surface of older rock, which is not only smoothed and striated, but exhibits typical *roches moutonnées*. In the last named instance there is some possibility that the boulder bed is older than Talchir,[2] but excluding this, there is evidence enough that glaciers must have descended to low levels in Talchir times.

The coal bearing rocks of the Karharbári coalfield were originally assigned to the Barakar group in the publications of the Geological Survey, on account of their mineral character and their position immediately above the Talchir beds. The examination of the Karharbári fossil flora has, however, shown that, whilst all the species known to be found in the Talchir beds are represented, one of them (*Gangamopteris cyclopteroides*) being the commonest fossil of the Karharbári beds, many of the common Damuda fossils are rare or wanting, and several very remarkable species are found which have not hitherto been detected in the Damuda series. The peculiar excellence of the coal, and its superiority to that obtained from the majority of the Damuda seams, have led to extensive mining operations in the Karharbári field, and it has consequently been possible to obtain good collections of the fossil plants.[3] It has also been noticed that the coal of Karharbári differs in structure from that of the Damuda series generally, and a partial re-examination of the field appears to justify the inference that there is also a slight distinction between the

[1] W. T. Blanford, *Records*, XX, 49, (1887).

[2] *Supra*, p. 106.

[3] These have been chiefly collected by Mr. I. J. Whitty, Superintendent of the East Indian Railway Company's collieries at Karharbári.

Karharbári and Barakar sandstones, although it is as yet uncertain whether a passage may not eventually be found between the Karharbári group and the Barakars. The palæontological evidence hitherto obtained tends, however, to connect the former with the Talchir group, and it appears best, for the present, to keep the Karharbári rocks distinct from the overlying Damúda series, under the name of the coalfield in which they were first distinguished.

The rocks of the Karharbári group consist almost solely of sandstones, grits, and conglomerates, with seams of coal. Very little shale occurs, the little which exists being associated with the coal seams. The sandstones are mostly white, grey, or brown, and felspathic, often gritty and conglomeratic, from containing large fragments of felspar and pebbles of quartz. The chief distinction between the constituents of the grits and conglomerates forming the Karharbári group, and those which make up so large a portion of the Barakars, is that in the former, and especially in the coarser grits and conglomerates, a large proportion of the fragments of felspar and quartz are angular or subangular, whereas in the Barakars the pebbles are, as a rule, particularly well rounded. The coal of Karharbári is rather dull coloured and tolerably homogeneous in structure, the layers of very bright jetty coal, which are so conspicuous in the Damuda seams, being in general few and ill marked. The seams appear to be somewhat variable in thickness, but to undergo very little change in composition throughout the small field in which they are found. Some of the seams, both in the Barakar and Ránígánj subdivisions of the Damuda series, furnish fuel equal in quality to that extracted at Karharbári, but they are much more distinctly laminated.

The Karharbári beds rest with apparent conformity on the Talchirs, but the former completely overlap the latter in places, within the limits of the little Karharbári field, and the mineral characters of the two groups are strongly contrasted. In the west of the Karharbári basin the Talchirs attain a thickness of about 500 or 600 feet, whilst, within a distance of less than four miles to the eastward, the Karharbári beds rest upon the gneiss. It is probable that the highest rocks seen within the coalfield may be of Barakar age, and there is some slight appearance of the Karharbári beds being overlapped by these higher strata, but the overlap is not clear. The whole thickness of the Karharbári group is probably about 500 feet.

Outside the limits of the Karharbári field the Karharbári group has been recognised, on palæontological grounds, at Mohpáni in the Narbadá valley, in the Dáltonganj coalfield and, with considerable degree of probability, in Hutar and Rewá. The coal in a seam, lying very little above the Talchir group, in the Ránígánj field resembles the Karharbári coal in mineral character, and it is probable that this group will be found to have a wider distribution than is now known, and to be represented in all those sections where the Talchirs are described as conformable to the beds above them.

The only fossils which have so far been formed in the Karharbári beds are plants of which the following is a list[1]; those species distinguished by an asterisk having also been found in the Damuda beds, while a dagger marks the forms which have been found in the underlying Talchirs:—

EQUISETACEÆ—
Schizoneura, cf. *meriani*.
* *Vertebraria indica*.

FILICES—
*†*Gangamopteris cycloptervides*.
,, (?) *buriadica*.
,, *major*.
,, *obliqua*.
† ,, *angustifolia*.
,, cf. *spathulata*.
*†*Glossopteris communis*.
* ,, *damudica*.
,, *decipiens*.
* ,, *indica*.

FILICES,—*contd.*
Sagenopteris (?) *stoliczkana*.

CYCADEACEÆ—
Glossozamites stoliczkanus.
†*Næggerathiopsis hislopi*.

CONIFERÆ—
Euryphyllum whittianum.
Voltzia heterophylla.
Albertia, sp.

SEEDS—
Samaropsis, sp.
Cardiocarpum, sp.
Carpolithes milleri.

The Talchir-Karharbári groups are succeeded by a great series of beds, the Damuda series, which was first examined and described in the coalfields of the Dámodar valley. Nearly all the coalfields of the Indian Peninsula owe their mineral wealth to the presence of these beds, the Karharbári being the only other important coal bearing group, and the quantity of valuable minerals contained in the rocks of the Damuda series is probably greater than that of all the other rock groups of India combined.

The Damuda series in Bengal has been found to consist of three subdivisions, known in ascending order as the Barakar group, Ironstone shales, and Ráníganj beds. The first and lowest is also found in the Son, Mahánadí, Narbadá, and Godávari valleys, the upper subdivisions being represented by groups differing in mineral character from the Bengal beds. In the Sátpura area the Damuda subdivisions are known as the Barakar, Motúr, and Bijori groups, and in the Godávari valley, above the Barakar group, there also the only coal bearing formation, a single member of the upper Damuda beds occurs, and is known as the Kámthí group. A similar arrangement prevails in the Mahánadí and Bráhmaní area, only two Damuda subdivisions being found, which appear to correspond to those of the Godávari region.

The mineral characters and geological relations of all these different groups must be described separately. It is sufficient for the purpose at

[1] *Pal. Ind.*, series xii, III, pt. i, (1879-81); IV, pt. 2, (1882).

Trizygia speciosa, Royle.

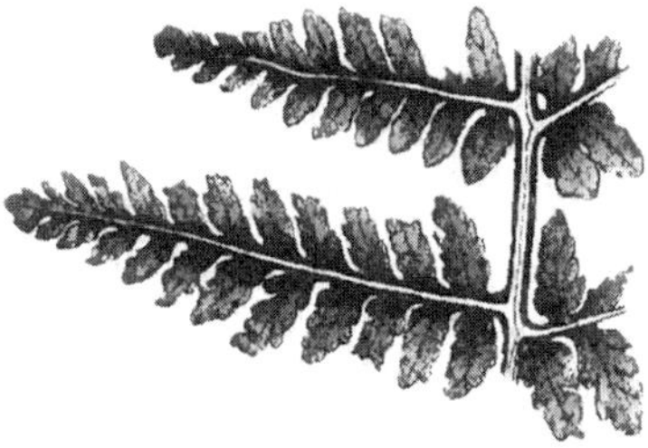

Sphenopteris polymorpha, Fstm.

Stem of Schizoneura gondwanensis,

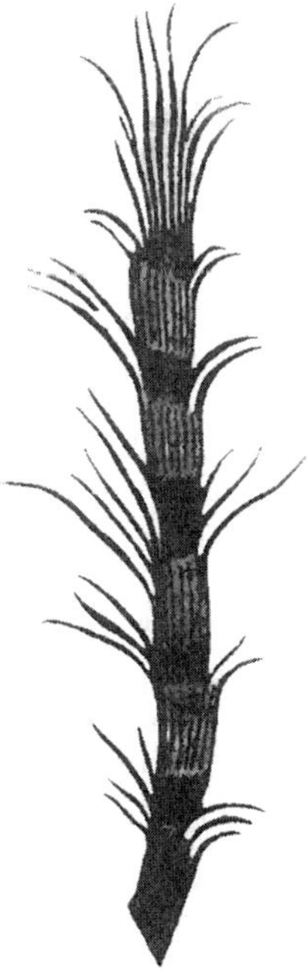

Phyllotheca indica, Bunb,

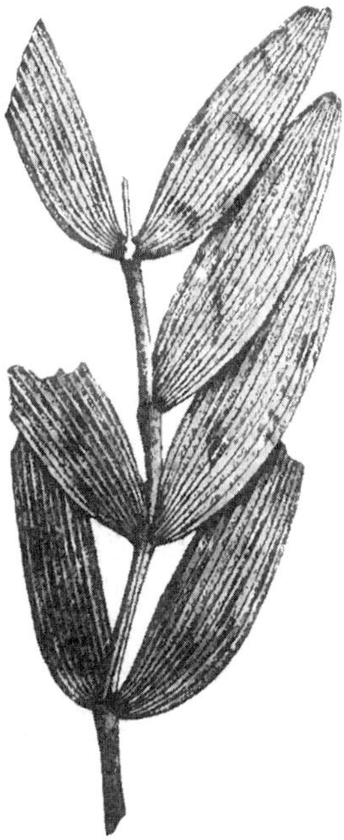

Schizoneura gondwanensis, Fstm,

Vertebraria indica, Royle.

Calcutta Phototype Co.

DAMUDA PLANTS.

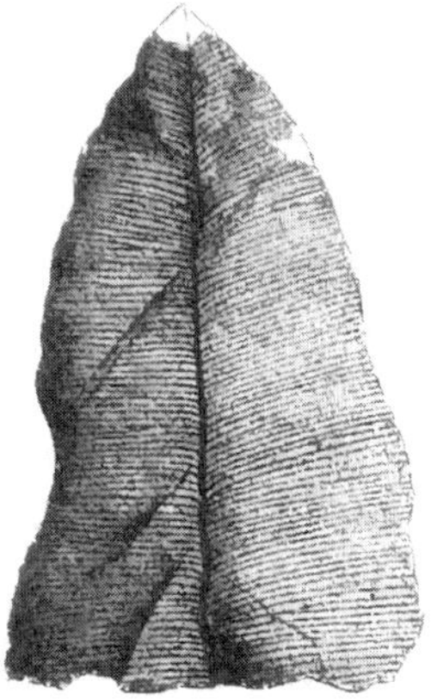

Macrotæniopteris Danæoides, Royle.

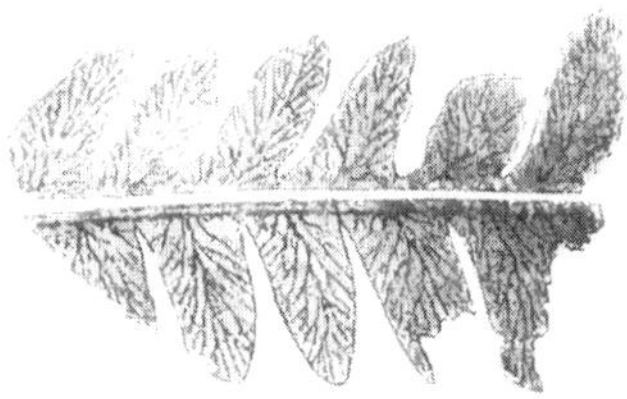

Alethopteris lindleyana, Royle

Glossopteris communis, Fstm

Glossopteris angustifolia, Bgt

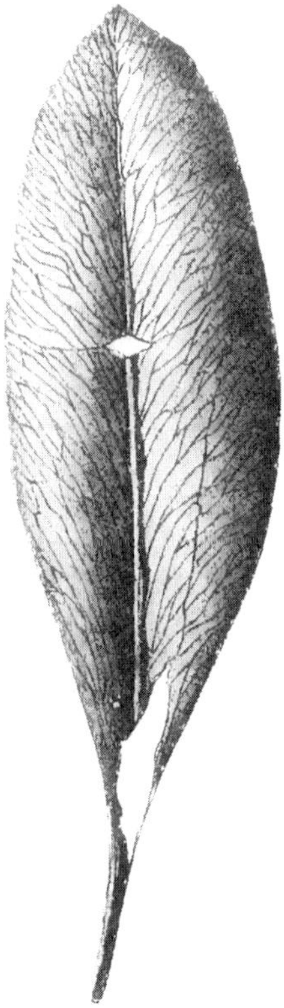

Glossopteris conspicua, Fstm.

Calcutta Phototype Co.

DAMUDA PLANTS.

present to note that all consist of sandstones and shales, with more or less ferruginous bands, and that some contain coal. Slight unconformity between the different groups has been noticed in places, and the Barakar beds are frequently unconformable to the Talchirs. The whole thickness of the Damuda series is 8,400 feet in the Ráníganj field, and about 10,000 feet in the Sátpura basin. It thus constitutes the most important portion of the Gondwána system.

The examination of the floras of the different groups of the Damuda series has shown that there are slight differences, but as nearly all the species of the Barakar group, and without exception all the more abundant forms, are also found in the Ránígan j group, they have been united in one list given below.[1] The letters *l.*, *m.*, *u.* prefixed signify that the species have been found in the Lower, Middle, and Upper, or Barakar Ironstone or Ránígan j beds, respectively; a dagger that it is also known from beds of Karharbári age, and an asterisk that it ranges into rocks newer than the Damuda series:—

EQUISETACEÆ—

* † l. u. *Schizoneura gondwanensis.*
u. *Phyllotheca indica.*
u ? „ *robusta.*
l. u. *Trizygia (Sphenophyllum) speciosa.*
* † l. u. *Vertebraria indica.*

FILICES—

l. *Cyathea*, cf. *tchihatcheffi.*
l. u. *Sphenopteris polymorpha.*
u. *Dicksonia hughesi.*
u. *Alethopteris whitbyensis.*
u. „ *lindleyana.*
u. „ *phegopteroides.*
u. *Pecopteris affinis.*
u. *Merianopteris major.*
* l. m. u. *Macrotæniopteris danæoides.*
u. „ *feddeni.*
u. *Palæovittaria kurzi.*
u. *Angiopteridium*, cf. *McClellandi.*
l. „ *infarctum.*
* † l. m. u. *Glossopteris communis.*
l. „ *intermittens.*
u. „ *stricta.*
u. „ ? *musæfolia.*
* † l. m. u. „ *indica.*
l. u. „ *browniana.*
l. u. „ *intermedia.*
m. u *Glossopteris retifera.*
m. u. „ *conspicua.*
u. „ *ingens.*
u. „ *divergens.*
* † l. m. u. „ *damudica.*
* l. m. u. „ *angustifolia.*
u. „ *leptoneura.*
u. „ *formosa.*
u. „ *orbicularis.*
u. *Gangamopteris anthrophyoides.*
u. „ *whittiana.*
u. „ *hughesi.*
† l. m. „ *cyclopteroides.*
u. *Belemnopteris wood-masoniana.*
u. *Anthrophyopsis*, sp.
l u. *Dictyopteridium*, sp.
u. *Sagenopteris* (?) *longifolia.*
u. „ *polyphylla.*
u. *Actinopteris bengalensis.*

CYCADEACEÆ—

u. *Pterophyllum burdwanense.*
l. *Platypterigium balli.*
† l. m. u. *Noeggerathiopsis hislopi.*

CONIFERÆ—

† u. *Voltzia heterophylla.*
u. *Rhipidiopsis densinervis.*
l. „ *gingkoides.*
l. *Cyclopitys dichotoma.*
*† l. u. *Samaropsis*, cf. *parvula*

[1] *Pal. Indica*, series XII, III, pt. ii, iii, (1880-81), IV, pt. ii, (1885).

Although there is little difference between the floras found in the various subdivisions of the Damuda series, the characters and relations of the minor groups require separate notice, and of these groups the lowest and the most important is the Barakar. This group derives its name from a river which traverses the western portion of the Ráníganj coalfield, and then falls into the Dámodar within the limits of the field.[1] In the higher portion of its course the Barákhar river receives the streams which drain the Karharbári coalfield.

The Barakars have an equally extensive range with the Talchirs, and consist of conglomerates, sandstones of various kinds, shales and coal. The sandstones are often coarse and felspathic, a variety of frequent occurrence being rather massive, white or pale brown in colour, soft at the surface and not much harder below, consisting of grains of quartz and *decomposed* felspar. The weathered surface of this sandstone frequently exhibits small projecting knobs, apparently due to calcareous concretions. One of the most striking distinctions between the sandstones of the Talchirs and those of the overlying formations consists in the felspathic constituents of the former being, as a rule, undecomposed, while in the Damuda series the grains of felspar are almost invariably converted into kaolin.

Besides the whitish felspathic sandstone, another typical Barakar rock is a conglomerate of small, well rounded, white quartz pebbles. These are sometimes found scattered over the surface and serve to indicate the presence of the conglomerate, where it is not exposed in section. The matrix of the conglomerate is usually white sandstone.

It must not be supposed that white is the only colour of the Barakar sandstones. Brown, red, yellow, and other tints are to be found, and predominate in many places. The whitish felspathic sandstone is however a typical rock, preserving its character in localities as far apart as Ráníganj in Bengal and Chándá in the Central Provinces, being well developed in the Godávari valley, but it is subordinate and forms but a small portion of the group to the eastward. Here the greater portion of the Barakar rocks consists of shales, grey, blue or black, frequently micaceous, and more or less sandy, occasionally associated with argillaceous iron ore, and often containing seams of coal. Not unfrequently the shaly beds are interstratified with hard flags.

The coals of the Barakar group vary greatly in quality and character in the different coalfields. They all, however, agree in having a peculiar laminated appearance, due to their being composed of alternating layers of bright and dull coal, the former purer and more bituminous than the latter, which, in many cases, is shale rather than coal. The best coals are

[1] *Memoirs*, III, 212, (1863).

those in which the bright layers predominate, but nearly all seams hitherto discovered are somewhat inferior to average European coal of the carboniferous formation, and there is a general tendency to variation in the thickness and quality of each seam within short distances. At the same time excellent fuel has been obtained from some Barakar seams. Some coal beds are of immense thickness, single seams (including partings of shale) amounting to as much as 35 feet in the Ránīganj coalfield, 50 feet near Chándá, and no less than 90 feet at Korba in Biláspur. Some of the Barakar coal exhibits a peculiar spheroidal structure, and round balls of various sizes, up to more than a foot in diameter, break away from the mass when the coal is mined. So thoroughly are these rounded that they were taken at first for rolled fragments, derived from some older formation.[1]

In places the Barakars rest quite conformably upon the Talchirs, and the two groups appear to pass into each other. In general, there is an abrupt change in mineral character, but the only case which has hitherto been found in which there is clear evidence of denudation having removed portions of the lower beds, during the deposition of the higher group,[2] is in the Rámgarh coalfield, where rolled fragments derived from the Talchirs have been found in the beds of the Barakar group. The Barakars, however, overlap the underlying Talchirs in many places and rest upon the metamorphic rocks, and in some coalfields, as in that of Ránīganj, there appears to be overstep as well, the highest beds of the Talchirs disappearing first, as if they had suffered from denudation. It yet remains to be seen whether representatives of the Karharbári beds do not intervene in those cases in which there is an apparent passage between the Barakar and Talchir groups.

The Barakars appear nowhere to exceed the thickness of 3.300 feet, a development which they attain only, so far as is known, in the Jhariá field. In no other field, except Rámgarh, do they exceed 2,000 feet.

Above the Barakar group in the Ránīganj and a few other fields of the Dámodar valley, there is found a great thickness of black or grey shales,[3] with bands and nodules of clay ironstone (carbonate of iron, mixed with clay), some of which is of the carbonaceous variety known as black band. Towards the base these beds become more sandy, and interstratifications of sandstone occur amongst them. The shales disintegrate slowly, and consequently the tract covered by this group is barren, and frequently elevated, but the rocks are not as a rule well exposed on the surface, although their presence is indicated by fragments of ironstone being scattered about.

[1] See *Jour. As. Soc. Beng.*, XVII, 59, (1848); XVIII, 412, (1844); XIX, 75, (1850).

[2] *Memoirs*, VI, 113, (1867).

[3] *Memoirs*, III, 40, (1863).

The greatest thickness attained by the ironstone shales is about 1,500 feet in the Bokáro coalfield, and they are nearly as thick in the Ráníganj field. As a rule, they are quite conformable to the underlying Barakars, the slight unconformity, which has been observed in places, is very possibly local, but one case[1] has been noticed where a break in time may be indicated.

Fossils are not common, and most of the species recorded were obtained from the South Karanpurá coalfield.

The highest group of the Damuda series, in the Dámodar valley, derives its name of Ráníganj from the principal town of the mining district of Bardwán, and comprises a great thickness of coarse and fine sandstones, with shales and coal seams.[2] The sandstones are moderately coarse, as a rule in thick massive beds, white or brown in colour, and obliquely laminated. They are usually more or less felspathic, the felspar being converted into kaolin. Bands of rather calcareous, fine, hard, yellow sandstone, often weathering out at the surface in nodular fragments, are common and characteristic of the group. Conglomerates are of rare occurrence. Shales form a much smaller portion of this group than they do in the Dámodar area of the subjacent Barakars. They are sometimes black and carbonaceous, sometimes bluish grey, and occasionally red or brown, more or less mixed with sand or stained by iron, and small bands of argillaceous ironstone occasionally occur, though they are not common. The coal is composed of alternately bright and dull layers, as in the Barakars.

This group is of considerable thickness in the Ráníganj field, being as much as 5,000 feet from top to bottom where fully developed, and it is possible that this is less than the original thickness, for the next group in ascending order rests upon the denuded surface of the present. The Ráníganj group diminishes in thickness in the other fields to the westward, and appears to be represented by groups of different mineral character beyond the limits of the Dámodar drainage.

As a general rule, the Ráníganj beds are conformable to the ironstone shales, but the higher group oversteps the lower, and rests on the Barakars, in the Bokáro coalfield, near Hazáribágh.

No animal remains have been found in the rocks of this group, but plants are abundant and comprise nearly all those in the Barakar groups, besides a number of species that are not known from any lower horizon.

The lithological distinction of the threefold division of the Damuda series, and the overlying Pánchet group, which will be noticed further on, has only been recognised with certainty in the coalfields of the Dámodar

[1] *Memoirs*, III, 42, (1863).

[2] *Memoirs*, III, 46, (1863).

valley. The Ráníganj and Pánchet groups have been recognised, palæontologically, in south Rewá, though they have not been mapped, but in the Mahánadí and Godávari drainage areas the Barakars are overlaid by a great series of beds which have been described in different areas by various names.

In the Sátpura ranges, south of the Narbadá, the lower Gondwána beds, above the Barakars, have been divided into two groups,—the Motúr and Bijori. This area has not had the same attention paid to it as that of Bengal, the Godávari valley, and Orissa, and the classification of the beds above the Barakars must be regarded as purely provisional.

The Motúr group[1] derives its name from a village of that name situated about 12 miles south-south-east of Pachmarhí, on the dividing ridge between the valleys of the Denwa which runs into the Táwa, a tributary of the Narbadá, and the Kanhán, which is a tributary of the Godávari. The village is on the road from Badnúr and Chhindwárá to Pachmarhí, and was at one time used as a sanitarium.

The beds of this group somewhat resemble the Pánchets of Bengal in mineral character. They consist of thick, coarse, soft, earthy sandstones, grey and brown, sometimes with red and mottled clays and calcareous nodules. Shales occur, but they are usually sandy and very rarely carbonaceous. It is probable that the Motúr group is unconformable to the Barakars. No collections of fossils have hitherto been made from the beds of the Motúr horizon.

The highest members of the Damuda series in the Sátpura region are exposed in the upper Denwa valley, at the southern base of the Mahádeva or Pachmarhí hills. For the rocks of this horizon the name of Bijori has been proposed,[2] from a small village rendered famous by being the locality whence the only distinctly vertebrate fossil, except *Brachiops*, yet obtained from the Damuda series, was procured.

The rocks of the Bijori horizon are characteristically Damudas, and comprise shales, occasionally carbonaceous, micaceous flags and sandstones.

Nothing definite is known of the relations between the Bijori and Motúr groups, nor has the thickness of either been determined, but the greater portion of the 3,000 to 4,000 feet of beds, intervening between the Motúr beds and the base of the Pachmarhí sandstone, may be assigned to the Bijori group.

The most important fossil hitherto found in the Bijori beds is the specimen already referred to, which is the skeleton of a Labyrinthodont allied to *Archegosaurus*, described by Mr. Lydekker under the name of *Gondwanosaurus bijoriensis*.[3]

[1] *Memoirs*, X, 161, (1873).

[2] *Memoirs*, X, 159, (1873).

[3] *Pal. Indica*, series iv, I, pt. 4, (1885).

Besides the labyrinthodont, the following plants have been identified:—[1]

EQUISETACEÆ—	FILICES,—*contd.*
Schizoneura gondwanensis.	*Glossopteris damudica.*
Vertebraria indica.	„ *retifera.*
Trizygia speciosa.	„ *angustifolia.*
FILICES—	*Gangamopteris*, sp.
Dicksonia, sp.	CONIFERÆ—
Glossopteris communis.	*Samaropsis*, cf. *parvula.*

The general facies of this flora corresponds best with that of the Ráníganj group in Bengal, with which it may be correlated in a general way, as long as exact contemporaneity of origin is not asserted.

In the Godávari valley, and in Chhatísgarh and western Orissa, the beds which overlie the Barakar group have been described under the names of Kámthí and Hingir, respectively, but in spite of some mineralogical differences, the two seem to represent each other so closely that they may be united under the first mentioned and older name.

The name Kámthí is derived from the military station so called, twelve miles north-east of Nágpur, and the station again derives its name from a village on the opposite side of the Kanhán river, where there is a famous quarry which has yielded a large number of fossils. The term Hingir is derived from a zamindári of that name situated north of Sambalpur.[3]

The typical Kámthí rocks consist of conglomerates, grits, sandstones, shales, and clays. The conglomerates contain pebbles of quartz. The grits are sometimes hard and silicious, so much so as to be quarried for quernstones, but usually they are soft and argillaceous. They are frequently stained by iron, and are often intersected by hard ferruginous bands of a dark brown colour. The sandstones are of every shade of colour, and vary greatly in character. They comprise fine grained micaceous beds, white in colour, with blotches and irregular streaks of red, and one of the most characteristic beds of the formation is a very fine argillaceous sandstone, hard, massive, and homogeneous, resembling a shale in structure, except that it exhibits no trace of lamination, yellow in colour below the surface, but becoming red when exposed. It passes into red shale. Another characteristic bed is a hard grey grit or sandstone, ringing under the hammer and breaking with a conchoidal fracture. The clays are red or green in colour, and chiefly prevail in the upper portions of the group.

These typical beds, with the exception of the clays, are chiefly developed near Nágpur. Elsewhere the Kámthís consist mainly of soft, porous

1 *Pal. Indica*, series xii, III, pt. ii, 17, (1880).
2 *Records*, IV, 50, (1871); *Memoirs*, IX, 305, (1872).
3 *Records*, VIII, 112, (1875).

sandstone, brown or white in colour, and conglomeratic in places, often with hard, ferruginous bands, and a few red shales. Here and there, however, a band of one of the characteristic rocks is met with towards the base of the formation.

The chief peculiarity, which distinguishes the Kámthí group from the Ráníganj and Bijori groups. is the absence of carbonaceous markings. In other Damuda groups, with the exception of the ironstone shales, the remains of plants generally retain a portion of their original carbon, but this appears very rarely to be the case amongst the Kámthís.

The thickness of the Kámthí group has not been determined, but it is undoubtedly considerable, probably 5.000 to 6,000 feet at least. The beds belonging to this group generally appear conformable to the Barakars, but it is extremely doubtful if the conformity is more than apparent, for the Kámthí beds overlap the Barakars in a most irregular manner, and the break in conformity between the two is well marked in places. The Hingir beds, both near Sambalpur and in the Tálcher coalfield, certainly rest unconformably in places on the Barakar group.[1]

The fossil plants of the Kámthí group comprise the following species :[2]

D.	*Phyllotheca indica.*		*Glossopteris musæfolia.*
D.	*Vertebraria indica.*		„ *leptoneura.*
	Pecopteris, sp.		*Gangamopteris hughesi.*
D. P.	*Glossopteris communis.*		*Angiopteridium,* cf. *macclellandi.*
D. P.	„ *indica.*	D.	*Macrotæniopteris danæoides.*
D.	„ *browniana.*	D.	„ *feddeni.*
D.	„ *damudica.*		*Næggerathiopsis hislopi.*
	„ *stricta.*		

In the foregoing list the letter *D* prefixed to the name of a species signifies that it is also found in the Damuda series of Bengal, and the letter *P* that the same species is known from the Pánchet group. Of the former, all are found in the Ráníganj group, but not all in the lower groups, one species *Angiopteridium macclellandi* has been found in the Rájmahál group of the upper Gondwánas. The character of the flora would lead us to regard it as homotaxial with the Ráníganj group or possibly newer.

In the neighbourhood of Mángli, a small deserted village lying at the northern extremity of the Wardhá Gondwána basin, about fifty miles south of Nágpur and thirty-five north-west of Chándá, some quarries have long existed, from which a very fine red and yellow sandstone is obtained and employed in building, chiefly for ornamental purposes and for carvings. The stone is precisely similar to that of Silewáda and other typical exposures of the Kámthí group, near Nágpur, and the coarser associated sandstones of Mángli differ in no way from the ordinary Kámthi grits.

[1] *Records,* VIII, 113, (1875).

[2] *Pal. Indica,* series xii, III. pt. ii, 19, (1880).

The quarries of Mángli have become well known by name to Indian geologists, and even to those of other countries, having furnished to Mr. Hislop the first Labyrinthodont amphibian fossil (*Brachiops laticeps*) detected in India[1] They have also yielded a species of *Estheria* and a few plant remains. The latter are so poor that very little dependence can be placed upon their determination. One is believed to be coniferous, and has been referred to *Palissya*;[2] another is a stem of a fern. The species of *Estheria* has been named *E. mangaliensis* by Rupert Jones.[3] A smaller variety closely resembles the *Estheria* found in the Pánchet group of Bengal and may be identical, but the identification is not quite certain, as the Pánchet fossil is so poorly preserved that some of the specific characters, depending upon the microscopical texture of the shell, cannot be ascertained.

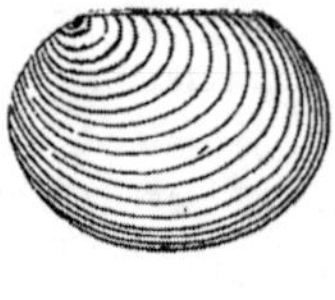

Fig. 11.—*Estheria mangaliensis*, Rupert Jones (enlarged 3 diameters).

The uppermost beds of the lower Gondwánas in the Sátpura range have been distinguished under the name of Almod from a village at the south base of the Pachmarhí escarpment. The rocks consist of sandstones with a few carbonaceous shales, from which no fossils have been obtained. Their relations to the groups above and below require further investigation. No unconformity has been traced and their sole importance comes from their position between the Mahádevas and Damudas, and the consequent possibility of their representing the Pánchets of Bengal.

The term Pánchet was originally applied to two groups of beds in the Ránígani coalfield.[4] It is now restricted to the lower of these groups, the upper Pánchets of the Dámodar valley being referred to an upper Gondwána age, and ascribed to the Mahádeva series. The name was derived from an important zamindári, which still comprises a large tract in the southern portion of the Ráníganj coalfield and formerly included much more, and the same name is that of a large hill, the basal portion of which consists entirely of Pánchet beds.

The great mass of this group consists of thick beds of coarse felspathic and micaceous sandstones, often of a white or greenish white colour, with bands of red clay from a few inches to twenty feet in thickness. The felspar, in the sandstones, is occasionally undecomposed, which is never the case in

[1] See *Quart. Jour. Geol. Soc.*, X, 472, (1854); XI, 37, (1855).

[2] Feistmantel, *Records*, X, 26, (1877). The identification seems doubtful, for Sir Charles Bunbury suggested the possibility of the same stem belonging to the Lycopodiaceous genus *Knorria*.

[3] *Pal. Soc.*, Mem. Foss. Estheriæ, p. 78, (1862).

[4] *Memoirs*, III, 30, 126, 132, etc., (1863).

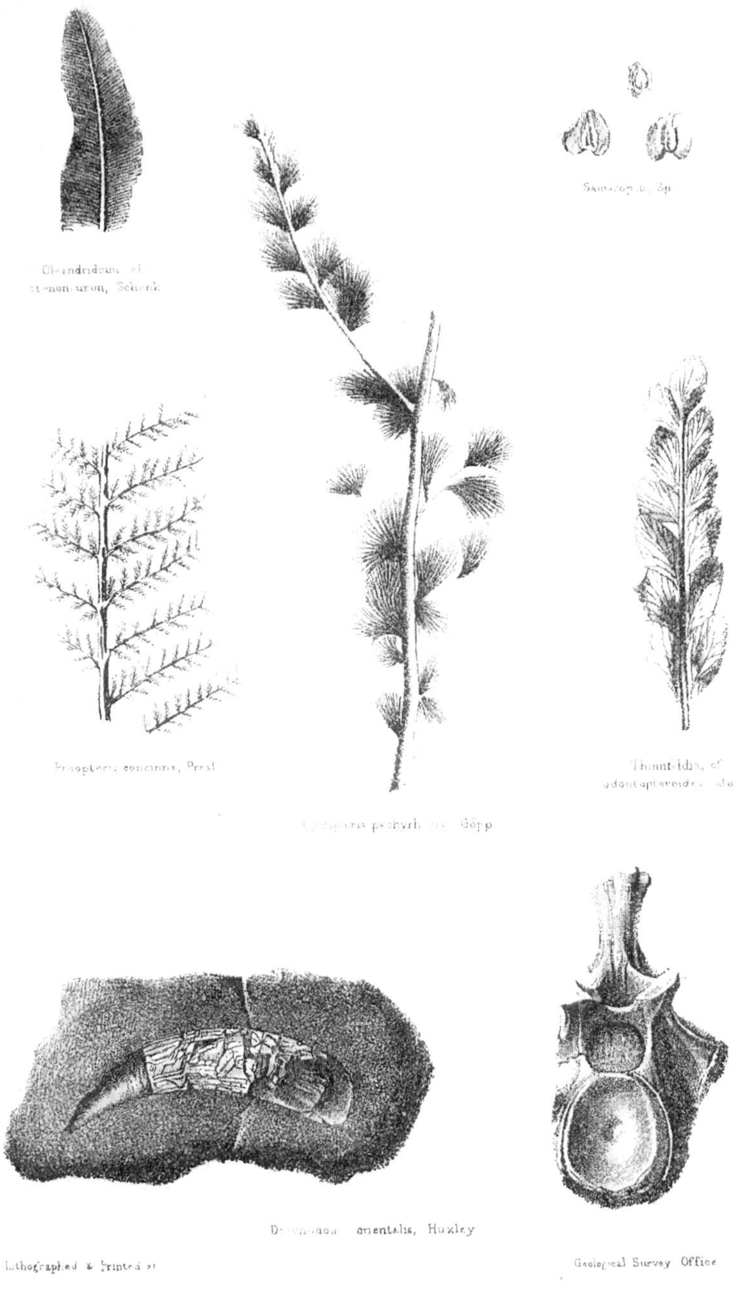

Lithographed & Printed at Geological Survey Office

PANCHET FOSSILS.

the Damudas. Conglomeratic beds sometimes occur in the upper portion of the group, but they are not common. At the base of the group grey and greenish grey sandstones and shales are usually found in very thin beds, and often highly micaceous. In places the greenish micaceous clays are met with higher in the group.

The Pánchet rocks are distinguished from the typical Damudas by the presence of red clay and the absence of carbonaceous shales, and by the sandstone being, as a rule, much more micaceous. But rocks of the Pánchet character are found in parts of India interstratified with the Damudas, as in the Motúr group.

The thickness of the present group in the Dámodar valley nowhere exceeds about 1,800 feet. It rests with slight, but distinct, unconformity upon the denuded surface of the Ráníganj group, and in some places the Pánchets completely overlap that group and rest upon lower beds, the unconformity being most marked in the Bokáro coalfield. Fragments of coal and shale, apparently derived from the Damudas, have occasionally been found in the conglomerates of the Pánchet group.

The most important remains of animals hitherto found in the lower Gondwána rocks have been derived from the Pánchets. In the upper portion of the group there is, in the Ráníganj coalfield, a well marked conglomeratic band containing reptilian and amphibian bones. These are isolated from each other and sometimes slightly rolled The specimens obtained have been examined and described by Prof. Huxley and Mr. Lydekker[1] and comprise the following forms, besides a few others whose affinities are doubtful. The plants were described by the late Dr. Feistmantel.[2] A dagger in the following list indicates species also found in the Damudas:—

ANIMALIA.

AMPHIBIA—
- Labyrinthodontia—
  - *Gonioglyptus longirostris.*
  - „ *huxleyi.*
  - *Glyptognathus fragilis.*
  - *Pachygonia incurvata.*

REPTILIA—
- Dicynodontia—
  - *Dicynodon orientalis.*
  - *Ptychosiagum* (*Ptychognathus*) *orientale.*
- Dinosauria—
  - *Epicampodon* (*Ankistrodon*) *indicus.*

CRUSTACEA—
- *Estheria mangaliensis* (?).

PLANTÆ.

EQUISETACEÆ—
- † *Schizoneura gondwanensis.*
- † *Vertebraria indica.*

FILICES—
- *Pecopteris concinna.*
- *Cyclopteris* (?) *pachyrhaca.*
- *Thinnfeldia*, cf. *odontopteroides.*
- *Oleandridium*, cf. *stenoneuron.*
- † *Glossopteris communis.*
- † „ *indica.*
- † „ *damudica.*
- † „ *angustifolia.*

CYCADEACEÆ—
- Scales.

CONIFERÆ—
- † *Samaropsis*, cf. *parvula.*

[1] *Pal. Indica*, series iv. Indian pretertiary Vertebrata.

[2] *Pal Indica*, series xii, III, pt. ii, pp. 51–56, (1880).

II.—THE UPPER GONDWÁNA SERIES.

The unfossiliferous beds of the lower part of the upper Gondwánas have, in the more recent publications of the Geological Survey, been generally referred to under the name of Mahádeva. This name, first applied to the sandstone of the Pachmarhí hills,[1] was subsequently extended so as to comprise all the beds of the Sátpúra basin above the Damudas of the lower Denwa valley, except the Jabalpur group, and has ultimately come to be a convenient, because indefinite, term to apply to the soft sandstones and conglomerates, of obscure stratigraphical position, seldom containing any fossils except fragments of carbonised wood, which there is good reason to believe belong to the lower part of the upper Gondwánas, though they cannot be assigned with certainty to any particular horizon.

The Mahádeva rocks consist chiefly of very thick massive beds of coarse sandstone, grit, and conglomerate. These are frequently ferruginous, or marked with ferruginous bands, as in the Kámthís. They are associated with clays, and occasionally with bands of impure earthy limestone. The sandstones form high ranges of hills, and often weather into vertical scarps of great height, forming conspicuous cliffs that contrast strongly with the black precipices of the Deccan traps and the rounded irregular masses of the more granitoid metamorphic rocks.

In the typical area of the Sátpura region the Mahádeva rocks attain a thickness of at least 10,000 feet, nine-tenths of which consist of coarse sandstone, grit, and conglomerate. They appear to be unconformable to the underlying Damudas, as the series overlaps the upper members of the lower Gondwána series.

In the Sátpura region the Mahádeva formation has been subdivided into three groups,—the Bágra, Denwa, and Pachmarhí,—each of which requires a few remarks.

The name of the Pachmarhí[2] group is derived from a village on the top of the hills of the same name, and the site of a sanitarium. The group consists of massive sandstone, whitish or brownish in colour, usually soft, often containing small subangular pebbles, and occasionally intersected by hard ferruginous bands. As a rule, the stratification is obscure, oblique lamination being common, and the different beds of which the group is composed exhibit great irregularity in superposition and often overlap each other. The hard ferruginous partings are most irregularly interspersed throughout the mass, usually as thin beds, though not always perfectly parallel to the planes of stratification. Sometimes the impregnation with iron is confined to pipes or nodules. Fragments of these ferruginous bands

[1] *Jour. As. Soc. Beng.*, XXV, 252, (1856); *Memoirs*, II, 183, 315, (1860).

[2] *Memoirs*, X, 155, (1873).

are often scattered in quantities over the surface, and serve to distinguish the outcrop of the Pachmarhí group from those of the underlying beds.[1]

The Pachmarhí group comprises, where thickest, 8,000 feet out of the 10,000 found in the Mahádevas of the Sátpúra hills.

The middle group of the Sátpura Mahádevas is named[2] after a stream which rises on the south side of the Pachmarhí range and, turning round the eastern end of the ridge, forms its northern boundary throughout, falling finally into the Táwa. The course of this stream, north of the Pachmarhí hills, is the area of the Denwa rocks, which, presenting a marked contrast to the massive Pachmarhí sandstone, are principally composed of soft clays, pale greenish yellow and bright red, mottled with white in colour, forming thick beds interstratified with discontinuous and subordinate bands of white sandstone, and very rare courses of earthy limestone. The sandstones are locally conglomeratic. In short, in mineral character the Denwa rocks are a repetition of the Motúr group in the middle of the Damuda series, and resemble the Pánchets of Bengal.

The thickness of these beds in the Denwa valley is about 1,200 feet. They appear in places to pass into the underlying group, although they are quite distinct in the typical area.

The Denwa group is the only one which can be correlated to those of other parts of India on palæontological grounds. Vertebrate remains have been found in it, and more abundantly in south Rewá, which show that it is the equivalent of the better known Kota-Maleri group of the Godávari valley. The name is consequently one which will probably drop out of use as the relations of the rock groups are more completely worked out.

The uppermost group was named Bágra[3] from a hill fort built upon it, where the river Táwa cuts its way through a spur of the Sátpura hills, south-east of Hoshangábád. It is largely composed of conglomerates, often coarse, frequently with a deep red sandy matrix. It is more calcareous than the other Mahádeva groups, and bands of calcareous sands and clays and limestones, sometimes dolomites, are of frequent occurrence. The group is very irregular in composition. The greatest thickness does not exceed 600 to 800 feet, and in places it overlaps the Denwa shales and rests directly on the Pachmarhí sandstones.

The generally unfossiliferous nature of the Mahádeva beds, their softness and ease of weathering, render their recognition with certainty a matter of difficulty. In the coalfields of the Dámodar valley some soft pebbly sandstones, which were formerly regarded as upper Pánchet, are now

[1] It should not be forgotten that similar ferruginous layers are found in the Kámthís.

[2] *Memoirs*, X, 153, (1873).

[3] *Memoirs*, X, 150, (1873).

regarded as very probably Mahádevas. Their relations to the underlying rocks are difficult to make out, the junctions being greatly obscured by pebbles and detritus derived from the newer grits, but there appears to be some unconformity.

In the Tálcher field and in Chhatísgarh the uppermost soft pebbly sandstones are believed to be Mahádevas, but have not been coloured as upper Gondwánas on the accompanying map, as their age has not been satisfactorily established, and the outcrops are small,

Mahádeva sandstones are found in the Narbadá valley, running out from under the scarp of the Deccan trap at Bárwai,[1] and further west certain sandstones, underlying the cretaceous of the Narbadá valley, are probably of Mahádeva age, but as they have been held to be cretaceous the question of their age will be discussed when dealing with that system.[2]

In the Rájmahál hills the lower Gondwánas are overlaid by a thick band of coarse sandstone, which was at first associated with the overlying beds, but has since been separated, as it is unconformable to them. It is lithologically very similar to some beds in the Dámodar valley, which are believed to be of upper Gondwána age, and possibly is a representative of them.

The Dubrájpur group, as this band of sandstones and conglomerates is called, takes its name from a village[3] in the Rájmahál hills, situated about forty miles north by east of Suri. The component beds are sandstones of several varieties, grits and conglomerates, for the most part ferruginous. Fine grained beds are not common, although shaly sandstones are occasionally met with. Most of the coarser beds are ferruginous, and one form of conglomerate, of frequent occurrence, consists of quartz pebbles in a ferruginous matrix. A precisely similar bed is found in the supposed Mahádeva beds of the Dámodar valley.

Along the western scarp of the Rájmahál hills the rocks of the Dubrájpur group rest partly upon the Damudas and partly upon the metamorphic rocks, the Damudas (Barákars) being repeatedly overlapped by the Dubrájpur beds in a manner which shows the two to be quite unconformable. The greatest thickness of the Dubrájpur group in the Rájmahál area does not exceed about 450 feet. Some specimens of a cycadeaceous plant (*Ptilophyllum*) were once found near the southern extremity of the hills in the uppermost beds underlying the Rájmahál trap, but there is some little doubt as to whether the fossiliferous band may not belong to the Rájmahál series itself.

[1] *Records*, VIII, 73, (1875).
[2] *Infra*, p. 253.
[3] *Pal. Indica*, series ii, I, 1, (1863); *Memoirs*, XIII, 198, (1877).

The Rájmahál series derives its name from a range of hills in Bengal, extending north and south from the Ganges to the neighbourhood of Suri in Bírbhúm, and, unlike the other members of the Gondwána system is confined to the neighbourhood of the eastern margin of the Indian Peninsula. Some species of fossil plants, identical with Rájmahál forms, have been found in other localities, but they are either isolated, or associated with plants belonging to a different flora.

In its typical locality the Rájmahál group of the Rájmahál series consists of a succession of basaltic lava flows or traps with interstratifications of shale and sandstone. The sedimentary bands are held to have been deposited in the intervals of time which elapsed between the volcanic outbursts, by the circumstance that the different bands of shale and sandstone differ from each other in mineral character, and also that the upper surface of the shaly beds has sometimes been hardened and altered by the contact of the overlying basalt, whilst the lower surface is never affected. The sedimentary bands are chiefly composed of hard white and grey shale, carbonaceous shale, white and grey sandstone, and hard quartzose grit.

Fig. 12.—Radiating columnar trap, Rájmahál hills.

The trap rocks are all dark coloured dolerites. They vary in character from a fine grained, very tough and hard rock (anamesite), ringing under the hammer, and with the edges of its fracture almost as sharp as those of a quartzite, to a comparatively soft, coarsely crystalline basalt. The latter usually contains olivine in large quantities. Many of the trap rocks

[1] *Jour. As. Soc. Beng.*, XXIII, 263, (1854); *Memoirs*, II, 313, (1860); XIII 209, (1877); *Pal. Ind.*, series ii, I, 1. (1863).

are amygdaloidal, the enclosed nodules usually containing some form of quartz, either agate, chalcedony, or rock crystal. Occasionally, but less frequently, zeolites are found, stilbite being the commonest, natrolite less abundant, and analcime has also been detected. It is not usual to find the cavities lined with green earth, as is so frequently the case amongst the amygdaloids of the Deccan trap. The basaltic flows above the sedimentary bands are, as a rule, compact.

Very little light is thrown on the source of the basaltic rocks by any observations within the Rájmahál area. Dykes are rare, and there is only one instance known of an intrusive mass which may mark the site of an old volcanic outburst. This is about 22 miles south south-east of Colgong on the Ganges, close to a place called Simra, where a group of small conical hills occurs, composed of pinkish trachyte, porphyritic in places, and surrounded by Damuda rocks. The surface of the ground is much obscured by superficial deposits, but there appears good reason for supposing that the core of a volcanic vent is here exposed. It appears not an unfrequent occurrence that the later outbursts from a volcano are more silicious than earlier eruptions, and that a volcanic core, even when the lava flows have been doleritic, should itself prove trachytic, when exposed by denudation. This may be due to the solution of the highly silicious metamorphic rocks through which the outburst took place by the molten lava remaining in the fissure after the eruption, and the consequent conversion of that lava from a basic into an acid rock.

Trap dykes and intrusions, believed to be of Rájmahál age, are abundant in the coalfields of the Dámodar valley, and both dykes and cores of basalt are common in the portion of Bírbhúm lying south-west of the Rájmahál hills. It is possible that the principal vents lay in this direction, or they may have been in the region now covered by the Ganges alluvium. The difficulty of determining the original source of eruptive rocks will be again illustrated in the case of the Deccan traps.[1]

The bedded basaltic traps of the Rájmahál hills, with their associated sedimentary beds, attain a thickness of at least 2,000 feet, of which the non-volcanic portion never exceeds 100 feet in the aggregate. They rest with general parallelism on the grits and coarse sandstones of the Dubrájpur group, but nevertheless several instances of overlap take place,

[1] Some doubt still attaches to the determination of the true age of the Rájmahál traps. Detailed examination has shown that a close lithological resemblance exists between them and the Deccan traps; *Records*, XX, 104, (1887); XXII, 226, (1889). This is not, however, in itself sufficient to prove their contemporaneity. On the other hand the examination of the stratigraphical relations of these traps in the field cannot be said to have been so close as to preclude the possibility that the supposed interbedded traps are really intruded along the planes of bedding. The correlation of the Rájmahál traps with the Sylhet traps is an important point, as (if it is correct) the former must be older than the Deccan trap period.

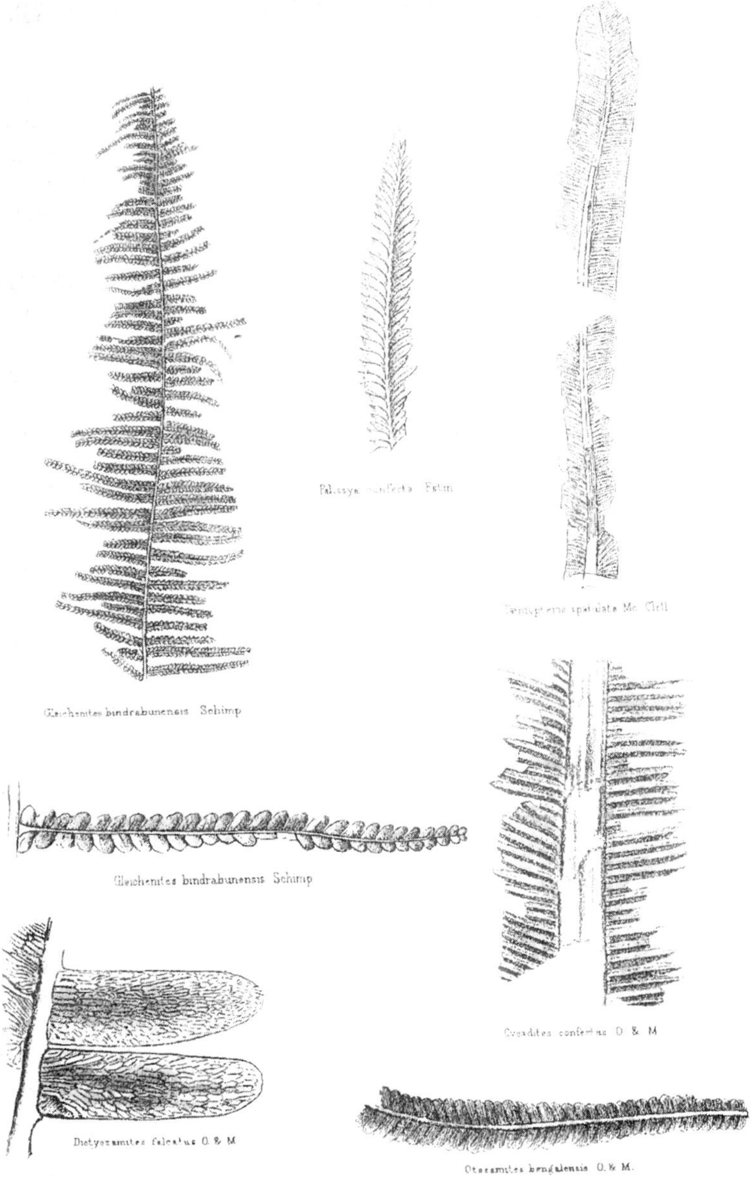

Lithographed & Printed at Geological Survey Office.

RAJMAHAL PLANTS.

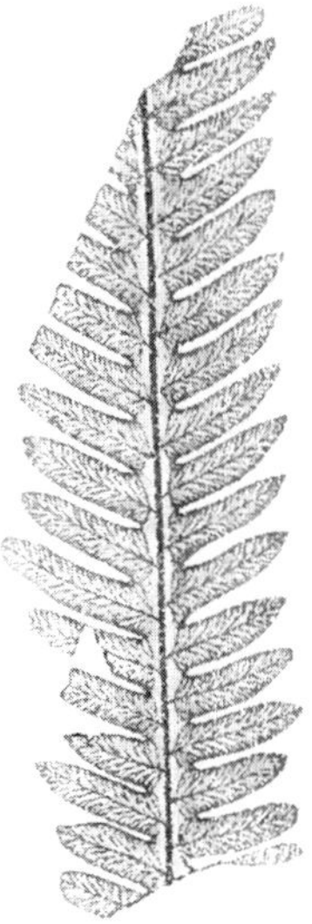

Alethopteris indica. O. & M.

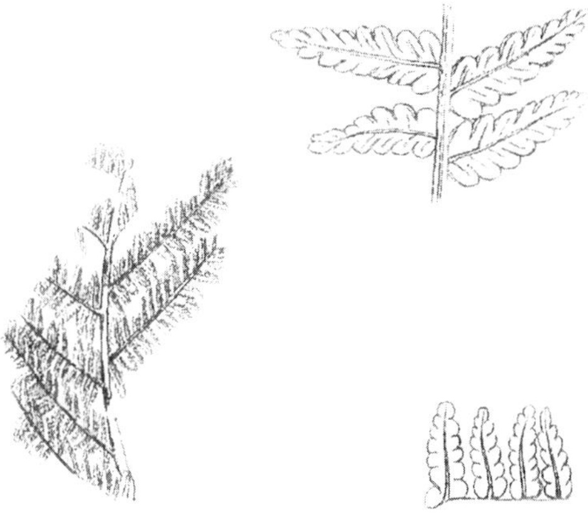

[illegible] Oldham.

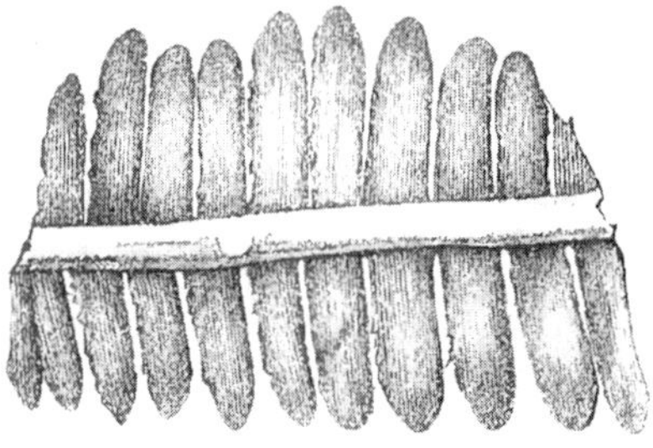

Pterophyllum rajmahalense. Morris.

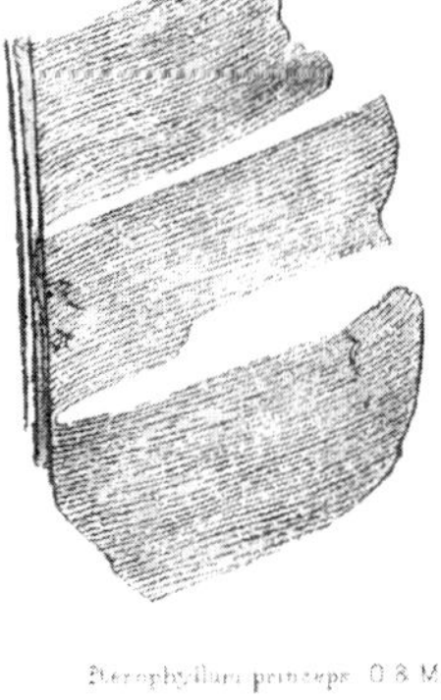

Pterophyllum princeps. O. & M.

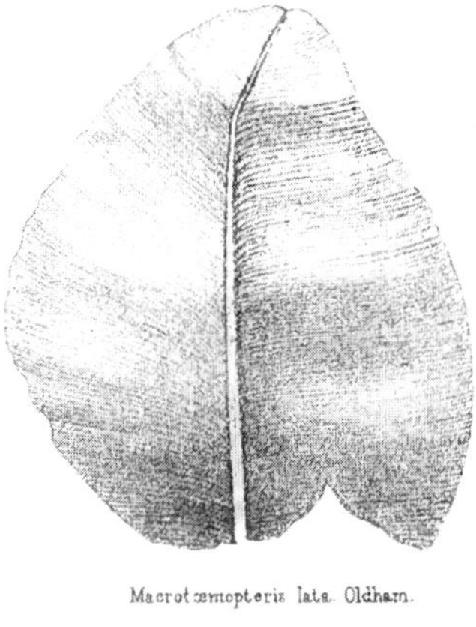

Macrotæniopteris lata. Oldham.

Lithographed & Printed at Geological Survey Office.

RAJMAHAL PLANTS.

and, in one locality at least, there is evidence of the Dubrájpur beds having been denuded before the deposition of the Rájmahál group.

From the extent of the area throughout which these dykes are developed, conclusions may be drawn as to the original limits of the volcanic action coincident with the period of deposition of the Rájmahál group. The number of trap dykes gradually diminishes in the coalfields of the Dámodar valley from east to west, until finally, in the Karanpurá field south-west of Hazáribágh, volcanic intrusions disappear almost entirely, and none appear to be known further west, until basaltic dykes of different age, which apparently are contemporaneous with the much newer Deccan trap, make their appearance. Outside of the coalfields it is difficult to distinguish the dykes belonging to the Rájmahál period from older eruptions, but there is not the same abundance of extensive basaltic intrusions in southern Monghyr, Házaribágh, and Chutiá Nágpur as in Bírbhúm. So far as can be judged, the region immediately north of the Rániganj coalfield was one of the foci of eruption and it is far from improbable that the bedded traps of the Rájmahál hills had originally a considerable extension to the south-west and south, though, as no single outlier has been preserved, it is impossible to feel sure of the inference. There is, however, considerable probability that a large tract in the Dámodar valley, including the whole Rániganj field may have been once covered with bedded traps.

The great difference of age between the Rájmahál group on the one hand and all the lower Gondwána rocks, including the Damudas and Pánchets, on the other, is well illustrated by the change in the flora and by the very much greater amount of disturbance to which the Damuda rocks have been subjected. The Rájmahál traps are almost horizontal, and no faults have been observed in them, while the dykes which abound in the Rániganj field,and are almost certainly of Rájmahál age, are newer than the faults of the coalfield.

By far the greater part of the Rájmahál fossils have been obtained from two bands of fine grained whitish or greyish shales—the upper 25 to 30 feet thick, the lower 10 to 15,—separated from each other by lava flow, and having other beds of trap, with intercalations of sandstone and shale, above and below.

The following is a list[1] of those hitherto described from this group :—

EQUISETACEÆ—
Equisetum rajmahalense.
FILICES—
Sphenopteris arguta.
FILICES,—*contd.*
Sphenopteris hislopi.
" *membranosa.*
Dicksonia bindrabunensis.

[1] The list is taken from O. Feistmantel, *Pal. Indica*, series ii, I, 143, (1877). In the case of *Thinnfeldia salicifolia*, *Dictyozamites falcatus* and *Echinostrobus indicus*, the original specific names have been restored in place of *P. indica*, *D. indicus* and *E. rajmahalensis*, as there does not seem to be sufficient reason for the change.

FILICES—*contd.*
Hymenophyllites bunburyanus.
Cyclopteris oldhami.
Thinnfeldia salicifolia.
Alethopteris indica.
Asplenites macrocarpus.
Pecopteris lobata.
Gleichenia bindrabunensis.
Angiopteridium macclellandi.
" *spathulatum.*
" *ensis.*
Macrotæniopteris lata.
" *crassinervis.*
" *ovata.*
" *morrisi.*
Danæopsis rajmahalensis.

CYCADEACEÆ—
Pterophyllum distans.
" *carterianum.*
" *morrisianum.*
" *medlicottianum.*
" *princeps.*
" *crassum.*

CYCADEACEÆ—*contd.*
Pterophyllum rajmahalense.
" *fissum.*
" cf. *propinquum.*
Zamites proximus.
Ptilophyllum acutifolium.
" *cutchense.*
Otozamites bengalensis.
" *abbreviatus.*
" *oldhami.*
Dictyozamites falcatus.
Cycadites confertus.
" *rajmahalensis.*
Cycadinocarpus rajmahalensis.
Williamsonia microps.
Ptilophyllum cf. *W. gigas.*

CONIFERÆ—
Palissya indica.
" *conferta.*
Chirolepis gracilis.
Cunninghamites dubiosus.
Echinostrobus indicus.

The first thing which must strike any one in looking over the above list is the great change in forms of life between the upper and lower Gondwána series, so far as we are yet acquainted with them. It is highly probable that intermediate beds may hereafter be found, but for the present there seems to be, in Bengal at least, just as great a break in the flora as in the stratigraphy. The most striking distinction is that the prevalent forms in the lower Gondwánas are *Equisetaceæ* and ferns of the *Glossopteris* type, *Cycadeaceæ* being rare, whilst in the upper Gondwánas, and especially in the Rájmahál group, *Cycadeaceæ* prevail, their individual abundance being so great that they frequently form the mass of the vegetation. In fact, the cycads, and especially *Ptilophyllum acutifolium* are just as abundant and characteristic in the Rájmahál group, as *Glossopteris* and *Vertebraria* are in the Damúdas.

The Rájmahál beds are represented along the east coast by a series of small outliers, most of them too small to deserve detailed notice here, which are interesting, as they appear to comprise some rock groups of later age than the Rájmahál, and contain marine fossils associated with the plants.

The most northerly of these is the Athgarh basin, a tract of sandstone, some twenty miles long from north to south, and eighteen miles from east to west close to the town of Cuttack, on the western margin of the alluvial plain. Some carbonaceous shale, occurring in the lower portion of the sandstones, has been supposed to indicate the presence of Barakar beds, but no Damúda fossils have been found. The relations of the remaining part of the rocks are very obscure, and they were believed to be of Kámthí age till the discovery of characteristically Rájmahál fossils in them.[1]

On the right bank of the Godávari, near Thalapúdi, about ten miles above Rájámahendri, a well marked belt of upper Gondwána beds commences, which extends for sixty miles, from the Godávari to beyond Golapilli west of Ellore. The width of this belt varies from ten to fifteen miles. There is a general dip to south-east or east-south-east at 5° to 10°, and the beds rest unconformably, throughout a considerable portion of their area, upon various members of the Kámthí group, but they overstep this group, both to the east and west, and rest upon a sloping floor of gneiss, which has the appearance of a plane of marine denudation formed after the deposition of the Kámthí rocks, as the latter rest upon a much more uneven surface of the metamorphic formations. This appearance of resting upon a surface which had been fashioned by denudation after the deposition of the lower Gondwána beds, quite agrees with the peculiar distribution of the Rájmahál group and its associates, which evidently were accumulated in a distinct area from that in which the Gondwána beds of the Godávari valley were deposited. To the south-east the upper Gondwána beds of the Ellore area disappear beneath the Cuddalore sandstones and the alluvial deposits of the Godávari delta, except west of Rájámahendri, where the Gondwánas are covered by outliers of the Deccan traps.

The rocks of the Ellore area are peculiarly interesting, because they appear to contain representatives of groups higher than the Rájmaháls, associated with beds in which the typical Rájmahál flora is well preserved. Dr. King, who surveyed the rocks of the Godávari district, classed the upper Gondwána beds in three subdivisions, thus distinguished in descending order:[2]

1. Tripetty sandstones.
2. Ragavapuram shales.
3. Golapilli sandstones.

The Golapilli sandstones consist of brown and red sandstones and conglomerates which form a broad plateau near Golapilli, capped by conglomerates and gravels, probably belonging to the Cuddalore sandstones.

The following plant fossils have been obtained from the Golapilli beds:—[3]

FILICES—
Alethopteris indica.
Pecopteris macrocarpa.

FILICES,—*contd.*
Angiopteridium ensis.
" *spathulatum.*

[1] Feistmantel, *Records*, X, 68, (1877); *Pal. Indica*, series xii, I, 187, (1879).

[2] *Records*, X, 56, (1880); *Memoirs*, XVI, 211, (1889).

[3] *Pal. Indica*, series ii, I, 163, (1877).

CYCADEACEÆ—
Ptilophyllum acutifolium.
„ *cutchense.*
Dictyozamites indicus.
Pterophyllum morrisianum.
„ *carterianum.*
„ *kingianum.*
„ *distans.*

CYCADEACEÆ—*contd.*
Williamsonia gigas.

CONIFERÆ—
Palissya conferta.
„ *indica.*
Cheirolepis, cf. *muensteri.*
Araucarites macropterus.

All the above, with the exeption of *Pterophyllum kingianum*, the *Cheirolepis* and *Araucarites macropterus* are characteristic Rájmahál forms.

Resting upon the Golapilli beds, in the neighbourhood of Ellore, there is found a thin band of white and buff shales, having a few interstratifications of sandstones towards the base, not more than 100 feet thick. No unconformity has been detected between these shales and the Golapilli sandstones, but there appears to be some difference in the flora, for, while the plants of the Golapilli standstones are all Rájmahál forms, except a few species peculiar to the beds, the flora of the overlying shales comprises, in addition to several forms common to the beds below, a few species allied to Jabalpur plants. The shales have been called Ragavapuram, from a village situated about twenty-six miles north-north-east of Ellore. A list of the plants will be found in the tabular statement on page 183.

With the plants are some marine shells, chiefly casts, amongst which are some *Ammonites*, apparently allied to middle jurassic species, the principal form being near *A. opis*, but distinguished by having the ribs simple throughout. *A. opis* belongs to the subgenus *Stephanoceras*, and to the group of *A. macrocephalus*, and is found in the Chári and Katrol beds of Cutch (callovian and oxfordian). Besides the ammonites, *Leda*, *Pecten*, *Gervillia*, etc., occur, the *Leda* being especially common and characteristic.

Above the shales just noticed there is another thin band of dark brown and red sandstones and conglomerates, chiefly ferruginous, with silicious and argillaceous bands, and beds of concretionary clay ironstone named from a pagoda called Chinna (little) Tirupati (Tripetty), which stands upon a scarp composed of them, about twenty miles north-north-east of Ellore. Towards the bottom these sandstones become softer and less ferruginous. In the main area, near Ellore, these Tripetty beds are only 40 feet in thickness.

The Tripetty beds in the main area have only yielded fossil wood, but from some outlying patches, near Innaparazpálayám about twenty-four miles north by east of Coconáda, supposed to belong to the same band, Dr.

King obtained two *Trigoniæ*,—*T. smeei* and *T. ventricosa*,—both of which are characteristic of the Umia beds of Cutch.

The sequence of upper Gondwána beds in the neighbourhood of Ellore is very instructive. The whole series rests unconformably on the Kámthís (lower Gondwána), and although the whole thickness of the upper Gondwána series is trifling, apparently not exceeding 200 or 300 feet, it comprises representatives of the Rájmahál and Umia groups, and of an intermediate formation. Yet these thin bands exhibit no marked unconformity. The middle group is overlapped at both ends, it is true, but there is no sign of any important break. It is clear that the country must have undergone very little disturbance in the interval between the deposition of the different groups, and, judging from this instance, it is impossible to argue from the small amount of discordance between successive subdivisions of the Gondwána series, that the period of time which elapsed between the different groups was of small amount. No notice would, in all probability, have been taken of the distinctions between the different beds at Ellore, but for the fossils, and many similar subdivisions might be practicable in such groups as the Kámthí or Pachmarhí if the strata were fossiliferous.

South of the Kistna river Mr. Foote has detected a threefold division of the Gondwána beds, similar to that of the Ellore region, and has distinguished the groups seen near Ongole as follows:—[1]

3. Pávulur sandstones.
2. Vemávaram shales
1. Budaváda sandstones.

Besides numerous remains of marine organisms, not yet determined, which have been obtained from the two lower groups, the Vemávaram shales have yielded a tolerably rich flora of a type similar to those of the Ragavapuram and Sripermatúr groups. To avoid needless repetition of names and to exhibit more clearly the relationships of the flora of these groups, they have been combined in the tabular statement on page 183, from which it will be seen that nearly half the Ragavapuram species are also found in the Vemávaram beds, and that the relationship of the floras to those of the Rájmahál and Jabalpur groups is very similar in each case. It is, therefore, tolerably certain that the similarity of grouping of the beds in the two areas is not merely accidental, but that the three groups in each region are respectively equivalents of each other.

The only Vemávaram fossil, apart from the plant remains, which has been determined, is a macrurous crustacean regarded by Dr. Feistmantel[2] as

[1] *Memoirs*, XVI, 69, (1880).

[2] *Records*, X, 193, (1877). The locality is there given by mistake as Sripermatúr. The correct locality is given by Mr. Foote in *Memoirs*, XVI, 63, (1879).

probably identical with the liassic *Eryon barroviensis*. There are, however, some important differences which make the specific identity of the two forms doubtful.

The upper Gondwána beds near Madras are divided into two groups,[1] the lower of which has been named from Sripermatúr, a town 25 miles west-south-west of Madras and a well known locality for fossil plants. The group is composed of white shales, containing plants, associated with sandstones, grits and micaceous sandy shales. Conglomerates occur, especially towards the base, where they are coarse and occasionally contain boulders of great size, but all the conglomerates are loose in texture and not compact. A boring recently put down at Place's garden near Madras has penetrated beds of carbonaceous shale, overlapped at the surface, which have raised hopes of finding workable coal. It is not at present known whether these belong to an outlier of the lower Gondwánas or not.

It is in the Sripermatúr shales that the fossils of the group are found. They consist of both animals and plants. The shells are ill preserved and have not been determined. They comprise two or three species of *Ammonites* and several lamellibranch bivalves. The *Cephalopoda* were regarded by Dr. Waagen[2] as resembling neocomian rather than jurassic forms, but the species cannot be determined, owing to the poor state of preservation, too much weight must not be attached to the opinion, though it is of interest in connection with the resemblance between the Rájmahál series and the Uitenhage series of South Africa, now regarded as neocomian in age. The occurrence of *Trigonia smeei* and *T. ventricosa*, both South African Uitenhage species in the outlier north of Coconada, has already been mentioned, and it is not necessary to make any further reference to this subject here, as it will be more fully treated of in the next chapter.

The Sripermatúr group is overlaid by a set of beds of coarse compact conglomerate, with intercalated sandstones and grits, which have been distinguished as the Sattavédu group, from a series of moderately elevated ridges of the same name, lying about thirty-five miles north-west of Madras. Only imperfectly preserved plant remains have been obtained from the Sattavédu beds. The junction with the Sripermatúr beds is ill seen, the groups appear to be conformable, and it is doubtful whether there is sufficient justification for the separation of the upper beds as a separate group.

The upper Gondwánas of the Trichinopoli district occur as narrow outcrops along the western edge of the cretaceous beds, which they separate from the gneissose rocks, being quite unconformable to both. The

[1] *Memoirs* X, 64, (1873).

[2] *Pal. Indica*, series ix, p. 236, (1875).

Tabular Statement showing the distribution of the fossil plants of the Ragavapuram, Vemávaram, Sripermatúr groups, and Utatúr beds of the Rajmahál series.[1]

[† Signifies that the species is also found in the Rájmahál group* in the Jabalpur group.

	Ragavapuram.	Vemávaram.	Sripermatúr.	Utatúr.
FILICES—				
Thinnfeldia subtrigona	...	*	...	...
Dichopteris ellorensis	*	...	...	..
* *Alethopteris whitbyensis*	...	*	*	...
,, *indica*	...	...	*	*
Pecopteris reversa	*	...	*	...
† *Angiopteridium spathulatum*	*	*	*	*
† ,, *macclellandi*	*	*	*	...
† *Macrotæniopteris ovata* (?)	...	...	...	?
CYCADEACEÆ—				
Anomozamites jungens	...	*	...	...
,, *lindleyanus*	...	...	*	...
† ,, *fissus*	...	*	...	...
Pterophyllum footeanum	...	*	*	...
† *Zamites proximus*	...	*	...	...
* *Podozamites lanceolatus*	*	...	...	...
† *Otozamites abbreviatus*	*	*	*	...
,, *rarinervis*	...	*	*	...
,, *bunburyanus*	...	*	*	...
* ,, *hislopi*	...	*	...	...
,, *parallelus*	...	*	...	...
,, *acutifolius*	...	*	...	...
,, *angustatus*	...	...	...	*
† *Ptilophyllum acutifolium*	*	*	*	*
† ,, *cutchense*	...	*	*	*
† *Dictyozamites indicus*	...	*	*	*
Cycadites constrictus	...	*	...	...
CONIFERÆ				
† *Palissya conferta*	...	...	*	*
† ,, *indica*	...	*	...	?
* ,, *jabalpurensis*	...	*	...	...
* *Araucarites cutchensis*	...	*	*	?
* ,, *macropterus*	...	...	*	...
Pachyphyllum peregrinum	...	*	*	...
,, *heterophyllum*	...	*	...	...
† *Echinostrobus rajmahalensis*	...	...	*	...
* ,, *rhombicus*	...	...	*	...
* ,, *expansus*	...	*	...	...
* *Taxites tenerrimus*	*	...	...	...
,, *planus*	*	*	*	...
Gingko crassipes	*	...	*	...

[1] *Pal. Indica*, series ii, I, 199, (1879).

most important of the outcrops is that near Utatúr, and the rocks consist chiefly of soft sandy clays and micaceous shales, with sandstones and a coarse conglomerate of rounded gneiss pebbles at the base.[1]

The Utatúr outcrops are the most southerly known to be of Gondwána age, but Mr. Foote has recorded some exposures of shales and conglomerates in the Madura district which closely resemble the beds of the coastal Gondwánas. No fossils were, however, found, so the identification is not fully established.

Reference has already been made to the resemblance between the floras of the Ragavapuram, Vemávaram, Sripermatúr groups, and of the Utatúr outcrops and, to save repetition, a separate list of fossils has not been given in each case, but the whole united in the tabular statement on the previous page. From this it will be seen that not only are there a certain number of forms common to two or more of the groups, but that in each case the flora is characterised by a large proportion of Rájmahál species, and a much smaller proportion of forms that are only known from the Jabalpur and Umia groups. Apart from the presence of Jabalpur species, and species peculiar to these outliers, their flora is distinguished from the true Rájmahál flora by the absence of broad leaved *Tæniopterideæ*, and the greater abundance of certain forms, such as *Angiopteridium spathulatum* and *Dictyozamites indica*, which though represented in the Rájmahál beds, are found only in a smaller proportion.[2] The presence of the genus *Macrotæniopteris* in the Utatúr beds, and the smaller proportion of purely Jabalpur species, may show that they are nearer in age to the Rájmahál group than the others, but, with this possible exception, we may take the groups as being at any rate approximately of the same age, intermediate between that of the Rájmahál and Jabalpur groups, as is shown in the tabular statement on page 156.

In the Godávari valley the principal representatives of the upper Gondwánas are the Kota and Maléri groups, more commonly referred to as the Kota-Maléri group, the name being derived from those of two villages long known to Indian geologists, the former by the discoveries of fish teeth and fossil fish by Dr. Walker and Dr. Bell[3] in 1851, the latter by the late Revd. S. Hislop's discovery of reptilian bones.[4] The village of Kota is on the left bank of the Pránhíta or Waingangá, about 8 miles

[1] H. F. Blanford, *Memoirs*, IV, 39, (1863); R. B. Foote, *Records*, XI, 247, (1878).

[2] *Pal. Indica*, series ii, I, 199, (1879).

[3] *Quart Jour. Geol. Soc.*, VII, 272, (1851); VIII, 230, (1852); IX, 351, (1853); X, 371, (1854).

[4] *Quart. Jour. Geol. Soc.*, XX, 280, (1864).

above its junction with the Godávari. Maléri is about 32 miles north-west of Sironchá.

The combined group, usually spoken of as a whole, is slightly, but distinctly, unconformable to the underlying Kámthís, and is divided into two subgroups, which were separately mapped by Dr. King.[1] The lower, or Maléri, consists essentially of bright red coloured clays, interbedded with soft, light coloured and open textured sandstones subordinate in thickness to the clays. The fossils are found in the red clays, coprolites being much the most abundant; besides them three species of *Ceratodus* and two genera of reptiles, *Hyperodapedon* and *Parasuchus*, have been distinguished.

The Kota subgroup, which overlies the Maléri, consists principally of coarse, loosely compacted sandstones, with some subsidiary bands of shale, and three very strong bands of limestone, from which all the animal remains have been obtained, the few plants being all derived from sandstone bands.

Though the distinction between these two subgroups is traceable in the field, they are so closely associated that the fauna may be treated as a whole and, as the few fossils of the Maléri subgroup have already been mentioned, there will be no difficulty in separating them in the subjoined list of the fauna and flora of the combined group:—

ANIMALIA.

CRUSTACEA—
Estheria kotahensis.
Candona kotahensis.

INSECTA—
Undetermined—

PISCES—
Lepidotus deccanensis.
„ *longiceps.*
„ *breviceps.*
„ *pachylepis.*
„ *calcaratus.*
Tetragonolepis oldhami.
„ *analis.*

PISCES.—*contd.*
Tetragonolepis rugosus.
Dapedius egertoni.
Ceratodus hunterianus.
„ *hislopianus.*
„ *virapa.*

REPTILIA—
Hyperodapedon huxleyi.
„ sp.
Pachygonia incurvata.
Belodon, sp.
Parasuchus hislopi.
Massospondylus, sp.

PLANTÆ.

FILICES—
Angiopteridium spathulatum.

CYCADEACEÆ—
* † *Ptilophyllum acutifolium.*
Cycadites, sp.

CONIFERÆ.
† *Palissya conferta.*
* „ *jabalpurensis.*
* † „ *indica.*
† *Cheirolepis*, cf. *muensteri.*
* *Araucarites cutchensis.*

[1] *Memoirs*, XVIII, 267, (1881).

In the list an asterisk prefixed to a species shows that it is also known from the Jabalpur group, a dagger that it is found in the Rájmahál group of the Rájmahál hills, or, in the case of *Cheirolepis muensteri*, in the Golapilli beds. With the exception of the last mentioned species, all the plants are found in the Sripermatúr or some of the other groups of the same age on the east coast. We may consequently regard the Kota-Maléri beds as somewhat later in age than the Rájmahál group, and nearly equivalent to the Sripermatúr, though possibly somewhat newer.

The palæontological relations of the animal remains will be treated in the next chapter, and all that need be noted here is the occurrence of the Pánchet form *Pachygonia incurvata*. The animal is believed to be specifically,[1] certainly generically, identical with that of the Pánchet group, and its presence here along with a flora which indicates a much newer age is remarkable.

The discovery of animal remains in the Denwa group of the Sátpuras and south Rewá has been already noticed. In the former area they are represented by scutes and vertebræ of *Parasuchus* and *Mastodonsaurus*, in the latter by remains of two species of *Hyperodapedon* and *Parasuchus*. The material is not sufficient for establishing those specific identifications which would alone allow us to assign the groups to the same horizon, but the resemblances are suggestive and the position of the Denwa group, below the Jabalpur and separated from it by the Bágra group, places it stratigraphically on very much the same horizon as is indicated by the plant fossils of the Kota-Maléri group.

Resting on the Kota-Maléri beds is a group, which was separated by Dr. King[2] under the name of Chikiála, from a village of that name situated close to their boundary, though actually upon the Kota sandstones. They extend along the eastern side of the Gondwána outcrop, with a width of eight to ten miles, for nearly seventy miles, from Rebni in the Wardhá basin to the reach of the Godávari below Enchapalli.

The group is composed of soft sandstones and heavy bands of conglomerates of white quartz pebbles. Clay bands and seams of shale are frequent and the group is very ferruginous, the iron ore is collected and worked into iron to a considerable extent.

The relations of this group to the Kota-Maléri are obscure, but there appears to be a slight unconformity. It was believed by Dr. King, on the ground of its lithological similarity, to represent the Tripetty sandstones of the coastal region—a correlation which is not inconsistent with its relations to the Kota-Maléri group.

[1] Lydekker, *Records*, X, 34, (1877); XV, 25, (1882).

[2] *Memoirs*, XVIII, 140, (1881).

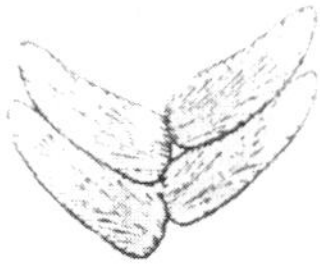

Otozamites gracilis Schimp.

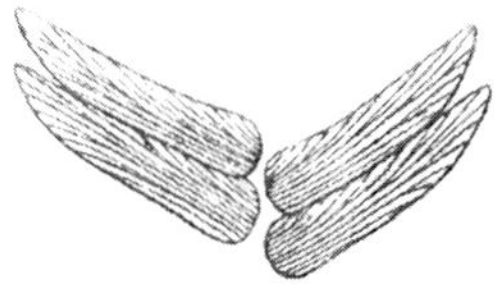

Otozamites hislopi, Oldh.

Brachyphyllum mammillare L. & H.

Brachyphyllum mammillare, L. & H.

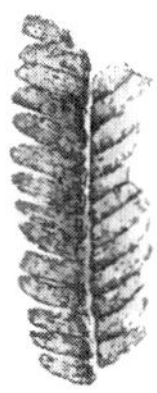

Otozamites gracilis, Schimp.

Otozamites hislopi Oldh.

Palissya indica, Oldh.

Palissya jabalpurensis Fstm,

Alethopteris medlicottiana Oldh.

Podozamites lanceolatus, Schimp

Podozamites lanceolatus
var. spathulatus Fstm.

Calcutta Phototype Co.,

JABALPUR FOSSIL PLANTS.

In the first account of the central portion of the Narbadá valley,[1] a group of rocks was distinguished as 'upper Damuda.' It was, however, pointed out at the time that this group was not only unconformable to the 'lower Damuda,' but that it contained a very different flora. When a true upper Damuda group was subsequently found in the Ráníganj coal-field, it became desirable to distinguish the Narbadá beds by a different name, and as they are well developed in the immediate vicinity of Jabalpur, they have been named from that town.

The Jabalpur group consists of clays, shales, and earthy sandstones, with some thin beds of coal. The clays and soft shales, which are the most characteristic beds of the formation, are pale coloured, usually white, pale lavender grey, or pale red. The sandstones are generally coarse and conglomeratic. Carbonaceous shales are met with in several places, and occasionally one or more thin bands of jet-coal, very different in character from the coal of the Damuda formation. Limestone is rare. At the base of the formation, when resting upon gneissic rocks, there is frequently found a coarse, compact sandstone, so hard and compact as almost to resemble a quartzite. It is often conglomeratic, and the matrix containing the pebbles consists of white earthy rock in a porcellanic condition. Occasionally, but rarely, this bed is calcareous.

The thickness of the Jabalpur group does not appear to have been determined with any accuracy. It is, however, of no great vertical extent, and so far as is known nowhere exceeds 1,000 feet. The relations of the Jabalpur group to the underlying Mahádevas have not been examined in detail, but they appear to be generally conformable.

The following is a list of the fossils found in this group, those found also in the Umia beds of Cutch being marked with an asterisk, whilst those met with in the Rájmahál group are distinguished by a dagger[2]—

FILICES—
- † *Sphenopteris*, cf. *arguta*.
- *Dicksonia*, sp.
- *Alethopteris lobifolia*.
- „ *medlicottiana*.
- „ *whitbyensis*.
- *Macrotæniopteris satpurensis*.
- *Glossopteris*. cf. *communis*.
- *Sagenopteris*, sp.

CYCADEACEÆ—
- *Podozamites lanceolatus*.
- „ *spathulatus*.
- „ *hacketi*.
- *Otozamites hislopi*.
- „ *gracilis*.
- „ *distans*.
- „ *angustatus*.
- *Pterophyllum nerbuddaicum*.

[1] J. G. Medlicott, *Memoirs*, II, 176, (1860). The Jabalpur formation was at this time not clearly distinguished in places from the Mahádevas, the former being supposed to be the lower; in reality the Jabalpur formation is not only newer than the Mahádeva, but it appears to be the latest member of the whole Gondwána series, with the possible exception of the Umia group of Cutch, or of some of the uppermost groups of the east coast. Further accounts of the Jabalpur group will be found in *Records*, IV, 75, (1871); and *Memoirs*, X, 142, (1873).

[2] *Pal. Indica*, series xi, II, 83, (1877).

CYCADEACEÆ—*contd.*

* † *Ptilophyllum cutchense.*
* † " *acutifolium.*
† *Williamsonia*, cf. *gigas.*
Cycadites, cf. *gramineus.*

CONIFERÆ—

* † *Palissya indica.*
" *jabalpurensis.*

CONIFERÆ—*contd.*

* *Araucarites cutchensis.*
Brachyphyllum mammillare.
* *Echinostrobus expansus.*
Echinostrobus rhombicus.
Taxites tenerrimus.
Gingko lobata.
Phönicopsis, sp.
Czekanowskia, sp.

It will be seen that nearly as many Rájmahál as Umia species, five of the former and six of the latter, are found in the Jabalpur group, so far as the flora has hitherto been determined. It should, however, be remembered that the known species of the Rájmahál flora are nearly fifty in number, while those of the Umia flora are much less numerous, about twenty-two.[1] Moreover, the Jabalpur beds are distinguished by a conspicuous want of many of the commonest and most characteristic Rájmahál plants, such as the broad leaved species of *Pterophyllum*.

On the whole, the Jabalpur beds are probably on nearly the same horizon as the Umia beds of Cutch, but possibly represent a period intermediate between the Umia and Rájmahál groups, though nearer to the former. At the same time the circumstance that no representative of the Jabalpur flora has yet been found on the east margin of the Indian Peninsula, to which the Rájmahál flora is confined, suggests that the distinction may be due to the beds having been formed in different botanical regions. Bearing in mind, however, the large amount of evidence which exists to show that the greater part, if not the whole, of India proper was a land area in Gondwána times, this idea of the country having been divided into distinct botanical regions is less probable than the theory of a difference in age between the Rájmahál and Jabalpur groups.

The plant bearing beds of the Umia group in Cutch are only mentioned here because of their relations to the uppermost beds of the Gondwána series. The name Umia is derived from a village about 50 miles northwest of Bhúj, the chief town of Cutch. The group will receive a fuller description under the head of the jurassic formations, and an account will there be given of its mineral character and animal fossils.[2]

The special interest of this group in connection with those just enumerated is due to the fact that beds containing plants, several of which are identical with those of the Jabalpur beds, are interstratified with rocks yielding marine fossils.

[1] Dr. Feistmantel enumerates twenty-eight in his Memoir, but some are only varieties and others stems not identified generically.

[2] *Infra*, p. 223.

The following is a list of the plants from the Umia beds,[1] a dagger indicating species found in the Jabalpur group :—

ALGÆ—
- (?) *Chondrites dichotomus.*

FILICES—
- *Oleandridium vittatum.*
- *Tæniopteris densinervis.*
- †*Alethopteris whitbyensis.*
- *Pecopteris tenera.*
- *Pachypteris specifica.*
- „ *brevipinnata.*
- *Actinopteris*, sp.

CYCADEACEÆ—
- †*Ptilophyllum cutchense.*
- † „ *acutifolium.*
- „ *brachyphyllum.*
- *Otozamites contiguus.*
- „ *imbricatus.*
- „ cf. *goldiæi.*
- *Cycadites cutchensis.*
- *Cycadolepsis pilosa.*
- *Williamsonia blanfordi.*

CONIFERÆ—
- *Palissya bhoojoorensis.*
- † „ cf. *indica.*
- „ cf. *laxa.*
- *Pachyphyllum divaricatum.*
- †*Echinostrobus expansus.*
- †*Araucarites cutchensis.*

At a somewhat lower horizon in the rocks of Cutch, a few plants have been found near a village named Narha, in the northern part of the province, in beds interstratified with the Katrol group, the *Cephalopoda* of which are considered by Dr. Waagen as corresponding to those of the Kimmeridge and Upper Oxford beds of Europe. These plants consist of the following species[2] :—

- *Sphenopteris*, cf. *arguta.*
- *Alethopteris whitbyensis.*
- *Otozamites*, cf. *contiguus.*
- *Araucarites cutchensis.*

The three last are apparently identical with species found in the Umia beds, whilst *Sphenopteris arguta* is an English lower oolite species, found also in the Jabalpur and Rájmahál groups. The *Alethopteris* and *Araucarites* are also Jabalpur forms. This evidence, so far as it goes, tends to show a great persistency in the flora, and it may indicate that the Jabalpur beds are a little older than the Umia group, since the connection of the flora found in the Katrol beds of Narha with that of the Jabalpur group is quite as strong as with the Umia plant fossils.

In northern Káthiáwár a series of soft white and ferruginous sandstones, with pebbly bands, is exposed, of which a few fossil plants have been obtained, which are comprised in the following list,[3] an asterisk and dagger indicating species that are also found in the Umia and Jabalpur groups respectively :—

FILICES—
- * † *Alethopteris whitbyensis.*
- *Pecopteris*, sp.
- *Tæniopteris*, sp.

CYCADEACEÆ—
- † *Podozamites lanceolatus.*
- * † *Ptilophyllum cutchense.*

CONIFERÆ—
- † *Palissya jabalpurensis.*
- † *Taxites tenerrimus.*
- * † *Echinostrobus expansus.*
- * † *Araucarites cutchensis.*

[1] *Pal. Indica*, series xi, II, 63, (1876).

[2] *Memoirs.* IX, 213, (1872); *Pal. Indica*, series xi, II, 80, (1876).

[3] *Memoirs*, XXI, 81-82, (1884).

The Káthiáwár beds have been regarded as the equivalents of the Umia group in Cutch, and the lithological resemblance and geographical proximity are certainly in favour of correlating them with the only group, of the more extended series, which appears to have been deposited under similar conditions, but four of the seven species of plants recognised are found both in the Umia and Jabalpur groups and three in the latter alone. The palæontological relationship is consequently closer with the more geographically remote beds, and if the Káthiáwár sandstones are the equivalents of the Umia group, they indicate a greater approximation in age between the latter and the Jabalpur group than a direct comparison of the two floras would necessarily imply.

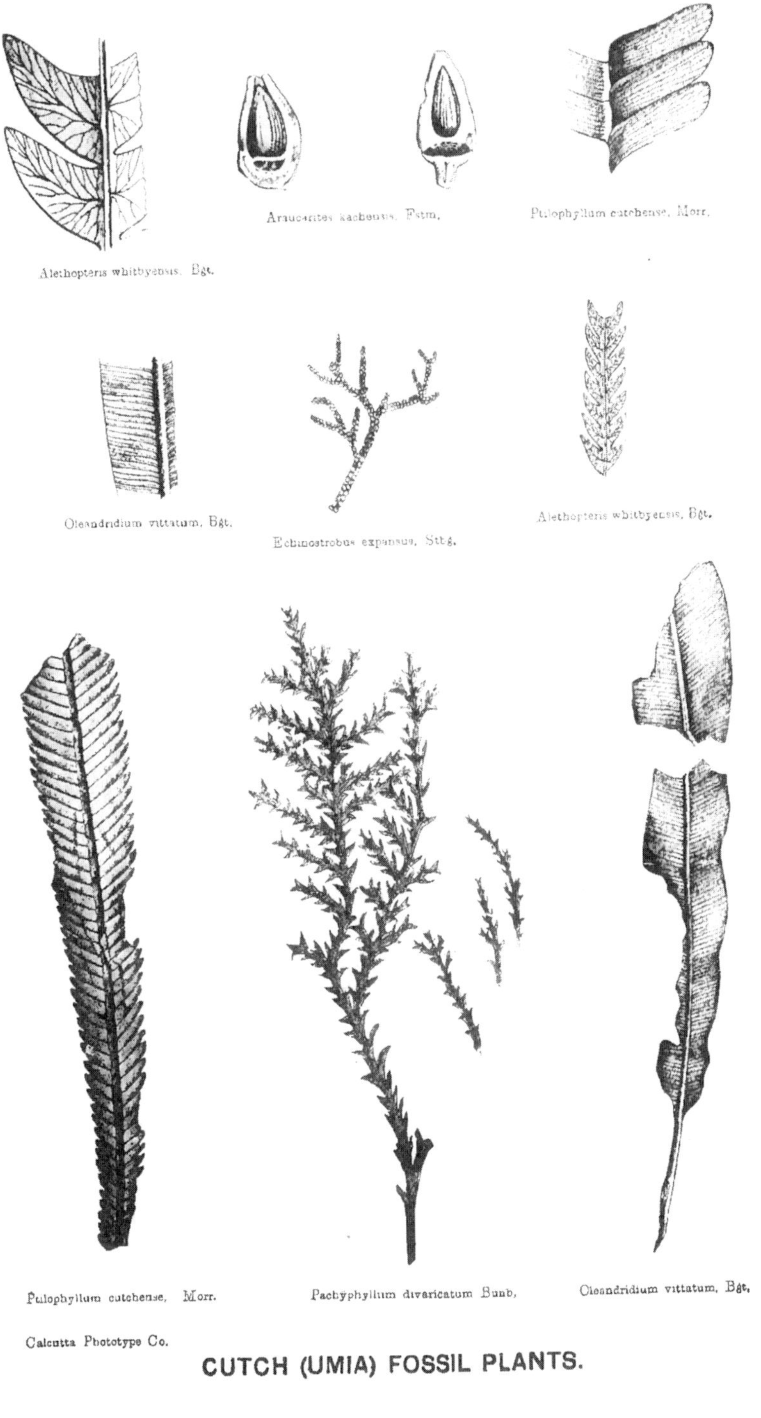

Calcutta Phototype Co.

CUTCH (UMIA) FOSSIL PLANTS.

CHAPTER VIII.

HOMOTAXIS OF THE GONDWÁNA SYSTEM.

Comparison of Gondwána and European floras—Palæontological relations of the animal remains—Asiatic representatives of the Gondwána system—Representatives of the Gondwána system in Australia and in South Africa—Correlation of the rock groups in the four continents—A former land connection between India and Africa—Bearing on the doctrine of the permanence of oceans—and on the hypothesis of secular variations in latitude.

In the preceding chapters all reference to the very interesting question of the age of the Gondwána system, as compared with the geological sequence in Europe, has been purposely omitted. The subject is one whose interest and complexity deserves a special treatment, and it would have been impossible to deal with its bearing on several of the most important problems of theoretical geology that are still unsettled, without interrupting the general description of the stratigraphy of the Gondwána system.

The writer of these pages is fortunate in the fact that the bitter controversy which long raged over this question, is now practically extinct. The history of this controversy would be an interesting one, showing, as it does, how truth is ultimately arrived at by the contact of conflicting opinions, each involving an element of falsehood and each containing a large amount of truth. But the purpose for which this book has been written, and that for which it will principally be consulted, is the statement of the present state of our knowledge, and a recapitulation of the past would be a task as uncongenial in its execution as unprofitable in its result. The main points for which the members of the Geological Survey have contended were accepted at last, even by the talented palæontologist who, alone among his colleagues, disputed them, and the last smouldering embers are mere minor and unimportant differences of opinion as to the exact position of certain individual groups of the Gondwána system.

The most obvious method of determining the age of the rock groups of the Gondwána system would be a direct comparison of the fossils they contain with those of Europe. Unfortunately this method leads to very unsatisfactory and inconclusive results.

If we take the flora of the Damuda series we find, according to Dr.

Feistmantel, the following species identical with, or allied to, European forms[1]:—

PERMIAN—

Macrotæniopteris feddeni, Fstm., allied to *M. abnormis*, Gutg.

TRIAS—

Schizoneura gondwanensis, Fstm., allied to *S. paradoxa*, Schimp.
Merianopteris major, Fstm., allied to *M. angusta*, Heer.

JURA OF RUSSIA AND SIBERIA—

Phyllotheca indica, Bunb., allied to *P. sibirica*, Heer.
P. robusta, Fstm., allied to *P. stschurowskii*, Schmalh.
Cyathea, cf. *tscihatcheffi*, Schmalh., probably identical.
Dicksonia hughesi, Fstm., allied to *D. saportana*, Heer.
Samaropsis, cf. *parvula*, Heer., probably identical.
Rhipidopsis, cf. *gingkoides*, Schmalh., probably identical.

while *Belemnopteris woodmasoniana* is said to closely resemble the living form *Hemionotis cordata* found in Southern India.

If we take the Rájmahál flora the evidence is no less conflicting. The following list gives the relationships with European forms as determined by Dr. Feistmantel[2]:—

PERMIAN—

Macrotæniopteris lata, O. & M., allied to *M. abnormis*, Gutbg.
Pterophyllum carterianum, O. & M., allied to *Pt. blechnoides*, Sandbg.

TRIAS—

Gleichenia bindrabunensis, Schimp., allied to *G. gracilis*, Heer.
Danæopsis rajmahalensis, Fstm., allied to *D. marantacea*.

RHÆTIC—

Equisetum rajmahalense, Schimp., allied to *E. muensteri*, Stbg.
Thinnfeldia salicifolia, O. & M., allied to *T. decurrens*, Schenk.
Alethopteris indica, O. & M., allied to *Asplenites rosserti*, Schenk.
Angiopteridium maclellandi, O. & M., allied to *A. muensteri*, Göpp.
Macrotæniopteris lata, O. & M., allied to *M. gigantea*, Schenk.
" cf. *propinquum*, Göpp., probably identical.
Pterophyllum fissum, Fstm., allied to *P. comptum*, L. & H., and *minus*, L. & H.
" *distans*, Morr., allied to *P. braunianum*, Göpp.
Pterophyllum princeps, O. & M., allied to *P. brauni*, Schenk.
Otozamites, cf. *brevifolia*, Brgt., probably identical.
Palissya indica, Fstm., allied to *P. brauni.*, Endl.
Cheirolepis, cf. *muensteri*, Schimp., probably identical.
" *gracilis*, Fstm., allied to *C. muensteri*, Schimp.

LIAS—

Equisetum rajmahalense, Schimp., allied to *E. liasinum*, Heer.
Angiopteridium macclellandi, O. & M., allied to *A. muensteri*, Göpp.
Cycadites rajmahalensis, Oldh., allied to *C. linearis*, Stbg.

[1] *Pal. Indica*, series xii, III, pt. ii, (1880). [2] *Pal. Indica*, series ii, I, pt. ii, (1877).

OOLITE—

Sphenopteris arguta, L. & H., identical.
Hymenophyllites bunburyanus, Fstm., allied to *Tympanophora racemosa*, L. & H.
Alethopteris indica, O. & M., allied to *A. whitbyensis*, L. & H.
Pterophyllum fissum, Fstm., allied to *P. minus*, L. &. H., a Rhætic species also.
Williamsonia, cf. *gigas*, Carr., probably identical.
Araucarites macropterus, Fstm., allied to *A. brodiei*, Sap.

These two floras were regarded by an eminent palæontologist as indicating a triassic and a liassic age respectively, but the most striking points about them are, *firstly*, the want of definite evidence of a difference of age corresponding to the great stratigraphical break and palæontological contrast between the two groups and, *secondly*, the extremely heterogeneous nature of the alliances exhibited by the flora. For the last of these an explanation will be found in the sequel, but it is also largely due to the absence of any true test of relationship in fossil plants. The shell of a marine mollusc, the test of a crustacean or the cup of a coral give real clues to the zoological position of the animal they once formed parts of, but leaves or, in the case of ferns, fronds, either barren or with the fructification too obscure to be determinable, are all that we have when dealing with fossil plants, and these, which have the least weight in determining the relationships of living plants, are often only imperfectly preserved. When a number of leaves are found, all showing the same shape and venation, there is a considerable probability that they belonged to the same species, but when small differences are observed, which lead to their being classed as belonging to distinct species, there is no certainty that they did not belong to plants widely separated from each other in all important characteristics, while plant remains that are classed under distinct genera or even families may have belonged, not merely to closely allied species but may have formed different parts of one and the same plant. It may consequently result that a plant fossil may most resemble one of a very different age, which possibly would prove to be widely distinct from it did we but know the whole of both.

The alliances of the Pánchet flora are shown in the following table[1]:—

TRIAS—

Schizoneura gondwanensis, Fstm., allied to *S paradoxa*, Schimp.

RHÆTIC—

Pecopteris concinna, Presl., identical.
Cyclopteris pachyrhaca, Göpp., identical.
Oleandridium, cf. *stenoneuron*, Schenk., probably identical.
Thinnfeldia odontopteroides, Morr, allied to *T. rotundata*, Nath.

JURA—

Samaropsis, cf. *parvula*, Heer.

In the last case, however, the resemblance is of little value.

[1] *Pal. Indica*, series xii, III, pt. ii, 51, (1880).

If we pass upwards to the Jabalpur and Umia groups we find a much more homogeneous flora, as is indicated by the following statement of the alliances with European fossil plants :—

ALLIANCES OF THE JABALPUR FLORA.[1]

LIAS—

Otozamites gracilis, Schimp., identical.

LOWER OOLITE—

Sphenopteris, cf. *arguta*, L. & H., probably identical.
Alethopteris lobifolia, Schimp., identical.
" *whitbyensis*, Göpp., identical.
Podozamites lanceolatus, L. & H., identical.
Williamsonia, cf. *gigas*, probably identical.
Cycadites? cf. *gramineus*, Heer., probably identical.
Araucarites cutchensis, Fstm., allied to *A. phillipsi*, Carr.
Brachyphyllum mammillare, L. & H., identical.
Echinostrobus expansus, Schimp., identical.

ALLIANCES OF THE UMIA FLORA.[2]

RHÆTIC—

Oleandridium vittatum, Schimp., identical.
Actincpteris, sp. allied to *A. peltata*, Schenk.

LOWER OOLITE—

Oleandridium vittatum, Schimp., identical.
Alethopteris whitbyensis, Göpp., identical.
Pachypteris specifica, Fstm., allied to *P. lanceolata*, Brgt.
Otozamites, cf. *goldiæi*, Brgt., probably identical.
" *imbricatus*, Fstm., allied to *O. brongniarti*, Sap.
Cycadites cutchensis, Fstm., allied to *O. zamioides*, Leck.
Williamsonia blanfordi, Fstm., allied to *W. sp.* Carr.
Taxites cf. *laxus*, Phill., probably identical.
Pachyphyllum divaricatum, Bunb., identical.
Echinostrobus expansus, Stbg., identical.
Araucarites cutchensis, Fstm., allied to *A. brodiei*, Carr.

UPPER JURA (Kimmeridge ?)—

Gycadolepsis pilosa, Fstm., allied to *C. hirta*, Sap.

Reviewing the evidence of the Gondwána plants, we find that the two most important of the floras, those of the Damuda and Rájmahál series, do not show a definite relation to any single horizon of the European sequence, nor do they show any distinct evidence of a difference of age. The flora of the Pánchet group has a much more defined relationship to the rhætic, and the beds might have been referred to this age on the evidence of the plants alone, were there not other considerations, to be detailed below pointing to an older date. The Jabalpur and Umia floras show a still

[1] *Pal. Indica*, series xii, II, 83, (1877). | [2] *Pal. Indica*, series xii, II, 63, (1876).

greater definiteness of relationship, in this case to the lower oolite, and though the latter group was described as newer than the former, the difference in age of the two is possibly not so great as to introduce any difficulty. The Cutch plant beds are, however, found resting on marine deposits whose *Cephalopoda* show that they are of uppermost oolitic age, and are overlaid conformably by beds containing upper neocomian ammonites;[1] there is consequently a direct conflict here between the evidence of the marine mollusca and the fossil plants, and the question of which is to be preferred arises.

This will not be discussed here in detail, but the explanation is to be found in the diversity of the forms of terrestrial life inhabiting distant regions of the earth at the present day. There is a much greater difference between the terrestrial faunas and floras of Africa, Australia and America than between the animals inhabiting the Atlantic, Indian and Pacific oceans, and it is a common circumstance to discover fossil remains of animals and plants, without any living representatives in neighbouring lands, but allied to forms still living in a distant region. Such was also the case during the Gondwána epoch, and, as will appear, the distinctions at its commencement were even more trenchant than at the present day.

If we turn from the plants to the animal remains found in the Gondwána system, the evidence is little less ambiguous. The *Gondwanosaurus* from the Bijori group belongs to the family *Archegosauridæ*, which in Europe is principally carboniferous and permian, though a specialised form ranges into the trias. The affinities of the Indian specimen are said to be permian. On the other hand, *Brachyops laticeps*, from the Mángli group, which is believed to be of about the same age as the Bijori group, is in Europe only allied to *Rhinosaurus*, a jurassic form.[2] Its nearest allies are to be found in the Karoo beds of South Africa.

The reptiles of the Pánchet group exhibit but little connection with European forms, and their connection with the South African fauna will be noticed further on.

In the Kota-Maléri and Denwa groups the genera *Belodon*, *Hyperodapedon*, and *Mastodonsaurus* are all represented in the upper trias of Europe, the first and last being also known from rhætic beds. *Parasuchus* belongs to the same typically mesozoic group of crocodiles with biconcave vertebræ as *Belodon*, and is placed with it, by Prof. Huxley,[3] in a section of the family which is almost confined to triassic rocks in Europe. Of the fishes represented in the Kota-Maléri group the genus *Lepidotus* ranges from the lias to the lower chalk, and the Kota species were regarded

[1] *Infra* p. 286.
[2] *Pal. Indica*, series iv, I, 13, (1885).
[3] *Quart. Jour. Geol. Soc.*, XXXI, 427, (1875).

by Sir P. Egerton,[1] as showing liassic or oolitic affinities, *Tetragonolepis* is only known from liassic beds, and *Dapedius* is also a liassic genus. *Ceratodus* is principally triassic, but species have been found in beds of later date, and the genus is still living in Australia.

Here the fauna of the Maléri group indicates an earlier age than that of the Kota, and in this agrees with the relative stratigraphical position of the two groups, but the fossils indicate a much greater difference in the age of the two rock groups than their intimate stratigraphical association suggests, and we have very much the same palæontological contradiction as there is between the land plants and marine animals of the Cutch jurassics.

From the coalfields of Tongking a fossil flora has been described[2] which contains certain Gondwána forms, and exhibits a much closer relation to a definite European horizon than any of the groups of that system. Out of a total of nineteen species, ten are found in the rhætic beds of Europe, and eight of the remainder in India; of the latter *Phyllotheca indica*, *Palæovittaria kurzi*, *Macrotæniopteris feddeni*, *Glossopteris browniana*, and *Noegerathiopsis hislopi*, are Damuda forms, and the Rájmahál and Sripermatúr groups are represented by *Angiopteridium spathulatum*, *Tæniopteris ensis* and *Otozamites rarinervis*. Here the evidence, so far as it goes, is distinctly in favour of regarding the Tongking beds as rhætic in age and intermediate between the Damuda and Rájmahál series in India, or more or less contemporaneous with the Pánchets, whose flora has also a rhætic facies.

Outside the limits of India proper, in north-western Afghánistán, a series of coal bearing sandstones, intercalated with marine beds, and having at its base a boulder bed precisely similar in character to that of the Talchirs has been described by Mr. Griesbach.[3] The general classification of the beds as adopted by him is as follows :—

Age.	Formation.	Localities.
Jurassic . . .	Densely red grits and sandstone, shales with plant remains.	Upper Almar stream near Pain-guzar; Astar-ab below Paisnáh. Khorak-i-Bala, north of the Kara Koh.
	Trap.	
	Dark bluish grey grits and sandstone; plant remains.	
	Ash-beds.	
	Sandstone and black alum shales with plant impressions; marine fossils.	Doáb north of the Kara Kotal.

[1] *Pal. Indica*, series iv, I, 2, (1875), "Ganoid Fishes from the Deccan."

[2] R. Zeiller: "Examen de la flore fossile des couches de charbon du Tongking, *Annales des Mines*, 8th series, II, 299, (1882). The details in the text are taken from *Pal. Indica*, series xii, IV, Introduction, pp. xv-xvii, (1886).

[3] *Records*, XIX, 239, (1816). The uppermost red grits were afterwards said to be neocomian. *Records*, XX, 94, (1887).

Age.	Formation.	Localities.
Upper Trias or Rhætic.	Light coloured sandstones and shales with coal seams.	Kotal-i-Sabz (north slope of Kara Koh), Shisha Alang.
Upper Trias — Upper.	Great thickness of marine sandstone, limestone, and shales with coal seams. *Schizoneura* sp., etc.	Chahil; Shisha Alang.
Upper Trias — Middle.	Brown sandstones and shales with coal seams. *Equisetites columnaris.*	Chahil, north slope of Kotal-i-Sabz.
Upper Trias — Lower.	Marine sandstone and limestone beds. *Halobia lommeli*, *Monotis salinaria.*	Chahil.
Permo-Carbon.	Altered shales (mica-schist, etc.) with graphitic and anthracitic seams. Clay shales with impure coal. The whole traversed by hornblendic granite.	Saighán; Ak Robát Kotal, north.
	Coarse conglomerate in greenish matrix, altered by granite.	Palú Kotal and gorge; Ak Robát.
	Massive dark limestone with brachiopod casts.	Ditto ditto ditto.

There can be little doubt that these beds are the equivalents of the Gondwána system of the Indian Peninsula and will be of the greatest importance in determining the age of the various members of that system, when they have been more fully studied. At present the subdivisions of the series of beds seen in Afghánistán cannot be correlated with those of the Gondwánas in detail, while as regards the correlation of the beds with the European sequence, it must be remembered that no fossils from this area have been critically determined, and the correlation, depending merely on a field determination of two or three species, may be upset by a fuller study of the fauna and flora as a whole. It is necessary to bear this in mind as the horizon indicated by Mr. Griesbach, for the coal bearing beds of Afghánistán, is higher than that which we shall have cause to regard as the horizon of the Barakar or Rániganj groups, their probable equivalents in India. The country in which these Afghán Gondwánas are developed is not open to detailed examination by Europeans, such information as is available having been obtained during rapid journeys through the country, and until a closer and more detailed examination of them can be made they have not the importance, from the present point of view, that their proximity to India, and the occurrence of beds containing marine fossils, intercalated with those which contain the fossil plants, would otherwise give them.

If, instead of looking to the west or the east, we turn to the south, we will find in Australia a series of beds which clears up the vexed question of the homotaxis of the Gondwánas in a wonderful manner.

From Bacchus marsh in Victoria three species of *Gangamopteris* have been obtained, but of this very limited fauna one species is identical with, and the other two allied to, Karharbári forms. This would in itself suggest a correlation of the Bacchus marsh with the Karharbári beds, and there is further evidence in the presence of large blocks of granite and pebbles of rocks that must have travelled long distances, imbedded in a fine grained matrix of mud.[1] Like the Talchir boulder bed, the beds are of glacial origin, and this, combined with the palæontological evidence, justifies us in regarding them as the equivalents of the Indian Talchirs.

By itself this would be of little importance, as the Bacchus marsh exposure is small and the beds cannot be palæontologically connected with any others of known age. But there is in New South Wales a much better and more complete section, which is in fact the standard one for Australian geology, so far as the period we are dealing with is concerned. The sequence has there been divided into the following groups :—

6. Wianamatta shales.
5. Hawkesbury sandstones.
4. Newcastle beds, or upper coal measures.
3. Upper marine beds with carboniferous fauna. }
2. Stony creek beds, or lower coal measures. } Muree beds.
1. Lower marine beds, with carboniferous fauna. }

The marine beds are important on account of the fossils they contain, and because they give us a fairly definite geological horizon to start from. Their equivalence to the glacial beds at the base of the speckled sandstone in the Salt range, and the close alliance of the fauna of these last named with that of the Australian marine carboniferous beds, has already been referred to,[2] and need not be recapitulated here.

The only palæontological evidence bearing on the correlation of the Bacchus marsh beds, with the sequence in New South Wales is the occurrence of *Gangamopteris angustifolia* in the Newcastle beds. But there is weightier evidence of a different character, which renders it certain that the marine beds, and not the overlying Newcastle beds, are the true equivalents of the glacial boulder clays of Bacchus marsh.

The marine beds, and especially those immediately associated with the lower coal measures of Stony Creek are composed of a fine grained matrix of sand or shale, enclosing numerous delicate *Fenestellæ* and bivalve shells with their valves still united, which had lived, died and been tranquilly preserved where they are now found, thus proving, as conclusively as

[1] Report on the geology of the district of Ballan by Richard Daintree, Melbourne, 1866, p. 10.

[2] *Supra*, p. 121.

the texture of the matrix in which they are preserved, that they could never have been exposed to a current of an great rapidity. Scattered through this matrix there are are numerous more or less subangular blocks of stone, of all sizes, ranging up to several feet in diameter, some of which exhibit most characteristically developed glacial striæ. It was not mere velocity of current that brought these fragments, and deposited them where they are, for to move even the smallest of them would require a current that would have swept away the matrix in which they are imbedded, and destroyed the delicate fossils with which they are associated. They must have been floated to their present position and dropped on to the bottom of a tranquil sea, and taking into consideration their abundance, as well as the distinct traces of glacial action that some of them exhibit, the only agency than can be appealed to is that of floating icebergs.[1]

We find then that the marine carboniferous deposits of New South Wales were formed during a period of exceptionally cold climate, and it is to the latter that we must look for an equivalent of the glacial beds of Bacchus marsh rather than to the overlying Newcastle beds, which indicate a more temperate climate at the time of their formation.

The lower coal measures consist of a comparatively thin band of sandstones and coal, intercalated between the lower and upper marine beds. The fact of this intercalation has been questioned, on account of the supposed mesozoic age of the flora obtained from them. But no one who had actually examined them in the field doubted the intercalation, and it is now too well established to be questioned. Under these circumstances it is important to see what are the plants which co-existed with a marine fauna of carboniferous type. The following is the list given by Dr. Feistmantel[2] :—

Phyllotheca australis.
Annularia australis.
Glossopteris browniana.
„ *clarkei.*
Glossopteris elegans.
„ *primæva.*
Nœggerathiopsis prisca.

At a glance the flora can be seen to be of the type of the lower Gondwánas, in India, and to differ totally from the European flora of corresponding age. The same alliance with the lower Gondwána flora is to be seen more conspicuously in the following list of the plants of the Newcastle beds:—

Phyllotheca australis.
Vertebraria australis.
Sphenopteris alata.
„ *flexuosa.*

[1] *Records*, XIX, 39, (1886); *Quart. Jour. Geol. Soc.*, XLIII, 190, (1887)

[2] O. Feistmantel, *Palæontographica*, supplement, 1878-79. A resumé—, with lists of fossils and notices of previous literature, is given by the same author in *Sitzungsber. K. böhm. Ges. Wiss.* 1887, pp. 55-77, and in *Jour. Roy. Soc. New South Wales*, XIV, 103, (1881).

Sphenopteris germana.	*Glossopteris reticulum.*
" *hastata.*	" *tæniopteroides.*
" *lobifolia.*	" *wilkinsoni.*
" *plumosa.*	*Gangamopteris angustifolia.*
Glossopteris browniana.	" *clarkeana.*
" *ampla.*	*Caulopteris* (?) *adamsi.*
" *cordata.*	*Zeugophyllites elongatus.*
" *elongata.*	*Nöggerathiopsis media.*
" *linearis.*	" *spathulata.*
" *parallela.*	*Brachyphyllum australe.*

Of these, *Gangamopteris angustifolia* and *Glossopteris browniana* are also found in India, the former in the Talchir and Karharbári groups, the latter in the Damuda series. Three species of *Glossopteris*, *G. linearis*, *G. ampla*, *G. parallela*, are represented in the Damuda flora by the allied forms *G. angustifolia*, *G. communis* and *G. damudica*. The Australian *Phyllotheca* is very closely allied to, and has been considered identical with, *P. indica* of the Damudas, *Vertebraria*, which is common in both the floras, is only known elsewhere by a distantly related species from the jurassic of Siberia. These specific relationships between the Damuda and Newcastle flora are strengthened by the general resemblance in the type of the floras as a whole. All the principal and more characteristic genera of the Newcastle beds are represented in both floras, and the genus *Glossopteris* in both cases includes about one-third of the total number of species.

The palæontological evidence would of itself be almost sufficient to justify the correlation of the Newcastle and Damuda series, and when we bear in mind that their position, relative to the underlying glacial beds, is similar in both cases, it becomes certain that the two coal bearing series must be more or less completely the equivalents of each other.

Above the Newcastle comes the Hawkesbury group of sandstones and shales. The few fossils that have been obtained from this and the overlying Wianamatta group are scarcely sufficient to establish their correlation with any particular division of the Gondwána system, none of the species being found in India except *Thinnfeldia odontopteroides*, a Pánchet form, but there is evidence of a recurrence of glacial conditions in the Hawkesbury beds which is worth noting. Large angular fragments of shale, similar to that interbedded with the sandstones, are found imbedded in a confused manner, with their original bedding planes lying at all angles, in a matrix of sand. They occur nearly always immediately above the shale beds, and are accompanied by well rounded quartz pebbles.[1] It is difficult to account for the facts that have been described without the agency of ice, in one form or another, but they are in no way comparable with the proofs

[1] C. S. Wilkinson, *Jour. Roy. Soc. New South Wales*, XIII, 105, (1880).

of glacial action exhibited by the marine beds. The evidence indicates the action of winter ice rather than of actual glaciers.

No evidence of the recurrence of glacial conditions has been recorded in India, but there is an indication of a return of cold in the undecomposed felspar, found in the sandstones of the Pánchet group. Undecomposed felspar is characteristic of the sandstones associated with the boulder beds of the Talchirs, and Prof. Green[1] has remarked on the extreme freshness of the felspar in the glacial beds of the same age in South Africa. The beds of glacial origin in the Indus valley, which have been supposed to be eocene[2] also contain an abundance of fragments of undecomposed felspar. Apart from these observations there is an inherent probability that sandstones containing undecomposed felspar would be found in cold climates. They mean that the disintegration of the parent rock from which the material was derived, together with the transport and final accumulation of the debris, went on at a greater rate than chemical decomposition of the constituent minerals, and this might be due either to extreme dryness, which would retard the rate of decomposition, or to an extreme severity of climate, which would accelerate the rate of disintegration.

Taking these considerations into account, it may well be that the undecomposed felspar of the Pánchet sandstones indicates a recurrence of a cold period, less severe than that of the Talchir, and comparable to that of the Hawkesbury group in New South Wales. Quite independent of this the stratigraphical position of the two groups would suggest their correlation, and the flora, which contains a mixture of purely lower Gondwána with upper Gondwána genera, is consistent with this conclusion.

Newer than the Hawkesbury beds a number of groups of plant bearing sandstones are found in different parts of eastern Australia, which probably represent the upper Gondwánas of India. Their relations to each other have not been fully worked out, and the recorded fossils indicate an admixture of specimens obtained from older beds. In any case they are of little importance in the present connection, and will not be further noticed.

In South Africa we again find a representative of the Gondwána flora which, though perhaps less valuable for the purposes of establishing the age of the Indian formation than the Australian beds, is otherwise of great interest and importance. The Karoo series, as it is now generally called, consists of a thickness of many thousands of feet of sandstones and shales, with interbedded coal seams. Like the Gondwánas, they were at first regarded as lacustrine in their origin, but are now looked upon by many geologists as having been deposited by rivers. Whether lacustrine or river deposits,

[1] *Quart. Jour. Geol. Soc.*, XLIV, 244, (1888). | [2] *Infra*, p. 346.

they were certainly formed by fresh water, and the coal resembles that of the Damudas in its laminated structure and in the absence of an underclay

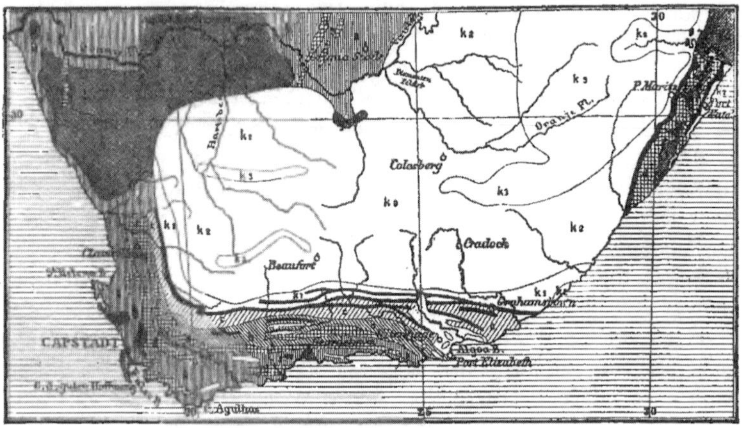

Fig. 13.—Geological sketch of South Africa.

G, granite and gneiss. *S*, silurian. *B*, porphyritic intrusions. *C*, carboniferous. *Black*, Ecca-conglomerate. *K*, Karoo formations and overlying trap. *White* (in the neighbourhood of Algoa Bay and Cape Agulhas) Uitenhage and younger marine deposits.

or even roots penetrating the underlying beds. More important than these mineralogical resemblances are the similarities shown by the fossil flora and fauna of the two countries, as will be described immediately. But, first, it will be well to notice briefly the distribution and relation to older rocks of the beds we are concerned with.

The Karoo formation occupies a very large tract of dry lands, in the interior of South Africa, forming the head waters of the Orange river and its southern affluents, and of the principal rivers which issue on the south coast. Northwards its extension has not been worked out, but rocks of a similar character have been observed in central Africa. Details of the palæontology of these regions, which alone would render them of importance in the present connection, are however wanting. Over the whole of the area in South Africa occupied by the Karoo system, the beds lie horizontally in general, except for slight disturbance of the lower beds along the southern margin, and form large, flat, desert, plains, known as Karoo, from which the system derives its name.

Between the Great Fish and St. John's rivers the Karoo beds extend to the east coast, but elsewhere they are separated from it by a series of ranges formed of more or less disturbed palæozoic beds. Among these the Bokkeveld beds contain a fauna of distinctly Devonian facies, overlaid by plant bearing beds, the Witteberge and Zuurberge quartzites, containing a flora allied to the carboniferous of Europe.

The subdivisions of the Karoo system have received a variety of names at different times from different authors, but generally it may be divided into three divisions as follows:—

3. Upper or Stormberg beds.
2. Middle or Beaufort beds.
1. Lower or Ecca and Koonap beds.

In dealing with the palæontology of this system we have the great advantage that the latest, and most complete, description of the plants is by Dr. Feistmantel,[1] the same distinguished palæontologist who examined and described the floras of India and Australia. According to him the specimens from the lower, or Ecca, group comprised the following species:

D. *Glossopteris browniana.*
K. *Gangamopteris cyclopteroides.* (a variety).
K. *Noeggerathiopsis hislopi.*

The middle or Beaufort beds yielded—

Schizoneura (?) *africana.*
Phyllotheca (?)
D. *Glossopteris browniana.*
D. „ *angustifolia.*
D. „ *communis.*
D. *Glossopteris stricta.*
D. „ *retifera.*
K. D. „ *damudica* (a variety).
Rubidgea mackayi, Tate.

the last being, according to Dr. Feistmantel, very probably the same as *Palæorittaria kurzi*, Fstm, from the Damudas.

In these lists the letter K signifies that the species occurs in the Karharbari group, D that it is found in the Damuda series, and a glance at the list will show that, with the exception of one doubtful species, all those recognisable are identical with Indian species, and there can be no danger in correlating the Ecca and Beaufort beds with the Talchir and Damuda.

Apart from the palæontology of these beds, the lithological character of the lower or Ecca group would suggest its correlation with the Talchirs, for it contains a boulder bed, composed of blocks of stone of various sizes, imbedded in a fine grained matrix, precisely similar to that of the Talchirs, and now generally regarded as owing its origin to glacial action. The Ecca group also resembles the Talchirs in containing beds which closely resemble volcanic traps or ashes.[2] There can be no doubt that these glacial beds were formed during the same cold period which has left such conspicuous traces throughout India and Australia, and, taken in conjunction with the fossils, they leave no room for doubt that the Ecca beds and

[1] Uebersichtliche Darstellung der geologisch-palæontologischen Verhältnisse sud Afrikas I Theil; Die Karooformation und die dieselbe unterlagernden schichten. *Abhand. K. böhm. Ges. Wiss.*, VII, Band 3, (1889).

[2] A.H.Green, *Quart. Jour. Geol. Soc.*, XLIV, 244, (1888).

Talchirs were deposited contemporaneously. The Devonian fauna, and carboniferous flora, underlying them have been already referred to, and as far as can be judged from the published description, there is no great discordance, implying a long lapse of time, between the beds in which they are found and the base of the Karoo system.

Besides the plants, and much better known than them, are the numerous reptilian remains, which have been found in the Beaufort beds. These, too, show a very distinct connection with the much more limited lower Gondwána fauna. The Mángli labyrinthondont, *Brachyops laticeps*, Owen is closely related to *Micropholis stowii*, Hux., from the Beaufort beds. The aberrant genus *Dicynodon* is represented by no less than thirteen species in South Africa, and is not known elsewhere except from the Pánchet group, and from reputed triassic beds in North America, while *Ptychosiagum orientale* resembles *P. declive*, Owen, so closely that the specific distinction is difficult.[1]

The age of the Beaufort beds, as compared with the European sequence, cannot be said to be definitely established, the opinions expressed by different palæontologists, and by the same one at different times, having varied, but the general consensus appears to be that the reptilian fauna represents a triassic age. They have generally been regarded as the equivalents of the Pánchet group, on the strength of the resemblances between the reptiles of the two, but the conjecture is open to question. The genus *Dicynodon* ranges into the next succeeding rock group, which also contains one plant identical with a Pánchet species, and agrees better in stratigraphical position than the underlying Beaufort group.

The uppermost group, of the Karoo system, again contains reptilian remains, among which is one species of *Dicynodon*, and the following plants, besides some undetermined equisetaceous stems :—

A. *Sphenopteris elongata.*	*Anthrophyosis*, sp.
A. *Thinnfeldia odontopteroides.*	*Alethopteris*, sp.
A. „ *trilobata* (?).	A. *Podozamites elongatus.*
A. *Tæniopteris carruthersi.*	„ sp.
A. „ *daintreei.*	*Baiera schencki.*

In this list the letter A denotes that the species has been found in Australia, in the so called mesozoic beds of Victoria, Tasmania or Queensland, whose exact position in the New South Wales sequence has not been fully established. The two species of *Sphenopteris* have been found in South America, and only one species, *Thinnfeldia odontopteroides*, a Pánchet form, has been found in India. The probability that these beds represent the Pánchet group has just been referred to, and the presence of this plant makes the suggestion more probable.

[1] R. Lydekker, *Records*, XXIII, 19, (1890).

It will be seen that the flora of the Stormberg beds is as distinctly allied to the Australian one as those of the lower groups are Indian, but Indian affinities reappear in the next succeeding rock group.

In the neighbourhood of Algoa bay a set of fossiliferous shales and sandstones, known as the Uitenhage series, of no great thickness and not found far from the coast, bears much the same relation to the Karoo system of the interior, as do the coastal outliers of the Rájmahál series to the lower Gondwánas of the interior of the Peninsula. Like these, they are of marine origin and have yielded marine fossils, which were at one time believed to indicate a lower oolitic age,[1] but have since been shown to be neocomian,[2] at least so far as the upper beds are concerned.

The plant remains that have been described appear to have been procured from the lower beds of the series.[3] Unfortunately, Dr. Feistmantel did not live to complete the description of the plants of this series, and we have only the earlier description of a more limited collection by Prof. Tate, of which the following is a list[4] :—

FILICES—
Pecopteris atherstonei.
" *rubidgei.*
" *africana.*
" *lobata.*
Sphenopteris antipodum.
Cyclopteris jenkinsiana.

CYCADEACEÆ—
Palæosamia (*Otozamites*) *recta.*
" (*Podozamites*) *morrisii.*
" *rubidgei.*
" (vel *Pterophyllum*) *africana.*

CONIFERÆ—*Arthrotaxites.*

Of these, *Pecopteris lobata* is a Rájmahál form, *P. atherstonei* and *P. rubidgei* are allied to *Alethopteris indica*. *Cyclopteris jenkinsiana* is allied to *C. oldhami*, and the *Arthrotaxites* is very like the Rájmahál *Echinostrobus indicus*.

The alliances of the Cycads are vague and appear to be rather with European lower oolite than with Rájmahál forms, but *Pecopteris lobata* appears to be the commonest plant, and *C. jenkinsiana* is also abundant. On the whole, however, the flora is distinctly related to that of the Rájmahál group, though the resemblances are not sufficient to establish a contemporaneity of origin.

Having detailed the facts, so far as they bear on the subject in hand, we can now proceed to the discussion of the age of the different rock groups.

[1] R. Tate, *Quart. Jour. Geol. Soc.*, XXIII, 169, (1867).

[2] Holub, u. Neumayr, *Denks., k. k. Akad. Wien.*, XLIV, 267, (1882).

[3] *Quart. Jour. Geol. Soc.*, XXIII, 147, (1867). See also G. W. Stow, *Quart. Jour. Geol. Soc.*, XXVII, 497, (1871).

[4] *Quart. Jour. Geol. Soc.*, XXIII, 144, (1867).

To begin with, we find, in Africa, India, and Australia alike, certain beds containing abundant and conspicuous traces of glacial action. The plant remains show that in South Africa, in the Indian Peninsula, and in Victoria these are of approximately the same age, and marine fossils show the same with regard to the beds in New South Wales and the Salt range. The deposits in every case were formed during a period of great cold, which was succeeded by a much more temperate climate, and it is almost impossible to doubt that this wide spread change of climate must have been due to some far reaching, if not cosmic, cause. It is consequently justifiable to use these glacial deposits for the purpose of correlation, and to conclude that the boulder beds of the three continents were formed contemporaneously.

In this way we at once find in the marine fossils of New South Wales and the Salt-range a means for determining the homotaxis of the Talchir group. The former were once regarded as lower carboniferous in age, but the bulk of the lower carboniferous species they contain range through the whole of the epoch. The absence of the group of *Productus giganteus*, and the presence of the genus *Strophalosia*, point to a newer horizon, and they are now looked upon as upper carboniferous or somewhat newer.[1] The reasons for ascribing a similar age to the Salt-range boulder beds have already been given.

The Barakar group in India and the Beaufort beds in South Africa agree so closely in their stratigraphical relations to the glacial boulder beds, and in their fossil plants, that they are clearly equivalent to each other. In the case of the Newcastle beds of Victoria, though the palæontological agreement with the Barakar group is less close, the stratigraphical relations are equally intimate, and there can be little doubt that there is no great divergence in the homotaxis of the two groups. The stratigraphical connection, between the Newcastle beds and the underlying, marine carboniferous deposits, is too close to allow of any great interval of time, and the Newcastle beds and Barakars cannot well be newer than permian. In the Salt range a similar stratigraphical position is occupied by the lower division of the Productus beds which are separated from the boulder beds of Talchir age by the speckled sandstones. If not the equivalent of the Barakar group, they cannot be much older. Their age has already been established as corresponding to the permo-carboniferous of the Ural mountains, and it is noteworthy that the flora of this age in eastern Russia is as essentially and typically palæozoic as the contemporaneous Barakar or Karharbári flora is mesozoic.[2]

[1] W. Waagen, *Jahrb. K. K. Geol. Reichs. Wien*, XXXVII, 163, (1887); *Records*, XXI, 106, (1888). See also *Pal. Indica*, series xiii, IV, 153, (1890).

[2] See *Pal. Indica*, series xiii, IV, 175, (1890).

As we ascend the sequence the evidence gets less satisfactory. The probable equivalence of the Pánchet group with the Hawkesbury beds of Victoria and the Stormberg beds of South Africa has already been referred to. The fauna of the Beaufort beds, once regarded as permian, is now more usually looked on as triassic, but the evidence is not conclusive. In the Salt range no certain equivalent of the Pánchets can be detected, though it is possible that the base change of fauna at the top of the Ceratite beds was due to a change of ocean currents caused by, or the cause of, that change of climate which is indicated in the Pánchet and Hawkesbury beds. This suggestion would harmonise with the conclusions that have been drawn from the study of the animal remains of the Pánchet group.

The age of the Rájmahál group must remain uncertain till the marine fossils of the outliers on the east coast have been determined. It is older than the Tripetty sandstones from which *Trigonia smeei* and *T. ventricosa* were obtained, both of which occur in the Umia group of Cutch, and the last named also in the neocomian beds of the Uitenhage series. The cephalopoda of the Sripermatúr group were believed by Dr. Waagen to have a neocomian facies, but the specimens are too ill preserved for very great value to be attached to this determination.

Indirectly we can form some sort of a guess at the age of the Rájmahál group, for the plant beds of Cutch overlie beds containing a marine fauna which represents an upper oolitic (portlandian) horizon, and underlie a bed of ferruginous oolite of neocomian or aptian age. The lower oolitic facies of the flora has been mentioned, but in view of the uncertainty that attaches to palæobotanical evidence, when large distances intervene, and the distinctness of that afforded in the present case by the marine fossils, it is impossible to regard the Cutch plant beds as older than upper oolite while they may verge into the neocomian period. The Jabalpur group, which is closely related to the Cutch plant beds, becomes, consequently, middle oolite at the oldest, and the Rájmahál series ranges backwards from that, throwing the Rájmahál group of the Rájmahál hills into the lower oolite or even the lias.

Taking everything into consideration, we may then accept the correlation indicated below as approximately representing the true interpretation of known facts. Two important reservations must, however, be made in this connection. In the first place the suggestion made by Mr. H. F. Blanford[1] in 1875, that the Talchir boulder bed was contemporaneous with the permian glacial deposits of England, has never been absolutely disproved, and as recent investigations have shown that the supposed lower carboniferous deposits of Australia are newer than they were formerly

[1] *Quart. Jour. Geol. Soc.*, XXXI, 528, (1875).

Tabular Statement of the probable equivalence of the Upper Palæozoic and Lower Mesozoic rocks of India, Africa and Australia.

Europe.	India.	New South Wales.	South Africa.
Upper Oolite . .	Umia.		
Middle Oolite . .	Jabalpur.		
Lower Oolite . .	Kota-Maléri.		
Lias . . .	Rajmáhál.		
Rhætic . . .	Mahádeva . .	Wianamatta.	
Trias . . .	Pánchet . .	Hawkesbury . .	Stormberg.
Permian . .	Rániganj . }	Newcastle . .	Beausfort.
Permo-Carboniferous	Barakar . }		
Upper Carboniferous	Talchir . .	Lower coal measures and associated marine beds.	Ecca.

considered to be, it is still possible that this may be the true equivalence, in which case the Pánchets might be of rhætic age, as their flora indicates. In the second place the correlation of the Rájmahál group with the lias is open to question. Dr. Waagen's statement regarding the neocomian facies of the Sripermatúr cephalopoda cannot in itself carry any weight, on account of the very imperfect and ill preserved material he had to deal with, but, taken in conjunction with the palæontological connection of the Rájmahál flora with that of the lower beds of the Uitenhage series, whose upper beds are now regarded as neocomian, it is strongly suggestive of a later age for the Rájmahál beds than is indicated in the table. The palæontological grounds on which the Rájmahál is considered older than the Jabalpur group have been referred to in the last chapter, where it was shown that, looked at from a purely local point of view, it is more probable that the difference between the two floras, and their admixture in the Kota-Maléri group and the upper members of the outliers on the east coast, is due to a difference in age than to a mere difference of situation These reasons are not, however, conclusive, and it seems possible that an examination of the marine fauna of the east coast outliers would show that the Rájmahál beds are newer than has been supposed, and that the differences between the Rájmahál, Jabalpur, and Cutch floras do not indicate successive periods of time so much as divergent conditions of soil or climate, existing in different parts of the continent on which the beds of the Gondwána system were deposited. Further, there are grounds for taking the Umia group out of the jurassic system and placing it at the base of the cretaceous, since its stratigraphical connection with the overlying beds, containing upper neocomian ammonites, is as close as with the underlying Katrol group, whose age is regarded as upper oolite (Oxford). Though it is improbable in either case that the Cutch groups are exactly contemporaneous with the groups they have been referred to, on the strength of their marine fauna, there is ample room between the possible limits of error to allow of the Umia group being either lowermost neocomian or uppermost oolite, and the occurrence of Uitenhage neocomian *Trigoniæ* in the Tripetty beds, which are probably, and the Katrol beds which are certainly, slightly older lends some support to a correlation with newer beds than that adopted.

It will be seen from this that there are grounds for shifting the whole of the Gondwána groups a step higher in the sequence than the positions they occupy in the tabular statement, but the commencement will still remain in the palæozoic era, whence the system ranges throughout the lower half of the mesozoic era well into its upper portion.

Apart from these elements of uncertainty regarding the exact correlations of the different rock groups, the tabular statement does not truly represent their relations of the different rock groups in point of time. It has been necessary to space the rock groups equally, as it is quite impossible to determine what relation the periods they respectively represent may bear to each other, though it is quite certain that they are far from being equal or nearly so. Apart from this, and with the reservations already made, the table may be taken as representing the nearest approach to the truth which is at present possible of attainment, and equivalence adopted as that which appears to agree best with the known facts, taken as a whole.

The comparison of the South African and Indian Gondwána floras is of less importance and interest from the point of view of establishing their homotaxis than as indicating a former distribution of land and sea very different from what now exists. Naturalists have before now appealed to a former land area stretching across what is now the Indian Ocean, to explain certain relationships between the living fauna of the Indian Peninsula on the one hand, and South Africa and Madagascar on the other, and the name Lemuria, given to this suppositious continent, is familiar to many. The hypothesis has of late years been discredited, at least in the form in which it was first propounded, and for the purpose it was originally intended to serve, and the most distinguished of the authors who have treated this subject, Dr. A. R. Wallace, has not only denied the necessity of appealing to any land connection in order to explain the peculiarities in the distribution of living animals, but has declared[1] that the

[1] "Island Life," ed. 1880, p. 418.

fauna and flora of the Mascarene islands, lying between the two continents, is such as to preclude the possibility of their being the remnants of an ancient continent, as was supposed by the believers in Lemuria. Dr. W. T. Blanford has, however, not only shown that the facts are not fully stated in Dr. Wallace's book, but has shown that the actual distribution of certain genera of birds, fishes, reptiles, and land mollusca, are strongly suggestive of a stretch of dry land having formerly extended from Southern India to Madagascar.[1] The question is a complicated one, but even if Dr. Wallace's conclusion is granted, it no ways justifies the much wider inference he has drawn in support of the somewhat popular hypothesis of the permanence of continental and oceanic areas, and is quite consistent with the existence of an Indo-African continent in pretertiary times.

The facts that have been detailed regarding the fauna and flora of the Karoo system show that there is a closeness of relation, amounting to identity and, extending throughout the whole of the Talchir and Damuda periods, which is inexplicable unless there had been a continuous land communication along which the plants could freely migrate between two areas.[2] And the conclusion is vastly strengthened when we remember that throughout the greater part, if not the whole, of this period, a very different type of flora was flourishing in Europe and North America.

Whether the comparative absence of Indian forms in the flora of the Stormberg beds indicates a break up of this land connection and the establishment of free communication with what is now Australia, it is difficult to decide, but if broken up, the presence of the Indian forms in the Uitenhage series suggests that the connection must once more have been established. It is true that only one species is identical in the two areas, but after allowing for the uncertainty of the alliances of fossil species of plants, the connection between the two floras seems to be real, and the differences are such as would naturally follow from a difference in their age.

The indications of a former Indo-African land area do not cease with the Gondwána epoch. From a study of the jurassic fauna of the world, Neumayr came to the conclusion that a land barrier must have stretched from Africa to India during that period, separating two distinct faunas.[3] This conclusion was especially founded on the study of the neocomian fauna of the Uitenhage series, and has lately received a strong confirmation in the identification of four species of *Belemnites* from

[1] *Proc. Geol. Soc. Lond.*, 1890, Presidential, address p. 88.

[2] Dr. A. R. Wallace ("Island Life," p. 398) speaks of the "fragmentary evidence derived from such remote periods" and the futility of the notion that "a similarity in the production of widely-separated continents at any past epoch is only to be explained by the existence of a direct land communication." As may be seen from what has gone before and what is to follow, he hardly appears to have appreciated the full weight of the evidence. The subject has been treated of by Dr. W. T. Blanford in his presidential address to the Geological Society, 1890.

[3] *Denkschr. k. k. Ak. Wiss. Wien*, L, 132, (1885).

Madagascar.[1] Three of these belong to the group of *Notocœli*, a group which is typical of his equatorial fauna, while the fourth belongs to the *Hastati*, a group which is distinctly southern in Europe. The only *Belemnite* in the Uitenhage beds is not only different from any of the Madagascar forms, but belongs to the group *Absoluti*, which is typical of the boreal regions in the northern hemisphere. The inference from this is that the neocomian beds of northern Madagascar were deposited in an extension of the tropical sea, while those of the extreme south of Africa were formed in a different, probably colder ocean.[2]

This barrier does not seem to have been absolutely continuous throughout the jurassic period, or there may have been a mode of communication round the north of the Peninsula of India by which some migration took place, and so the presence of a few Cutch species, which are also found on the east coast of India and in South Africa, is accounted for.

In cretaceous times the evidence is even stronger. The fauna of the cretaceous (cenomanian) beds of Bágh is closely allied to that of Arabia and Europe, but is as distinct from that of the cretaceous beds in Trichinopoli as is possible in the case of two homotaxially equivalent faunas. But the Trichinopoli cretaceous fauna is very closely allied to that of the Khási hills and of South Africa, showing that these areas, which are separated from each other by distances much greater than that which divides the Trichinopoli and Bágh exposures, were parts of one marine province, and the difference of the fauna from that of the lower Narbadá valley can only be explained by the existence of a land barrier, separating the sea in which the Trichinopoli, Khási, and South African fauna lived, from that in which the Narbadá, Arabian, and European cretaceous beds were deposited.[3]

We see then that throughout the later part of the palæozoic and the whole of the mesozoic era, there was a continuous stretch of dry land over what is now the Indian Ocean, which finally broke up and sank beneath the sea in the tertiary period.

This conclusion has an important bearing on the generally, though not universally, accepted doctrine of the permanence of continental and oceanic areas. It is claimed, by many geologists of eminence, that the deep oceanic areas of the present day have been oceans throughout the whole of the period represented by the sedimentary formations of the geological sequence, and that, if we except small volcanic islands rising from the depths of the ocean, the dry land of every geological period was confined to the present dry land and the shallower parts of the sea surrounding it.

[1] *Quart. Jour. Geol. Soc.*, XLV, 333, (1889).

[2] *Neues Jahrb. Min. Geol.*, 1890, Band I, p. 1; W. T. Blanford, *Quart. Jour. Geol. Soc.*, XLVI, proceedings, p. 98, (1890). The evidence is good for the distinctness of the marine areas, but the climatic inferences are vitiated by the possibility of there having been extensive alterations of latitude since these beds were deposited (*vide infra*).

[3] See pp. 247, 252.

The soundings that have been made in the Indian Ocean are not so numerous as to preclude the possibility of there being a bank connecting India and Africa, which would allow a bridge of dry land having existed without imperilling this popular theory, but there are no indications of the existence of such a bank, and the Indian Ocean is generally regarded by the supporters of the hypothesis as one of the original oceanic areas of the world. So the conclusion we have to draw from known and accepted facts is in conflict with an hypothesis which has much to be said in its favour and has the support of many of the most eminent geologists of the day.

But this hypothesis is, consciously or unconsciously, to a large extent bound up with, and based on, ideas relative to the constitution of the earth as a whole, which represent it as either solid throughout, or at any rate as having a solid crust whose thickness is very considerable in proportion to the whole diameter of the earth. And one of the consequences that follow on this theory of the constitution of the earth is, that there cannot have been any great changes in the direction of the earth's axis of revolution, or changes of latitude of places on its surface caused by the shifting of the superficial crust over the internal core.

Here again we find the facts in conflict with a generally accepted hypothesis, which is, however, being gradually discredited in later years. Whatever may be the cause of these cold periods, of which two are now well established in the geological history of the world and several more are less completely indicated,[1] there can be no doubt that their effects will be more widespread, more extensive, and their traces more conspicuous in high latitudes than in low. Yet the remains of this carboniferous glacial period so conspicuous in India, Africa, and Australia, all lie within, or only just beyond, thirty degrees from the equator. The furthest from the equator lie in latitudes where the last glacial period, of pleistocene times, has left but few traces at low altitudes, and those of a somewhat doubtful character, while most of the remains are in latitudes to which the ice of the pleistocene glacial period never penetrated, and many are well within the tropics.

At the same time the corresponding deposits not merely of the temperate clime in Europe and America, but even within the Arctic circle, in which one would expect the traces of this cold period to be more abundant, more extensive, and more conspicuous, are almost free from traces of glacial action. Boulders of rock imbedded in fine silt have been found, and some have shown a striation believed to be due to the action of ice, but they are sporadic and indicate that the carboniferous beds of England and Europe, in which they were found, lay in a latitude which was near the limit which the floating icebergs could reach before melting.

[1] J. Croll; "Climate and Time," London, 1875, Chap. XVIII. See also pp. 106, 346, *h. l.*

If we compare this with the very distinct evidence that glaciers descended to low altitudes in the Pengangá valley and the great desert of Rájputána, the contrast is not only striking but inexplicable, unless there has been a very considerable change of the position of the earth's surface relative to the present position of the poles.

There is independent evidence that similar changes of latitude are actually taking place at the present day, in the records of all the principal observatories of Europe and America.[1] The justice of the conclusion has been questioned and the variations have been ascribed to instrumental errors and errors of observation, but their consensus is so strong that this appears to be out of the question, and their accuracy has to a great extent been confirmed by a series of observations recently made at Honolulu.[2] In fact, the only grounds for questioning the possibility of changes of latitude on the earth's surface are based on those hypotheses regarding the constitution of the earth which have been referred to. These hypotheses, however, are not based on direct observation, but on mathematical reasoning of a very brilliant nature which, to quote Prof. Huxley's well known simile, is a mill that grinds very fine, but can only grind what is put into it. In the present case the fundamental data on which the investigations were based are uncertain, and the conclusions must consequently be questionable. This view has frequently been verbally accepted, and more recent investigations have gone far to modify the earlier ones, yet such is the glamour of genius, and such the natural tendency of human nature to mistake precision for accuracy, and to prefer a definite statement to a guarded inference, that the idea of the general solidity of the earth, with its consequences, that very extensive changes in its form, or the position of its surface relative to its axis of revolution, are impossible, and the limitation of the period which has elapsed since the earth's surface had cooled sufficiently to support life and to admit of the deposition of ordinary sediments, to a comparative few millions of years, have exerted an influence on the speculations of physical geology none the less important because usually unconscious.

It has already been stated that some of the more recent mathematical investigations have tended to show the uncertainty of the earlier conclusions, and in one of the most recent and philosophical of these,[3] the Revd. O. Fisher has almost returned to the old idea of a thin crust lying on a molten interior. According to him the actual solid crust of the earth is

[1] Asaph Hall, *Am. Jour. Sci.*, 3rd series, XXIX, 223, (1885); R. D. Oldham, *Geol. Mag.*, 3rd decade, III, 300, (1886); G. C. Comstock, *Am. Jour. Sci.*, 3rd series, XLII, 470, (1891). Flinders Petrie's observations ("Pyramids and Temples of Ghizeh," London, 1883, p. 125) are important, as showing a distinct change of azimuth, and probably of latitude, since the pyramids were built.

[2] *Proc. Roy. Soc.*, L., 227, (1891).

[3] O. Fisher, Physics of the Earth's Crust London, 1st edn., 1881; 2nd edn., 1889.

comparatively very thin, not exceeding 25 to 30 miles in thickness, and rests on a magma which, in its essential characters, must be regarded as a fluid. What may be the condition of the great mass of matter forming the central core of the earth is unknown and immaterial, but if Mr. Fisher's theory of the earth's crust is correct, it not only allows of changes of latitude having taken place, but renders it exceedingly improbable that they have not taken place. In this respect it agrees better with the observed facts than the more usually held hypothesis, and we will see, when the Himálayan range is dealt with, that this is not the only respect in which a confirmation of his theory can be derived from the facts of Indian geology.

CHAPTER IX.

MARINE JURASSIC ROCKS.

Cutch—Western Rájputána—Salt range—Himálayas—Afghánistán—Doubtful jurassics of western Garhwál.

With the exception of the fossiliferous upper Gondwána beds of the east coast, referred to in a previous chapter, no marine beds of jurassic age are known to occur in the peninsular area proper, and even as regards these it is still uncertain whether they should not be regarded as more recent in their origin. But in the debateable tract lying east of the Indus and west of the Arávallis, marine jurassic rocks attain a large development in Cutch and in western Rájputána.

The jurassic area of Cutch[1] may be considered as occupying a number of post-tertiary islands, now connected by alluvial flats. The largest of these islands, that forming the western and central portion of Cutch, is about a hundred and twenty miles long, from east to west, by about forty broad. To the north-east of it is the district of Wágad, another ancient island nearly fifty miles from east to west, and, excluding alluvium and "Rann," twenty-five miles broad. Farther north four isolated masses of hills, chiefly composed of lower jurassic rocks, extend in a line nearly a hundred miles in length from east to west. These are the so called islands in the Rann,[2] Patcham, Kharir (Khurreer or Kurreer), Bela and Chorar.

[1] The account of this province is taken partly from a report by Mr. Wynne, *Memoirs*, XI, pp. 1—293,(1872), and partly from manuscript notes by the late Dr. Stoliczka. The *Cephalopoda* have been determined by Dr. Waagen, and described in the *Pal. Indicas* series ix, (1873-75). It should, perhaps, be noticed that Dr. Waagen's views of specific distinction differ from those of many palæontologists, and that, as he points out, several of the forms described by him as species might by other naturalists not be considered to rank higher than varieties.

[2] The "Rann" of Cutch is an immense tract surrounding the province on all sides, except the south, and consisting of barren salt marsh periodically overflowed by sea water. This tract, which is evidently an ancient sea basin now filled up by alluvial deposits, will be further described in a subsequent chapter on post-tertiary and recent deposits.

A few smaller islands also occur, but none of them are of sufficient size to be worthy of notice.

None of the rocks found in Cutch and the adjoining islands are of older date than jurassic. In one spot some limestones, containing upper neocomian *Cephalopoda*, are found resting upon the jurassic series, the uppermost group of which may perhaps itself be of intermediate age, and belong in part to a lower neocomian horizon.[1] In general, the upper jurassic beds disappear to the south beneath the Deccan traps, but marine tertiary beds (nummulitic) overlap the traps and rest upon the older series in many parts of the country, both traps and nummulitic beds being quite unconformable to the jurassic formations.

The lowest beds are seen dipping to the south in the Rann islands, and are locally exposed in an anticlinal which runs along the northern edge of the province, the intervening synclinal being, for the most part, concealed beneath the Rann. From the anticlinal near the Rann there is a general dip, varying in amount, to the southward. The greater portion of the series is, however, repeated twice in consequence of a great fault, which runs from east to west along the northern scarp of the Chárwár range of hills south of Bhúj.

By the earlier observers, including Mr. Wynne, the jurassic series in Cutch was simply divided into a lower and an upper group, the former chiefly marine, the latter apparently fresh water for the most part, though, as was shown clearly by Mr. Wynne, no marked line of division can be drawn, for not only is there an absence either of unconformity or of any marked break in lithological character between the two subdivisions, but marine beds are occasionally found interstratified with the upper, and plant beds with the lower group. The examination of the *Cephalopoda* by Dr. Waagen indicated the probability that representatives of several European jurassic groups existed in Cutch, and Dr. Stoliczka, re-examining the beds with the aid of Mr. Wynne's geological map and his own knowledge of palæontology, found no difficulty in distinguishing four subdivisions, the three lower of which had been included in the inferior or marine group of previous observers, whilst the upper comprised the higher fresh water beds, with the uppermost marine strata. The names of the groups proposed by Dr. Stoliczka, with their homotaxial equivalents amongst European formations, are exhibited on the opposite page.

The whole thickness of the Cutch jurassic series has been estimated by Mr. Wynne at 6,300 feet, of which 3,000, or very nearly half, belong to

[1] Dr. Stoliczka unfortunately did not live to publish the results of his examination of Cutch, but from his rough field notes it appears probable that the upper neocomian bed of Ukra (*infra*, p. 286) is conformable to the underlying Umia beds. He does not precisely state, however, what are the relations of the upper bed to the lower at this spot.

Classification of the Jurassic series in Cutch.[1]

CUTCH.		EUROPE.	
GROUPS.	BEDS.	ZONES.	GROUP.
UMIA	1. Beds with *Crioceras* and *Ammonites* of the *rhotomagensis* group.	......	CRETACEOUS. *Upper Neocomian.*
UMIA	2. Sandstones and shales with *Cycadeæ* and other plants.	?	? WEALDEN.
UMIA	3. Sandstones and conglomerates with marine fossils, *Ammonites* (*Perisphinctes*) *eudichotomus*, *frequens*, *Trigonia smeei*, *T. ventricosa* etc.	Upper Tithonian . Lower Tithonian .	UPPER JURASSIC. *Tithonian and Portland.*
KATROL	4. Sandstones and shales with *Am.* (*Phylloceras*) *ptychoicus*, *A.* (*Oppelia*) *trachynotus*, *A.* (*Perisphinctes*) *torquatus*, *pottingeri*, etc.	Zone of *A.* (*Perisph.*) *mutabilis*. Zone of *A.* (*Oppelia*) *tenuilobatus*.	UPPER JURASSIC. *Kimmeridge.*
KATROL	5. Red ferruginous and yellow sandstones (Kantkot sandstones) with *Am.* (*Stephanoceras*) *maya*, *A.* (*Aspidoceras*) *perarmatus*, *A.* (*Perisphinctes*) *virguloides*, *leiocymon*.	? Zone of *A.* (*Pelt.*) *bimammatus*. ? Zone of *A.* (*Pelt.*) *transversarius*.	MIDDLE JURASSIC. (Middle oolite). *Oxford.*
CHÁRI	6. Oolites (Dhosa oolite) with *Am.* (*Stephanoceras*) *polyphemus*, *A.* (*Perisphinctes*) *indo-germanus*, *A.* (*Aspid*) *perarmatus*, *babeanus*, *A.* (*Pelt.*) *arduennensis*, etc	Zone of *A.* (*Amaltheus*) *cordatus*. Zone of *A.* (*Amaltheus*) *lamberti*.	MIDDLE JURASSIC. (Middle oolite). *Oxford.*
CHÁRI	7. White limestones with *Am.* (*Pelt.*) *athleta*, *A.* (*Oppelia*) *bicostatus*, etc.	Zone of *A.* (*Pelt.*) *athleta*.	MIDDLE JURASSIC. (Middle oolite). *Kelloway.*
CHÁRI	8. Shales with ferruginous nodules, with *Am.* (*Perisph.*) *obtusicosta*, *anceps A.* (*Harpoceras*) *lunula*, *punctatus*, etc.	Zone of *A.* (*Perisph.*) *anceps*.	MIDDLE JURASSIC. (Middle oolite). *Kelloway.*
CHÁRI	9. Shales with calcareous bands and locally with golden oolite: *Am.* (*Steph.*) *macrocephalus*, *tumidus*, *bullatus*, *A.* (*Oppelia*) *subcostarius*, *A.* (*Perisph.*) *funatus*, etc.	Zone of *A.* (*Steph.*) *macrocephalus*.	MIDDLE JURASSIC. (Middle oolite). *Kelloway.*
PATCHAM	10. Light grey limestones and marls with *Am.* (*Oppelia*) *serriger*, Corals and Brachiopoda, etc.	......	MIDDLE JURASSIC (Lower oolite). *Bath.*
PATCHAM	11. Yellow sandstones and limetones, with *Trigoniæ*, *Corbulæ*, *Cucullеæ*, etc.	......	MIDDLE JURASSIC (Lower oolite). *Bath.*

[1] Waagen, Pal. *Indica*, series IX, Introduction, (1875).

the uppermost group alone, the thickness of the other groups not having been estimated separately. It must be remembered that the base of the whole series is not exposed, and that the upper beds had suffered from denudation before they were covered by the traps.

The Patcham group is thus named from its occurrence in the island of Patcham in the Rann. The lowest beds are exposed on the northern scarp of a range of hills which runs east and west through all the Rann islands from Patcham to Chorar. The rocks composing the range dip south at a low angle, the crest of the hills and the surface of their southern slopes being formed of a thick massive bed of yellowish sandstone and limestone, which contains *Corbula pectinata*, *Astarte compressa*, a *Trigonia* closely resembling *T. interlævigata*, *Cucullæa virgata* and other fossils.[1] Below the massive bed come shales and sandstones, all more or less calcareous, containing a *Rhynconella*, near *R. concinna*, *Lima*, *Gervillia*, a small *Exogyra*, etc. The lowest bed seen in Patcham island is calcareous sandstone abounding in the small *Exogyra*. The same lower beds are seen in Koari Bet, a small islet north-west of Patcham, and on the northern flank of the range, in Kharir, Bela, and Chorar, the top of the range in all cases consisting of the yellow calcareous rock. The thickness of this portion of the beds is at least 500 feet.

Besides forming the range of hills in the islands of the Rann, the Patcham limestone is exposed at four places in Cutch itself,—at Jarra, Kira hill near Chári, Jura hill, and in Halamán hill near Lodai—all situated along the northern edge of the main province of Cutch, near the borders of the Rann. In all these places they appear as inliers, exposed at the crest of an anticlinal, and surrounded on all sides by higher beds. At Jarra, about fifty miles north-west of Bhúj, there is a bed of white limestone containing *Scyphia*, a *Terebratula*, and small *Rhynconellæ* and, immediately above it, a bed of corals. These rocks do not appear to be equally well exposed elsewhere. They are at the base of the Chári group and were considered by Dr. Stoliczka as the uppermost beds of the Patcham group of Cutch.

The lower portion of the Patcham group has yielded no *Cephalopoda*, and the higher beds only eight species, all of which are rare. One is *Nautilus jumarensis*, the others are *Ammonites*, of which one belongs to the sub-genus *Oppelia*, three to *Stephanoceras*, and three to *Perisphinctes*. One *Stephanoceras* is a variety of *Ammonites macrocephalus*, the typical form of which is abundant in the next higher subdivision, and both the other species of *Stephanoceras* pass likewise into the lower beds

[1] As only the *Cephalopoda* of the Cutch beds have been properly compared, it is possible that some of the identifications of other fossils may require modification. Only those are mentioned which are in all probability correctly determined.

of the Chári group. With the exception of *A. macrocephalus*, the only species found also in European rocks is *A.* (*Oppelia*) *serriger*, which was originally described from upper bathonian beds. So far as the *Cephalopoda* are concerned, it would be difficult to correlate the Patcham group with any subdivision of the European oolites, but the Patcham *Brachiopoda*, which, however, have not been thoroughly compared, and the position of the beds immediately beneath the strata containing *A. macrocephalus* in abundance, have induced Drs. Stoliczka and Waagen to refer the group to the horizon of the Bath oolite (bathonian).

The next group in ascending order derives its name from the village of Chári, situated close to the borders of the Rann, about thirty-two miles north-west of Bhúj. This village has been known since the time of Captain Grant, the earliest geological explorer of Cutch, as an admirable locality for fossils, and especially for *Cephalopoda*, of which large numbers are found in the calcareous sandstones exposed around Kira hill.

The Chári group is composed of four subdivisions, each marked by its mineral characters and by the fossils it contains. The group, as a whole, is much more shaly than any of the other subdivisions; but it contains hard bands of limestone or calcareous sandstone forming ridges, which are usually distinguished by characteristic forms of *Ammonites*.

The lowest of the four zones or subgroups consists of shales, usually of a grey colour, with occasional bands of golden oolite, and sometimes nodular shaly limestone. The rock called golden oolite (which is not peculiar to India, but which is also found in the jurassics of Europe, and at about the same horizon) is very characteristic and easily recognised. It is a rather coarse grained limestone, composed of calcareous grains, which are coated with a very thin ferruginous layer and are surrounded by a matrix of carbonate of lime, so that the stone has much the appearance at first sight of a rock with golden coloured mica. In places, as at Chári itself, the golden oolite is thick and conspicuous, but it is locally distributed and often wanting. The most characteristic fossils of these lowest Chári beds are *Ammonites* (*Stephanoceras*) *macrocephalus*, and allied forms of the sub-genus *Stephanoceras*.

Above the *macrocephalus* beds come dark shales, often black, with ferruginous bands and concretions. Sometimes, however, the nodules are of white limestone, and the shales are locally sandy, and associated with sandstones, but the beds appear to preserve their lithological characters in general thoroughout Cutch. The chief palæontological peculiarity of this subdivision is the extreme abundance of a *Terebratula*, considered by Sowerby a variety of the cretaceous *T. biplicata*. Planulate *Ammonites* (*Perisphinctes*) are also very common. The shales not unfrequently contain remains of plants, but no distinct impressions have been found.[1]

The next subdivision, in ascending order, is a very thin band, sometimes only 20 to 30 feet thick, of whitish or grey shale, with bands of limestone, which are generally white, but occasionally yellowish or brown. Usually this band may be recognised easily by its colour and by its presence beneath the Dhosa oolite. The most characteristic fossil is *Ammonites* (*Peltoceras*) *athleta* and, in north-western Cutch, the shell of this mollusc is usually changed into black calcspar.

The uppermost Chári subdivision, or Dhosa oolite, is the most characteristic of all both lithologically and palæontologically. It is of no great thickness, though more developed than the *athleta* beds, and consists of grey, reddish, or brown oolite, sometimes sandy and often nodular. *Cephalopoda* are extremely abundant. And it abounds in many places in a *Terebratula* closely allied to the cretaceous *T. sella*, and referred to that species as a variety by Sowerby.

The Chári beds are exposed in several places in Cutch, but they nowhere occupy a large area. They are found resting upon Patcham beds in the southern part of Patcham and Kharir, and in Kakindiya and Gángta, two small islands south-east of Kharir, forming only the axis of a quaquaversal anticlinal on the latter, but none are exposed in Bela or Chorar, though a small area exists in the extreme north of Wágad. In these outcrops the subdivisions are less well marked than to the southward, and the two characteristic *Terebratulæ* have not been noticed. In the mainland of Cutch, the Chári group occupies two series of inliers. One of these series is scattered at intervals along the northern anticlinal range. The rocks appear at three places west north-west of Chári, again around Kira hill, near Chári, the typical locality, they extend nearly twelve miles from east to west around the Patcham beds of Juria hill, north of Bhúj, and are found in two more outcrops farther east around Halamán hill, where they extend more than six miles, and they again appear a mile farther east. Another series of outcrops occurs in the Chárwár range, south of Bhúj. Here the Chári beds are brought up at intervals along the southern side of the great fault; they are greatly disturbed and cut up by cross faults, but the different bands can be easily recognised,—the Dhosa oolite with *Terebratula sella*, var., the white *athleta* beds, and the band with *T. biplicata*, var., being always conspicuous.

The cephalopodous fauna of the Chári group comprises a hundred and twelve species, of which thirty-seven are European. The relations between those found in the different subdivisions and the corresponding Kelloway

[1] In Dr. Stoliczka's field notes he mentions having at one locality found fragments of quartz and of a limestone derived from one of the lower groups, probably from the Patcham beds, cemented together in the rock at this horizon. This may indicate an unconformity.

and lower Oxford groups in Europe, are the following, according to Dr. Waagen:—

In the lowest Chári subdivision, or *macrocephalus* beds thirty-one *Cephalopoda* have been found, *viz.* two species of *Belemnites*, three of *Nautilus* and twenty-six of *Ammonites* (*Phylloceras* 2, including *A. disputabilis*, *Lytoceras* 1, *Oppelia* 1, *Harpoceras* 1, *Stephanoceras* 13, including *A. macrocephalus*, and *Perisphinctes* 8). Three are common to this subdivision and the upper Patcham beds, whilst none are known to range into higher strata. Sixteen species, or rather more than one-half, are found in Europe, all, except two, belonging exclusively to the beds with *A. macrocephalus* (Lower Kelloway).

In the next subdivision, the dark shales with *Terebratula biplicata*, twenty-seven *Cephalopoda* are found, *viz.* three *Belemnites*, one *Nautilus*, one *Ancyloceras*, and the remainder *Ammonites* (*Phylloceras* 2, *Oppelia* 3, *Harpoceras* 5, *Stephanoceras* 1, and *Perisphinctes* 11). Six of these range into higher beds, whilst seven are European, and of these latter five are only found in the beds with *A. anceps* (Middle Kelloway).

The *Athleta* beds have yielded twenty species, three *Belemnites* and seventeen *Ammonites* (*Phylloceras* 1, *Amaltheus* 2, *Oppelia* 2, *Harpoceras* 2, *Peltoceras* 1, *Aspidoceras* 2, and *Perisphinctes* 7); five of these are common to the next lower subdivision, and two to the Dhosa Oolite. Eight are European, six being peculiar to the zone of *A. athleta* (Upper Kelloway).

In the Dhosa oolite there are thirty-four *Cephalopoda*, *viz.* four *Belemnites*, one *Nautilus*, and twenty-nine *Ammonites* (*Phylloceras* 2, *Harpoceras* 1, *Peltoceras* 5, *Aspidoceras* 4, *Stephanoceras* 8, and *Perisphinctes* 9). Four of these range into higher and three into lower beds. Eight are found in Europe, the most important being *Am.* (*Aspidoceras*) *perarmatus*, and seven of these belong exclusively to the zones of *A.* (*Amaltheus*) *lamberti* and *A.* (*Amalth.*) *cordatus* (Lower Oxford). Other fossils, especially *Terebratula sella*, are abundant in this group.

The Katrol group, which rests upon the uppermost subdivision of the Chári beds, is of considerable thickness. It consists of sandstones of various kinds, white, brown, pinkish grey, etc., and shales usually grey or reddish, but sometimes very dark coloured, like those of the *Am. anceps* zone. Ferruginous nodules and concretions sometimes occur in the shales which prevail towards the base of the group, the upper portion being chiefly sandstones. On the whole, however, shales predominate.

These beds form two belts in Cutch proper. The first occurs in the anticlinal along the Rann and extends for nearly eighty miles, surrounding the inliers of the Patcham and Chári groups, and extending to a considerable distance beyond them. The exposure of Katrol rocks varies in breadth being, where broadest, nearly ten miles wide. The second belt is in the Chárwár range, south of the great fault. This tract is about thirty-five miles from east to west, but nowhere more than two miles broad. Besides this the greater part of Wágad is occupied by beds apparently belonging to the same group. The rocks are very similar in mineral character, consisting of a coarse and fine grey, pinkish and white sandstones above, and grey or yellowish shales below, but the *Cephalopoda* found are almost all distinct, and appear to indicate a lower horizon. From their development around the town of Kantkot, these Wágad beds have received the name of Kantkot sandstone.

The *Cephalopoda* of this Kantkot sandstone are nineteen in number, four *Belemnites* and fifteen *Ammonites* (*Phylloceras* 1, *Aspidoceras* 2, *Stephanoceras* 5, *Perisphinctes* 7). Four of these *Am.* (*Aspidoceras*) *perarmatus*, *A.* (*Stephanoceras*) *maya*, *fissus*, and *opis*, are also found in the Dhosa oolite of the Chári beds, whilst only one species, *Belemnites grantianus* (*B. kantkotensis*), is common to the Kantkot bed and the Katrol group in Cutch proper. Thus the Kantkot beds appear by their cephalopodous fauna allied more closely to the uppermost Chári beds than to the Katrol group. Three species only of the Kantkot *Cephalopoda* are European, *A.* (*Asp.*) *perarmatus*, *A.* (*Per.*) *plicatilis*, and *A* (*Per.*) *martelli*, and only one of these, the last, is limited to a single zone, that of *A.* (*Pelt*) *transversarius* (Upper Oxford) in Europe, the other two ranging lower. Several forms are, however, allied to upper oxfordian species.

The Katrol group proper has yielded twenty-six species of *Cephalopoda*, four *Belemnites* and twenty-two *Ammonites* (*Phylloceras* 2, *Lytoceras* 1, *Haploceras* 2, *Oppelia* 4, besides an *Aptychus*, *Harpoceras* 1, *Aspidoceras* 5, *Perisphinctes* 7). Only one of these species, *Bel. grantianus*, is found with certainty in any other group in Cutch. Four species are found in Europe, all belonging to the beds of the Kimmeridge group, with *A.* (*Asp.*) *acanthicus*. By far the most characteristic and abundant of the *Cephalopoda*, is a non-canaliculate Belemnite, *B. katrolensis*. The commonest *Ammonites* are *A.* (*Oppelia*) *kachhensis*, *A.* (*Per.*) *pottingeri*, *A.* (*Per.*) *katrolensis*, and *A.* (*Per.*) *torquatus*.

Imperfect plant remains are common in the Katrol group, as they are in many of the lower beds of Cutch, but in one instance near the village of Narha, as has already been mentioned in the description of the Gondwána system several remains of plants, whose relations have already been discussed on a previous page[1] were found by Mr. Wynne, in shales interstratified with the Katrol beds and distinctly inferior in position to some of the marine bands of the group.

[1] *Supra*, p. 189.

The Umia group derives its name from a small village in western Cutch, rather more than fifty miles north-west of Bhúj. Taken as a whole, this group appears to equal in development all the other jurassic beds together, being, according to Mr. Wynne's estimate, upwards of 3,000 feet thick. It is the equivalent of the upper jurassic group of Mr. Wynne's Memoir. As a rule, it consists of sandstones of various kinds, and more or less sandy shales. The sandstones are very often soft and white or pale brown, sometimes variegated, and very generally distinguished by thin bands of hard black or brown ferruginous grit. Occasionally the sandstones are variegated with pink, red, and brown, they are often very argillaceous and tend to decompose into a loose sandy soil, which covers and conceals the rocks over a great part of the country. In a few instances carbonaceous shale occurs, and in one locality a thin seam of bright jetty coal. A few thin hard bands of sandstones are met with, some being so hard as to be almost a quartzite. There is a marked resemblance in the beds of this group to some of the upper Gondwána strata of Central India: there are the same soft argillaceous sandstones and sandy shales and the same hard ferruginous gritty bands.

Towards the base of the Umia group there is a thick band of calcareous conglomerate, hard and grey, sometimes ferruginous, associated with sandstones and shales. In this conglomerate and in some associated beds marine fossils are numerous. Throughout all the rest of the group plant remains are common, but they are not often sufficiently well preserved to be identified. Marine fossils are very rare, but *Trigonia smeei*, the most typical fossil of the group, has been found in places, as near Vigori, forty miles north-west of Bhúj, in beds near the top of the group and well above the horizon at which most of the plant fossils have been obtained.

The beds of the Umia group are covered unconformably by the Deccan traps and by tertiary rocks, except in one place, where they underlie the upper neocomian (aptien) beds of Ukra hill in north-western Cutch.[1]

The surface occupied by the rocks of the Umia group corresponds in magnitude with the thickness of the formation, and embraces nearly, if not quite, half of the jurassic area in Cutch. In Cutch proper these beds extend throughout the province from the western extremity near Lakhpat to the eastern end beyond Bachao, forming a great plain south of the irregular range of hills along the edge of the Rann. They also extend round each end of the range, especially to the eastward, where the bottom Umia beds extend north of the hills about twenty miles along the edge of the Rann. The main belt of Umia beds is from eight to twelve miles across on an average. These rocks lap round the western end of the Chárwár range, where the great east and west fault to which the range is due appears to die out, and they cover another plain, nearly fifty miles in length from east

[1] *Infra*, p. 286.

to west and about eight miles broad, south of the Chárwár range. They also form the western portion of Wágad.

The plant remains of the Umia group and their relations have already been described in the chapter relating to the Gondwána system. It was there shown that twenty-seven species had been identified, of which ten are either common to the lower oolitic beds of Yorkshire or represented by very closely allied forms. Bearing in mind that the plant beds overlie the portion of the group which has furnished *Cephalopoda*, it is remarkable to find that the latter exhibit a very decided upper oolitic facies. They are eleven in number,[1] *viz.*—

Belemnites grantianus (*kantkotensis*).	*A.* (*Perisphinctes*) *bleicheri.*
Belemnites, 2 sp. indet.	*A.* (*Per.*) *occultefurcatus.*
Am. (*Haploceras*), cf. *tomephorus.*	*A.* (*Per.*) *eudichotomus.*
A. (*Aspidoceras*) *wynnei.*	*A.* (*Per.*) *frequens.*
A. (*Perisphinctes*), cf. *suprajurensis.*	*A.* (*Per.*) *denseplicatus.*

Of these eleven species, one (*Belemnites grantianus* var. *kantkotensis*) is found in lower beds in Cutch, and the two other forms of *Belemnites* are closely allied to the Katrol species *B. claviger* and *B. katrolensis*, and may be identical. All the eight *Ammonites* are restricted in Cutch to the Umia group, and two of them (*A. tomephorus* and *A. eudichotomus*) are tithonian species, found in the uppermost jurassic beds of southern Europe, whilst *A. bleicheri* and *A. suprajurensis* are found in the Portland strata of northern France, and *A. occultefurcatus* is barely distinguishable from another Portland species, *A.* (*Perisphinctes*) *boidini.* The connection between the *Cephalopoda* of the Umia group and the forms found in the uppermost jurassic beds of Europe is consequently very marked, and Dr. Waagen states that the same marked similarity exists between the lamellibranchiate bivalves of the same beds in the two regions.[2]

The *Cephalopoda* are, however, rare and exceptional in the Umia group, and they form by no means so important a portion of the fauna as in the other groups. The commonest Umia fossils are two species of *Trigonia* (*T. smeei* and *T. ventricosa*), the latter being also found in the neocomian rocks of South Africa, whilst a very closely allied form (*T. tuberculifera*) occurs in cretaceous beds in Southern India. The occurrence of these *Trigoniæ* in upper Gondwána strata near Rájámahendri has already

[1] Nine, according to Dr. Waagen (*Pal. Indica*, series ix, 225, 232), but he appears to have overlooked two forms—*Belemnites kantkotensis* (*grantianus*), stated at page 4 to have been found in Umia beds, and the specimens from the same group doubtfully referred to *B. claviger* on p. 7. These very trifling and unimportant oversights are not noticed in order to call attention to a trivial error, but because the relations of the Umia group are of considerable importance and have been disputed. In consequence of the great importance of this group, the evidence upon which its relations to the upper jurassic beds of Europe are based is given in full.

[2] *Pal. Indica*, series ix, 225, (1875).

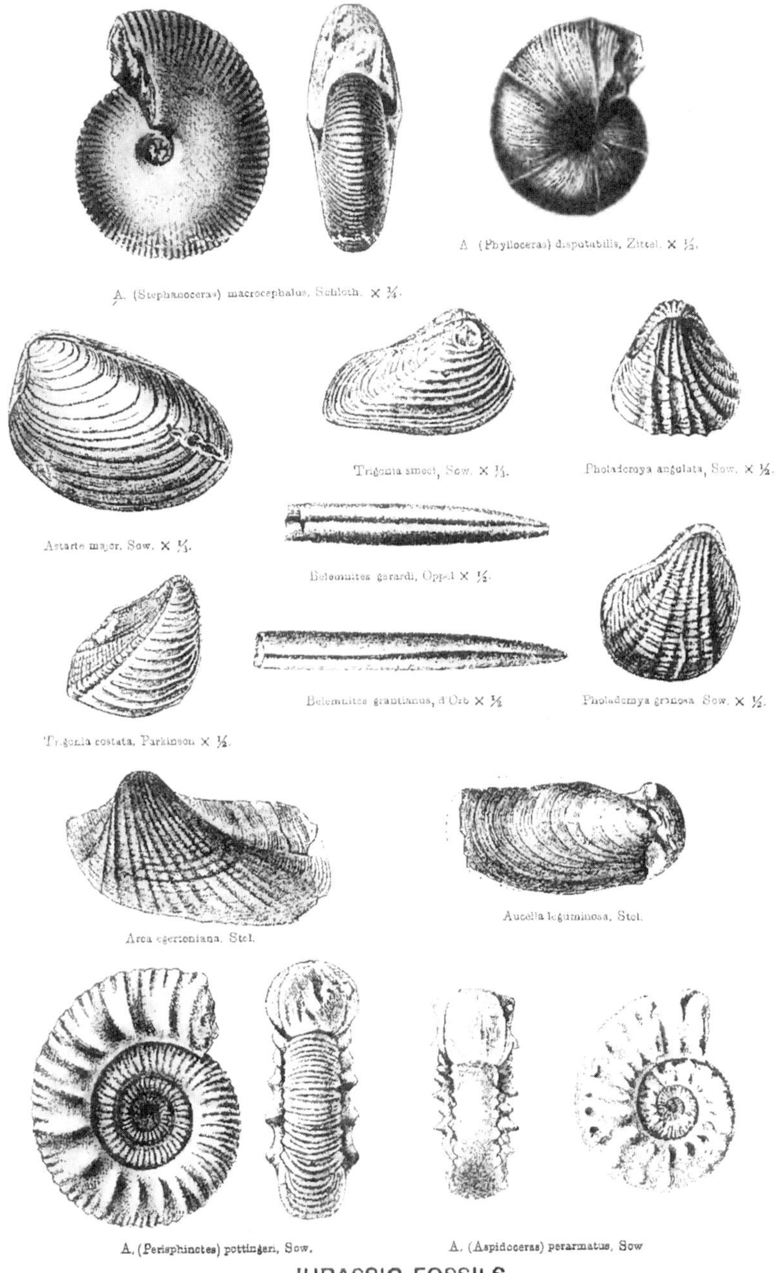

JURASSIC FOSSILS.

Calcutta Phototype Co,

most characteristic beds being whitish or greyish sandstone, very fine and been noticed. Other forms of common occurrence in the Umia group are species of *Trigonia* allied to *T. vau, clavellata* and *gibbosa*, *Astarte major*, a *Gervillia*, a peculiar *Gryphæa*, intermediate in form between *G. dilatata* and *G. vesicularis*, *Goniomya*, etc. Some of these range into lower groups also. A portion of the jaw of a *Plesiosaurus*, also obtained from these beds, is said to be undistinguishable from that of *P. arcuatus*, Owen, from the lower lias of England.[1]

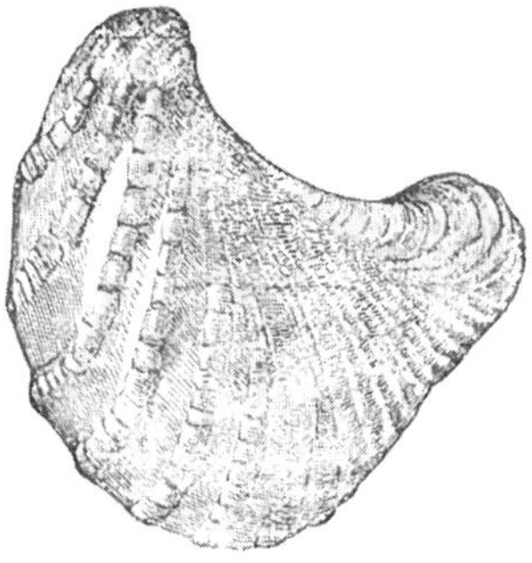

Fig. 14.—*Trigonia ventricosa*, Kraurs, natural size.

Owing to the circumstance that, with the exception of the *Cephalopoda*, the large collections of fossils made in Cutch by Messrs Wynne, Fedden, and Stoliczka have not hitherto been examined and compared, the distribution of many of the most characteristic species has not been definitely ascertained. Amongst the forms which are most abundantly preserved in the lower groups of the Cutch jurassic series are species of *Pleurotomaria*, *Pholadomya granosa*, *Ph. angulata*, *Ph. inornata*, *Corbula lyrata*, *C. pectinata*, *Nucula cuneiformis*, *Cucullæa virgata*, *Trigonia costata* and *Ostrea marshii*.

On the next page is given a table which shows the general result of Dr. Waagen's examination of the jurassic *Cephalopoda* found in Cutch. The correspondence, not only with the European jurassic rocks as a whole, but with the different groups into which they are divided, is remarkable, and greater than is known in any other Indian formations, the only other series of Indian rocks of which the fauna has been sufficiently examined to justify the comparison, the cretaceous series of Southern India, showing much less close agreement in the distribution of the fauna, and especially of the *Cephalopoda*, with the corresponding groups in Europe. The only remarkable instance in which the *Cephalopoda* of the Cutch jurassics differ from their representatives in the jurassic rocks of Europe, is in the prevalence in the Indian area of the *macrocephali* at a higher horizon than in Europe. In Cutch they abound in the Dhosa oolite and Kantkot sandstone, the other *Cephalopoda* of which are of Oxford, and in the latter case of upper Oxford types, whilst in Europe they are not known above the base of the Kelloway group. As will, however, be shown in the next chapter, some of the cretaceous forms of *Ammonites* found in Southern India show a remarkable resemblance to the jurassic forms of

[1] Lydekker, *Records*, XXII, 50, (1889).

macrocephali, and in this instance they are associated with species allied to *Ammonites* characteristic of even older European deposits.

Name of group.	Name of subdivision.	Total number of *Cephalopoda*.	Species peculiar to group.	Species ranging into higher beds.	Species ranging into lower beds.	Common to European jurassics.	Remarks.
Umia	Marine beds	11	9	...	2	4	Two of the European species occur in Portland beds of northern France and two in Tithonian beds of southern Europe.
Katrol	Katrol beds proper	27	26	1	1	4	All the four European species belong to the zone of *Am. acanthicus* (Kimmeridge).
	Kantkot beds	19	14	1	4	3	......
Chari	Dhosa Oolite (*Terebratula sella* beds).	34	27	4	3	8	Seven characteristic of the zone of *A. transversarius* (Lower Oxford) of Europe.
	Athleta beds	20	13	2	5	8	Six characteristic of the zone of *A. athleta* in Europe.
	Anceps beds, with *Ter. biplicata*	27	21	6	...	7	Five of the seven species found also in Europe are peculiar to the beds with *A. anceps*.
	Macrocephalus beds	31	28	..	3	16	Fourteen of the sixteen exclusively found in beds with *A. macrocephalus* in Europe.
Patcham	Upper	8	5	3	...	1 or 2	......
	Lower	...	...	...	...	...	

The occurrence of jurassic rocks in the desert tract to the north of the Rann of Cutch has been known for many years. A few species were obtained from the country immediately north of the Rann by Sir H. Pottinger, but the only tracts which have yet been explored by a geologist lie further north, near Bálmer, in Jaisalmer and in Bíkaner. Five distinct groups of rocks have been recognised and named as follows :—

5. Abur beds.
4. Parihar sandstones.
3. Bedésar group.
2. Jaisalmer limestones.
1. Bálmer sandstones.

The Bálmer rocks consist of sandstones, grits, and conglomerates, the compact, and a still finer rock approaching a compact shale, white, but veined and blotched with purple. These beds must attain a considerable thickness, but only the lowest are well exposed, the upper strata being probably softer. The lower members of the group are well seen at Bálmer itself, where they rest upon the Maláni volcanic rocks, and in some hills near Naosir, about thirty miles farther east. Fragmentary plant remains are common, but nothing sufficiently well preserved for determination has been found, and no remains of animals have been detected in the beds.

East and south-east of Jaisalmer, beneath the marine jurassic beds of the next group, a considerable thickness of white, grey, and brown sandstones is exposed, interstratified with numerous bands of hard black and brown ferruginous sandstone and grit. Towards the base are some soft argillaceous sandstones, streaked and blotched with purple, and closely resembling the Bálmer beds, except that they are less hard. These rocks probably belong to the Bálmer group. They have a lithological resemblance to the Umia group of Cutch and to some of the Gondwána beds of the Central Provinces. The only fossils found, except fragments of leaves, were some pieces of dicotyledonous fossil wood

The sandstones and limestones of the Jaisalmer group rest upon the beds last described, and consist of thick bands of compact buff and light brown limestone interstratified with grey, brown, and blackish sandstone, with some conglomerate. The limestone forms conspicuous scarps close to the town of Jaisalmer, and it is highly fossiliferous, containing amongst other species *Terebratula biplicata*, *T. intermedia*, *Pholadomya granosa*, *Corbula lyrata*, *C. pectinata*, *Trigonia costata*, *Nucula cuneiformis*, *Pecten lens*, and *Nautilus kumagunensis*. *Ammonites* (*Stephanoceras*) *fissus* has been obtained from the neighbourhood, but very possibly from a different horizon, for in Cutch it belongs to the Dhosa oolite and the Kantkot sandstone (both Oxford), whilst *Nautilus kumagunensis* is only found at a lower horizon in the beds with *Am. macrocephalus* at the base of the Chári group. *Terebratula biplicata* in Cutch is chiefly characteristic of a rather higher horizon than that of the *macrocephalus* beds. There can, however, be but little hesitation in referring the Jaisalmer limestones to the age of the Chári group.

The Jaisalmer limestones are overlaid by a group of purplish and reddish sandstones, with thin layers of black vitreous ferruginous sandstone. Some beds of a red calcareous sandstone contain fossils which have not been determined. Some of them closely resemble forms from the Katrol group of Cutch. These sandstones have been distinguished as the Bedésar group[1] and are overlaid by the Parihar group of soft, white felspathic sandstones, which weather into a clean sugary sand, largely composed of

[1] R. D. Oldham, *Records*, XIX, 158, (1886).

subangular fragments of transparent quartz. The uppermost beds of the jurassics are sandstones, shales and limestones, among which there is one very conspicuous fossiliferous band, known locally as Abur, a name which has been applied to the village where it is quarried, and formerly referred to in the Survey publications as the ammonite bed of Kuchri.[1] It is a thin bed of buff coloured limestone, weathering red, and abounding in yellow coloured ammonites of three or four species. None can be safely identified with any Cutch species, though one form is very near *A.* (*Stephanoceras*) *opis*, which is common to the Dhosa oolite and Kantkot sandstone of Cutch.

Jurassic beds are again found in the western half of the Salt range and in its trans-Indus continuation, where the lower beds rest upon the triassic rocks, and consist of sandstones of varying colour, succeeded in ascending order by limestones, clays, and soft white sandstones, then come bands of hæmatite, several feet in thickness, and thinner layers of golden oolite, precisely similar to the rock of Cutch, the upper portion of the group consisting of coarse brown sandstones, yellow marls, white sandstone and hard grey limestone bands. The sandstones are often conglomeratic and the limestones are most largely developed to the westward. Small layers and patches of bright jetty coal occur in places towards the base of the group and west of the Indus near Kálabágh, but there is nothing like a seam of coal. The patches of coal appear to be merely carbonised fragments of drift wood.

The Salt range jurassic beds are not found east of the neighbourhood of Naushahra. They begin to appear a little further west than the triassic ceratite strata and, increasing much in thickness, continue to the Indus. West of that river the same rocks re-appear in the Maidáni (Chichāli) hills, where they are well developed and more fossiliferous than in the Salt range. They are well seen in the Chichāli pass, and extend round the curve of the range further to the southward than the Productus limestone does, disappearing beneath the tertiary rocks about six miles south of Mulakhel,[2] but are again found in the Shaikh Budín hills.

Until the fossils of the Salt range jurassic beds are examined in detail, it is not possible to say exactly what members of the jurassic series are represented. Dr. Waagen has shown that there is a close connection between the Salt range oolitic beds and those of Cutch,[3] but that the Spiti shales of the Himálayas contain a very different fauna. The Kelloway portion of the Chári beds is distinctly represented in the Punjab, and some of the higher jurassic groups also. *Cephalopoda* are scarce, except west of the Indus, where *Ammonites* and *Belemnites* occur rather more

[1] W. T. Blanford, *Records*, X, 16, (1877).

[2] Fleming, *Jour. As. Soc. Beng.*, XXII, 278, (1853).

[3] *Pal. Indica*, series ix, 236, (1875).

abundantly, especially in the fine section of jurassic beds exposed in the Chicháli pass.

The jurassic rocks of the Himálayas have long been known by their fossils, which had acquired a sacred character and become objects of commerce. They are represented in the central Himálayas by a series of dark grey and black shales, known as the Spiti[1] shales, whch contain numerous calcareous concretions, a large proportion of which are formed round fossils. The thickness of these shales is about 300 feet, and their original extension has been very much restricted by the great denudation they have undergone.

The Spiti shales are said to lie conformably on the underlying limestones, which were regarded by Dr. Stoliczka as lias, and the presence of passage beds, in the shape of a small thickness of clayey slates, is indicated. In view of the probability that the beds regarded as lias are in reality uppermost trias, and of the abrupt change of lithological character which takes place at the base of the Spiti shales everywhere except in Spiti, it seems more natural to suppose that there is an unconformity which has not yet been recognised.

The Spiti shales are overlaid by about 600 feet of a light yellow silicious sandstone, known as the Giumal sandstone, which was regarded by Dr. Stoliczka[2] as upper jurassic in age, but has more recently been classed as neocomian.[3]

Jurassic rocks are known to occur north of Nepál, characteristic fossils having been brought from that direction by traders, and the Spiti shales have been recognised to the north of the Karakoram range in one direction, and in Hazára in the other. In Hazára they are perfectly typical and are recognisable both lithologically and palæontologically. In the Sirban mountain they rest unconformably on a surface of triassic limestones, eroded and pierced by boring molluscs. They are conformably succeeded by sandy and calcareous beds, abruptly overlaid, but with no observed unconformity, by a sandstone containing a lower cretaceous fauna.

Further south the Spiti shales have not been recognised with certainty, but in the Suláimán range some black shales, overlaid by sandstones, are found below the cretaceous limestone of the Takht-i-Suláimán[4] and a similar section is said to be observable in the eastern termination of the Safed Koh,[5] but the identification lacks the support of fossil evidence. In the southern part of the Hazára district jurassic rocks are represented by a

[1] F. Stoliczka, *Memoirs*, V, 85, (1865).

[2] *Memoirs*, V, 113, (1865).

[3] C. L. Griesbach, *Memoirs*, XXIII, 80, (1891).

[4] *Records*, XVII, 184, (1884).

[5] *Records*, XXV, 81, (1892).

band,[1] composed almost entirely of *Trigonia ventricosa*, associated with layers containing *Ammonites*, *Gryphæa* and *Belemnites*, whose age relative to the Spiti shales is undetermined.

This will be the most convenient place to mention certain fossiliferous beds, underlying the nummulitics of the outliers in south-western Garhwál, whose chief interest lies in the fact that they contain the only pretertiary organic remains that have been recognised in the outer Himálayas south of the main snowy range. They were originally discovered by Mr. Medlicott [2] in the Tál valley immediately east of the Ganges and have consequently been frequently referred to as the Tál beds.

They are described by Mr. Middlemiss as grits or quartzites, frequently calcareous and passing into limestone in places. The fossils are mostly fragmentary, but among them he considered that corals, belemnites lamellibranchs and gasteropods were represented, the whole indicating a probably jurassic age. Subsequent critical examination, however, failed to discover any specimens determinable with sufficient accuracy to indicate the age of the rocks. Judging from their stratigraphical position they are probably mesozoic, but beyond this nothing can be decided.

[1] A. B. Wynne, *Records*, XII, 125, (1879).
[2] *Memoirs*, III, pt. ii, 69, (1864).
[3] *Records*, XVII, 161, (1884); XVIII, 73, (1885).

CHAPTER X.

MARINE CRETACEOUS ROCKS OF THE INDIAN PENINSULA.

Cretaceous rocks of Southern India—Relations to cretaceous rocks of Assam and South Africa—Cretaceous rocks of the lower Narbadá valley—Relations to cretaceous of Europe—Contrast to cretaceous of Southern India—An Indo-African land connection—Doubtfully cretaceous sandstones of the Narbadá valley and of Káthiáwar.

The occurrence of cretaceous rocks in Southern India was first observed in 1840 by Mr. Kaye of the Madras Civil Service, who, in company with Mr. Brooke Cunliffe and others, collected a large series of fossils, which were examined by Professor E. Forbes. The rocks near Pondicherri had, however, some years before attracted the notice of Mons. E. Chevalier, but no account of them was published until after the appearance of Mr. Kaye's description. A collection of fossils from the neighbourhood of Pondicherri was examined by A. D'Orbigny, and referred to an upper cretaceous age. Forbes, on the other hand, referred the beds of Trichinopoli and Viruddháchalam to the age of the upper greensand or gault, and the Pondicherri beds to the neocomian. It was shown by Mr. H. F. Blanford that beds of two ages exist near Pondicherri, and he, following Professor Forbes, considered the lower of these or Valudayur beds neocomian and older than any of the Trichinopoli rocks, but the thorough examination of all the Southern Indian fossils by Dr. Stoliczka has proved that the general homotaxis is middle and upper cretaceous, and that the neocomian and oolitic forms, which led to a portion of the beds being originally classed as lower cretaceous, are less numerous than the middle cretaceous species with which they are associated. It was also found that the fauna of the Valudayur beds had more species than was at first supposed in common with the lowest group of the Trichinopoli area, and the two were consequently considered identical. The *Cephalopoda* of the lower beds comprise several species found in the gault of Europe, and the number was at first supposed to be larger than it proved on subsequent closer investigation, but as there are scarcely any representatives of gault forms amongst the very numerous and beautifully preserved *Gasteropoda* and *Lamellibranchiata* (*Pelecypoda*), the whole of the Southern Indian beds were finally referred by Dr. Stoliczka to an age not older than the upper greensand of England (cenomanian), and ranging thence to the upper chalk (senonian).

The rocks of cretaceous age in Southern India[1] occupy, with relation to older and newer formations, a very similar position to that of the outcrops of upper Gondwána beds farther to the northward. The cretaceous beds occur in the great plain which extends along the Coromandel coast from the north of the Godávari to Cape Comorin. They rest to the west upon the gneiss, or occasionally upon small patches of the upper Gondwána (Rájmahál) beds, they have a low dip to the eastward, and are covered up on the east by pleistocene beds, known as Cuddalore sandstones, and by the alluvium of the sea coast. The cretaceous beds are exposed at the surface in three detached areas, separated from each other by the alluvial deposits of the Penner and Vellar rivers. The southern and largest of these areas, between the Vellar and Coleroon rivers, is in the Trichinopoli district, and known as the Trichinopoli area. North of Vellar ar two much smaller exposures near Viruddháchalam and Pondicherri respectively, and named from those towns.

The Trichinopoli area extends about twenty-five miles from north to south, and for about the same breadth where widest, but it is very irregular in form. South of the Coleroon (the principal outlet of the river Cauvery) no cretaceous beds have been traced, and the southern boundary of the cretaceous area north of the Coleroon is chiefly formed by gneiss. To the northward the cretaceous rocks disappear beneath the alluvium of the Vellar river and re-appear north of the river at Viruddháchalam forming the Viruddháchalam area, in which only the highest cretaceous group is exposed, and even this is only visible at very few points. It occupies a tract of country about fifteen miles long from north-north-east to south-south-west by about five broad, with gneiss to the west and tertiary Cuddalore sandstone to the east. There is a second break in the rocks at the Penner river, and alluvium extends to the neighbourhood of Pondicherri, causing an interval of about twenty-five miles in the belt of cretaceous rocks before they re-appear near Valudayur, ten miles west by north from Pondicherri. Here they occupy a small tract of country about twelve miles long from north-east to south-west, by six miles broad, and only separated from the sea on the east by a band of Cuddalore sandstones two to four miles wide. To the west is a narrower strip of Cuddalore sandstone, beyond which the country consists of gneiss.

[1] For a complete description of the geology by Mr. H. F. Blanford, see *Memoirs*, IV, pp. 1-217, (1862). The fossils are described and figured in four volumes, comprising Series i, iii, v, vi, and viii, (1861-73) of the "*Palæontologia Indica*," all by Dr. F. Stoliczka, with the exception of the *Belemnites* and *Nautili*, which are by Mr. H. F. Blanford. Some additional notes on the *Cephalopoda* are published in the *Records*, I, 32, (1868), and on the fossils generally, by Mr. R. B. Foote, in *Records*, XII, 159, (1879).

In all three areas there appears to be a low dip to the east, the lowest beds appearing at the western boundary and higher groups succeeding in regular order to the eastward. Many of the dips seen in the rocks are, however, deceptive, being due to oblique lamination or false bedding, which prevails extensively throughout the series and especially in the southern portion of the Trichinopoli area. In the Viruddháchalam and Pondicherri areas the rocks are ill seen, and the dips are less distinct, but there appears every probability that the same low dip prevails in the Pondicherri or Valudayur area; the direction is, however, south-east rather than east.

The series is divided into three groups named, in descending order Ariyalúr, Trichinopoli, and Utatúr. The following table taken from the Palæontologia Indica[1], exhibits Dr. Stoliczka's final views as to the representation by these groups of the European cretaceous subdivisions:

South India.		England.	France.	Germany.
Ariyalúr Group.	Zone of *Nautilus danicus* and *Ammonites ootacodensis*, *Ostrea pectinata*, and *O. ungulata*, *Gryphæa vesicularis*, *Inoceramus cripsii*, *Crania ignabergensis*.	Upper chalk	Senonian .	Ober Quader.
Trichinopoli Group.	Zone of *Ammonites peramplus*, *Pholadomya caudata*, *Modiola typica*, *Ostrea diluviana*, *Rhynconella compressa*.	Lower chalk	Turonian .	Mittel Quader.
Utatúr Group.	Zone of *Ammonites rostratus* and *rotomagensis*, *Inoceramus labiatus*, *Exogyra suborbiculata* (*Gryphæa columba*), and *Terebratula depressa*.	Chalk marl and upper greensand.	Cenomanian or Tourtia.	Unter Quader Unterer Quadersandstein, and Unterer Pläner.

The Utatúr group derives its name from a large village twenty miles north-north-east of Trichinopoli. The beds composing the group are chiefly argillaceous; fine silts, calcareous shales, and sandy clays, frequently concretionary and more or less tinted with ochreous matter, prevail throughout the group, and in the southern portion of the area constitute almost the entire bulk of the deposit. North of the parallel of Utatúr, limestone bands become intercalated in the lower or western part of the group and sands, grits, and conglomerates in the upper or eastern part, these changes in mineral character being accompanied by a great enrichment of the

[1] *Pal. Indica*, series viii, IV, Introduction, p. ii, (1873). As there are several slight errors, or misprints, the proof of the original was probably not corrected by Dr. Stoliczka.

fauna in the first case and an impoverishment in the latter. Conglomerates are of very rare occurrence in the lower beds. Gypsum occurs in most of the argillaceous strata, and is to a certain extent characteristic of the group. The dips are often irregular, and apparently due to the original deposition of the beds on shelving banks. This irregularity of dip renders it impossible to form any trustworthy estimate of the thickness attained by the group as a whole; it may, however, be roughly estimated as probably not less than 1,000 feet.

At the base of the Utatúr group there are, in several places, large masses of coral reef limestone, resting sometimes on the upper Gondwána plant beds, more frequently on the gneiss, and occasionally on the lowest beds of the Utatúr group itself. The rock is a nearly pure pale coloured limestone, compact and homogeneous, but often with a flaggy structure, and frequently irregularly banded with white streaks, which, on weathered surfaces, exhibit the corals of which they are composed. The mass of the rock also sometimes abounds in corals, but more frequently no organic structure can be traced. In lithological character this rock precisely resembles the coral reef limestone of the present day, as described by Darwin, Dana, Jukes, and other observers.

The usual position of this limestone is at the base of the Utatúr group, resting upon older rocks. The coral reefs appear to have been frequently exposed to denudation during the deposition of the later Utatúr beds, amongst which, in places, calcareous bands are found, apparently derived from the waste of the reefs. The coral limestone now remains in the form of small isolated patches, scattered along the western and southern margins of the cretaceous beds. In one locality, however, close to the village of Caligudi, on the southern boundary of the cretaceous area and twenty miles north-east of Trichinopoli, by far the largest outcrop of the limestone in the area occurs at the base of the Trichinopoli group. This outcrop is of considerable breadth, and extends, with one or two breaks, for about six miles. From an examination of all the circumstances, however, it has been satisfactorily ascertained that this outcrop also belongs to the Utatúr group, and that the Trichinopoli group rests unconformably upon it.

The coral reefs appear to have been scattered over the sea bottom in shallow water, and probably along the coast, at the commencement of the period during which the cretaceous deposits of Southern India were formed. The remaining beds of the Utatúr group were probably deposited in water of moderate depth, and some of them appear to have accumulated on submarine banks, possibly formed in tidal channels. Hence the false bedding so prevalent in the rocks. The coarser constituents of the rocks to the northward appear to indicate that the current which brought the sediment flowed from that direction, and the occurrence of littoral forms of mollusca in greater abundance throughout the northern parts of the area may be accounted for in the same manner. The beds in the southern portion of the Utatúr area appear to have been formed of fine silt, deposited in a bay where the force of the current was less than to the northward, and the fossils which occur are mostly the remains of pelagic animals, such as *Belemnites*, a few *Ammonites*, chiefly of *Cristati* group, or else peculiar forms of *Vermetidæ* (*Tubulostium discoideum* and *T. callosum*), which probably lived in the mud. The *Ammonites* and *Nautili*, which are numerous to the northward, are scarce in the southern portion of the area. Cycadeaceous fossil wood, sometimes bored by *Teredo* and other *Pholadidæ*, abounds in certain parts of the group. On the whole, there appears every reason to believe that the Utatúr beds were formed in the neighbourhood of a coast line.

The distribution of the Utatúr beds in the Trichinopoli district is very simple. They form the western portion of the cretaceous area throughout, their outcrop being in general from three to five miles broad, except to the northward, where it diminishes in consequence of the beds being overlapped by those of the next group, till, at the village of Olapádi in the northern portion of the tract, the breadth of the Utatúr outcrop does not exceed half a mile. At the extreme northern point of the area, both the Utatúr and Trichinopoli groups are completely overlapped by the uppermost subdivision.

The Utatúr beds are not represented in the Viruddháchalam area, but they re-appear, as already mentioned, near Pondicherri. Here the beds formerly classed as the Valudayur group, and considered neocomian by Forbes, but which were shown by Stoliczka to contain several species of fossils common to the Utatúr group, consist chiefly, like the strata near Utatúr, of argillaceous beds, sandy shales, and sands, with occasional bands of limestone and calcareous concretionary nodules. Conglomerates occasionally occur amongst the lowest beds seen, but the most characteristic band is composed of dark grey, compact limestone in large nodules, sometimes highly fossiliferous, *Baculites vagina* being the commonest fossil.

The area occupied by the Utatúr or Valudayur beds near Pondicherri extends from Valudayur for about nine miles to the north-east and is about four miles broad. The beds are not seen to rest upon any older formation. The country north and south is covered with alluvium. To the eastward the Utatúr beds disappear beneath the Ariyalúr group, and to the westward beneath the Cuddalore sandstones of Tiruvakarai (Trivicary). The beds to the westward appear to be the lowest, and there is a dip to the eastward.

The fauna of the Utatúr group is very rich, no less than 297 species of *Invertebrata* having been described from it. It has yielded an especially large number of *Cephalopoda*, 109 species, 95 of which have not been met with in the Trichinopoli or Ariyalúr group. Of these 109 species, 27 are known to occur in Europe, or elsewhere out of India, and although the majority are distinctly and characteristically middle cretaceous forms, three are, in Europe, neocomian species, *viz. Nautilus neocomiensis*, *Ammonites velledæ*, and *A. rouyanus*, whilst no less than nine are found in the gault, several of the latter ranging, however, into the upper greensand (cenomanian). Amongst the forms which are not European, the most remarkable are three species belonging to the section of *Ammonites* known as *globosi*, which, amongst European rocks, are especially characteristic of the triassic period. A very large proportion of the *Cephalopoda* were collected in the neighhourhood of two villages, Odiam and Maravatúr, on the road from Perambalúr to Arialúr, and about twelve miles north-east of Utatúr.

The *Gasteropoda* comprise only 43 species, a number far inferior to that found in each of the other groups, and the majority of these are littoral forms. The *Lamellibranchiata* (*Pelecypoda*) are 79 in number, the *Brachiopoda* 9, *Echinodermata* 10, and *Corals* 42, with one species of sponge and one annelid. The forms found also in other countries belong almost without exception to the upper greensand (cenomanian) or higher groups, thus presenting a remarkable difference from the *Cephalopoda*, in which gault forms are so largely represented. The only fossils of much importance, besides the *Mollusca*, are the corals, which, from the prevalence of reefs at the base of the group, are superbly represented, no less than 42 species, belonging to 23 genera, being known to occur.

The Trichinopoli, or middle group, of the Southern Indian cretaceous series derives its name from the district of Trichinopoli, to which it is, so far as present exploration extends, entirely restricted. To the south it consists chiefly of sands and clays, very irregularly bedded, with a few bands of limestone and some conglomerates, and it differs lithologically only in one important respect, which will be described presently, from the Utatúr group. North of the parallel of Utatúr regular bands of shell limestones become intercalated in the lower beds of the deposit and, to the northward, the whole group is composed of regularly stratified alternations of sand, sandy clays, and shales, with bands of shell limestone, calcareous grit and conglomerate.

A peculiarity by which both the Ariyalúr and Trichinopoli beds in the southern part of the cretaceous area are distinguished from the Utatúr is the occurrence of granite pebbles in considerable quantity in the gravels and conglomerates of the two former, whilst none are found in the lower subdivision. In the Utatúr group the materials of the few conglomeratic or gravelly beds which occur are derived either from the gneiss or from the coral reef limestone, whilst in the two upper groups conglomerates are more frequently met with, and loose masses of unstratified gravel and beds of rolled pebbles, almost entirely composed of granitic materials and resembling the shingle of a sea beach, are of common occurrence. The source of the granite pebbles was evidently the broad belt of granitic rocks which forms the southern boundary of the cretaceous area, dividing it from the alluvium of the Cauvery throughout the greater portion of its extent, and the necessary inference is, that this band of rock was in all probability beneath the sea during the deposition of the Utatúr beds, and that it was elevated above the water in the interval between the Utatúr and Trichinopoli ages.

The Trichinopoli beds are, even more characteristically than the Utatúrs, the littoral deposits of a shallow sea. This is proved, not only by the frequent occurrence of coarse sediment and the great irregularity of the deposits in part of the area, but by the abundance of fossil wood, almost exclusively exogenous and apparently cycadeaceous. Trunks of trees are met with of great size, as much as three feet in diameter and sixty feet in length, much of the wood being perforated by boring mollusca.

The shell limestone of Garudamangalam, east of Utatúr, and other places is a very fine, hard, bluish grey, translucent rock, usually abounding in beautifully preserved shells, both *Gasteropoda* and *Lamellibranchiata*, which retain their original polish, and occasionally even the colouration of their surfaces. This rock, known as Trichinopoli marble, is largely quarried for ornamental purposes, and has yielded a considerable proportion of the fossils found in the group. The limestone occasionally contains pebbles of granite or fragments of fossil wood, either of which is sufficient to distinguish it, even when it is unfossiliferous, from the Utatúr limestones.

The beds of the Trichinopoli group are unconformable to the Utatúrs, upon which they rest throughout the greater part of the area, the evidence of unconformity not being confined to overlap, but depending chiefly upon the proof afforded, by the rocks at the southern edge of the area, that the Utatúr beds had been disturbed and faulted, probably at the period of upheaval of the granitic band already mentioned, before the deposition of the Trichinopoli formation. Elsewhere also, the Trichinopoli beds rest in places upon a denuded surface of Utatúrs. There is also a great change in the fauna. In the southern portion of their range the Trichinopoli beds rest partly upon the coral reefs, which have been already shown to be some of the lowest beds of Utatúr age, and partly on the metamorphics, a considerable portion of the boundary being formed by the granitoid rocks so frequently mentioned already.

The present group, like the Utatúr, is so irregularly bedded, and the dips seen are so frequently those of original deposition, that no trustworthy estimate of the thickness can be formed. The general inclination is to the eastward, the average breadth of the outcrop is nearly the same as that of the Utatúr beds, and the same minimum thickness, *viz.* 1,000 feet, may be assumed; the general dip of the bedding in the more regularly stratified portion of the group to the northward is, however, lower than in the underlying group, averaging about 6°. The beds thin out greatly to the northward, and are at length completely overlapped by the Ariyalúrs.

It has already been stated that the Trichinopoli group is confined, so far as is at present known, to the Trichinopoli area. Within that area it forms a belt east of that formed by the Utatúr group, and extending similarly from south-south-west to north-north-east. The Trichinopoli outcrop is, however, broader in the southern half of the area, where it is about four miles across, than in the northern half, where it is in no place more than two miles wide. It thins out and disappears completely about two miles south of the place where the Utatúrs are similarly overlapped by the Ariyalúr beds. Along the southern boundary of the Utatúr area, several outliers of Trichinopoli beds are found, resting partly on the Utatúrs and partly on the gneiss, and occasionally overlying the faulted boundary between the two formations. These small outliers, one of which, south of Tirupatúr, forms the south-western corner of the whole area, are composed of coarse sands and conglomerates, usually unfossiliferous, but occasionally containing *Chemnitzia undosa* and other characteristic Trichinopoli fossils, and the materials of which they are formed are derived chiefly from the metamorphic rocks, but partly from the denudation of the Utatúr beds.

The fauna of the Trichinopoli group, although not quite so rich as that of the Utatúr beds, affords a full illustration of the life existing at the period, 186 species of *Invertebrata* having been described from these beds by Dr. Stoliczka. The *Cephalopoda* are comparatively poorly developed, only 23 species having been detected, and of these but 10, of which four are European, are in India peculiar to the group. All the *Cephalopoda* identified belong to the two genera *Nautilus* and *Ammonites*, the non-discoid Ammonitoid genera, such as *Anisoceras*, *Scaphites*, *Turrilites*, etc., so largely represented in the Utatúr group, as well as the *Belemnites*, so abundant in the lower subdivision, being apparently wanting in the Trichinopoli beds. The *Rotomagenses* Ammonites, so characteristic of the lowest cretaceous subdivision in Southern India, are also wanting in the higher groups, with one doubtful exception. A few forms, usually associated with older strata, still survive, however, such as *Ammonites menu*, belonging to the *Armati* (a jurassic group), *A. koluturensis* of the *Macrocephalus* group, allied to such oolitic species as *A. macrocephalus* and *A. herveyi*, and *A. theobaldianus*, one of the *Planulati* allied to upper jurassic forms, such as *A. biplex.* Most of the types found are, however, characteristically upper cretaceous.

On the other hand, *Gasteropoda*, comprising 86 species, are much more abundant than in the Utatúr group, *Lamellibranchiata*, comprising 66 species, being rather less numerous. There are but 5 *Brachiopoda* and 6 corals, whilst no *Echinodermata* have been recognised. The *Gasteropoda* include several siphonostomate genera, rare in the older rocks, and not found in the Utatúr beds, the number increasing greatly in the next higher subdivision, that of Ariyalúr. The whole fauna exhibits a mixture of upper and middle cretaceous forms, and appears fairly to represent the lower chalk of England or the turonian of continental geologists.

The name of the highest group of the Southern Indian cretaceous series is derived from the town of Ariyalúr, which is situated nearly in the middle of the comparatively large expanse of Ariyalúr beds in the Trichinopoli district. The country occupied by the beds of this group is much covered with cotton soil, and sections are even rarer than in the two lower cretaceous subdivisions.

The Ariyalúr beds are more sandy than the two lower groups and more uniformly bedded, the beds being thick and homogeneous, consisting principally of white unfossiliferous sands and grey argillaceous sands, with casts of small fossils. Beds of calcareous grit and nodular calcareous shales are found towards the base, and again in the upper portion of the group, constituting two highly fossiliferous zones, separated by a considerable thickness of deposits, in which fossils are rare or wanting, although some interesting remains of a *Megalosaurus* were found in one of the beds. A band of flints is associated with the uppermost beds. There is a marked difference between the fossils of the upper and lower zones in Trichinopoli, and it appears very probable that further examination of the rocks, now that the fossils have been compared and determined, would justify the separation of this group into two—a probability which was pointed out by Mr. H. F. Blanford at the time of the original survey, although not shown on the map nor applied in the discrimination of the fossils, because of the doubts which remained as to the distinction of the two subdivisions in the Pondicherri area, where the fossils of both upper and lower Ariyalúr beds appear to occur together. Conglomerates are of rare occurrence in the Ariyalúr group, though a coarse bed is found in places near the base, and there is but little irregularity in the bedding, except close to the southern boundary. The constituents of the Ariyalúr beds were derived chiefly from the metamorphic rocks, amongst others from the granitic band to the southward, but a portion of the sediment must have been furnished by the waste of some of the older cretaceous groups, probably the Utatúrs.

The above description of the lithological characters is principally taken from the beds near Ariyalúr, but it is also to a great extent applicable to the rocks seen near Viruddháchalam and Pondicherri. In both localities the Ariyalúr deposits are chiefly represented by sands or sandy clays, and by beds of arenaceous limestone or calcareous sandstone at the base of the group. The strata appear to thin out to the northward, and it is far from clear whether the uppermost fossiliferous zone extends in that direction, although some of its characteristic fossils, such as *Nautilus danicus*, occur abundantly near Pondicherri. It has not, however, been hitherto found practicable to determine whether a distinct upper zone exists near Pondicherri or whether representatives of the upper fauna occur in beds of lower horizon than those in which the same species are found near Ariyalúr.

There is consequently some obscurity concerning the relations of the beds belonging to the Ariyalúr group amongst themselves, and this difficulty is complicated by the circumstance that there is in many places an apparent passage from the Trichinopoli group into the Ariyalúr beds, the rocks being similar in mineral character near the junction, and the fossils being chiefly forms which appear to range from one group into the other. It is highly probable that further examination of the ground, which, as has been already noticed, is so much concealed by superficial accumulations that the different groups can frequently only be traced by their fossils, would show either that the number of groups or of palæontological zones must be increased, or else, in some cases, that fossils supposed to have been procured from the Trichinopoli group, have really been derived from the Ariyalúr, and *vice versá*.

The area occupied by the Ariyalúr beds in the eastern portion of the Trichinopoli tract amounts to about 200 square miles, or more than that covered by both the other subdivisions together, the outcrop where broadest, near Ariyalúr, is about sixteen miles wide and extends for twenty-six miles from north to south.

The Ariyalúr beds also occupy the greater portion of a tract sixteen miles long by five miles broad near Viruddháchalam, and another about twelve miles long from south-west to north-east, by two miles broad, west of Pondicherri, whilst a very small exposure of them occurs close to the coast ten miles north of Pondicherri, and another still smaller three miles farther north.

The lowest fossiliferous zone is found resting upon the Trichinopoli beds, throughout the western portion of the Ariyalúr area in the Trichinopoli district, and the same zone appears to be also represented in the Viruddháchalam and Pondicherri exposures. The great bulk of the outcrop in all three tracts appears to consist of the thick sands, with but few determinable fossils, forming the middle portion of the formation, whilst the upper fossiliferous beds are only seen north of Ariyalúr, near the villages of Sainthoray, Niniyur, and other places farther north, in the long strip of cretaceous rocks which forms the north-eastern extremity of the Trichinopoli area.

Although the thickness of the Ariyalúr group can be estimated with a nearer approach to probability than in the case of the two lower cretaceous formations, the estimate is still far from accurate. The dip of the beds is very low, rarely exceeding two or three degrees, the general inclination being north-east, and the whole of the beds in all probability do not exceed 1,000 feet in Trichinopoli. Near Viruddháchalam they appear to be very thin, and in the neighbourhood of Pondicherri they are too obscurely exposed for any estimate of their thickness to be attempted. There is an apparent diminution of thickness to the northward, as in the other groups, but this attenuation appears to be greatest near Viruddháchalam, and takes place less rapidly farther north, even if the beds are not thicker in that direction.

The Ariyalúr beds, as has been already stated, frequently appear to pass into the Trichinopoli group at their base. They overlap the lower groups however, both to the north and south, and there is, in places, an appearance of unconformity where they rest upon the Trichinopoli beds, nor is it easy to understand the very rapid diminution in the thickness of the latter to the northward without supposing that they had been partially denuded in pre-Ariyalúr times.

As was noticed in the description of the Utatúr group, the Ariyalúr beds rest upon the Utatúrs for a distance of rather more than two miles in the northern part of the Trichinopoli area, and still farther north the former were deposited directly on the gneiss. They also rest on the gneiss throughout the whole breadth of their outcrop in the south of the Trichinopoli area, and in the Viruddháchalam cretaceous tract, whilst in the neighbourhood of Pondicherri they are deposited to the eastward on the Valudayur representatives of the Utatúr group, and to the westward no beds are seen beneath them, the Cuddalore sandstones covering the boundary completely. Throughout the Trichinopoli and Viruddháchalam areas the Ariyalúr beds disappear to the eastward beneath the Cuddalore sandstones, which are unconformable to the cretaceous beds, and the latter are covered up by alluvial deposits, intervening between the three areas, in the valleys of the Vellar and Penner rivers, and also to the north of the Pondicherri area.

The Ariyalúr beds appear to have been chiefly deposited in a tranquil sea of small depth, although the deposits are less characteristically littoral than those of the Trichinopoli group, and the evidence of the neighbourhood of land afforded by the occurrence of fossil wood is less abundant.

The invertebrate fauna of the Ariyalúr group exceeds in richness even that of the Utatúr beds, no less than 365 species having been detected in the uppermost subdivision of the cretaceous rocks of Southern India. The *Cephalopoda* comprise 36 species, *Gasteropoda* 138, *Lamellibranchiata* 117, *Brachiopoda* 12, *Bryozod* 23, *Echinodermata* 26, *Anthozoa* 10, *Foraminifera* 1, and *Vermes* 2. It is highly probable that this large number may be due partly to the circumstance that the Ariyalúr deposits comprise two groups differing somewhat in age. The lower fossiliferous beds, from which the bulk of the fossils have been procured, correspond very fairly with the senonian beds of France and the upper chalk with flints of England. From this horizon all the *Cephalopoda* found in the formation have been derived, with the exception of *Nautilus danicus*, which was only observed in the upper beds of Niniyur, etc., in the Trichinopoli area, although some specimens were obtained, apparently from a lower horizon, near Pondicherri. The fauna of these upper beds will be noticed separately; the following remarks apply to the remainder of the group.

In the Ariyalúr beds, as in the lower subdivisions, there are some forms of *Cephalopoda* which are in Europe characteristic of older beds. These comprise two gault species of *Nautilus*, *N. bouchardianus* and *N. clementinus*, *Ammonites menu*, found also in the lower groups, and belonging to the jurassic section of *armati*, *A. velledæ*, a lower and middle cretaceous form in Europe, two *macrocephali*, *A. deccanensis*, and *A. arrialoorensis*, and one of the *Planulati*, *A. theobaldianus*. Very few older forms occur in the other classes of mollusca, and the great majority of the species common to Europe are found in the upper cretaceous beds of England, France, and Germany.

The most striking peculiarity of the Ariyalúr fauna is the great abundance of *Gasteropoda*, and especially of the carnivorous prosobranchiate forms, which, as is well known, appear to replace the *Cephalopoda* of the older periods in tertiary and recent seas. Several genera not previously known from cretaceous beds have been detected in the Ariyalúr group, and the *Cypræidæ* and *Volutidæ* are especially well represented. The *Lamellibranchiata* are also very numerous, whilst all the *Bryozoa* and the great majority of the *Echinodermata* hitherto found in the cretaceous beds of Southern India have been obtained from the highest subdivision. Lower forms of animals are but poorly represented. Amongst the *Vertebrata* the only important species is a *Megalosaurus*,[1] of which a tooth was found in the middle beds of the deposit, together with a number of bones, which, however, could not be extracted in a sufficiently perfect state for determination. The tooth closely resembles that of *M. bucklandi*, found in the Stonesfield slate and Portland oolites of England, and the occurrence of this genus in the upper cretaceous beds of India is of

[1] *Memoirs*, IV, 128, 139, 186;—*Records*, X, 41, (1877).

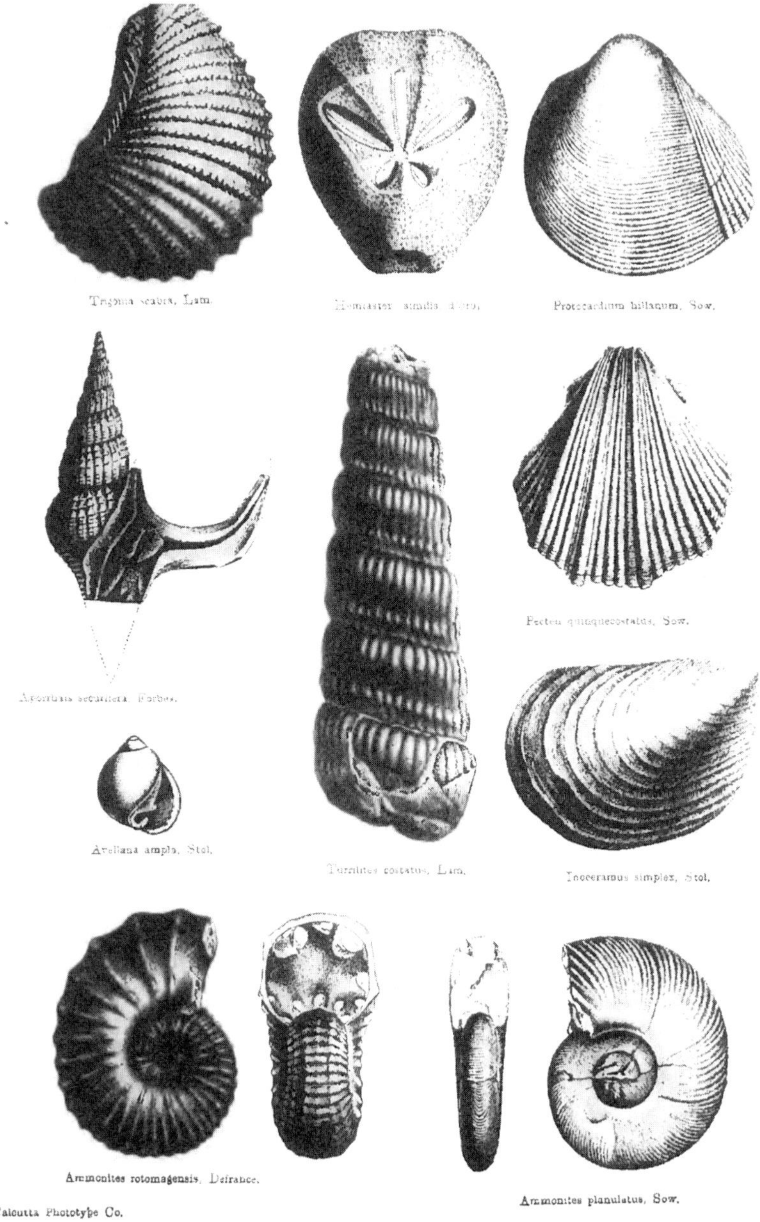

Calcutta Phototype Co.

CRETACEOUS FOSSILS.

peculiar interest, because it only ranges from the lias to the wealden in Europe. In this instance, as in several others, the land fauna appears to have differed more from that which inhabited distant parts of the earth than the marine fauna did.

The fauna of the uppermost Ariyalúr beds found at Niniyur and other places to the north-east of Ariyalúr, comprises very few species which are found in the lower portion of the group. Some of the fossils found most abundantly, such as *Nautilus danicus* and *Orbitoides faujasi*, are characteristic of the uppermost cretaceous deposits of Maestrich, Aix la Chapelle, and the Danish Island of Rugen (Danien of D'Orbigny). No other Cephalopod except *Nautilus danicus* occurs in the Niniyur beds, whilst the characteristically mesozoic genera *Inoceramus*, *Radiolites*, *Trigonia*, *Trigonoarca*, and *Leptomaria*, which are abundantly represented in the lower portion of the Ariyalúr group, are entirely wanting in the uppermost fossiliferous zone, where the only important mesozoic genus is *Nerinæa*. On the other hand, however, no typically tertiary forms make their appearance except carnivorous Gasteropoda, and these are not more numerous in proportion than in the lower zone, although some additional forms are represented.

Besides the fossils characteristic of each group, there are a few species which are found throughout the whole series. The most important of these are *Nautilus huxleyanus*, *Ammonites planulatus* (cenomanian, gault), *Ampullina bulbiformis* (turonian, senonian), *Gyrodes pansus*, *Solariella radiatula* (senonian), *Vola quinquecostata* (upper and middle cretaceous), *Ammonites menu*, Forbes, is also supposed to be found in all three subdivisions, although there is some doubt about the Utatúr beds, and a rare *Lucina*, *L.* (*Myrtea*) *arcotina*, has also been procured from all the groups. Some of these fossils, although found throughout the series, are especially characteristic of one subdivision, as in the cases of *Nautilus huxleyanus* and *Solariella radiatula*.

A large number of forms are common to two groups. The table on page 245 exhibits the number of each class of *Invertebrata* found in the different formations, and the proportion found also in Europe, or common to two or more groups. The *Vertebrata* are represented by 17 species of fishes and one saurian, but the remains are of the most fragmentary description, consisting in most cases of single teeth, and it is not certain from which group some of the specimens were originally derived.

Adding these few vertebrata to the numbers given in the table, we have a grand total of nearly 800 species of animals from the cretaceous deposits of Southern India. Much time was devoted to the collection of the fossils, and their exhaustive examination by Dr. Stoliczka has furnished the best evidence extant for the correlation of any Indian fossil fauna with that of European rocks of corresponding age. Of the whole invertebrata 16·36 per cent. consist of forms known to occur in cretaceous beds in Europe. The great majority of these are middle or upper cretaceous (cenomanian to senonian), but there are amongst the *Cephalopoda* several forms which, in Europe, have only been found in lower beds (neocomian and gault), whilst a few are representatives of European jurassic forms, and three species of *Ammonites* belong to a triassic section of the genus. The general facies of the cephalopodous fauna found in the lowest group, that of Utatúr, approximates to that of the European gault, but nearly all the species of the other classes of mollusca found in the same beds belong to a higher horizon, cenomanian (upper greensand), or even higher.

The whole of the cretaceous rocks of Southern India appear to have been formed in shallow water, in the neighbourhood of a coast line, and it is possible that the relative elevations of the country have undergone but little change since cretaceous times. Then, as now, there was higher ground to the westward, and the ancient coast line appears to have been approximately parallel to the present, although farther to the west. We have thus in the cretaceous formation a confirmation of the evidence, already afforded by the lower mesozoic deposits, that the Indian peninsula is a land area of great antiquity.

Amongst the descriptions by Dr. Stoliczka, of the cretaceous fossils from Southern India, the following species of *Lamellibranchiata* are included from Sripermatúr, twenty-five miles west-south-west of Madras, already mentioned as the typical locality for a group of the upper Gondwána series :—

Sphæriola, sp. indet.	*Pseudomonotis fallaciosa.*
Hippagus emilianus.	„ *inops.*
Yoldia obtusata.	*Lima oldhamiana.*
Trigonoarca galdrina.	*Pecten arcotensis.*

Two of these, *Yoldia obtusata* and *Trigonoarca galdrina*, are also found in the Ariyalúr group of the Trichinopoli district. but the identification of the Sripermatúr species referred to the *Yoldia* is slightly open to doubt. *Trigonoarca galdrina* is, however, a well marked form, and it belongs to a characteristically cretaceous genus.

The specimens were collected by the late Mr. Charles Oldham before the country was properly examined, and there appears some slight doubt as to the precise beds from which they were obtained. Some of the specimens were from Sripermatúr itself, others from Rajah's Choultry. The only cretaceous fossils found by Mr. Foote, who mapped the country in the Sripermatúr neighbourhood, occurred in water-worn blocks of grey or greenish grey gritty sandstone, resting loosely on the surface of jurassic

Table showing the Distribution of Invertebrate Fossils in the Cretaceous Rocks of Southern India.

	1	2	3	4	5	6	7	8	9	10	11	12	13	14	15
	Total number of species found in Utatúr group.	Peculiar to Utatúr group in India.	Species in column 1 found also in Europe.	Total number of species found in Trichinopoli group.	Peculiar to Trichinopoli group in India.	Species in column 5 found also in Europe.	Total number of species found in Ariyalúr group.	Peculiar to Ariyalúr group in India.	Species in column 8 found also in Europe.	Species common to Utatúr and Trichinopoli groups.	Species common to Utatúr and Ariyalúr groups.	Species common to Trichinopoli and Ariyalúr groups.	Species common to all three groups.	Total species in South Indian cretaceous series.	Total found also in Europe.
Cephalopoda	109	95	25	23	10	4	36	20	5	4	7	6	3	146	38
Gasteropoda	43	36	5	86	59	6	138	113	10	3	1	1	3	237	30
Lamellibranchiata	79	69	8	66	51	5	117	106	10	6	2	7	2	243	29
Brachiopoda	9	8	3	5	2	1	12	8	3	...	1	3	...	21	9
Bryozoa	...	...	...	...	...	...	23	23	4	...	...	...	...	23	4
Vermes	1	1	...	...	...	...	2	2	2	...	...	...	...	3	2
Echinodermata	10	10	4	...	...	...	26	26	4	...	...	...	...	42	8
Anthozoa	42	42	2	6	5	3	10	9	...	...	...	1	...	57	5
Spongiozoa	1	1	1	...	...	...	...	...	...	...	...	...	...	1	1
Foraminifera	...	...	...	...	...	...	1	1	1	...	...	...	...	1	1
TOTAL	294	262	48	186	127	19	365	308	39	13	11	18	8	774	127

beds near Sripermatúr.[1] The origin of these boulders could not be traced, and the fossils cannot now be found; amongst the forms obtained were four or five species of *Ammonites*, some *Belemnites*, etc.

As will be presently shown, there is a great difference between the fauna of the cretaceous rocks in Southern India and that of the deposits of similar age on the Narbadá, but on the other hand many of the fossils of the Trichinopoli area are found in the cretaceous rocks of the Khási hills, to the north-east of Bengal, between Assam and Sylhet. So many species indeed are common to the Trichinopoli and Khási deposits that it is probable that the two regions were part of the same marine area. The cretaceous rocks of the Khási hills are almost unquestionably identical with those extending throughout the hill ranges south of Assam and the same strata are probably represented in Arakan. The description of these rocks will be best deferred to a subsequent chapter,[2] but the palæontological results may be noticed here.

In a small collection of fossils Dr. Stoliczka[3] recognised the following species. The highest fossiliferous band, about 200 feet below the edge of the cliff at Maosmai, a coarse sandy limestone, contains small *Lamellibranchiata*, a *Cellepora*, and echinoderms; a finer rock is principally made up of an *Astrocænia*, allied to *A. decaphylla.*

From about the middle of the series, above Mahádeo, in a stream under Laisophlang, in a soft, ochreous, glauconitic sandstone these fossils were found:—

Nautilus (? *N. elegans*).
Nautilus, with a central siphon; fragments.
U.T. A *Ammonites planulatus.*
U. „ *dispar.*
U. „ *orbignyanus.*
A. „ ? *pacificus.*
U. *Anisoceras indicum.*
U. „ *subcompressum.*
U. A. *Baculites*, near *B. vagina.*
T. A. *Alaria papilionacea.*
T. A. *Rostellaria palliata.*
T. *Gosavia indica.*
T. A. *Cerithium inauguratum.*
T. *Tritonidea requieniana.*
T. *Hemifusus cinctus.*
Phasianella.
Turritella.
Euspira.
Dentalium.
Janira, near *J. fleuriausiana.*
Exogyra matheroniana.
Spondylus striatus.
Modiola typica.
T. *Cardita jaquinoti* (*orbicularis*).
Cardium.
Terebratula, near *T. carnea.*
Turbinolia.
Hemiaster.
Holoster
Brissus.

The facies of this group rather resembles that of the Utatúr beds of Southern India.

[1] *Memoirs*, X, 61, (1873).
[2] *Infra*, p. 295.
[3] Stoliczka, *Memoirs*, VII, 181, (1869); in the lists the letters U, T, A prefixed signify that the species is found in the Utatúr, Trichinopoli, and Ariyalúr groups respectively.

From the well known fossil locality about two miles from Tharia, on the fourth crosscut taken by the footpath between the zigzags of the road to Cherra Punji, or the first below the Devil's Bridge, the following were named :—

A. *Nautilus lævigatus*
U. A. *Baculites vagina.*
A *Cypræa globulina.*
U. „ *pilulosa.*
T. A. *Rostellaria palliata.*
T. A. *Alaria tegulata.*
T. „ *glandina.*
A. *Lyria crassicostata.*
A. *Volutilithes septemcostata*
T. *Tritonidea requieniana.*
T. *Lathirus reussii.*
A. *Pseudoliva subcostata.*
A. *Turritella pondicherriensis.*
T. A. „ *multistriata.*
U. A.? *Mitreola citharina.*
A. *Euspira lirata.*
T. A. *Gyrodes pansus.*
A. *Gibbula granulosa.*
A. *Nerita divaricata.*
A. *Euptycha larvata.*
A *Actæon curculio.*
Pecten septemplicatus.
Janira quadricostata.
A. *Gryphæa vesicularis.*
Spondylus striatus.
Pecten, near *P. rugosus.*
Inoceramus
T. *Rhynconella compressa.*
Terebratula, sp., probably *T. biplicata* and *T. carnea.*
Ananchytes, *Brissus* } several species, but distinct from any described.

Nearly all the fossils of this list occur also in the Ariyalúr group of Southern India, but there are a number of species in the Tharia beds which appear to be peculiar, and most of them new. It may be worth recording that the observer who collected these fossils considered the latter locality to be lower in the series than the former, which would be remarkable, if true, seeing that the Ariyalúr group is newer than the Utatúr.

Before quitting the subject of the Trichinopoli and Assam cretaceous beds, it is necessary to notice the very remarkable resemblance between a portion of their fauna and the species found in certain strata in South Africa.[1] In the description of the Gondwána system, and again in the account of the upper jurassic beds of Cutch, the remarkable affinities between Indian fossil plants and animals and the forms found in South African beds were repeatedly noticed, and there is a similar connection between the cretaceous formations in the two regions. In some deposits found resting upon the Karoo beds on the coast of Natal, 22 out of 35 species of mollusca and echinodermata collected and specifically identified, are identical with forms found in the cretaceous beds of Southern India, the majority being Trichinopoli species. Amongst the South African fossils are some of the commonest and most characteristic fossils of the Southern Indian cretaceous deposits, namely *Ammonites gardeni* (Ariyalúr), *A. kayei* (Utatúr)

[1] C. L. Griesbach, *Quart. Jour. Geol. Soc.*, XXVII, 60, (1871). Some of the fossils were described by Baily, *Quart. Jour., Geol. Soc*, XI, 454, (1855).

Anisoceras rugatum (Utatúr), *Pugnellus uncatus* (Trichinopoli), *Fasciolaria rigida* (Trichinopoli), *Chemnitzia undosa* (Trichinopoli), *Euchrysalis gigantea* (Trichinopoli and Ariyalúr), *Solariella radiatula* (all three groups), *Avellana ampla* (Trichinopoli), *Turritella multistriata* (Trichinopoli and Ariyalúr), *Pecten* (*Vola*) *quinquecostatus* (all three groups) and *Cardium hillanum* (Trichinopoli group). There is also some slight indication of a representation of the different Indian zones.

From the cretaceous rocks of Madagascar six species of cretaceous fossils were examined by Mr. R. B. Newton [1] in 1889, of which three, *Alectryonia* (*Ostrea*) *ungulata*, *A. pectinata* and *Gryphæa vesicularis* are also found in the Ariyalúr group, the other three species being typical neocomian belemnites, from a different locality and evidently from rocks of an older date.

The South African beds are clearly coast or shallow water deposits like those of India; the great similarity of forms certainly suggests continuity of coast line between the two regions, and thus supports the view that the land connection between South Africa and India, already shown to have existed in both the lower and upper Gondwána periods, was continued into cretaceous times. It is very surprising to compare the middle cretaceous fauna of Southern India with that of the distant beds of Natal, and then with the widely differing forms found in beds of the same age in central India and southern Arabia.

The marine cretaceous formations found in the western portion of the Narbadá valley have been commonly known as Bágh beds, from the town of Bágh, situated about 90 miles west by south of Indore and 35 miles west-south-west of Dhár. The town is not on cretaceous rocks, though they are well developed in the neighbourhood. The occurrence of cretaceous fossils near Bágh was discovered by Colonel Keatinge [2] in 1856, but the existence of fossiliferous limestone in this part of the Narbadá valley had been known for a long time, although the exact locality had not been ascertained. The circumstance that blocks of limestone, containing fragments of *Bryozoa* and other fossils, had been employed in building the houses of Mándogarh, a city now in ruins about twenty miles south of Dhár, first attracted attention, and it was mainly owing to an ingenious and happy suggestion by Dr. Carter [3] that attention was attracted to the neighbourhood of Bágh, where limestone had been observed in 1818 by Captain Dangerfield.

[1] *Quart. Jour. Geol. Soc.*, XLV, 333, (1889).
[2] *Jour. As. Soc. Beng.*, XXVII, 116, (1858).
[3] *Jour. Bom. Br. Roy. As. Soc.*, V, 238, (1857). Geological papers on Western India, p. 685.

The cretaceous rocks of the lower Narbadá valley[1] occur chiefly along the edge of the Deccan traps, and intervene between the latter and the metamorphic rocks. West of Bágh the outcrop of the cretaceous beds may be traced with a few interruptions to the neighbourhood of Baroda. East of Bágh they only occur in places around the inliers of older rocks.

As a general rule, the Bágh beds are composed of a calcareous rock above and of sandstone below, but the character of each portion of the formation varies. Commencing to the eastward, the first place where marine cretaceous beds are known to occur is in the neighbourhood of Bárwai, on the Narbadá, nearly due south of Indore. Here some conglomerates, more or less calcareous, and sandstones containing marine shells, represent the cretaceous formation, and in one place are seen to be distinctly unconformable to an outlier of Mahádeva conglomerate belonging to the upper Gondwána series. From the neighbourhood of Bárwai the whole Narbadá valley is composed of trap for nearly 50 miles to the westward. Lower rocks re-appear near Mándogarh, between which place and Bágh the cretaceous beds are found, forming a narrow fringe to the traps, around several inliers of Bijáwar and metamorphic rock.

The Bágh fossiliferous beds are divided into three zones, all calcareous underlaid by a variable thickness of conglomeratic and sandy beds. The fossiliferous zones are known as (1) the nodular limestone, (2) the Deloa and Chirákhán marl, (3) the coralline limestone.

The nodular limestone group is the most extensive of the three, being found in all the exposures, but at the eastern and westernmost outcrops the peculiar nodular character is wanting. It is an argillaceous, whitish, compact and generally nodular limestone.

The Deola and Chirákhán marl is a soft band, never more than 10 feet thick, which is chiefly interesting as having yielded the majority of the fossils. The coralline limestone is the rock of which Mándogarh is built it is yellow or red in colour, and consists chiefly of small fragments of *Bryozoa* shells, etc. The freshly broken surface has a somewhat granular mottled appearance, and the fossils are not conspicuous, except on the weathered surface.

The two upper groups do not extend so far westwards as the lowest, the most western known exposure of the coralline limestone being at Umráli, near Ali.

The total thickness of these groups united does not exceed 60 or 70 feet at its maximum, and they thin out to the northwards, attaining their greatest thickness along the southern limit of the exposures.

The fossil fauna obtained from these beds is an interesting one, though

[1] The description of these rocks is based principally on W. T. Blanford, *Memoirs*, VI, 163, (1869), and P. N. Bose, *Memoirs*, XXI, 35, (1884).

amounting to only forty forms in all. Of these the late Prof. Duncan determined the following:[1]—

	Nodular limestone.	Deola marl.	Coralline limestone.	
LAMELLIBRANCHIATA—				
Neithea alba[2]	...	*	...	*P. quinquecostata* in Europe (gault to lower chalk); Africa, Utatúr, Trichinopoli and Arialúr.
Pecten quadricostatus	...	*	...	
BRACHIOPODA—				
Rhynconella depressa	...	*	...	Europe (lower greensand).
BRYOZOA—				
Escharina, sp.	...	*	...	
Eschara, sp.	...	*		
ECHINODERMATA—				
Cidaris namadicus	...	*	*	Lebanon.
Salenia traasii	...	*	...	
Cyphosoma cenomanensis	...	*	*	Europe (cenomanian).
Orthopsis indicus	...	*	...	
Echinobrissus goybeti	...	*	...	Lebanon.
Nucleolites similis	...	*	*	Europe (chloritic marl).
Hemiaster cenomanensis	*	*	*	Lebanon, Europe (cenomanian).
H. similis	*	*	*	Europe (cenomanian).
VERMES—				
Vincularia, sp.	...	*	...	
Serpula plexus	...	*	...	
CORALLIA—				
Thamnastræa decipiens	...	*	...	Europe (neocomian to gosau).

To which Mr. Bose[3] has added the following:—

	Nodular limestone.	Deola marl.	Coralline limestone.	
CEPHALOPODA—				
Ammonites guadeloupi	*	*	...	Europe, Trichinopoli.
GASTEROPODA—				
Fulguraria elongata	*	*	...	Europe (cemonanian and senonian); Trichinopoli.
Lyria granulosa	*	...	...	Africa; Utatúr and Trichinopoli.
Fasciolaria rigida	*	*	...	Trichinopoli.
Triton, sp.	*	*	...	
Natica, sp.	*	*	...	
Cerithium, sp.	*	*	...	
Turritella, sp.	*	*	...	
LAMELLIBRANCHIATA—				
Ostrea leymerii	*	...	...	Ariyalúr, Europe (neocomian).
O. arcotensis	...	*	...	
O. sp.	...	...	*	
Radula obliquistriata	*	*	...	Ariyalúr.
Plicatula multicostata	...	*	...	Trichinopoli.
Modiola archiaci	*	...	...	Europe (neocomian).
Inoceramus concentricus	*	...	...	Europe (gault).
I. coquandianus	*	...	...	Europe, (gault).
I. multiplicatus	...	*	...	Trichinopoli.

[1] *Quart. Jour. Geol. Soc.*, XXI, 349, (1865); *Records*, XX, 81, (1887).

[2] This and the following species were regarded by the late Dr. F. Stoliczka as but doubtfully distinct from each other and from *P.* (*Vola*) *quinquecostata*.

[3] *Memoirs*, XXI, 37, 40, 43, (1884).

	Nodular lime-stone.	Deola marl.	Coral-line lime-stone.	
Inoceramus, sp.	*	...	...	
Pinna laticostata	...	*	...	Ariyalúr, Europe (neocomian).
Arca securis	*	...	...	
Cardium altum	*	*	...	Utatúr.
C. hillanum	...	*	...	Trichinopoli.
Venus, sp.	*	*	...	
Panopæa arcuata	*	...	...	Europe (neocomian).
BRACHIOPODA—				
Rhynconella plicatiloides	...	*	*	Trichinopoli, Aryalúr.
R., sp.	...	*	*	
POLYZOA—				
Ceriopora dispar	*	*	*	Ariyalúr.

In the most recently published account of these beds an attempt has been made,[1] to correlate the three zones with the three great groups of Southern India and to the European groups ranging from albian (gault) to senonian (chalk). To say nothing of the improbability of the greater portion of the cretaceous period being represented by some 50 feet of fossiliferous beds, there is no palæontological evidence in favour of the supposition. Seven at least of the species of the lowest bed range into the Deola and Chirákhán marl, and the same number of species are found common to this and the coralline limestone. Three species (*Ceriopora dispar*, *Hemiaster cenomanensis*, and *H. similis*) range through all three beds. Considering the limited number of species found in the upper and lowermost beds the palæontological resemblances are greater than would be consistent with a range of time from cenomanian to senonian. Nor do the palæontological relations of the fauna to those of other parts of the world bear out the supposition. None of those forms which have been determined by the late Professor Duncan are found in Europe in beds of other than cenomanian or turonian age, and of the forms identified by Mr. Bose with European species of albian and cenomanian age, an identification confessedly rough, at least half are forms whose specific identity or the reverse is difficult for any one but a practised palæontologist to determine. Omitting those species whose specific identity is open to doubt and confining ourselves to those forms which have been determined by Professor Duncan, all those which are found in Europe occur there in beds of cenomanian (upper green sand) age and most are characteristic of it, consequently the cretaceous rocks of the Narbadá valley must closely correspond to the Utatúr group of Southern India.

In contrast to the relation between the cretaceous faunas of South Africa and Southern India may be noticed the divergence between the

[1] *Memoirs*, XXI, 48, (1884).

latter and that of the Narbadá valley. Among the fossils identified by Professor Duncan, *Neithea alpina* and *Pecten quadricostatus*, were regarded by Dr Stoliczka as doubtfully distinct from *P. quinquecostatus* found in the Utatúr group, but with this exception no other species is found in the cretaceous of Southern India, and even in this case the identification is one on which palæontologists are not thoroughly agreed. *Thamnastrea decipiens* is replaced by a closely allied form *T. hieroglyphica*, and the two *Hemiasters* by remotely allied forms. The more recent additions to the fauna made by Mr Bose have yielded thirteen species apparently identical with Southern Indian forms, but it is probable that this number will be reduced when the fossils come to be more critically examined, and five of the thirteen are wide ranging species found also in Europe or South Africa. Even accepting the identifications, this number out of a total of forty distinct forms is a much smaller proportion than obtains in the case of the south African and Trichinopoli cretaceous beds.

Another contrast between the Madras and Narbadá valley cretaceous beds is the very small proportion of European forms found in the former and the large proportion in the latter. Of the eight species of *Echinodermata* four are also found in Europe and two more in the Lebanon, and of other orders *Neithea alpina*, *Pecten quadricostatus* and *Thamnastraea decipiens* are found in Europe. To these Mr. Bose has added eight species also found in Europe. In the intervening area, two small exposures of cretaceous rocks are known at Ras Fartak and Ras Gharwen on the south-east coast of Arabia, from which small collections, comprising but thirteen species have been examined.[1] Yet, three of these are also found in the Bágh beds, and no less than ten in Europe. It is evident from this that there must have been tolerably free communication between the seas in which these different exposures were deposited, and that they belong to one of the great marine provinces of the cretaceous epoch.

With the South African cretaceous fauna the relationship of the Bágh fauna is of the slightest, only two species, *Pecten quinquecostatus* and *Lyria granulosa*, being common to the two regions. Of these the former is a species of almost world-wide range, and the latter is also found in the cretaceous of Southern India.

The contrast between the faunas of Narbadá and South Indian cretaceous is consequently as strongly marked as the relationships of the former to the European fauna and of the latter to that South Africa. Yet the distance which separates the two Indian exposures is but 750 miles, only half the distance which separates the lower Narbadá exposure from Arabian localities, and a much smaller fraction of the distance to the European localities or between Trichinopoli and South Africa.

[1] P. M. Duncan, *Quart. Jour. Geol. Soc.*, XXI, 349, (1865). The Arabian localities were originally described, and the fossils collected, by Dr. Carter, *Jour., Bombay Br. Roy. As. Soc.*, IV, 71, (1853), and "Geological Papers on Western India," p. 603, (1857).

These apparently anomalous relationships and divergences between the cretaceous faunas are easily explicable by, and are indeed proof positive of, the supposition that dry land stretched continuously from India to Africa during the cretaceous period and formed a barrier between two distinct marine provinces.

Besides the fossiliferous beds of the lower Narbáda valley already noticed, a lower division has been described under the name of the Nimár sandstone,[1] whose age is open to doubt. The frequent presence of a band of conglomerates and sandstones underlying the nodular limestone has already been noticed. In fact, it is only at or near Kachbaoda, in the Man valley, that its absence is recorded As a rule, the thickness is small in the eastern exposures, but they begin to thicken rapidly south-westwards of Bágh.

In the Hatni valley they have a thickness of nearly 200 feet; near the deserted city of Ali they are 500 feet thick, and in the inlier south of Kawant they amount to over 700 feet. With all its variation in thickness this sandstone preserves the same general type, of conglomeratic beds and conglomerates at the base, overlaid by fine grained sandstones and shaly beds.

The age of this sandstone is doubtful. It has been regarded as lower cretaceous, and in favour of this supposition there is the absence of any observed unconformity between it and the beds it underlies. Mr. Bose classed it as neocomian on the strength of an oyster bed, composed of a species which was identified with the European *O. leymerii*, but even if the identification were correct the evidence is not sufficient to establish the age of the bed, and there is some doubt whether the oyster band is really conformable to the underlying sandstone or not.[2]

The sandstone contains no recognisable fossils, the uppermost beds show crustacean and annelid tracks on their surface; some indeterminable fragments of bone were found and fragments of carbonised driftwood are said to occur. In this, as well as in lithological facies, the Nimár sandstone agrees with the Mahádevas of the Dhár forest area. In Káthiáwár there is the same absence of beds intervening between the upper Gondwánas and the cretaceous as would be implied by a Mahádeva age for the Nimár sandstones, but the absence of any observed unconformity, the constancy of the nodular limestone beds, and the fact that the thickening of the sandstones is in the same direction as that of the upper beds lend some support to the supposition that they are cretaceous. In the absence of more conclusive evidence their true age must remain a matter of doubt.

In Káthiáwár a series of sandstone beds known as the Wadhwán sand-

[1] *Memoirs*, XXI, 23, (1884).

[2] *Memoirs*, XXI, preface to pt. ii, p. vii, (1884.)

stones are found between the Umia group and the overlying Deccan traps.[1] They are composed of brick red or dull reddish brown sandstone with some argillaceous beds, at the top of which are in places cherty beds or thin bands of limestone, recalling the rocks of Bágh. In the neighbourhood of Wadhwán, there is a thin band of drab coloured, tough, sometimes gritty or chalcedonic limestone, containing marine fossils, chiefly Bryozoa, a few small corals and a portion of a flattened, keeled ammonite, resembling the cretaceous *A. guadaloupae*, in a matted mass of broken indeterminable shells. In other localities *Ostrea* and *Natica* were found, but no fossils sufficiently well preserved to determine the precise age of the beds. The general facies of the fossils, as well as the lithological character of the rock they are preserved in, suggest the correlation of these beds with the cretaceous of Bágh, and their direct superposition on the Wadhwán sandstones corresponds to the relation between the Bágh beds and the Nimár sandstones, whose probably upper Gondwána age has already been indicated. The relation of the Wadhwán sandstones to the overlying trap appears to be very similar to that of the Bágh cretaceous, there being a distinct erosion unconformity.

[1] F. Fedden, *Memoirs*, XXI, 87, (1884).

CHAPTER XI.

DECCAN TRAP.

Extent—Nomenclature—Petrology—Igneous formations—Sedimentary beds—Lametá group—Infra and inter-trappeans of Rajamahendri—Inter-trappeans of the main area—Subaërial origin of the traps—Foci of eruption—Age of the Deccan trap—Probable conditions during their formation.

In the last chapter the cretaceous rocks of the Peninsula were described, and the present should, in the ordinary course, be devoted to the rocks of the same age in the extra-peninsular area. There are, however, two very good reasons for departing from this course, the first being the intimate relation which subsists between the cretaceous and tertiary rocks of a large portion of extra-peninsular India, necessitating their being dealt with together; the other is the presence, immediately above the cretaceous beds in the Peninsula, of a series of volcanic rocks, forming one of the most prominent and widely spread of all the rock systems found in the Peninsula.

In superficial area the Deccan traps are only exceeded, within the limits of peninsular India south of the Indo-Gangetic plain, by the metamorphic series and, although the traps are far inferior in thickness to the Vindhyan and Gondwána formations, their remarkable horizontality, throughout a great part of the region covered by them, enables them to conceal all older rocks. Some faint idea of the extensive area occupied by this formation may be gained from the fact that the railway from Bombay to Nágpur, 519 miles long, never leaves the volcanic rocks until it is close to the Nágpur station, and that the traps extend without a break from the sea coast at Bombay to Amarkantak at the head of the Narbadá, and from near Belgáum to north of Goona Even this extent, great as it is, by no mean represents the whole area originally occupied by the formation; for outliers are found east of Amarkantak as far as Jamírá Pát in Sargúja, to the south-east a small outcrop occurs close to Rájámahendri, whilst to the westward the series is well developed in Káthiáwár and Cutch, and is even believed to be represented, though only by two very thin bands west of Kotri, in Sind. We have, therefore, proof of the existence of this volcanic formation throughout nearly ten degrees of latitude and sixteen of longitude, whilst the area covered in the Peninsula of India can be little less than 200,000 square miles. It is probable that the limits mentioned very nearly correspond to the original boundaries of the volcanic rocks, because the high level laterite, which rests conformably upon the uppermost traps of the Deccan, is found to the southward, eastward and northward, resting on rocks older than the volcanic series, and if, as will be shown to be probable in a later chapter, this laterite was formed at a date shortly subsequent to the cessation of the igneous outbursts, it may be inferred that the lava flows never extended to the localities (such as Gwalior, Rewá, etc.) in which the laterite is found resting immediately upon Vindhyan, transition, or metamorphic rocks.

In adopting the name of Deccan[1] trap for this great volcanic formation, the Geological Survey has been guided partly by old usage, partly by the circumstance that the term 'trap' was originally applied to similar horizontally stratified lava flows. Some geologists have condemned the term on account of the loose manner in which it has been used for a great variety of igneous rocks, but it is difficult to replace it, and in the present case, at all events, it is employed in a well defined sense.

In consequence of its geological structure, the volcanic region of central and western India is distinguished by marked peculiarities of scenery, and the characters of the surface are widely different from those found in other parts of the Indian Peninsula. Great undulating plains, divided from

Fig. 15.—Hill composed of Deccan trap, near Harangaon, north of Nimáwar, Narbadá valley.

each other by flat topped ranges of hills, occupy the greater portion of the country and the hillsides are marked by conspicuous terraces, often

[1] It is scarcely necessary to state that the Deccan (Dakshin) comprises that part of the Indian Peninsula which is south of the Vindhyan range.

traceable for great distances, and due to the outcrop of the harder basaltic strata, or of those beds which resist best the disintegrating influences of exposure. In some parts of the area great scarps are found, some of those in the Sahyádri range being 4,000 feet in height, all conspicuously banded with horizontal terraces.

The vegetation of the trap area differs no less conspicuously from that which is found on other formations, the distinction in the dry season being so marked that, especially when taken in connection with the form of the surface, it enables hills and ranges of trap to be distinguished at a distance from those composed of other rocks. The peculiarity consists in the prevalence of long grass and the paucity of large trees,[1] and in the circumstance that almost all bushes and trees, except in the damp districts near the sea, are deciduous. The result is that the whole country presents, except where it is cultivated, a uniform straw coloured surface, with but few spots of green to break the monotony during the cold season, from November till March, whilst from March, when the grass is burnt, until the commencement of the rains in June, the black soil, black rocks, and blackened tree stems present a most remarkable aspect of desolation. During the rainy season, however, the country is covered with verdure, and in many parts it is very beautiful, the contrast afforded by the black rocks only serving to bring into relief the bright green tints of the foliage.

Throughout the trap area the prevailing rock is some form of dolerite or basalt, but there is a large amount of variety in the characters presented by different beds. Some are excessively compact, hard, and homogeneous, the crystalline structure being so minute as to be detected with difficulty (anamesite), others are coarsely crystalline, and these frequently contain olivine in considerable quantities, and one variety is porphyritic, containing large tabular crystals of glassy felspar, white or green in colour. Many of the basalts again are soft and earthy, evidently in most cases, and probably in all, from partial decomposition. The most striking peculiarity is, perhaps, the great prevalence of amygdaloid, in which the nodules, chiefly containing zeolite or agate, sometimes form the principal part of the rock. These nodules are very often coated with glauconite (green earth), and the prevalence of this mineral is highly characteristic. Almost throughout their range, the Deccan traps may be recognised by the occurrence of the amygdaloidal basalts with green earth, or of the porphyry with crystals of glassy felspar.

[1] The want of large trees is partly due to the wanton destruction to which the forests of India have been exposed for ages through reckless cutting, to equally reckless clearing for temporary cultivation of a rude kind, and perhaps more than all, to the practice of annually burning the grass at the commencement of the hot season.

Exfoliating concretionary structure is common in the softer forms of basalt, which have undergone some amount of decomposition, but it is never seen in the hard compact beds. Frequently the hard unaltered spheroidal cores of concentric nodules, which may easily be mistaken for rolled fragments, are to be found scattered over the surface of the bed, from which they have weathered out. Columnar structure is less common, though it is occasionally seen, a fine example being shown in the following woodcut. In some cases this structure has been observed in

Fig. 16.—Radiating basaltic columns in a dyke near Gújri, north-west of Maheswar, Narbadá valley.

the compact basaltic flows; it is frequently seen in the lowest flow, a very thick one, west of Hoshangábád, in the Narbadá valley, and in one of the lower flows in Málwá, but the appearance is often confined to intrusive dykes, as in the example illustrated. Trachytic rocks are extremely rare, and have hitherto only been found in intrusive masses.

Beds of volcanic ash are common, so common indeed in places as to form a very considerable proportion of the strata, and they appear to be much more prevalent towards the upper part of the series.[1] They often differ but little in appearance from the basaltic lavas with which they are interstratified, but, on close examination, their brecciated structure can always be readily detected, and the blocks of scoriæ which they contain

[1] Possibly due to the upper part of the series being chiefly preserved near the old volcanic foci. Ashes are found interstratified with the lower beds on the Narbadá, west of Baroda, where remains of ancient volcanic cores also occur.

generally weather out on exposed surfaces and remain in relief, precisely as on old volcanic cones. Magnificent examples are to be seen on most of the higher portions of the Sahyádri or Western Gháts and on the high peaks around Poona, formerly used as hill forts; well marked instances occur also in Bombay and Salsette.[1] Very frequently a thin bed of ash intervenes between two basaltic flows. Occasionally pumice is found in the ash beds, the interstices being, however, all filled up by the same process as that by which vesicular lava has been converted into amygdaloid. Here and there, throughout the traps, beds of red bole occur; they are usually only a foot or two thick, but occasionally more. Sometimes the bole contains scoriæ, and in this case it frequently covers the upper surface of a basaltic flow, into which it appears to pass. In some instances the bole is so uniformly stratified that it has the appearance of having been deposited from water.[2]

In a few instances bands of very homogeneous structure and of a pale lilac colour, formed of an apparently argillaceous rock resembling bole in texture and so perfectly laminated as to exactly simulate shale, have been found interstratified with the basalts. This is especially the case at a large hill called Páwagarh, 2 000 feet high, near Baroda, and similar beds are said to occur in Káthiáwár; they have also been noticed east of Surat. The occasional occurrence of glassy felspar crystals in these beds and the circumstance that some of the harder basalts at times weather on their exposed edges into a somewhat similar soft lilac rock, render it possible that these shaly strata result from the alteration of trap. At the same time it is far from improbable that some of them may be consolidated volcanic mud, composed of fine lapilli washed down and deposited by water.

No crystallised pyroxene has been observed, except locally in some of the ash beds, and the only felspar which occurs in distinct crystals appears to be the form of orthoclase (gassy felspar) which is found in the porphyritic rock already mentioned. Olivine and magnetite are common, the former occurring as translucent yellowish grains, the latter in minute crystals, too small, as a rule, to be recognised by the naked eye

[1] Amongst the best examples are the rocks in which the Keneri caves of Salsette are cut; some beds on the Kamatki ghât between Poona and Mahábaleshwar; and a conspicuous bed at the lower gateway of the fortress of Singarh near Poona. Ash-breccias also occur in Bombay Island at Flag-staff hill and Rai hill, Parel, and in the neighbourhood of Sion fort. It must not be supposed from these examples that the rock is rare. It is found almost throughout the trap country, but it is much less common towards the base of the traps.

[2] Sir C. Lyell has shown that bands of red clay interstratified with the lavas of Etna have been formed from the crust of the lower lava flow, decomposed into clay and then baked and reddened by the heat of the overlying flow, or where "volcanic sand has been showered down from above and washed over the older lavas by torrents and floods;" *Phil Trans.*, 1858, p. 711. Similar beds appear to be characteristic of subaerial lava flows; Judd, *Quart. Jour. Geol. Soc.*, XXX, 227, (1874).

but easily detected, if abundant, by the effect of the rock upon the magnetic needle. Magnetic iron sand derived from the traps is frequently found in the streams which traverse the rocks. With the tabular felspar crystals small scales of red mica are found.

Secondary minerals of various kinds, which have been formed since the consolidation of the volcanic strata, are found in the greatest abundance in some of the flows, especially in the amygdaloidal, and in some of the more earthy and decomposed traps. These minerals not only form the nodules of the amygdaloid, but they are found lining cracks and hollows, the finest crystals being always in geodes or cavities, some of which are as much as two or three feet across, and even larger hollows lined with crystals are said to have been found. The commonest minerals are quartz (either crystalline or in the form of agate, bloodstone, jasper, etc.,) and stilbite, next in abundance are apophyllite, heulandite, scolecite (poonahlite[1]), laumonite and calcite; thomsonite, epistilbite, prehnite and chabasite also occur, but they are rare. The great prevalence of glauconite or green earth has already been noticed.

The crystalline quartz is occasionally, though rarely, amethystine; it but seldom occurs in crystals which exceed an inch in diameter, and the larger crystals are not often transparent. The form known as trihedral quartz, in which the terminal pyramid of each quartz crystal consists of three planes instead of six, or in which three planes are very much more developed than the other three, is of common occurrence. The agates occur chiefly in geodes or nodules, large and small; many are finely banded, and, after being coloured by heating, are cut into ornaments.[2] Jasper and heliotrope or bloodstone occur chiefly in flat plates, which appear to have been formed in cracks, and agate is sometimes met with of apparently similar origin. Stilbite is very common, though less so than quartz; one magnificent variety consists of large orange or salmon coloured crystals, often two or three inches in length, usually compound or in sheaf like aggregations, but occasionally in large flat prisms terminated by a four sided pyramid. Apophyllite is the finest of all the Deccan trap minerals. It generally occurs in four sided prisms with terminal planes, a form which closely resembles the cubical crystals of the isometric system, the double pyramid, with replacements of the secondary prismatic faces and terminal planes, so characteristic of this mineral in other localities, being chiefly typical of small crystals in the Deccan traps. The

[1] Two other mineral species besides poonahlite have been described from the Deccan traps. One of these is hislopite, Haughton, *Phil. Mag.*, 4th series, XVII, 16, (1859), which appears to be calcite coloured by glauconite (green earth) and the other, syhedrite, Shephard, *Am. Jour. Sci.*, 2nd series, XL 110, (1865) is stilbite, coloured in the same manner.

[2] Most of the stones cut for ornaments are either procured from rivers or from the tertiary gravels derived from the denudation of the traps.

colour of the Deccan apophyllite is usually white, more rarely pink or green, some crystals are perfectly transparent, and one of the most magnificent associations of minerals to be found anywhere is seen when, as occasionally happens, perfectly clear vitreous crystals of apophyllite, of large size, are inserted on a mass of orange stilbite. Some apophyllite crystals are as much as three or four inches across. The other minerals are less deserving of notice, but very beautiful long acicular crystals of scolecite with exquisitely formed pyramidal terminations are of occasional occurrence, and fine crystals of white heulandite are not unfrequent. The glauconite is usually amorphous, but occasionally forms an aggregate of crystalline scales, and a massive mineral, which, if not green earth, is closely akin both in appearance and composition, occasionally occupies small cavities completely.

One of the most remarkable characters of the Deccan traps is their persistent flatness or near approach to horizontality throughout the greater portion of their area. This is conspicuous throughout the Sahyádri range, over the whole of the Bombay Deccan, from Khándesh to Belgáum and Sholápur, throughout southern Berár and the north-western portion of the Haiderábád territory, in many parts of the Sátpura range between the Narbadá and Tápti, and on the Málwá plateau north of the Narbadá. Where exceptions occur, as in the western Sátpura and Rájpipla hills and along the coast near Bombay, the disturbance is shown to be of later date from its affecting contemporaneous or newer beds of sedimentary origin. The only departure from absolute horizontality to be seen in the lava flows of the Deccan is frequently no more than may be due to the lenticular form of the beds, but usually there is a very low dip discernible, seldom exceeding 1°, and fairly constant over large areas. This circumstance tends to show that even this small amount of inclination may be due to disturbance, because if the dips represented the original angle at which the lava flows were consolidated, they would be found to radiate from the original volcanic vents. Nothing of the kind has, however, been traced.

The separate lava flows are, as a rule, of no great thickness. The average in the two sections of the Bhor and Thal Gháts, measured on the railway lines, is apparently 64 and 87 feet respectively, but really less, because the distinction between the flows can in most cases only be recognised by lithological characters, and where, as must frequently be the case, two or more beds of similar appearance and composition occur together, they must often be confounded and measured as one. Many of the more amygdaloidal beds appear to be made up of several smaller flows from six to ten feet thick, distinguished by being highly amygdaloidal above, less so in the middle, and traversed towards the base by long cylindrical vertical pipes filled with zeolite.[1] But even supposing that these apparent distinctions are accidental, some well marked crystalline flows in each section do not exceed 15 feet in thickness.

Hitherto only the igneous portion of the Deccan series has been described, but volcanic rocks, although they form the great mass of the formation, do not compose it exclusively, for sedimentary bands, frequently fossiliferous, have been found in several places interstratified with the lava flows, and have become widely known and described as intertrappean beds. There is also found in many places, at the base of the whole series, a small group of limestones, sandstones and clays, known as the Lametá group, from its occurrence at Lametá Ghát, on the Narbadá, near Jabalpur.

The intertrappean beds have been found in two distinct portions of the Deccan series, first close to the base, throughout the greater portion of the enormously extensive circuit of the volcanic area, and, secondly, in the highest portion of the traps, only known to occur close to the coast in Bombay Island and the immediate neighbourhood.[2] A rough classification of the whole series is presented in the following sections:—

	Approximate thickness in feet.[3]
1. Upper traps, with numerous beds of volcanic ash and the intertrappean sedimentary deposits of Bombay	1,500
2. Middle traps, ash beds numerous above but less frequent towards the base, no sedimentary beds known	4,000
3. Lower traps, with intertrappeans of Nágpur, Narbadá valley, etc., volcanic ash of rare occurrence or wanting	500
4. Lametá or infratrappean group	20 to 100

The whole thickness, as will be shown presently, is probably considerably greater than 6,000 feet in the neighbourhood of Bombay, but the rocks gradually thin out in other directions. At Bombay the upper limit of the series is not seen. It is highly probable that near Surat and Baroda the trap may have been even thicker than near Bombay, but the upper portions have been greatly denuded, and it is extremely difficult here, as in most other places, to estimate the thickness with any accuracy. In Cutch the traps are about 2,500 feet thick, whilst in Sind they have

[1] Bearing in mind that amygdaloidal basalt must have been originally vesicular lava, and that what are now nodules of quartz or zeolite were originally air or steam bubbles, it is easy to understand that the upper portion of a lava flow, having been more vesicular originally than the lower portion, would be characterised by a prevalence of amygdaloid. The vertical tubes must also have been originally filled with air or vapour, perhaps expelled from the underlying stratum by the heated mass flowing over it.

[2] The reasons for considering the Bombay traps higher in the series than the others will be explained subsequently.

[3] The thickness given is little more than a guess, except in the case of the lower traps and Lametás. The other figures are minimum estimates of the vertical extent of the series, where fairly developed.

dwindled down to two bands at different horizons, each less than 100 feet thick. Throughout the greater portion of their area, no higher beds, except laterite or post-tertiary deposits, are found resting upon them, and it is impossible to form any accurate estimate of their original development. In the extreme south of the volcanic area, near Belgáum, their thickness has been estimated by Mr. Foote to be 2,000 to 2,500 feet. On the plateau of Amarkantak, at the eastern extremity of their main area, they are about 500 feet thick, but farther east in the outlier on the Máin Pát in Sargúja, not more than 300 to 400, whilst to the south-east near Rájámahendri they are represented by a thin outlier, in which from 100 to 200 feet of basalt may be exposed.

Before proceeding further it will be necessary to give a fuller description of the sedimentary formations, and in accordance with the system adopted throughout this work, the Lametá group as the lowest will first receive attention.[1] Formerly this group was supposed to be a representative of the Mahádeva group of the Gondwána system, but further examination has shown that the Mahádevas are much more ancient, and that the Lametá beds are so closely associated with the lowest trap that they must be considered as part of the same series. The origin of the name has already been mentioned, and it has been stated that the group consists of limestones, sandstones and clays. The limestones are the most characteristic and persistent beds, they frequently occur alone, and they form the upper portion of the group when other beds are associated with them. Occasionally the limestone is pure, but it is commonly full of sand and small pebbles, so as to form a calcareous grit rather than a limestone, and as a rule it contains an abundance of masses, sometimes irregular, sometimes more or less lenticular in form, of segregated chert. Some of the small pebbles frequently consist of red jasper, the occurrence of which is very characteristic. This gritty limestone, with its included chert nodules, is found over a very extensive tract of country in the Central Provinces, and appears to be rarely absent throughout any large area in which the base of the traps is exposed.

The bed which, after the limestone, is most commonly found in the Lametá group, is a rather fine porous earthy sandstone, usually of a greenish colour. The clays are red or green, and are very frequently sandy or marly; sometimes they contain nodular carbonate of lime. They are of local occurrence and appear but rarely to extend over any considerable area. All these beds pass into each other; the limestone is not unfrequently merely the sandstone cemented by carbonate of lime,

[1] For details, see *Quart. Jour. Geol. Soc.*, XVI, 154, (1860); *Memoirs*, II, 196, (1860); VI, 216, (1869); IX, 315, (1872); XIII, 87, (1877); *Records*, V, 88, 115, (1872).

the marls are an argillaceous form of the limestone, and, except where the characteristic gritty limestone is the sole representative of the formation, there is, as a rule, a frequent change of character in the beds, both horizontally and vertically. This is usually the case where the thickness exceeds 20 or 30 feet, but where the group is only represented by a thin band, either the gritty limestone or the earthy greenish sandstone is commonly found alone.

The Lametá group is quite unconformable to all the various older formations upon which it rests, from the metamorphics to the Jabalpur group. As a rule, the lowest flows of trap are conformable to the infratrappean beds, but in some instances distinct unconformity has been detected, especially in one case near Jabalpur,[1] and it is highly probable that closer examination would show that such cases are common, and that in many localities where Lametás are wanting their absence is due to denudation in pre-trappean times. At the same time the denudation appears to have been local, not general, patches occurring here and there, whilst in the intervals between them the trap rests upon a formation older than Lametá, but at such an elevation as to show that the absence of the infratrappean bed is not due to the ground having been above the water in which the Lametás were deposited. It is impossible that the Lametás can ever have been co-extensive with the base of the trap, because the surface on which the latter rests is extremely uneven, and many portions of it must have been above the level at which the infratrappean beds were deposited. It will, however, be necessary to recur to this subject, when discussing the relations of the trap series as a whole to the older formations.

It is unnecessary to give a list of localities at which the Lametá group has been observed. It is principally developed in the Central Provinces, around Nágpur, Jabalpur, etc. It has not been found in the southern Maráthá country, but elsewhere along the boundary of the volcanic area from the Godávari valley to Bhopál and Indore, it is rarely absent over any considerable area. As a rule, owing to its small vertical development, it only covers small portions of the surface, and it usually forms a narrow fringe to the trap country. In the western Narbadá valley it has been recognised and described as lying unconformably on the cretaceous Bágh beds.[2]

The Lametá group is, as a rule, singularly unfossiliferous, the principal fossils which have been found in it, consisting of some bones of a large Dinosaurian reptile, *Titanosaurus indicus*,[3] allied to *Pelorosaurus* of the wealden and *Cetiosaurus* of the Bath oolite. These fossils occur near Jabalpur, and similar bones, together with coprolites and some chelonian

[1] *Records*, V, 115, (1872).

[2] Bose, *Memoirs*, XXI, 46, (1884).

[3] Lydekker, *Records*, X, 38, (1877).

remains, were found at Pisdura about eight miles north of Warorá in the Chándá district.[1] In the last named locality some of the characteristic fresh water mollusca of the intertrappean beds, such as *Physa prinsepii*, are associated with the bones, and the same shells have also been found in beds at the base of the trap in one or two other localities; for instance, a *Paludina*, apparently identical with *P. deccanensis*, an intertrappean fossil, was found by Mr. Hislop at Nágpur,[2] *Melania* and *Corbicula* have been met with in infratrappean beds near Ellichpur in Berár,[3] and *Physa prinsepii* in a similar position at Todihal, 15 miles north-north-east of Kaládgi in the southern Maráthá country.[4] But it is by no means clear, in those localities, where fresh water shells are found in beds beneath the trap, with the exception of Nágpur, that an intertrappean bed has not overlapped the edge of the underlying lava flow, so as to rest upon an older rock, which may be either Lametá or any other more ancient formation and in the particular case of Pisdura, where all the fossils are found scattered on the surface of a field consisting of red Lametá clay, there is always a possibility that *Physa prinsepii* and similar fossils may have come from some small unnoticed intertrappean band, concealed beneath the deep surface soil. At the same time it is by no means improbable that the *Physa* and other shells are really derived at Pisdura from the Lametá beds, and that this group consequently is not much older than the volcanic beds which overlie it.

The only other noteworthy occurrence of fossils in the Lametá group is that of some fish remains at Dongargaon, six miles east by south, and Dhamni, nine miles east by north, of Warorá.[5] The species have not been described; one of the fish found was considered by Sir P. Egerton allied to the *Sphyrænodus* (a cycloid acanthopterygian) of the London clay, but according to Mr. Smith Woodward, only differs from *Belonostomus*, an upper cretaceous genus, in the prominence of the vertical foldings of the teeth.[6]

Leaving the question of the mode of origin of the Lametá group to be discussed hereafter, and deferring for the moment the description of some beds with marine fossils found at the base of the traps near Rájámahendri, the next group which requires notice is that comprising the fresh water beds interstratified with the lower traps in many parts of India, and especially in parts of the Central Provinces, northern Haiderábád, Berár, and the states north of the Narbadá valley. Throughout these tracts of country, and beyond them almost throughout the great trap area, there

[1] *Quart. Jour. Geol. Soc.*, XVI, 163, (1860); *Memoirs*, XIII, 88, (1877).

[2] *Quart. Jour. Geol. Soc.*, XVI, 167, (1860).

[3] *Memoirs*, VI, 283, (1869).

[4] *Memoirs*, XII, 193 (1876).

[5] *Quart. Jour. Geol. Soc.*, XVI, 163, (1860).

[6] *Records*, XXIII, 24, (1890).

are found here and there, near the base of the volcanic formations, and in no case so far as has hitherto been recorded, at a greater height than from three to five hundred feet above the base, thin bands of chert, limestone, shale or clay, often abounding in fossils of fresh water or terrestrial origin.

Perhaps the most common form of the intertrappean bands, or that which is most conspicuous, is a compact, blackish, cherty rock, a kind of lydian stone It is clear that this rock has been originally a silt, and has been hardened, either by the outpouring of igneous rock over it or by chemical infiltration, the former being the more probable, because it very frequently happens that the upper portion of the bed only is cherty, the lower portion being a soft earthy shale. Other forms of intertrappean bands are a dark or pale grey limestone, often earthy and impure, but rarely gritty, like the characteristic Lametá bed. Not unfrequently the sedimentary bed is composed of volcanic detritus, whether removed by denudation from solid basalt, or consisting merely of the loose products of eruptions, such as lapilli, it is difficult to say. Red and green clays or bole are also found, often associated with other intertrappean rocks.

As a rule, the sedimentary beds interstratified with the lava flows are distinguished from those underlying the whole volcanic series by the absence of pebbles and sand, but occasionally, though rarely, sandy and even pebbly beds are found at some distance above the base of the trap. In the south Maráthá country most of the intertrappean beds are sandstones and conglomerates. One peculiar detrital form of intertrappean accumulation has hitherto only been described from the country north of the Narbadá and south of Chhota Udaipur on the banks of the Karo, a tributary of the Hiran river.[1] The lower beds of the trap series here consist of conglomerates, sandstones, and sandy grits, sometimes resting on a stratum of basalt, but occasionally on the Bágh cretaceous beds, which underlie the volcanic formations. Occasionally the sandstone or conglomerate appears to be chiefly composed of detritus derived from the metamorphic rocks, but volcanic fragments, usually in the form of rolled pebbles of basalt, can always be found by search, and in many parts the bed becomes a mass of rolled volcanic fragments, often mixed with unrolled scoriæ. At times, indeed, the rock is a conglomeratic ash, in which rolled fragments of metamorphic rocks and of basalt occur together Hornblende and pyroxene crystals have been found in these conglomeratic ashy beds, which are in some places as much as 200 feet thick. In some instances the conglomerates appear to have accumulated in hollows, like river beds, but in any case the abundance of rolled pebbles and boulders of trap is important as a proof that denudation took place in the interval between successive lava flows.

[1] *Memoirs*, VI, 327, (1866).

With the exception of the detrital accumulations which have just been mentioned, the intertrappean bands rarely exceed a few feet, from three to about twenty, in thickness, and they frequently do not exceed half a foot. In many places two or more sedimentary beds occur at different levels in the same section, and the different bands are in some cases dissimilar in mineral character. Thus, at Mekalgandi[1] Ghát in the Sichel hills, south of the Pen Gangá river, on the old road from Nágpur to Haiderábád, a locality famous as being one of the first at which the intertrappean fossils were detected by Malcolmson, the following beds are observed in section:—

1. Trap.
2. Cherty bed containing *Unio*, *Cypris*, etc.
3. Trap.
4. Limestone containing *Cypris* and fragments of small mollusca.
5. Trap.
6. Calcareous grit, containing broken shells (Lametá).
7. Metamorphic rocks.

A single intertrappean bed can but rarely be traced for more than three or four miles without interruption; it then usually dies out. At the same time it is rare to go over any large tract near the base of the traps without finding some sedimentary bands interstratified, and occasionally one is found to be much more extensive than usual. Thus, an instance is recorded by Mr. J. G. Medlicott[2] in Sohágpur, east of Jabalpur, in which an intertrappean bed was traced for nearly 25 miles.

It would take up too much space to enumerate all the localities at which the lower sedimentary intertrappean beds have been observed. They have been noticed in several places in the southern Maráthá country; they are commonly found near the base of the trap flows almost throughout the great and irregular line of boundary extending from the Godávari to Rájputána, and they occur even in small outliers, for instance, at Máin Pát in Sargúja; they have been detected by Mr. Rogers[3] to the westward at Dohad, about 75 miles north-east of Baroda, and still farther west in Cutch, by Mr. Fedden[4] of the Geological Survey.

The abundance of fresh water and terrestrial animals and plants in the intertrappean beds has been the principal reason for the comparatively large amount of notice which these thin bands of rock have attracted. The mollusca are very abundant and are occasionally exquisitely preserved in the cherty layers, the commonest species being forms of *Physa* and *Lymnea*, whilst *Unio*, although abundant locally, is of comparatively rare occurrence. *Paludina*, *Valvata*, and *Melania* are far from uncommon.

[1] Mucklegundy pass of Malcolmson.
[2] *Memoirs*, II, 201, (1859).
[3] *Quart. Jour. Geol. Soc.*, XXVI, 122, (1870).
[4] *Memoirs*, IX, 58, 240, (1872).

Land shells are very seldom found, but they have been detected[1] in one case at least. Entomostracous crustaceans are very nearly as common as mollusca, all hitherto found belonging to the genus *Cypris*. The other remains of animals hitherto detected have consisted of insects, fishes, and reptiles, all of which are fragmentary. Plant remains abound, but leaves are rare, seeds and fragments of wood being more common and the most abundant vegetable fossils are the seed vessels of *Characeæ*, of which one species has been described under the name of *Chara malcolmsoni*.

The plants have not been described, with the exception of the *Chara*. Those collected near Nágpur are said by Mr. Hislop to comprise about fifty species of fruits and seeds, twelve of leaves, and five kinds of woods, the only forms mentioned are endogens and angiospermous exogens. The relations of the fossils will be discussed in the sequel, together with the fauna of the other intertrappean deposits.

The whole of the mollusca and crustacea are fresh water forms; no marine species have been detected associated with them, except in the beds near Rájámahendri, of which a description will be given in the next paragraph. The insects and plants, with the exception of *Chara* a fresh water form, are of terrestrial origin. The general prevalence of the pulmoniferous mollusca *Physa* and *Lymnea* appears to indicate that the water was shallow, as these forms live partly at the surface. *Cypris*, too, is commonly found in shallow marshes.

The outcrops of trap near Rájámahendri are so remote from any other exposure of the Deccan volcanic series, being about 210 miles distant from the nearest point of the great Deccan area north-west of Sironchá, that some doubt would remain as to the identification, despite the similarity of mineral character, had not some of the typical fresh water fossils of the Deccan intertrappean beds been discovered in the Rájámahendri area. The Rájámahendri outcrops occur on both banks of the Godávari,[2] and consist of an interrupted narrow band of volcanic rocks, chiefly earthy dolerite and amygdaloid of the usual character, extending altogether for about 35 miles from east-north-east to west-south-west. Traps are seen at Káteru on the left bank of the Godávari just north of Rájámahendri itself, and extend rather more than ten miles to the east-north-east, resting upon metamorphic rocks whenever lower beds are

[1] *Memoirs.*, II, 213, (1859); several forms were referred to the terrestrial genus *Achatina*. Some similar fossils from a French deposit had been placed in the same genus, but it appears more probable that the Indian shells are of fresh water origin and belong to *Lymnea* or to some allied type.

[2] The intertrappean beds were discovered originally by General Cullen and Dr. Benza, and collections of the fossils were made by Lieutenant Stoddart and Sir W. Elliot, and described by Mr. Hislop, *Quart Jour. Geol. Soc.*, XVI, 161, 176, (1860). The infratrappean band was first noticed by Dr. King, *Records*, VII, 159, (1874). See also *Memoirs*, XVI, 324, (1880); *Quart. Jour. Geol. Soc.*, X, 471, (1854).

seen. On the right bank the volcanic rocks appear in two areas, divided by small alluvial valley: the larger extends for about ten miles to the westward from Pungadi, 7 miles west of Rájámahendri, and the smaller occurs a few miles still farther west. In these outcrops the beds of the volcanic series rest upon the Rájmahál rocks of the Ellore region. In both cases the strata overlying the trap are Cuddalore sandstones and all the beds alike have a low dip to south or south-east. The whole thickness of the volcanic series at this locality, as already mentioned, nowhere appears to exceed about 200 feet, and in places it is not more than 100.

At the base of the traps, and intervening between the basalt flows and the underlying jurassic sandstone, about 50 feet of sandstone, white, yellowish, or greenish in colour, are exposed near the village of Dúdkúr, 12 miles west of Rájámahendri. The upper portion is calcareous, and on the top there is a band, about six inches to two feet thick, of sandy limestone abounding in marine fossils, the most abundant of which is a *Turritella*, apparently identical with *T. dispassa* of the cretaceous Ariyalúr group. If not identical, the two species are very closely allied. A *Nautilus*, about fifteen *Gasteropoda*, and eleven *Lamellibranchiata* accompany the *Turritella*, but not a single species, except *Turritella dispassa*, has been recognised as identical either with the cretaceous beds of southern India or with the eocene fossils of the nummulitic group. The collections have not, however, been sufficiently compared to enable the species to be determined with certainty. Only one single species, too, *Cardita variabilis*, has been recognised as occurring also in the overlying intertrappean bed. Although the whole facies is tertiary, there is a remarkable absence of characteristic genera,[1] and the chief distinction from the cretaceous fauna of the upper beds in Southern India is simply the want of any marked cretaceous form The fauna is distincly marine.

It is difficult to say whether this bed should be referred to the Lametá group or not. The mineral character is similar, but all known Lametá outcrops are so distant that the identification is somewhat doubtful. The distinctions between the fossils of the Bágh beds and those of the infratrappeans of Dúdkúr and Pungadi appear too great to be attributed solely to the existence of a land barrier between the two areas; it is difficult to suppose that the two formations can be of the same geological age, and the Bágh beds are probably more ancient than the Pungadi infratrappeans. The balance of evidence is rather in favour of referring the latter to cretaceous times than to tertiary, and they may be considered of intermediate age, as will be shown to be probably the case with the Lametás.

[1] Amongst the genera identified are *Rostellaria* several forms of *Muricidæ*, a *Volutilithes* near the tertiary *V. torulosa*, *Natica*, *Turritella*, *Dentalium*, *Cytherea* or allied genera (three sp.), *Cardita* (four sp.), *Corbis*, *Pectunculus*, *Cucullæa* and *Ostrea*.

Upon the fossiliferous limestone described in the last paragraphs a flow of basalt is superposed, varying in thickness from about 30 to about 100 feet. There is an appearance of slight unconformity where the volcanic rock rests upon the sedimentary bed, the surface of the latter being slightly uneven, as if denuded, and the upper fossiliferous infratrappean zone is occasionally wanting. The variation in thickness of the basalt stratum may be due to its having been poured out upon an uneven surface, but it is not quite clear whether this unevenness was due to disturbance of the sedimentary beds before the outburst of the traps. That the denudation of the underlying formations can have been only partial is shown by the fact that they may be traced between three and four miles, the upper portion alone being locally absent.

On the left bank of the Godávari, near Rájámahendri itself, the infratrappean band has not been observed. The thickness of the lower flow of basalt cannot be clearly ascertained, but it is not less than 40 or 50 feet, and is probably more. Above this lower flow on both banks of the Godávari there is found a sedimentary band, twelve to fourteen feet thick at Káteru, where it only extends for about half a mile, and about two to four feet thick in the Pungadi direction, where it has been traced for about ten miles. The intertrappean bed consists of limestone and marl, portions of which abound in fossils. Numerous quarries, which have been opened near both Pungadi and Káteru, have afforded good opportunities for obtaining fossils, which are difficult to extract from the argillaceous limestone when it is first quarried but weather out on exposure. About 30 or 40 feet above the fossiliferous limestone of Káteru, another sedimentary bed, consisting of yellow calcareous shale, is seen in one place. It is very thin, and no fossils have been found in it.

The most marked feature of this fauna is its distinctly estuarine character.[1] *Tympanotonus*, *Pirenella*, *Cerithidea* and *Potamides* are all brackish water forms. *Hydrobia* is an estuarine genus, and the fossil called *Hemitoma* closely resembles a species of *Acmæa* found living in creeks in the deltas of Indian rivers. The shell described as *Cerithium multiforme* appears to be a *Tympanotonus* or *Pirenella; C. leithi* has the characteristic form and sculpture of a *Cerithidea*, and *C. stoddardi* is, at least, as much allied to *Potamides* as to *Cerithium* proper. Some of the shells referred to *Cytherea* agree best with the typical forms of the genus (*C. meretrix*), many species of which abound in backwaters and at the mouths of rivers, and Mr. Hislop has remarked the similarity between *Corbula oldhami* and a Brazilian species belonging to the estuarine genus *Azara*. There is a complete absence of pelagic shells such as the *Cephalopoda*, no *Echinodermata* or corals are found, and, above all, four species *Physa prinsepii*, *Lymnea subulata*, *Paludina normalis* and *Corbicula ingens* are characteristically fresh water

[1] For a list of the species that have been obtained see *Memoirs*, XVI, 233, (1880).

forms; the first three of these are of comparatively rare occurrence, but the *Corbicula* is common, and the last named may perhaps have lived in brackish water, as its near ally *Cyrena* does at the present day, whilst the purely fresh water shells were washed down by rivers, this view being quite in accordance with the theory that the intertrappean beds of Rájámahendri were deposited in brackish water, which was supplied with fresh water by streams, but was also in communication with the sea.

The mollusca, however, cannot be considered as very characteristic of age. They were compared by Mr. Hislop with the nummulitic fauna of western India, but, as he points out, no forms appear to be identical, and although *Natica dolium*, *Turritella affinis* and an unnamed *Cerithium* found in the tertiaries of Sind and Cutch resemble *N. stoddardi*, *T. prælonga* and *C. stoddardi*, the intertrappean forms are more closely allied to the cretaceous *N.* (*Mammilla*) *carnatica*, *T. elicita* and *Cerithium vagans* than to the eocene species mentioned,[1] while other forms might easily be shown to be affined to those occurring in the cretaceous rocks of Southern India. In the case of *Turritella prælonga* and *T. elicita* the affinity is very great. The shell called *Vicarya fusiformis* appears not to be really congeneric with *V. verneuilli*, the type of the genus,[2] and the latter has now been found to be miocene, not eocene. On the whole, it may be safely asserted that no tertiary alliances of any value have been detected amongst the intertrappean Rájámahendri fossils, and that their relations are rather with the upper cretaceous rocks of Southern India, although the connection is not strong.

In the islands of Bombay and Salsette, and probably farther north on the same line of coast, the traps have an inclination of from 5° to 10° to the westward. The islands are separated from each other and from the mainland to the north by tidal creeks and alluvial flats, whilst the expanse of water forming Bombay harbour lies between them and the mainland to the eastward. In the islands of the harbour, and on the hills between Thána and Kalyán north of the harbour, the same westwardly dip is displayed, but further to the eastward, from Kalyán to the Sahyádri range, the traps are horizontal.

About 2,000 feet of horizontal beds are exposed on the flanks of Mátherán hill, and a still greater thickness farther to the east in the hills near the Bhor Ghát and close to the Great Indian Peninsula Railway line between Bombay and Poona, but it is impossible to say how far the lowest strata, exposed at the base of the hills, are above the bottom of the series, as no lower beds than the traps are seen. Owing to the numerous breaks in

[1] When Mr. Hislop wrote, the South Indian cretaceous fossils had not been described.

[2] This was pointed out by Mr. H. M. Jenkins, *Quart. Jour. Geol. Soc.*, XX, 58, (1864).

the section, it is difficult, without closer measurements than have hitherto been made, to estimate the precise thickness of the rocks dipping to the westward near Bombay, but taking the average dip at 5°, the whole thickness would be nearly 7,000 feet. This is a minimum estimate, as the average dip is probably higher and the thickness consequently greater. From 1,200 to 1,500 feet of rock are exposed in Bombay island, so that it is evident that the lowest beds seen on the island are higher in the series than the highest flows seen on the Sahyádri mountains to the eastward, although some of the higher portions of the range are 4,000 feet above the sea.

The intertrappeans of Bombay are entirely confined, so far as is known, to these higher beds, no sedimentary rocks having hitherto been found amongst the middle portions of the Deccan trap series and it is manifest that the Bombay fresh water beds belong to a very different horizon from that to which the intertrappeans of Nágpur and the Narbadá valley must be assigned. The most important bed is that which underlies the basalt of Malabar hill and Worlee hill, forming the broken ridge along the western or sea face of the island; this stratum is consequently immediately beneath the highest lava flow known to occur anywhere throughout the trap area, for the rocks, as already stated, dip to the west, and no beds higher than those of Bombay have been discovered. It must, however, not be forgotten that the coast north and south of Bombay has not hitherto been examined with sufficient care to make it quite certain that no higher beds occur.

This intertrappean bed on the east side of Malabar hill is more than one hundred feet thick in places, and consists principally of soft grey, greyish blue, brown, and brownish yellow earthy shales, with occasional harder bands, some of which are black and carbonaceous. The greater portion of the bed is evidently formed of volcanic detritus, whether lapilli washed down by water, or sand produced by the disintegration of lava flows, it is difficult to say, very possibly both may have contributed to the formation of the rock. At the top of the deposit the shale occasionally becomes hardened and silicious, as if by the action of the overlying basalt. The black carbonaceous shale is locally highly bituminous and sometimes contains small layers of a coaly substance and fragments of mineral resin. Impressions of vegetables abound, although they are but seldom well preserved, and remains of animals are common, the best known being skeletons of small frogs and carapaces of *Cyprides*.

Besides this thick sedimentary band, several thinner beds have been found at lower horizons amongst the lava flows and ash beds of Bombay island. They are, however, very thin and, except one which is seen in the quarries of Nowroji hill south of Mazagaon, they are difficult to detect; indeed, the circumstance of their occurrence has only become known through the careful scrutiny of local geologists, who, living in the town, could take advantage of any excavations for buildings, tanks, roads, etc.. to examine the strata exposed. According to Dr. Buist there are five or six sedimentary beds below the thick band of Malabar hill, but fossils have only been found in that exposed at Nowroji hill, where *Cyprides* occur. All these bands consist of shaly beds.[1]

The fossils found at Bombay are tolerably numerous, but hitherto only the *Vertebrata* appear to have received more than a superficial notice. The remains of a fresh water tortoise, *Hydraspis leithi* (*Testudo leithi*, Carter) belonging to the *Emydidæ*, and of a frog, *Rana pusilla*,[2] considered by Dr. Stoliczka an *Oxyglossus*,[3] have been found, the latter in abundance, while some bones of a larger frog have been obtained. The *Arthropoda* are represented by three species of *Cypris*, one of which, *C.* (*cylindrica*), is also found in the intertrappean deposits of the Deccan; another species has been called *C. semimarginata* by Dr. Carter, the third is unnamed. *C. semimarginata* is the most generally diffused, but the other forms also occur in great numbers. Only fragments of insects have been found. Mollusca are rare, and the few specimens hitherto procured have been in poor condition, they have been referred to *Melania* and *Pupa*, but with some doubt, and none of the characteristic Deccan forms have been detected. The plant remains comprise stems, leaves, seeds, and perhaps roots, but very little has been determined, except that endogens and angiospermous exogens are represented.

The life represented by the species named is clearly that of a shallow marsh. The frogs occur in large numbers, and their bodies have evidently been deposited near the spot where they died, as the whole skeleton is found perfect. In some cases, as was noticed by Dr. Stoliczka, the skeleton has been dragged along the surface of the shale in which it is imbedded, and he suggests with great probability that this was done by wind. The tortoise is a marsh or river form, the nearest living ally, according to Dr. Gray,[4] being a genus found in fresh water in South America.

After the description of the various sedimentary formations intercalated

[1] For fuller description of these beds see Carter, *Jour. Bo.. Br. Roy. As. Soc.*, IV, 161, (1853), and Geological Papers on Western India, p. 128; Buist, *Trans. Bo. Geogr. Soc.*, X, 195, (1852); Wynne, *Memoirs*, V, 193,(1866); VI, 385, (1869). It must not be forgotten that Dr. Carter's views as to the relations of the sedimentary beds differ essentially from those stated in text, with which all other observers agree.

[2] Owen, *Quart. Jour. Geol. Soc.*, III, 224, (1847).

[3] *Memoirs*, VI, 387, (1867). Dr. Stoliczka shows that the form agrees well with *Oxyglossus* and with no other known existing genus. At the same time, as some of the principal characters by which genera of frogs are distinguished are not preserved in the skeleton, the Bombay frog may have differed greatly from recent *Oxyglossi*. From the species of true *Rana* it is distinguished by the want of vomerine teeth, the large head, and short hinder limbs.

[4] *Ann. Mag. Nat. Hist*, 4th series, VIII, 339, (1871).

with the traps or underlying them, the next point for consideration is the mode of origin of the trap rocks themselves. Their volcanic character is sufficiently proved by their composition. Precisely similar rocks occur amongst the lavas poured out from recent volcanoes, whilst nothing of the same kind has ever been known to be deposited from water. But the first difficulty which arises, and it is one of very great importance, is to account for the persistent horizontality of the beds. Two observers certainly, Jacquemont[1] and Adolph Schlagintweit,[2] have considered that the traps are unstratified, but after the evidence already mentioned as to the differences in mineral character between successive bands, the frequent occurrence of vesicular structure on the upper surface of flows, the presence in abundance of beds of volcanic ash, and the repeated interstratification in the same localities of sedimentary layers, it is unnecessary to refute this view. A much more common opinion, and one which has been supported by numerous excellent geologists, from Newbold downwards, is that the Deccan traps are of subaqueous origin, and it is necessary to show why this opinion is untenable.

In all cases of subaqueous eruptions the ejected masses consist of substances very similar to the lava, ashes, scoriæ and lapilli of ordinary subaerial volcanic outbursts, but these materials being thrown out into the water are reduced by the sudden cooling to the condition of a fine powder, which is dispersed and deposited in layers in the same manner as ordinary detritus, so as to form what are known as stratified tuffs. With these tuffs ordinary marine deposits are necessarily intercalated, and both these and the tuffs are usually fossiliferous, the very destruction of life in the waters of the sea, caused by the heat and gases which are evolved during eruptions, encouraging the preservation of those portions of the organism which are not liable to destruction from the temperature of boiling water or the process of decomposition. Now, the volcanic ashes, already described as occurring in great abundance amongst the higher beds of the Deccan traps, are not, as a rule, stratified in the manner in which beds deposited from water would be. Although they occur in strata, intercalated with basaltic lava flows, these ash beds themselves have no internal lamination, except in a few rare instances in which they are chiefly composed of bole, and may have been formed in the small pools of fresh water so common in volcanic areas. Above all, not a trace of a marine organism has ever been found in any ash bed, or in any rock intercalated with the traps, except in the intertrappean and infratrappean formations of Rájámahendri, where the lava has evidently been

[1] "Voyage dans l'Inde," 4°, Paris, 1841, III, 504, etc.

[2] "Report of the Proceedings of the Officers engaged in the Magnetic Survey of India," No. I, p. 6.—Reisen in Indien und Hochasien. Vol. I, p. 141, (1869).

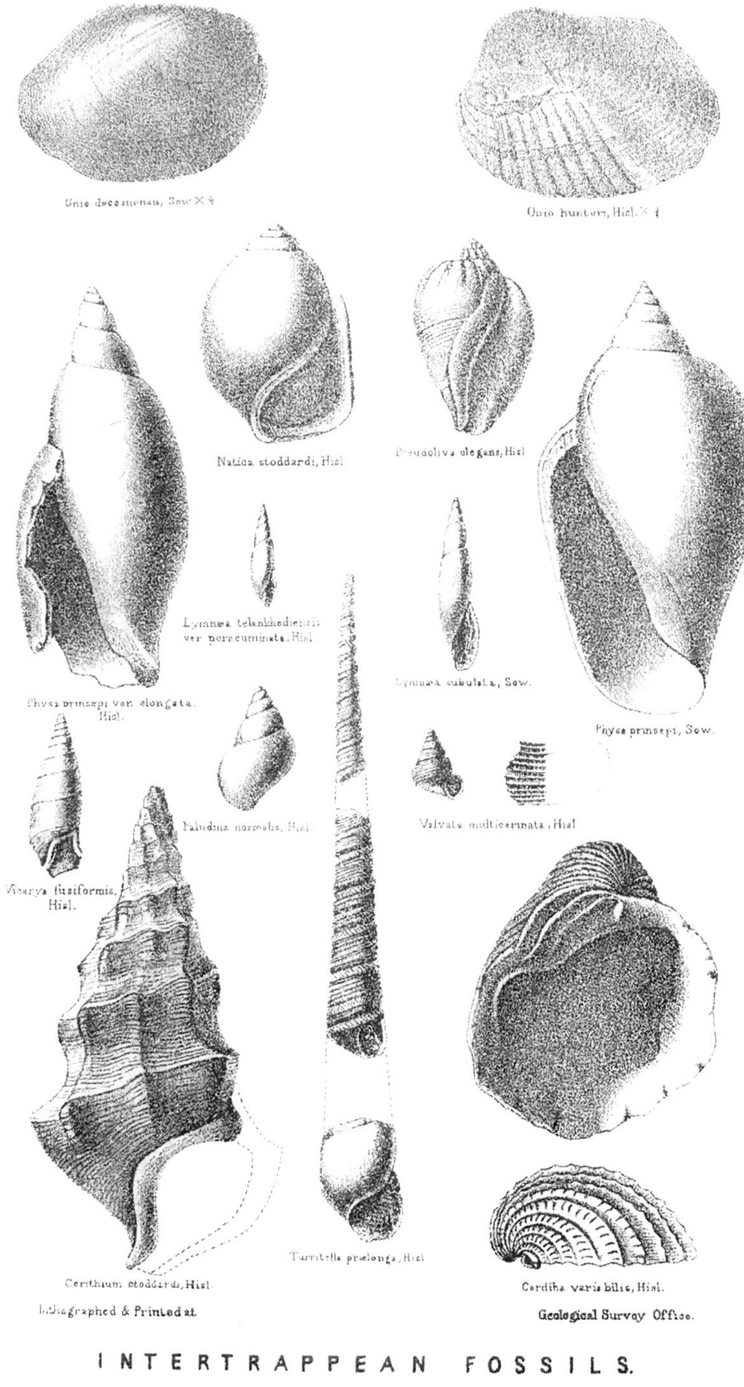

Lithographed & Printed at Geological Survey Office.

INTERTRAPPEAN FOSSILS.

poured out on the coast. It may be thought that the prevalence of volcanic conditions would destroy all life in the sea, and thus the absence of marine fossils in the traps may be explained, but even if this view were conceded, and it is entirely opposed to all that is known of recent submarine volcanic action, there must have been a great destruction of life at the commencement of the volcanic epoch, and some traces of the animals destroyed should have been preserved.

The evidence afforded by the characters of the traps and the absence of marine fossils is, therefore, opposed to the hypothesis of a submarine origin, and the relations of the lowest lava flows to the underlying rocks are strongly antagonistic to the idea that the volcanic outbursts were subaqueous. The surface of the older rocks upon which the traps rest is in many parts extremely uneven, the basalt filling great valleys, sometimes as much as 1,000 feet in depth, whose form shows that they were excavated by subaerial erosion. Admirable examples are seen between Bhopál and Hoshangábád, where the Deccan traps rest upon an extremely uneven surface of Vindhyan rocks.[1] It is true that this uneven surface might have been formed above the sea and then depressed beneath the water, but in that case we should expect to find aqueous deposits of considerable thickness at the base of the volcanic rocks, as periods of depression are always favourable for the accumulation of sediment. It is precisely in this uneven ground that no deposits whatever are found at the base of the traps and the general absence of any infratrappean deposit has been noticed in the south Maráthá country, where also the surface upon which the traps rest is very irregular.

Where the underlying formation consists of the cretaceous Bágh beds, these are, as a rule, apparently conformable to the volcanic series, and it might be thought that in this tract of country the traps were submarine. But every here and there a spot is found where the cretaceous rocks are wanting, and where the level of the infratrappean surface shows that their absence is due to denudation.[2] In some cases where the Bágh beds are not more than 30 or 40 feet thick, the denudation which has removed them has only extended over a small area, and has scarcely affected the harder rocks beneath, and from the small area, often only a few yards wide, over which the cretaceous rocks have been removed, it is evident that the denuding agent was subaerial erosion. It has also been ascertained that the Bágh beds had been locally disturbed to a small extent, besides having suffered from denudation, before the commencement of the volcanic outbursts.

Lastly, the circumstance that, with the single exception of the estuarine intertrappean band of Rájámahendri, every fossiliferous sedimentary bed

[1] *Memoirs*, VI, 240, 242, etc., (1867).

[2] *Memoirs*, VI, 212, 300, 313, etc., (1867).

intercalated with the Deccan traps is unmistakably of fresh water origin is a conclusive proof that all those lava flows which are associated with such sedimentary beds are not submarine. We have thus not only a complete absence of all proof of submarine origin,[1] but clear and unmistakable evidence that the traps were in great part of subaerial formation.

Another favourite idea with many writers, and especially with Mr. Hislop and Dr Carter, has been that the lower traps were poured out in a vast, but shallow, fresh water lake extending throughout the area over whieh the 'intertrappean limestone formation' extends.[2] This hypothesis involves the existence of a lake of enormous size, several hundreds of miles in length and breadth, but shallow throughout. It appears more probable that the lakes in which the Lametá group and the intertrappean beds were deposited were of moderate size, and that they were formed by unequal elevation of different parts of the area, prior to the volcanic outbursts, or by the obstruction of the drainage of the country by lava flows. The lake or lakes in which the Lametá beds were formed may have been more extensive, but it has already been shown that single sedimentary bands intercalated in the traps, can rarely be traced for more than three or four miles, and the character of the fauna, in the intertrappean formations, both of Central India and of Bombay, is in favour of the animals of which the remains are found having inhabited shallow marshes rather than deep lakes.

[1] It may appear to many geologists that an unnecessary amount of space and argument has been devoted to proving a very clear proposition, *viz.* that the Deccan traps are subaerial. The reason for giving the arguments at length is that a different view has been expressed by many geologists. A reference to the *Quart. Jour. Geol. Soc.*, XXX, 225, (1874), will show that the arguments used by Professor Judd to prove the subaerial origin of the volcanic rocks in the west of Scotland and north of Ireland, are precisely the same in many cases as those mentioned above. These views had been urged in the case of the Indian rocks (*Memoirs* VI, 145) some years before the publication of Professor Judd's papers.

[2] By both the writers named the intertrappean beds of Bombay were supposed to be identical with those of Central India, and both were under the impression that there was but a solitary fresh water bed which was deposited before any volcanic outbursts took place, which was then covered up by lava flows, and finally separated from the underlying rocks and broken up by a great sheet of intrusive basalt injected beneath it.

The geologists named would doubtless have modified their views had they been acquainted with all the facts now ascertained with regard to the Deccan traps and the associated sedimentary beds. The conception of a great sheet of intrusive basalt so injected between two formations that it always overlies the one and underlies the other, over an area of thousands of square miles, is quite untenable. It is a physical impossibility that an immense dyke should be injected horizontally for hundreds of miles instead of breaking through to the surface. Moreover, the fact that successive sedimentary beds, as in the case at Mekalgandi Ghát, mentioned on page 267, are often of different mineral composition, and the very frequent instances in which the upper surface of a sedimentary band is altered, whilst the lower is unchanged, prove that both lava flows and sedimentary intertrappean beds were regularly and successively formed, one above the other, as they now occur.

We are thus thrown back upon our original difficulty, the horizontality of the Deccan traps. It has been shown that this is not due to a subaqueous origin, whether marine or fresh water. At the same time the phenomenon cannot be said to have been thoroughly explained, because no such formation is known to be in process of accumulation at the present day. Many such masses of horizontal stratified traps are, however, found in various parts of the world, and though it is impossible, for want of recent examples, to demonstrate the circumstances which cause their formation in place of volcanic cones, there is abundant evidence that such traps were a common form of volcanic accumulation in past times, and that similar stratified lava flows were not confined to any particular epoch, although several instances are known of about the same geological age as that attributed to the Deccan outbursts.

Assuming, therefore, as we are justified in doing, that the horizontal dolerites of western and central India precisely resemble modern lavas in everything except their horizontality and the extent of area which they have covered, it remains to be seen what evidence there is of the sources from which this enormous accumulation of molten materials was poured out. The original cones and craters, if any ever existed, must have been the first portion of the volcanic area to suffer from denudation, and it is easy to conceive that subaerial erosion, from eocene times to the present, would have more than sufficed to remove every trace of the loose material of which volcanoes are chiefly composed. Still, it is surprising that the inclined beds forming the slopes of a volcanic cone should, in no single case which has been recognised, have been preserved by being encased in subsequent outbursts of harder materials. Possibly the tendency of great lava streams to sweep away all loose volcanic materials may suffice, in those cases in which large quantities of lava are poured out, to prevent volcanic cones from forming.

When, however, we look for other evidence of the neighbourhood of igneous outbursts, we find dykes and irregular intrusions abundant in some localities, rare or absent in others, whilst the presence of volcanic ash throughout a large portion of the trap area has already been noticed. The ash beds, especially when, as usually happens, they form a coarse volcanic breccia, containing blocks several inches in diameter, cannot have accumulated far from volcanic vents, although they may have been transported to a much greater distance floating on the surface of molten lava than they could have been ejected from the volcano.

A much closer examination of the Deccan area than has hitherto been practicable will be requisite before the distribution of dykes and ash beds can be considered as even approximately known. So far as the country has hitherto been examined, both appear to prevail much more largely in the region near the coast, from Mahábaleshwar to the neighbourhood of

Baroda, than in other parts of the trap area. It is, of course, very often difficult to recognise dykes amongst rocks of precisely similar mineral character, much closer search being needed than is requisite in order to detect volcanic intrusions amongst sedimentary formations. It is only where dykes are large and numerous that attention is likely to be directed to them.

There is one tract of country in which dykes are peculiarly large and abundant. This is in the Rájpipla hills, north-west of Surat. In this country, over a considerable area, very large, parallel, or nearly parallel, basalt dykes are found, sometimes not more than two or three hundred yards apart, the general direction being east by north to west by south. The traps are much disturbed, and frequently dip at considerable angles.

To the southward of the Tápti, along the line of the Sahyádri, and its neighbourhood, in western Khándesh, the northern Konkan, and the intervening small native states, a tract whose geology is unknown, it is probable that dykes may continue numerous for a considerable distance, as their number and size in the Konkan, north-east of Bombay, are especially noticed by Mr. G. T. Clark,[1] but intrusions are far from abundant in the lava flows exposed in the higher country east of the Gháts. At the same time the frequent occurrence of ash beds in the higher traps around Poona and Mahábaleshwar sufficiently attests the neighbourhood of the old volcanic vents.[2]

North of the Rájpipla hills and of the river Narbadá, and west of Baroda, trap dykes are not so common as in the Rájpipla hills, but intrusive masses occur. One of these, forming Matapenai or Karali hill, about 14 miles south-west of Chhota Udaipur, is a mass of grey trachyte or trachy-dolerite, containing enormous masses of granite, evidently derived from the metamorphic rocks through which the mass, when molten, passed on its way to the surface. The silicious character of the intrusion in this, and some other cases, is perhaps due to the fusion of quartzose metamorphic rock in the basic dolerite. Another trachytic core was noticed near the village of Padwani, 18 miles east of Broach. The occurrence of fragments derived from the metamorphic rocks in intrusive dykes is by no means an uncommon occurrence.

It is only natural that far better evidence of volcanic foci is to be

[1] *Quart Jour. Geol. Soc.* XXV, 164, (1869).

[2] Mr. Clark has described what he regards as a series of vents along the course of the Barwi river above Kalyán. They are said to be crater-like hillocks, in whose neighbourhood the traps lie in small streamlets, crossing and overlapping one another. These supposed vents lie along the region where the traps have undergone a maximum of denudation, believed by Mr. Clark to be due to the less degree of homogeneity of the accumulations in the neighbourhood of the foci, as compared with the more evenly bedded traps and ashes at a distance; *Records*, XIII, 69, (1880). The only reason for doubting those conclusions is the peculiar form assigned to the hillocks; as a rule, the material filling the neck of an old volcano is harder than the surrounding stuff, and would not form a hollow surrounded by a ridge, but stand out as a solid mass.

found outside the trap area, or in the inliers of older sedimentary rocks, than amongst the lava flows themselves, and it may therefore be useful to point out very briefly the distribution of such intrusive masses so far as the country is known. Commencing to the north-west, no trap dykes have been found in Sind, where, however, the deposits of older date than eocene cover an exceedingly small area. In Cutch intrusive masses of basalt and dykes of large size abound throughout the jurassic rocks, and some of the former rise into hills of considerable size.[1] In Káthiáwár the Girnár hill is said to be formed of the dioritic core of an ancient volcano, and intrusions are abundant throughout this district. Throughout the northern edge of the trap country in Rájputána, Gwalior, and Bundelkhand, dykes are rare or wanting, but they abound in some of the areas of older rocks exposed in the Narbadá valley, and they are especially common in the Gondwána tract, south of the river, in the neighbourhood of the Mahádeva hills. Farther to the eastward they are less numerous, but some occur throughout the upper Son valley, and they gradually die out in Sargúja and Palámau, only 200 miles west of the ground in which the older lava flows of Rájmahál age are seen, and within less than 100 miles of the Gondwána basins in the upper Dámodar valley, which are traversed by basalt dykes, probably of the same age as the Rájmahál traps. Passing southwards from Jabalpur and Mandlá however, there is a total absence of volcanic intrusions amongst the Vindhyan and Gondwána formations of Nágpur and Chándá, and none have as yet been noticed in the neighbourhood of the Pránhíta and Godávari between Chándá and Rájámahendri. The country south of the Godávari and north-west of Haiderábád is still imperfectly known, but in the south Maráthá country, and the Konkan near Vengurla, the few dykes which have been observed traversing the unmetamorphosed azoic strata are but doubtfully connected with the Deccan traps. Ashes, moreover, are much less abundant in this region, amongst the Deccan flows, than they are further north.

We have thus abundance of evidence of the former existence of volcanic foci in Cutch, the Rájpipla hills, and the lower Narbadá valley, and probably in the neighbourhood of the Sahyádri range east and north-north-east of Bombay, whilst there is every probability that vents extended to the eastward as far as south Rewá and Sargúja, but there is no evidence of any having existed in the Nágpur country or to the south-east. Yet, as the traps are found represented at Rájámahendri, it appears probable that they once extended over all the Godávari valley, though it is quite possible that the Rájámahendri outlier may have been originally isolated and derived from a centre which has not been discovered.

[1] One of these, called Denodhar, was described originally as a volcano, *Geol. Trans.*, 2nd series, V, 315, (1840), and the statement that it is an extinct crater has been repeated in numerous geological works. The hill is very probably the basaltic core of a pre-nummulitic volcano. Its crateriform appearance is due to denudation.

It is, however, very clear that the lava flows must have extended to an enormous distance from the vents through which the molten material was poured out. Trap dykes are rarely solitary; they are generally abundant in the neighbourhood of volcanic foci, and the country in the neighbourhood of Nágpur and Chándá has been examined so closely that the improbability of any intrusions of igneous rock having been overlooked is much greater than in most parts of the country. It is known that the comparatively moderate outbursts from existing volcanoes flow to great distances from their source, while the occurrence of the Deccan traps in immense horizontal sheets of but slight thickness, but of great horizontal extension, shows that they must have been extremely fluid when poured out, and it is difficult to form an accurate idea of the distances to which they may have flowed before consolidating. Further observations are necessary before all the sources of the great volcanic series of Western India can be said to have been even approximately determined.[1]

The question of the geological age to be assigned to the Deccan volcanic outbursts has been left to the last, because it was desirable to precede it by a full statement of all the facts upon which a conclusion may be founded. The evidence to be recapitulated is twofold, that founded on the affinities of the fossils found in the various intertrappean rocks, and that derived from the relations of the stratified traps to beds above and below them. It is, of course, clear that the traps rest upon cretaceous beds and are overlaid by nummultitics, and the only question is whether the lava flows are cretaceous or eocene.

The most important clue to the correlation of the volcanic rocks with the known series of fossiliferous deposits might be expected to be obtained from the marine beds associated with the volcanic formations at Rájámahendri. This, however, as has been already shown in the description of the sedimentary beds, proves of little service. So far as is hitherto known, the relations of both the infratrappean and intertrappean faunas are with the cretaceous rather than with the tertiary beds, but the points of connection, in the latter case especially, are quite insufficient to decide the affinity of the formations.

Turning to the fresh water fauna of the intertrappean beds, the question arises, as in the case of the Gondwánas, of the amount of dependence to be placed upon terrestrial animals and plants as a guide to geological age. In the case of the Gondwána formations it has been shown that forms

[1] Sir A. Geikie (*Nature*, 4th Nov. 1880) has suggested that the Deccan traps, like the great lava flows of the Pacific slopes of North America, were poured out from fissures, and not from volcanic cones. This would, to a great extent, account for the horizontality of the traps, but the ash beds must have been thrown out from vents, round which cones would accumulate.

characteristic of particular epochs in Europe occur, in a very different position, in the geological sequence in India, and it is, therefore, necessary to be cautious in accepting conclusions founded upon slight evidence. There is a very marked similarity between some of the fresh water mollusca of the Deccan intertrappeans and species found in some beds of plastic clay age (lower eocene) occurring at Rilly-la-Montaigne in northern France,[1] one species, *Physa gigantea*, from the latter locality being considered by some palæontologists identical with the Indian *P. prinsepii*. This identification is, however, to say the least, extremely doubtful, and the fauna of the Rilly beds appears more recent than that of the Deccan intertrappeans. Other resemblances between the plants and fish of the intertrappean beds and those of the London clay have also been indicated, and a *Physa*, said to be allied to *P. prinsepii*, has been found in the nummulitic rocks of the Himálayas, but even the generic identification in the latter case is far from certain.[2]

This evidence only suffices at the most to show an approximation between the age of the Deccan trap and the lower eocene, and is quite insufficient to prove whether the former should be classed as uppermost secondary or lowest tertiary. But the closest relationship of the intertrappean fauna is with that of the Laramie group in North America. According to Dr. Neumayr[3] the following species are very closely allied, if not identical:—

Intertrappean.	Laramie.
Physa prinsepii	*P. copei*.
,, ,, var. *elongata* .	*P. disjuncta*.
Acella attenuata	*A. haldemani*.
Paludina virapai	*Hydrobia anthonyi*.
Unio carteri	*U. gonionotus*. *N. gonioumbonatus*.
Corbicula ingens	*C. cleburni*.

The Laramie group is regarded as intermediate in age between the cretaceous and eocene—an age which would harmonise the conflicting alliances of the fauna of the inter and infra trappean beds near Rájámahendri.

The relations between the traps and the underlying cretaceous beds of the lower Narbadá valley have been already described. There is a general, though slight unconformity, due to subaerial denudation of the underlying beds. In a very few localities the latter appear to have been disturbed

[1] *Mem. Soc. Geol. de France*, 2nd series, III, 265, (1848). The genera found at Rilly-la-Montaigne are *Cyclas*, *Ancylus*, *Vitrina*, *Helix*, *Pupa*, *Clausilia*, *Megaspira*, *Bulimus*, *Achatina*, *Auricula*, *Cyclostoma*, *Paludina*, *Physa*, *Valvata*. The majority of these genera are terrestrial forms.

[2] D'Archiac and Haime, Description des animaux Fossiles du groupe nummulitique de l'Inde, Paris, 1853, p. 277.

[3] *Neu. Jahrb. Min. Geol.*, 1884, Bd. I, p. 74; *Records*, XVII, 87, (188).

before the formation of the lowest traps. Between the highest volcanic beds and the nummulitic rocks of Surat and Broach the break appears to be much greater; not only do the tertiary rocks rest upon a largely denuded surface of the traps, but they are in a great measure composed of materials derived from the disintegration of the lava flows, the lowest tertiary beds being frequently coarse conglomerates of rolled basalt fragments, whilst beds, hundreds of feet in thickness, are chiefly composed of agates derived from the traps. This, however, although it proves that great denudation of the volcanic rocks took place during the deposition of the nummulitic formations, does not necessarily imply a great break and an interval of disturbance prior to the commencement of the tertiary epoch, because the traps, being of subaerial origin, were, unlike most sedimentary rocks, subject to erosion from the period of their formation. In this case, however, the unconformity is distinctly marked, and appears to show a great break in the sequence. The lowest tertiary beds near Surat contain fossils which appear to be a mixture of middle and lower eocene forms (Kirthar and Ránikot).

Farther to the westward, in Cutch, the rocks at the base of the tertiary group resting upon the trap are locally conformable, and they have even been considered[1] to be partially volcanic, but, as will be shown in the next chapter, there can be no doubt that a break, marked by unconformity, exists between the two series. It appears most probable, too, that the lowest tertiary beds are really composed of detritus derived from the volcanic rocks, as all appear to be of sedimentary origin, and no instance has been noticed of intercalation with the lava flows. The great difficulty of distinguishing between volcanic ash and the detritus of igneous rocks when mixed with ordinary sediments, especially where the rocks are much decomposed, as in Cutch, is too well known to require comment. The beds immediately resting upon the traps are of older date than the nummulitic limestone. The trap rests unconformably on neocomian and jurassic beds.

In Sind the very thin representatives of the Deccan traps may, of course, only represent a small portion of the period during which the volcanic rocks were in process of accumulation further to the eastward. One band rests conformably upon beds which, according to the most recent palæontological investigations, are intermediate in age between the cretaceous and lower eocene, whilst a second bed of trap is found about 700 feet lower, interstratified with sandstones.

It will be seen, therefore, that whilst it is clear that the Deccan traps were poured out in the interval between middle cretaceous and middle eocene, the evidence tends to show that the lowest volcanic outbursts were at the oldest of uppermost cretaceous age. That an immense period

[1] Wynne, *Memoirs*, IX, 66, (1872).

of time was occupied by the accumulation of the successive volcanic outbursts is manifest; long intervals must have elapsed between successive flows in all those cases in which fossiliferous sedimentary beds are intercalated, for these intervals were sufficient to enable lakes to be formed and stocked with life, and in other cases for rivers to cut beds in the lava flows, and to fill up those beds with gravel and sand.

Recapitulating the whole evidence, so far as it is presented to us by the observations hitherto made, we find that in times subsequent to middle cretaceous, a great area of the Indian Peninsula formed part of a land surface, very uneven and broken in parts, but to the eastward apparently chiefly composed of extensive plains, which, by some slight changes of level preceding the volcanic period, were converted into lakes. There is much probability that springs charged with silica were common either at this epoch or shortly after. The lakes had apparently been drained, and the deposits, which had accumulated in them, had locally been subject to denudation before the first outbursts of lava took place. These occurred at considerable intervals, small and very shallow lakes or marshes being formed in the meantime by the interruptions to the drainage produced by lava flows, or by changes of level accompanying the volcanic eruptions. In these lakes a rich fauna of fish, mollusca, entomostracous crustacea and water plants existed, whilst a varied and probably a rich vegetation occupied the surrounding country. There is evidence of the existence of insects and of reptiles, whether terrestrial or aquatic has not been determined, but hitherto no remains of mammals or birds have been found—a circumstance which by no means proves that they did not exist. Fresh flows of lava filled up the first lakes, and covered over the sedimentary deposits which had accumulated in the waters, but these very flows, by damming up other lines of drainage, produced fresh lakes, so that several alterations of lava and sedimentary beds were produced in places. Gradually the lakes seem to have disappeared, whether the lava flows succeeded each other so rapidly that there was no time for the accumulation of sediment in the interval, or whether, as is more probable, the surface had been converted into a uniform plain of basalt by the enormous lava streams which had been poured out it is difficult to say, but no further traces of life have hitherto been found until towards the close of the volcanic epoch. It is possible that at the end, as at the commencement, of the period, the intervals between eruptions became longer, and the animal and vegetable life, which may have been seriously diminished or altogether driven out of the country during the rule of igneous conditions, resumed its old position, but a great change had taken place in the long interval, the old lacustrine fauna had died out, and the animals and plants which now appeared in the country seem to have differed from those which had formerly occupied it. Lastly, in the north-western portion of the area, parts of the volcanic country were depressed beneath the sea, and marine tertiary deposits began to be formed from the detritus of the extinct volcanoes and their products. A great tract of the volcanic region, however, appears to have remained almost undisturbed to the present day, affected by subaerial erosion alone and never depressed beneath the sea level though probably for a time at a lower elevation than at present.

CHAPTER XII.

CRETACEOUS ROCKS OF THE EXTRA-PENINSULAR AREA.

Absence of break between the cretaceous and tertiary—Isolated exposures of lower cretaceous beds—Cretaceous rocks of Sind and neighbouring areas—Suláimán range—Afghánistán—Himálayas—Assam—Burma—Doubtfully cretaceous beds of Tenasserim.

We have already seen that the great distinction between the palæozoic and mesozoic of Europe does not hold good in India, and that the interval is bridged by rock systems which include beds both of older mesozoic and newer palæozoic age. In the last chapter we saw that the interval between mesozoic and cainozoic is similarly bridged, in the Indian Peninsula, by the Deccan traps, and we will now find that in extra-peninsular India, in Sind and Balúchistán on the west, as in Assam on the east, it is similarly impossible to separate the secondary and tertiary eras, on any but purely palæontological grounds, as they are connected by a continuous series of deposits ranging from cretaceous to tertiary, which, in the intervening country, is not only carried through the tertiary epoch, but is extended into recent times.

It is not meant by this that there is on any one section a complete and conformable sequence extending from cretaceous to recent. The sections indeed are all imperfect, and unconformable breaks are found in all, but the unconformity of one section is bridged by a continuous series of deposits on another, and there is no widespread, universal break which would give a natural line of demarcation for the separation of the rocks below from those above them. In many ways it would, consequently, be more natural to group the cretaceous and tertiary beds together for descriptive purposes, but this would necessitate a system of description that would obscure some important relations between the tertiary and especially the upper tertiary beds of widely separated areas, and it will be best to take advantage of the recognised division between tertiary and secondary, and confine our attention for the present to the cretaceous rocks.

Before proceeding to the description of the more important and complete exposures it will be well to notice some isolated occurrences of lower cretaceous beds.

In Cutch there is a thin bed of ferruginous oolitic rock which occurs at the base of the Deccan traps forming Ukra the hill, seven miles south-east of Lakhpat in north-western Cutch, and rests upon beds of the Umia group. The outcrop is very ill seen, and nothing has been definitely ascertained as to the degree of conformity between the cretaceous bed and the underlying formation, but there appears to be no marked contrast between them.[1]

The following three *Cephalopoda* have been obtained from this locality, *Ammonites martini*, *A. deshayesi*, *Crioceras australe*. The two former of these occur in the lower greensand (Neocomian) of Europe, and are most characteristic of the upper portion; the third has been found in cretaceous beds of Australia, whose exact horizon is not known.

In the Chicháli pass, in the trans-Indus continuation of the Salt range, the jurassic beds are conformably overlaid by a band of tough black sandy clay, full of *Belemnites*, *Ammonites*, etc., among which Dr. Waagen recognised the lower neocomian form *Perisphinctes asterianus*.[2]

This fossiliferous band is overlaid by a band of soft yellowish unfossiliferous sandstones which increases in thickness to the west. This was at first regarded as cretaceous, but it is said to contain pebbles of alveolina limestone in the westerly exposures.[3]

In the Sirban mountain near Abbottábád, the jurassic sandstones are capped by a bed, 10 to 20 feet thick, of a similar, but much harder, sandstone, which weathers of a rusty brown colour and is commonly crowded with fossils. These include *Ammonites* of the groups *Cristati* and *Inflati*, *Ancyloceras*, *Anisoceras*, and *Baculites*; *Belemnites* are abundant, and the general facies of the fauna is that of the gault.[4] They are succeeded by a group of thin bedded unfossiliferous limestones which may belong either to the cretaceous period or to the nummulitics by which they are overlaid.

The only locality in Sind where beds of older date than eocene have been identified is in a range of hills running due south from the neighbourhood of Sehwán, and generally known to Europeans as the Lakhi range,[5] from the small town of Lakhi near the northern extremity. South-west of Amri, on the Indus, a number of very dark coloured hills are seen in this range, contrasting strongly with the cliffs of grey and

[1] The only published account of this bed is in the *Pal. Indica*, series ix, pp. 245-47, (1875). No account of the locality was ever printed by the discoverer, Dr. Stoliczka, and his note books contain scarcely any details on this particular point.

[2] W. Waagen, *Pal. Indica*, series ix, p. 245, 1875.

[3] A. B. Wynne, *Memoirs*, XVII, 242, (1880).

[4] W. Waagen, *Memoirs*, IX, 342, (1872).

[5] This range has no general name, different portions being known by a number of local terms. It is one of the ranges which combine to form the Hala range of Vicary and other writers, and the name is the less inappropriate in this case as there is an unimportant pass through the chain known as the Hala Lak. Different portions of the range are known as Tiyún, Kara, Eri, Surjana, etc.

whitish nummulitic limestones behind them. These dark hills consist of cretaceous beds, but the lowest member of the series is only exposed in a single spot, at the base of a hill known as Barrah, lying about ten miles south-west of Amri. The whole range here consists of three parallel ridges, the outer and inner, composed of tertiary rocks, while the intermediate one consists of cretaceous beds, faulted against the lower eocene strata to the eastward and dipping under them to the westward. Close to the fault some compact and hard whitish limestone is found, the lower portion pure; the upper portion, often containing ferruginous concretions, is sandy, gritty, and forms a passage into the overlying sandstones. The base of this limestone is not seen, the whole thickness exposed is a little over 300 feet, and the length of the outcrop does not exceed half a mile. The limestone is fossiliferous, containing echinoderms and mollusca, but it is so hard and homogeneous that nothing obtained from it can be easily recognised, except one fragment of a hippurite. This fossil is, however, of great importance, because it shows that the white limestone may very probably be an eastern representative of the hippuritic limestone, so extensively developed in Persia, and found, in numerous localities, from Teheran to east of Karman in longitude 58°, just ten degrees west of the Lakhi range in Sind.[1] The precise position of the Persian hippuritic limestone in the cretaceous series has not been determined, but the European formation, which is very similar and probably identical, is of the age of the lower chalk (turonian).

The sandstones resting on the hippuritic limestone occupy a considerable tract around Barrah hill, and extend for about three miles from north to south. They are also seen at Jakhmari to the northward, and in one or two other places in the neighbourhood. They are gritty and conglomeratic, frequently calcareous, and contain a few bands of shale, usually of a red colour. The prevailing colour on the weathered surfaces is dark brown or purple, many of the beds being highly ferruginous. On the top of the sandstones is a thick bed of dark coloured impure limestone, containing oyster shells, and occasionally large bones, apparently reptilian, but none have been found sufficiently well preserved for identification.

In one place a bed of basalt, about 40 feet thick, has been found interstratified in the sandstones, and it is possible that the band may exist elsewhere, but it has hitherto remained undetected. The position of this bed of basalt on the face of a hill called Bor, about 13 miles north of Ránikot, is at an elevation of 300 or 400 feet above the base of the sandstones, and about twice as much beneath the main band of interbedded trap, to be described presently.

These sandstones are overlaid by soft olive shales and sandstones,

[1] W. T. Blanford, Eastern Persia, London, 1876, II, pp. 457, 485.

usually of fine texture. The sandstone beds are thin, and frequently have the appearance of containing grains of decomposed basalt or some similar volcanic rock, or else fine volcanic ash. A few hard bands occur, and occasionally, but rarely, thin layers of dark olive or drab impure limestone. Gypsum is of common occurrence in the shales.

The olive shales are highly fossiliferous, the commonest fossil being *Cardita beaumonti,*[1] a peculiar, very globose species, truncated posteriorly, and most nearly allied to forms found in the lower and middle cretaceous beds of Europe (Neocomian and Gault). This shell is extremely abundant in one bed, about 200 or 250 feet below the top, but is not confined to this horizon. *Nautili* also occur, the commonest species closely resembling *N. labechei* of Messrs. D'Archiac and Haime, but differing in the position of the siphuncle. This form appears undistinguishable from *N. bouchardianus,* found in the upper cretaceous Ariyalúr beds of Pondicherri and at a lower cretaceous horizon in Europe.

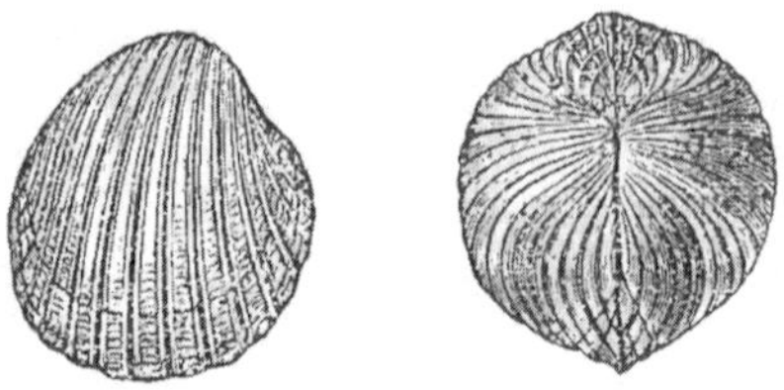

Fig. 17.—*Cardi'a beaumonti,* DArch, and Haime.

A second *Nautilus* resembles *N. subfleuriausianus* (another eocene Sind species) in form, and is also allied to some cretaceous types. Several *Gasteropoda* occur, especially forms of *Rostellaria, Cypræa, Natica,* and *Turritella* but none are very characteristic. Two forms of *Ostrca* are common, one of them being allied to the tertiary *O. flemingi* and to the cretaceous *O. zitteliana,* but distinct from both. The only mollusc which certainly passes into the Ránikot beds is *Corbula harpa.*

In the lower part of the beds with *Cardita beaumonti* some amphicœlian vertebræ were found, which Mr. Lydekker has ascertained to be crocodilian. All amphicœlian crocodiles are mesozoic, and the present form must be one of the latest known. So far as it is possible to form an opinion from very fragmentary materials, the vertebræ in question appear more nearly allied to the wealden *Suchosaurus* than to any other form hitherto described. It has, however, been already shown, when writing of the Gondwána flora, that the distribution of *Reptilia* in past ages was not the same in India as in Europe.

Only the corals and echinoids of the *Cardita beaumonti* beds have as yet been criticically examined, the former by the late Prof. P. M. Duncan, the latter by the same palæontologist with the assistance of Mr. P. M Sladen. The results obtained are not very definite so far as the correlation of

[1] D'Archiac and Haime, Description des Animaux fossiles du groupe Nummulitique de l'Inde. Paris, 1853, p. 253, pl. xxi, fig. 14.

the beds is concerned. Among the corals the genera *Caryophylla*, *Smilotrochus* and *Litharœa* are cretaceous, but also range into and through the tertiary, and the *Smilotrochus blanfordi* very closely resembles *S. incurvus* of the Italian eocene.[1] The echinoids are equally indefinite as regards their relations, being neither distinctly cretaceous, nor definitely eocene, and the general facies is such as to indicate an age intermediate between these two periods.[2]

Mention has already been made of one bed of basalt intercalated in the sandstones above the hippuritic limestone and a much more important band of the same igneous rock has been traced, resting upon the *Cardita beaumonti* beds, throughout a distance of twenty-two miles from Ránfkot to Jakhmari, about seventeen miles south of Sehwan, wherever the base of the Ránikot group is exposed. The thickness of this band of trap is trifling, and varies from about 40 to about 90 feet. Apparently in some places the whole band consists of two lava flows, similar in mineral character except that the upper is somewhat ashy and contains scoriaceous fragments; the higher portion of each flow is amygdaloidal, and contains nodules of quartz, chalcedony and calcite, and in places the nodules are surrounded by green earth, as is so frequently the case with the Deccan traps. Another characteristic accessory mineral, common also in the traps of the Deccan and Málwá, is quartz with trihedral terminations. The basaltic trap of the Lakhi hills is apparently of subaerial origin, although it rests conformably on the marine (or estuarine) *Cardita beaumonti* beds. There is nothing in the igneous bed to indicate its having consolidated otherwise than in the air, and the structure differs altogether from that of subaqueous volcanic tuffs.

The evidence that this band of basaltic rock is interstratified and not intrusive is ample; throughout the whole distance the trap is found in precisely the same position, between the lowest beds of the Ránikot and the highest of the *Cardita beaumonti* groups, and apparently perfectly conformable to both. The close resemblance in mineral character and the similarity of geological position, at the base of the tertiary beds, show that this band must in all probability be a thin representative of the great Deccan trap formation, and the occurrence of a second bed at a lower horizon, interstratified with the passage beds between cretaceous and tertiary, tends strongly to confirm the inference drawn from the relations of the traps to the cretaceous and tertiary rocks of western India, that the great volcanic formation must be classed as intermediate in age between those two eras.

In Balúchistán the section of cretaceous beds is more extensive than

[1] *Pal. Indica*, series xiv, I, pt. ii, p. 25, (1880). [2] *Pal. Indica*, series xiv, I, pt. iii, p. 28, (1882).

that seen in the one small exposure in lower Sind, and none of the horizons except the uppermost have been identified in the two regions.

In the neighbourhood of Quetta and the country to the east of it the lowest rock known is a massive limestone of great thickness. The lower portion of this, as seen at the head of the Bolan pass and at Sariab, is of a pale cream colour, the upper portion is more or less dark grey in colour. Fossils are not very abundant, but in places it exhibits sections of *Hippurites*, *Inoceramus* and corals on the weathered surface. The exact age of this limestone is undetermined, but it is regarded as lower cretaceous.

The massive grey limestone is succeeded by a series of dark grey or black shaly beds, often containing an admixture of volcanic ash, overlaid by red and green mottled shales and thin bedded limestone, capped by white limestone. The thickness of these beds appears to be about 1,000 feet, about 200 at the top of which consists the white limestone, but there are great variations in thickness of both members, owing partly to the manner in which the soft shaly beds have yielded to compression, and partly to the removal of the upper beds by denudation previous to the deposition of the next succeeding group. Locally the shaly beds of this group are abundantly fossiliferous, but except a few fragments of ammonites the only fossils that have yet been found are belemnites, mostly belonging to the section *Dilatati*, and including one very broad and flattened form; owing to the abundance of these fossils the group has been referred to, in previous publications of the Survey, as the Belemnite beds.[1]

The age of this group is clearly secondary, yet on some sections the white limestone at its summit contains numerous specimens of *Nummulina* and *Alveolina*, which are usually regarded as indicative of a tertiary age, thus introducing an anomaly which is repeated in the next succeeding group.

According to any local system of classification the next succeeding group of strata would be separated from those just described and united to the overlying Nummulitic beds, with which it is perfectly conformable, for there is a slight but distinct unconformity at the top of the Belemnite beds. The unconformity is unaccompanied by any recognisable want of parallelism of stratification between the beds below and above it, but is marked by a considerable degree of erosion and a complete change of fauna, none of the belemnites having been found in the overlying group.

The Dunghan group as this is called, from a hill of the same name east of Spintangi, is an important and interesting one. In the neighbourhood of Hurnai it is essentially a limestone formation and caps the bare

[1] *Records*, XXV, 19, (1892).

hogbacked hills east of the Hurnai route to Quetta. To the south of the road which connects the Spintangi railway station with Thal Chotiáli, the lower beds become argillaceous, and the argillaceous element more and more replaces the calcareous till, in the hills east of Khattan, the group has become essentially a shale group in which the calcareous element is quite subordinate. A similar change takes place in the hills west of the Bolan pass, but the country there has been less fully examined.

In the hills inhabited by the Mari tribe the lowest beds of the group are usually unfossiliferous, grey, green, and purplish shales, overlaid by about 1,000 feet of grey shales, many beds being so profusely fossiliferous as to become impure limestones. Above these shales there is a band of 100 to 200 feet thick, composed principally of more or less impure sandstones, capped by a limestone composed almost entirely of oysters (*Exogyra?*), but containing also a few *Nautili* and other fossils. The oyster bed is separated by some 600 feet of beds, on some sections of limestone on others grey shale, from a peculiar band of pseudo-breccia regarded as the base of the nummulitic series in this district.

The fauna of this group is a peculiar one; nummulites are abundant, but associated with them are *Crioceras*, *Baculites* and *Ammonites*, while fully half the echinoderms belong to the order *Echinoconidæ*, and an oyster resembling *O. carinata* is not uncommon. *Cardita beaumonti* was not found, but the admixture of characteristically cretaceous forms with an abundance of nummulites points to the group being intermediate in age between the cretaceous and eocene periods, and consequently equivalent to the *Cardita beaumonti* beds of Sind. The suggestion is supported by the frequent occurrence of beds of impure volcanic ash, immediately above the sandstones and oyster bed, and of ash beds and even basaltic trap apparently interbedded with the uppermost beds of the Dunghan group in the Bolan pass. It is natural to suppose that they represent the same horizon as the trap above the *Cardita beaumonti* beds in Sind.

It is not known how far the grouping adopted in the country east of Quetta holds good for the rest of Balúchistán, as this country has never been geologically examined with any thoroughness. Some particulars of observations made on his journeys through Balúchistán, have been recorded by Dr. Cook, in which it appears to be possible to recognise the various rock groups mentioned above. He describes[1] the cretaceous rocks as consisting of "more or less compact, fine grained, red and white limestone, interleaved with slabs and veins of chert; the limestone generally containing fine microscopic specks, and the upper part one or two massive strata of an excessively hard limestone abounding in *Orbitoides*, *Orbitolina* and *Operculina*, the lower strata becoming

[1] *Trans. Med. Phys. Soc., Bombay*, VI, 101, (1860).

argillaceous, shaly and containing (rarely) ammonites." These beds are underlaid by a dark blue fossiliferous limestone containing *Rhynconella*. There is, however, some uncertainty about the section, for near Khelát the white limestone appears to underlie the fossiliferous shales.

There is some indication in this description of the massive limestone, belemnite beds, and Dunghan group further east, and with one exception the fossils recorded would accord with the identification. *Ammonites*, *Ceratites*, *Crioceras*, and *Belemnites* have all been found in the eastern area, and *Scaphites* might well accompany them, but if the identification of *Orthoceras*, which has been referred to in a previous chapter,[1] was correct, it must belong to an older set of beds than cretaceous, and suggests that the section is more extensive and less simple than Dr. Cook's descriptions would indicate.

In the Suláimán range, west of Dera Ghází Khán, the cretaceous rocks, so far as they are exposed, comprise two well marked stages. The lower consists of dark grey limestones, occasionally sandy or shaly, passing downwards into dark to bluish grey, often nodular, calcareous shales. The limestone abounds in indistinct fossils, especially foraminiferæ, and in the underlying shales a cephalopod belonging to the *Ammonitidæ*, *Inoceramus* and two species of *Exogyra* resembling cretaceous forms have been found.[2]

These beds, of which about 1,500 feet are exposed, are overlaid by about the same thickness of sandstones, generally white or pale coloured, brown, greenish or purplish grey. No fossils have been found in the sandstones, and no unconformity has been detected between them and the overlying beds, while they overlie a bed of pseudo conglomeratic limestone exactly resembling that found at the base of the eocene beds of eastern Balúchistán.

Further north, in the neighbourhood of the Takht-i-Suláimán, the same pale sandstone and underlying shales and limestone are found underlaid by some hundreds of feet of massive grey limestone, showing sections of *Inoceramus* and corals on the weathered surface, which exactly resembles the massive limestone of the Quetta neighbourhood, and is probably of the same age. West of the Suláimán range, in the direction of the Zhob valley, this massive limestone is underlaid by a great thickness of green and grey slaty shales, intercalated with beds of sandstone and a few of limestone, from which no fossils have as yet been obtained.

Before passing on to the cretaceous beds of Afghánistán, it will be interesting to notice that the rocks just described are the source of the petroleum of eastern Balúchistán and of the Suláimán range. The Balúchistán petroleum is a thick, black, tarry maltha, traces of which are very

[1] *Supra*, p. 143.

[2] *Memoirs*, XX, 217, (1883).

frequently found in the Dunghan group. At Khattan there is a natural oil spring, where the petroleum issues along with an abundance of hot sulphurous water, and for seven years past an attempt to work this oil for profit has been in progress, but the quantity obtained has not been sufficient to prove remunerative. It has been supposed that the greater abundance of the oil was in some way connected with the unusual profusion of organic remains in the Dunghan group at this locality, but the connection is by no means clear. A precisely similar oil is found in the Bolan pass near Kirta, and in the Robdar valley, south of Bíbi Nání, it issues from the limestones below the Belemnite beds, that is to say, from rocks much older than the Dunghan group. No rocks older than this limestone are exposed in the Bolan pass, and it is impossible to say whether there are any profusely fossiliferous beds underground, analogous to those of the Dunghan group at Khattan, but the widespread presence of traces of a similar oil, even where fossils are rare, appears to indicate that the concentration of the oil at Khattan, Kirta, and the other localities, has no connection with the greater or less profusion of organic remains at those spots.

In the country round Kandahar,[1] and between it and the Khwája Amrán range, the cretaceous system is represented by a great thickness of hard grey limestone, usually unfossiliferous but locally containing an abundance of *Hippurites*, corals, etc. Near Kandahar this limestone is underlaid by a series of beds showing the following sections in descending order:—

3. A shaly sandstone, made up more or less of trappean material.
2. Bright green and intensely red shales with thin sandstone bands of trappean substance.
1. Coarse and thick conglomerate, almost entirely made up of pebbles of trap and cemented by a trappean, though calcareous matrix.

The description of these beds agrees well with certain beds seen near Kach and Hamadun on the Hurnai route to Quetta, which were formerly regarded as cretaceous. More recent examination has shown, however, that they are nummulitic and consequently cannot be representative of the Kandahar beds, unless one of the sections has been misinterpreted.[2]

Associated with the cretaceous limestone there are intrusive rocks, both basic and acid, and bedded traps. The latter are basaltic in character and overlie the limestone; very few details have been recorded, but one of the original foci of eruption was supposed to have been recognised about four miles west of Kandahar.[3] These bedded traps are newer than the trappoid conglomerates described above, which occur below the limestone.

The intrusive basic rocks are said to be lithologically similar to the bedded ones, but they have undergone a serpentinous change, and contain

[1] C. L. Griesbach, *Memoirs*, XVIII, 42 (1881).

[2] R. D. Oldham, MS. report, (1891).

[3] *Memoirs*, XVIII, 52, (1881).

veins and lumps of bright green and yellowish chrysotile where in contact with the limestone. The intrusions are numerous and vary in size down to mere wafer like strings.

The acid intrusive rocks are varieties of quartzsyenite, sometimes porphyritic in the hills crossed by the Maiwand pass west of Kandahar, which occur in veins and dykes varying from several hundreds of feet to quarter of an inch in thickness. In the larger syenitic masses numerous veins of a porphyritic rock with crystals of pink orthoclase imbedded in a fine grained felsitic matrix are found. Similar syenitic intrusions were observed at Dabrai and on the western side of the Khwája-Amrán pass. The syenitic intrusions are of somewhat older date than the basaltic, as the former are penetrated by dykes of the latter.

In Afghán-Turkistán the cretaceous system is well developed, and covers a large area of ground in which the older rocks only appear as inliers here and there. The lower cretaceous beds, consisting of about 800 feet of sandstones and shales with earthy limestones at the top, are said to be conformably underlaid by the red neocomian grits[1] at the top of the plant bearing series. The upper cretaceous is formed by about 1,800 to 2,000 feet, thickening to 4,000 in the sections near Balkh, of white thick bedded limestone with occasional sandstone bands. No defined subdivisions were recognised, but the limestones may be divided into three zones[2]:—

3. Chalk with flints.
2. Concretionary earthy white or brownish white limestones, occasionally dolomitic.
1. Hard white splintery limestones.

In the north-west Himálayas the cretaceous system, apart from the possibly cretaceous Giumal sandstones, is represented by a few small patches left on the tops of some of the hills in Spiti. They were named the Chikkim series by Dr. Stoliczka, who described it as consisting of a maximum thickness of about 500 feet of bluish or greyish white limestone, weathering white, with occasional earthy bituminous bands, overlaid by about 200 feet of grey or darkish unfossiliferous marly shale. The limestone yielded several fragments of *Rudistes* and numerous *Foraminiferæ*.[3]

Precisely similar limestones were observed further east by Mr. Griesbach in Hundes. They exhibit no features calling for special notice, and the fossils collected have not yet been described.

To the north hippuritic limestone has been observed in the Lokhzung range, north of the Lintzihang plain,[4] and at Sanju, on the road from

[1] *Supra*, p. 196.

[2] C. L. Griesbach, *Records*, XIX, 253, (1886).

[3] F. Stoliczka, *Memoirs*, V, 116, (1865).

[4] Drew: Jummoo and Kashmir Territories, London, 1875, p. 343.

Leh to Yarkand, Dr. Stoliczka recorded the presence of coarse grey calcareous sandstones and chloritic marls, some beds being almost entirely composed of the middle cretaceous *Gryphæa vesiculosa*.[1]

The occurrence of cretaceous beds on the shores of the Namcho lake, about 75 miles north of Lhasa, is proved by specimens of *Omphalia trotteri*, which were brought from that locality by one of the native explorers of the Trigonometrical Survey in 1876.[2]

In the Assam range the cretaceous rocks occur both on the plateau, where they lie nearly horizontal, and along the southern edge, where they are bent down to a steep dip in a monoclinal flexure. They thin out in a marked manner to the northward on the section south of Shillong, having a thickness of about 600 feet at the edge of the scarp, while ten miles further north, near Surarim, there is only about 100 feet. Still further north there are some small outliers which lie in hollows on the surface of the Shillong quartzites marking the position of pre-cretaceous valleys.

It is in these little primitive basins on the plateau that the cretaceous coal is found, one of them, a tiny coal basin at Máobehlarkár between Surarim and Mauphlong, having for years supplied the station of Shillong. The mineral itself has a persistent character throughout the whole cretaceous area. It is remarkable as being less of a true coal than is that of the overlying nummulitic group; the texture is compact and splintery, with a smooth conchoidal pasture, and the coal gives a dull wooden sound when struck. It has the additional peculiarity of containing numerous specks and small nests of fossil resin.

The most persistent member of the cretaceous series is known as the Cherra sandstone, about 200 feet of coarsish hard rock, unfossiliferous except for some vague stem-like vegetable impressions, which comformably underlies the nummulitic limestones. The next most constant member is the basal conglomerate, whose larger components are almost all derived from the neighbouring Shillong quartzites, and are generally subangular. Varying in thicknes from 20 to 100 feet, it everywhere forms the base of the series, but whether it represents a definite geological horizon is doubtful.

In the Maobehlarkár coal basin the basal conglomerate and the Cherra sandstone are in contact, but at the south scarp of the plateau they are separated by glauconitic sandstones, overlaid by a pale fine grained sandstone, often containing broken plant remains, and in places marine fossils.

[1] F. Stoliczka, *Quart. Jour. Geol. Soc.*, XXX, 572, (1874); *Records*, VII, 50, (1874); Scientific Results of the Second Yarkand Mission, Geology, p. 22, (1878).

[2] *Records*, X, 21, (1877).

These beds, about 400 feet in thickness, thin out to the northward by an original limit of deposition, and it has been noticed [1] that the matrix of the basal conglomerate, at the different levels, partakes of the nature of the corresponding horizon below Cherra, and on all the sections there is more or less of a transition, by interstratification, between it and the particular bed which happens to overlie it. The only point tending to cast a doubt on its being a marginal form of the successive sandstone beds is the frequent occurrence of carbonaceous matter in the rock immediately above, but this is not conclusive.

Where the beds bend over, and are exposed with a high dip in the low ground south of Tharia, the basal conglomerate is represented by a coarse felspathic ochrey sandstone, while the overlying beds, having a thickness of about 1,200 feet, consist of pale grey shales, locally nodular, calcareous, or ferruginous, with some thin layers of earthy limestone or sandstone. The whole series, besides being thicker than that exposed on the plateau north of the uniclinal axis, is earthy in character, instead of sandy, implying a greater distance from the margin of the sea.

The marine fossils of the cretaceous rocks of the Khási hills have already been mentioned,[2] and need not be further referred to here. No fossils have as yet been found west of the Khási hills.

In the Gáro hills the cretaceous attains a considerable development as an arenaceous series, containing important coal seams in places. The sandstones of the plateau are horizontal and rest on a more deeply eroded and irregular surface than those of the Khási hills. At the western end of the range the sandstones lap round the end of the Turá gneissic ridge, and the original relations of the rocks are nowhere better seen than here. The spur on which the station of Turá stands, some 2,000 feet below the crest of the ridge has a midrib of gneiss, with sandstone on both sides, through which the streams have again excavated their channels. There is but little disturbance in this locality, and it is plain that the ridge must have stood as it does now when these sandstones were laid down.

East of the Khási hills, throughout the south-east portion of the Jaintia hills from the neighbourhood of Jowai eastwards, cretaceous rocks are found at the surface, horizontal or nearly so, and to the eastwards pass conformably beneath the tertiaries near the Kapili (Kopili) river. Beyond this we have only isolated observations; the thin bedded sandstones at the falls of the Kapili are believed to be cretaceous, and typical cretaceous coal is associated with sandstone and some hard sandy limestone resting flatly on the gneiss in the Námbar and Doigrung valleys, near Golághát.

[1] H. B. Medlicott, *Memoirs*, VII, 171, (1869). | [2] *Supra*, p. 247.

The existence of cretaceous beds in the Arakan Yoma is only shown by the discovery of one species of mollusc in a single locality near Ma-í in the northern part of the Sandoway district of Arakan. The species found, *Ammonites inflatus*, is a characteristic cenomanian cephalopod, common in the Utatúr beds of Southern India. The only specimen obtained was picked up in the bed of a stream, and had evidently been derived from some shales in the neighbourhood. No other specimens nor other fossil of any kind could, however, be found.

What may be the extent of the cretaceous beds, and which strata should be referred to this group, are matters on which but little trustworthy information has been obtained. Mr. Theobald was disposed to consider that a peculiar, compact, light cream coloured, argillaceous limestone, resembling indurated chalk, sometimes speckled from containing sublenticular crystalline particles, belongs to the cretaceous system. This limestone has been traced at intervals from near Ma-í, about thirty miles north of Tongúp (Toungoop), to the neighbourhood of Sandoway, whilst somewhat similar limestone, though not so characteristic, may be traced to Keantali, some thirty miles farther south. The same limestone is found in the western part of Ramrí Island. Another peculiar formation is a greyish rather earthy sandstone, with a pisolitic structure in places, due to the presence of small globular concretions of carbonate of lime and iron. The concretions decompose and leave small holes, which impart to the earthy sandstone the aspect of an amygdaloidal trap. Like the limestone, this peculiar sandstone is traced from Ma-í to near Keantali, a distance of 94 miles, and if, as appears probable, these beds are really cretaceous, for both are closely associated with the shale from which the ammonite had apparently been derived, the rocks of this formation may be considered as extending at least the distance mentioned. The strata ascribed to the cretaceous group are less hardened and metamorphosed than the other rocks of the Arakan Yoma; they are of great thickness, and may include all the beds of the main range of the Yoma, as far south as Keantali. No rocks which can be referred to the Ma-í group have been detected east of the main Arakan range in Pegu. To the northward their range is unknown, but a limestone resembling that of the Ma-í group was seen in the hills east of Manipur.[1]

Apart from those just mentioned no rocks of cretaceous age are known to exist in Burma, though there is a probability that they may be represented in Tenasserim. On the Lenya river,[2] in the extreme south of the province, a bed of coal occurs, of very laminated structure and containing numerous small nodules of a resinous mineral, like amber. This peculiar

[1] *Memoirs*, XIX, 223, (1883); *Supra* p. 148. | [2] T. Oldham, *Sel. Rec. Govt. Ind.*, X, 48, (1856).

association of mineral resin is characteristic of the cretaceous coals in the Assam hills, and it is possible that the Tenasserim mineral is of the same age.[1] At the same time no palæontological evidence has been discovered, the rocks associated with the coal are soft clays and sands, having a more recent appearance than those accompanying the other coal seams of the Tenasserim province, and these other coal seams are, it is believed, not older than eocene. The coal occurs in an irregularly developed bed, varying from 1 to 5 feet, or rather more, in thickness, with thin layers of fine jetty coal between bands of hard black shale, and rests on clay with vegetable remains, and patches of jet coal. Thin coal laminæ are also found in the associated strata.

Below the rocks immediately associated with the coal are fine, whitish earthy sandstones and indurated clay, passing into marl, with some conglomerates. Above the coal is a series of soft muddy sandstones, marls, conglomerates and a few seams of carbonaceous matter. The whole may be 600 feet thick. The dip is considerable, about 35°, and the rocks have undergone disturbance and faulting. Nothing has been ascertained as to the relations of the coal bearing beds to other formations, indeed all that is known of the Lenya river coal is the result of a hurried visit to a locality very difficult of access.

[1] Mr. Bose (MS. Report, 1892) regards these beds as belonging to the Maulmain series, of Palæozoic age; the country is, however, singularly ill adapted to geological investigation and the examination was necessarily incomplete.

CHAPTER XIII.

TERTIARY DEPOSITS.

(*Excluding those of the Himálayas.*)

Quilon and Ratnágiri—Surat—Sind and Balúchistán—Cutch and Káthiáwár—Afghánistán—Kohát—Assam—Burma.

No tertiary beds are known in the Indian Peninsula except in the immediate neighbourhood of the coast, and if we exclude certain unfossiliferous sandstones, now regarded as subrecent though possibly of upper tertiary age, they are confined to a few small exposures on the west coast, the most southerly of which is near Quilon, in Travancore.

The earliest, and practically still the only, information published on the occurrence of tertiary beds in Travancore is comprised in some notes supplied by General Cullen to Dr. Carter, and published by the latter in his 'Summary of the Geology of India.'[1] Beneath the laterite of the neighbourhood of Quilon, at a depth of about 40 feet from the surface, grey fossiliferous limestone (or dolomite according to General Cullen) is found, partly compact and partly loose and rubbly. This limestone is exposed beneath a laterite cliff near the coast, four or five miles north-east of Quilon, and the same rock has been found in the neighbourhood of the town at a depth of about 40 feet in numerous wells, many of which were sunk or deepened by General Cullen for the purpose of ascertaining the presence of the limestone. Further south, near Warkalli, twelve to fourteen miles south of Quilon, the cliffs on the coast expose, beneath the laterite, beds of brightly coloured sand and clays with bands of lignite, abounding in fossil resin and iron pyrites, both in lumps of considerable size. The sandy beds overlie the lignites and clays.

The limestone contains marine shells in abundance, amongst which Dr. Carter recognised *Strombus fortisi*, *Cassis sculpta*, *Voluta jugosa*, *Ranella bufo*, *Conus catenulatus*, *Conus marginatus*, and *Cerithium rude*,

[1] *Jour. Bo. Br. Roy. As. Soc.*, V, 301, (1857); and Geological Papers on Western India, Bombay, 1857, pp. 740 and 743, footnote. This footnote is an addition to the original summary. The very small outcrop was not found during the geological examination of the Cochin neighbourhood. The locality had been incorrectly defined in the first instance, but its existence was subsequently verified by Mr. Logan; *Records*, XVII, 9, (1884).

besides species of several other genera resembling forms found in the tertiary beds of Sind and Cutch. A species of *Orbitolites* (?) was described by Dr. Carter as *O. malabarica*. All the mollusca identified belong to species occurring also in Cutch and Sind, and, so far as is known, in beds of later date than the nummulitic limestone. No plants appear to have been collected from the lignite beds.

Another deposit of obscure date and origin has been found beneath laterite at Ratnágiri (Rutnagherry) on the western coast.[1] White or blue clays with thin carbonaceous seams are found in various quarry and well sections near the town beneath a considerable thickness of laterite, 35 feet in one case. Some of the clay is said to be sandy or gravelly. Above the deposit is a layer of hard ironstone, about an inch thick, but said sometimes to be thicker. As in Travancore, fruits and leaves are found in the clay and lignite, together with mineral resin and pyrites. No specimens of the organisms found appear to have been collected. The beds are only a few feet thick, 27 in one section measured by Dr. de Crespigny, and rest unconformably upon Deccan trap.

There is but little evidence to connect this deposit with the Travancore beds, but, owing to some similarity of mineral character, the presence of lignite in both, and the circumstance that both underlie laterite, they have been classed together.

The tertiary rocks in Surat and Broach[2] are almost confined to two tracts of country, separated from each other by the alluvium of the river Kim, a small stream running to the sea from the Rájpipla trap area. The southern tract is the smaller, extending about ten miles north from the Tápti river and being about fifteen miles broad from east to west; the other area, between the Kim and Narbadá, extends about thirty miles from north-east to south-west, and is about twelve miles across where widest. In both the few good exposures of rock which occur are to the eastward.

At the base of the tertiary formations, north-east of Surat, are thick beds of ferruginous clay, assuming, where exposed, the characteristic brown crust and pseudo-scoriaceous character of laterite, from which they differ in no respect. These beds at first sight appear to be of volcanic origin, an idea which is strengthened by the neighbourhood of the traps on which they rest, but close examination has shown that they are really sedimentary deposits, although composed, in all probability, of materials derived from the disintegration and denudation of the trap. With them are interstratified beds of gravel or conglomerate, containing agate pebbles derived

[1] Carter, *Jour. Bo. Br. Roy. As. Soc.*, V, 626, (1857); Geological Papers on Western India, Bombay, 1857, p. 722, footnote; C. J. Wilkinson, *Records*, IV, 44, (1871).

[2] For a fuller description, see *Memoirs*, VI, 223-27 and 356-73, (1869).

from the traps and limestone, sometimes nearly pure, but more frequently sandy, argillaceous, or ferruginous, and abounding in nummulites and other fossils. The thickness of the whole can only be roughly estimated at between 500 and 1,000 feet.

These beds are well seen on the banks of the Tápti below Bodhán, a village eighteen miles east by north from Surat. They extend thence to the northward through Tarkesar to the Kim alluvium, and again north of the Kim to the neighbourhood of a village called Wágalkhor, about twenty-four miles north-north-east of Bodhán, and seventeen east by south of Broach. North of this they appear to be overlapped by higher beds.

The nummulitic limestones and their associates are distinctly unconformable to the underlying traps, and rest upon the denuded edges of the latter. Amongst the fossils found in the lower tertiary beds are *Nummulites ramondi*, *N. obtusa*, *N. exponens* (or *N. granulosa*), *Orbitoides dispansa* and some other species which are common in the Kirthar beds of Sind, together with *Ostrea flemingi*, *Rostellaria prestwichi*, and *Natica longispira*, which are particularly characteristic of the Ránikot group, and *Vulsella legumen* found in both. Some other fossils have been identified with species found at a higher horizon, but the identification appears doubtful. The nummulitic beds of Surat and Broach may safely be classed as eocene.

Above the limestones and lateritic beds there is found a great thickness of gravel, sometimes cemented into conglomerate, together with sandy clay and ferruginous sandstone, often calcareous. These higher beds are poorly exposed in the Tápti and Kim rivers, but they are well seen in the stream which runs past Ratanpur, east of Broach. Here they consist chiefly of sandstone, gravel, and conglomerate, with occasional beds of red and white clay and shales. The pebbles in the gravels and conglomerates consist chiefly of agates and quartzose minerals derived from the trap, and from some of these beds near Ratanpur, east of Broach, the agates and carnelians are obtained which have from time immemorial supplied the lapidaries of Cambay. At the base of the tertiary beds in this direction is a coarse conglomerate composed of large rolled fragments of basalt, but it is uncertain whether this bed belongs to the upper tertiary group or to the lower tertiaries, as it is not quite clear, owing to the few sections exposed, whether the lower eocene beds are completely overlapped to the northward, or merely represented by unfossiliferous beds of a different mineral character. Like the underlying beds, the higher tertiary strata have a steady dip to the westward, and the thickness of the whole tertiary series exposed near Ratanpur appears to be between 4,000 and 5,000 feet, but this estimate is based on a very imperfect exposure of the rocks. Of course, if, as appears possible, the lower beds are overlapped, the whole of this thickness consists of the upper members of the series.

No nummulites are found in these upper tertiary beds, and the few fossils discovered in them appear to differ from those in the nummulitic limestones below. The commonest organic remains are valves of *Balani*, which are also abundant in the Gáj (miocene) rocks of Sind. The abundance of *Balani* and the absence of *Nummulites* together form strong reasons for believing that the upper beds of Surat and Broach are of later date than eocene.

It is far from certain whether any pliocene beds are found in eastern Gujarát. They occur in Káthiáwár and on Perim Island in the Gulf of Cambay, and further search may detect them in Surat and Broach.

Tertiary deposits are found in the debateable ground west of the Arávallis which belongs structurally to the peninsular and stratigraphically to the extra-peninsular area, and in the extra-peninsular area they attain an immense development, both as regards their thickness and the area they cover. Taken as a whole, and ignoring local breaks in the continuity of deposition, they form a great system of deposits whose lower portion is a marine formation while the upper consists of fresh water subaerial deposits. The distinction is not absolute, nor can the line of demarcation between the two types be everywhere drawn on the same horizon, yet the distinction is a real and important one. Everywhere, from Sind on the one hand to Burma on the other, the eocene deposits are marine and the pliocene fresh water or subaerial, with the possible exception of the pliocene beds of the Irawadi valley, and wherever there is a continuous succession of deposits, there is a gradual transition from the one type to the other.

It will be well to commence the general description of the extra-peninsular tertiaries with those of Sind, as, owing to the completeness of the section there and the abundance and excellent preservation of the fossils of the various horizons, it may well be regarded as a type area for the rest of India. At the same time the tertiaries of those adjoining areas will be noticed where the Sind rock groups have been recognised with some degree of certainty, those of Cutch and Káthiáwar, where they are less distinctly represented, being taken separately. After noticing the tertiaries of Afghánistán and the western frontier, those of Assam, and finally Burma, will be described, the consideration of the Himálayan tertiaries being more conveniently postponed to the following chapter.

The great series of tertiary deposits of Sind has been divided into the following groups or subdivisions, whose approximate correlation with the European sequence is given—

Manchhar, 8,000—10,000 ft. . .	*Upper miocene to pliocene.*
Gáj, 100—1,500 ft. . . .	*Miocene.*

Nari, 500–6,000 ft. . . . *Upper eocene to lower miocene.*
Kirtar, 6,000—9,000 ft. . . . *Eocene.*
Ránikot, 2,000 ft. . . . *Lower eocene.*

The lowest group of the Sind tertiaries, which lies with perfect conformity on the *Cardita beaumonti* beds described in the last chapter, derives its name from a hill fortress of Sind Amirs, situated in the Lakhi range of hills, known as Ránikot and also as Mohan-kot, from the Mohan stream, which traverses the fortification.[1] The Ránikot group is much more extensively developed in Sind than the underlying cretaceous beds, for although it is confined to lower Sind, and although its base is only seen in the Lakhi range, north of Ránikot, its upper strata occupy a considerable tract of country.

All the lower portion of the Ránikot group, including by far the greater portion of the beds, consists of soft sandstones, shales and clays, often richly coloured and variegated with brown and red tints. Gypsum is of frequent occurrence. Some of the shales are highly carbonaceous and occasionally sufficiently pyritous to be used in the manufacture of alum. In one instance a bed of coal (or lignite), nearly six feet thick, was found, and a considerable quantity of the mineral extracted.[2] The quality was poor, the coal decomposed rapidly and was liable to spontaneous combustion owing to the quantity of iron pyrites present, whilst the deposit was found to be a small patch, not extending more than about 100 yards in any direction. The only fossils found in the lower portion of the Ránikot group, with the exception of a few fragments of bone, have been plants, some dicotyledonous leaves, hitherto not identified, being the most important. All the Ránikot beds, except towards the top of the group, have the appearance of being of fresh water, and are probably of fluviatile origin.

A variable portion of the group, however, towards the top, consists of highly fossiliferous limestones, often light or dark brown in colour interstratified with sandstones, shales, clays, and ferruginous bands. These are the lowest beds in Sind containing a distinctly tertiary marine fauna. The brown limestones are well developed around Lynyan, east of Band Vero and north-west of Kotri, and throughout the area of Ránikot beds near Jerruck and Tatta In this part of the country there appears to be a complete passage upwards into the overlying nummulitic limestone (Kirthar), but in the Lakhi range the upper marine beds of the Ránikot group are poorly represented or wanting, and it is evident that they were removed by denudation before the deposition of the Kirthar limestone, for the latter is seen at Hothian pass resting upon their denuded edges.

The greatest thickness of the Ránikot group in the Lakhi range, where alone the base of the group is visible, is about 2,000 feet, but generally

[1] *Memoirs*, XVII, 37, (1879). | [2] *Memoirs*, VI, 13, (1869).

the amount is rather less, about 1,500. It must, however, be recollected that in this locality some of the upper marine beds are wanting, and as these marine limestones and their intercalated shales, sandstones, etc., are 700 or 800 feet thick in places north-west of Kotri, it is evident that the original development of the group exceeded the 2,000 feet seen in the Lakhi range.

The fossils of the Ránikot group[1] indicate a lower eocene age, though cretaceous affinities are not wanting. The *Nautili* are all connected with cretaceous, rather than tertiary, types, a *Terebratula* is undistinguishable from *T. subrotunda*, one of the commonest upper mesozoic types. On the other hand, the presence of nummulites, and the general aspect of the mollusca, indicate a lower eocene age. The only fossils which have been critically examined are the corals and echinoderms. In neither case is the result decisive, but the corals comprise, out of a total of 50 species, 7 species identical with European eocene species and 5 closely allied to forms found on that horizon or in slightly newer rocks.

In the Mari hills of eastern Balúchistán the base of the strata regarded as nummulitic is marked throughout by a peculiar pseudo-conglomerate which has also been recognised in the Gáj river in Sind on the one hand and in the southern portion of the Suláimán range on the other. It has the appearance of being composed of subangular fragments of dark grey limestone, imbedded in a limestone matrix of paler colour, both matrix and pebbles containing numerous small nummulites, though no difference can be traced between the forms found in the two portions of the rock. The resemblance of this rock to a true conglomerate is especially striking in the sections near Khattan, but the similarity of the fossils found in the apparent pebbles and in the matrix, the comparative uniformity in thickness of this band, which lies among fine grained shales, the absence of any known rock from which the pebbles could be derived, and the presence of every gradation from the most conglomerate like form to a merely mottled limestone, all point to the structure being in some way of concretionary origin.[2]

Above the pseudo-conglomerate, which has been accepted as the base of the tertiaries, there comes a great thickness of green and grey shales with interbedded impure sandstones which are, as regards both their lithology and stratigraphical position, the equivalent of the Ránikot group in Sind, but owing to the fossils not having been examined and the homotaxis verified, they have as yet been provisionally described as the Gházij group.[3] Along the outcrop of this group, from Mach in the Bolan pass to

[1] A detailed list will be found in *Memoirs*, XVII, 197, (1879). The corals and echinoderms are described in *Pal. Indica*, series xiv, I, pts. 2 and 3, (1880-86).

[2] W. T. Blanford, *Memoirs*, XX, 149, (1883); R. D. Oldham, *Records*, XXIII, 94, (1890).

[3] *Records*, XXIII, 95, (1890).

Hurnai, coal seams are found near top, which attain a maximum thickness of about three feet and have proved of great economic importance in a country where fuel is so scarce. The distribution of the coal seams is peculiar. Besides the localities mentioned, coal has been found north of the Thal Chotiáli plain and in the Luni Pathán country to the east, in every case close to the western limit of the known exposures of this group, while to the eastwards the group ceases to be coal bearing. As the coal seams were doubtless formed in marshes near the margin of the sea, and as the only rocks known westwards of the present limit of the group are either older or very much newer, it would seem that the original western limit of deposition cannot have been far removed from the present limit of outcrop, at any rate, in the country east of Quetta.

Further north, in the Suláimán range, very similar shales are found immediately overlying the upper cretaceous sandstone, but they differ from what is seen in the southern sections in the prevalence of a red colour throughout the greater portion of the thickness of the shales.

In both these areas the relation of the lowest tertiary to the underlying beds is one of perfect conformity, and they form part of a continuous system of deposits with the upper cretaceous beds, as has been mentioned in the last chapter.

The Ráníkot group in Sind is overlaid by the Kirthar group, so called from the frontier range of hills of that name. Though inferior in thickness to several other subdivisions of the tertiary series in Sind, this group comprises by far the most conspicuous rock, the massive nummulitic limestone which forms all thé higher ranges in Sind. It forms the crest of the Kirthar range throughout, and all the higher portions of the Lakhi range, of the Bhit range south-west of Manchhar lake, and of several smaller ridges, and consists of a mass of limestone, varying in thickness from a few hundred feet in lower Sind to about 1,000 or 1,200 at the Gáj river, and probably 2,000, or even 3,000, farther north. The colour is usually pale, either white or grey, sometimes, but less frequently, dark grey, the texture varying from hard, close, and homogeneous, breaking with a conchoidal fracture, to soft, coarse and open. Ordinarily the nummulitic limestone is tolerably compact but not crystalline, and chiefly composed of *Foraminifera*, especially whole or fragmentary *Nummulites*; corals, echinoderms, and molluscs also abound, but the two latter very frequently only weather out as casts.

Throughout northern Sind, except near Rohri, no beds are seen beneath the Kirthar limestone. The remarkable range of low hills, surrounded by Indus alluvium, and extending for more than forty miles south from Rohri, consists of nummulitic limestone, having a low dip to the westward, and a considerable thickness of pale green gypseous clays, with a few bands of impure dark limestone and calcareous shale, is exposed beneath the limestone forming the eastern scarp of the hills, on the edge of the alluvial plain. No *Foraminifera* have been found in these beds, although *Nummulites* abound in the limestone immediately overlying. Several species of mollusca occur, but none are characteristic, and it is far from clear whether the green clays and their associates are merely thick bands intercalated in the limestone, or whether they belong to a lower group. Probably these argillaceous beds of the Rohri hills represent some of the marls, shales and clays forming the lower portion of the upper Kirthar group on the Gáj river.

In some places west of Kotri a band of argillaceous and ferruginous rock is found close to the base of the Kirthar group. It is mainly composed of brown hæmatite, weathers into laterite and appears to be found over a considerable area near Kotri and Jerruck.

It has already been mentioned that the Kirthar limestone rests unconformably on the Ránikot group in the Lakhi range. The Kirthar group here cannot be much more than 500 or 600 feet thick, and consists entirely of limestone. To the south-east, towards Kotri and Tatta, there is no unconformity between the Ránikot and Kirthar groups; on the contrary there is an almost complete passage between the two, and the limestone of the latter becomes much split up and intercalated with shales and sandy beds. This is even more the case further to the south-east in Cutch, where the whole group consists of comparatively thin beds of limestone, interstratified with shales. To the south-west the massive limestone dies out altogether, and although it is well developed in the southernmost extremity of the Kirthar range near Karchat, about 50 miles south of Sehwán, it disappears entirely within a distance of 12 or 14 miles, and is entirely replaced by shaly limestones, shales, and thick beds of sandstone in the ranges on the Hab river. Some rather massive beds of nummuliferous dark grey limestone, very different in character from the pale coloured Kirthar limestone, are found west of the Hab, but their precise position in the series is not known, and the rocks appearing from beneath the Nari group, in the place of the Kirthar limestone, consist of shales and sandstones, with some calcareous bands abounding in nummulites, and closely resembling, both in character and in the species of *Foraminifera* they contain, the nummulitic shales beneath the massive limestone on the Gáj river.

The most characteristic fossils of the Kirthar group are *Nummulites* and *Alveolina*, the extraordinary abundance of individuals rendering it usually easy to recognise even small fragments of the rock by the organisms preserved in it. Many of the species, and especially the *Foraminifera*, are characteristically eocene, and there can be no question that the nummulitic limestone of India is a continuation of the same formation in Europe. Several species pass from the Ránfkot beds into the Kirthar group; indeed, the palæontological differences between the two appear to be principally due to a change of conditions from the shallow muddy water of the Ránfkot to the deeper clear sea of the Kirthar beds.

The result of Messrs. Duncan and Sladen's examination of the echinoderms does not altogether bear out the conclusions regarding the relation between the Kirthar and Ránikot groups expressed above. They found no less than 63 out of 70 species being peculiar to the group, and the horizon of the remaining 7, which are supposed to have been obtained from the Ránfkot group, is very doubtful.[1] Caution is, however, necessary in applying the palæontological results obtained from a single order of animals, and the conclusions based on the general palæontology and stratigraphy of the two groups may be accepted in spite of this apparent contradiction.

In eastern Balúchistán the Kirthar limestone appears to be largely developed in the Mari hills and south of the Bolan pass, having been given the local name of Spintangi in this region. Between Hurnai and Quetta it has been very much reduced in thickness by denudation previous to the deposition of the Siwáliks. The relation of the Kirthar, or Spintangi, limestone to the underlying shales is one of perfect conformity by interstratification, and there is reason to believe that to a certain extent the Spintangi and Gházij, or Kirthar and Ránfkot, groups merely represent different conditions of deposition and are partly of contemporaneous origin.

The Spintangi limestone has frequently a nodular structure that makes it weather into an aggregate of rounded lumps, easily mistaken for a conglomerate; so much so that three practised geologists have each recorded the fact that, after crossing the boundary of the Siwálik conglomerate in the Bolan pass, they walked for some distance over the nummulitic limestone before discovering the change.

Another peculiarity of the group in the Mari hills is the occurrence of thick beds of gypsum, interstratified with the clear limestones and green shales. The thickness of these beds in the country east of Khattan is very considerable, one bed of 50 feet thick, besides four others aggregating 33, having been seen near Mámand.[2] Whether they were originally deposited as gypsum, or are due to the subsequent alteration of limestone beds, their occurrence among distinctly marine beds is not easy to account for. On the east flank of the Suláimán range the Kirthar group does not appear to be present in its characteristic form, but some thin beds of white nummulitic limestone overlying the shales regarded as the probable equivalent of the Ránikot group may represent it.

[1] *Pal. Indica*, series xiv, I, pt. 3, p. 245, (1884).

[2] *Records*, XXV, 24, (1892).

In western Rájputána two outcrops of nummulitic rocks are known, the larger one north-west of Jaisalmer, the smaller near Koilath, thirty miles west-south-west of Bikaner. The rocks represented are a white nummuliferous limestone, resembling that of the Kirthar group of Sind, and shaly beds, mostly grey and impregnated with salt, though a very fine grained pale buff coloured fuller's earth is also found and quarried for export under the name of Multani mitti. In Jaisalmer a bed of ferruginous lateritic rock, like that found near Kotri, is associated with the nummulitics. The rocks of these exposures resemble those of the Kirthar group as seen east of Sukkur, and there is good reason to suppose that they are of the same age and indicate an easterly extension of the nummulitic sea.[1]

The series of tertiary rocks above the Kirthar nummulitic limestone is superbly developed and very well seen in the hills on the frontier of upper Sind whose culminating ridge is known as the Kirthar. The names of the tertiary groups overlying the nummulitic formation have consequently been derived from places in this range, and the Nari group takes its title from a stream which traverses the lower portions of the range, where it is composed almost entirely of Nari beds, for a considerable distance, and issues from the hills west by north of Sehwan.[2] The present subdivision comprises at its base the uppermost bands of limestone containing *Nummulites*, the species *N. garansensis* and *N. sublævigata* being distinct from those so commonly found in the Kirthar subdivision, and the limestone itself is usually distinguished by its yellowish brown colour, and by being in comparatively thin bands, interstratified with shales and sandstones. Several other fossils, besides the nummulites, differ from those in the Kirthar beds. Not unfrequently, however, there is an apparent passage from the white or greyish white Kirthar limestone into the yellow or brown Nari rock, and the two groups appear always to be perfectly conformable, but no intermixture of the characteristic species of nummulites has been detected, and the division between the Kirthar and Nari beds can always be recognised by the fossil evidence.

In some places the lower Nari beds consist almost entirely of brown and yellow limestones, but more frequently the limestone bands are subordinate, dark shales and brown, rather thinly bedded, sandstone forming the mass of the rocks. The limestone bands are often confined to the base of the group, and always diminish in abundance and thickness above, although they are occasionally found as much as 1,500 feet above the top of the Kirthar group. These shales and fine sandstones, with occasional bands of limestone, constitute the lower Naris, and pass gradually into

[1] *Records*, XIX, 159. (1886).

[2] *Memoirs*, XVII, 49, (1879).

the coarser, massive, thick bedded sandstones that form the greater portion of the group, and attain a thickness of 4,000 or 5,000 feet. On the flanks of the Kirthar range a few bands of clay, shale, or ironstone, are interstratified with the sandstones, and bands of conglomerate occasionally occur. The Nari beds in their typical form extend throughout the eastern flank of the Kirthar range, and occupy a belt of varying width, from one or two miles to as much as ten miles in breadth, between the underlying Kirthar and the overlying Gáj beds.

On the western side of the Bhagotoro hills, four or five miles south of Sehwan, there is a break in the Nari beds, and some variegated shales, clays, and sandstones, richly tinted in parts with brown and red, which represent the upper Nari sandstones, rest unconformably on the denuded edges of the lower Nari limestones and shales. The break is evidently local. To the east of the Lakhi range the Nari beds are entirely wanting, and it appears very possible that they have never been deposited in this portion of the Indus valley. From the neighbourhood of Sehwan to Jerruck, the Manchhar beds rest with more or less unconformity on the Kirthar, a very faint and imperfect representative of the Gáj group occasionally intervening. But west of the Lakhi range, throughout lower Sind, the Nari beds are exposed almost wherever the base of the Gáj group is seen; they increase in thickness to the westward, and the Hab valley, from the spot where the river first forms the boundary of British territory to the sea, consists entirely of these strata. There is, however, no longer any such marked distinction between the subdivisions of the tertiary series as is found in the Kirthar range. The disappearance of the Kirthar limestone has already been mentioned, and with it the lower Nari limestones also disappear, so that it is no longer possible to draw a distinct line between the two groups. The two groups can still be traced, although the dividing line between them is obscured, as the calcareous shales, with the characteristic Kirthar nummulites below, and the massive Nari sandstones above, are still recognisable. Beds of brown limestone, too, full of *Orbitoides papyracea* or *O. fortisi*, occur in the Nari beds of the Hab valley, but instead of being found at the base, they appear in the middle of the group. Again, just as there is a difficulty in distinguishing the Naris from the Kirthars at their base, so the beds at the top of the former group can only be separated by an arbitrary line from the overlying Gáj beds. In the Kirthar range the upper boundary of the Nari group, although there is no unconformity, is distinct and definite, limestones with marine fossils of the Gáj group resting immediately upon the upper Nari sandstones. But in southern Sind bands of limestones or calcareous sandstone, with marine fossils, some of which are well marked Gáj species, occur in the upper part of the Nari group, whilst limestone bands with the Nari *Orbitoides papyracea* are found in the Gáj.

The sandstones, which form so large a portion of the Nari group in upper Sind, have hitherto proved destitute of animal remains, but the occasional interstratifications of shales and clays often contain fragments of plants, and some ill marked impressions, probably due to fucoids, have been found in the sandstones themselves. There appears a probability that these sandstones may be of fluviatile, and not marine origin, and although some species pass from the Kirthar, and even from the Ráníkot, group into the lower part of the Nari group, the fauna is chiefly distinct and marks a higher horizon. The most marked change is in the *Foraminifera,* because they are so abundant and characteristic, whole beds of limestone towards the base of the Nari group being entirely made up of three species, distinct from those occurring in the Kirthar group, *Nummulites garansensis, N. sublævigata* and *Orbitoides papyracea,* the last named frequently of large size, and reaching two or three inches in diameter. One of these species of *Nummulites* (*N. garansensis*) is of importance, because it occurs in Europe, as in Sind, in the highest strata characterised by the abundance of the genus, those beds being at the base of the miocene. *Nummulites sublævigata* is peculiar, so far as is known, to India.

Several of the molluscs and echinoderms of the Nari beds also, such as *Siliquaria granti, Solarium affine, Venus granosa,* and *Clypeaster profundus,* show distinctly miocene affinities, and some of these pass up into the Gáj group. At the same time there are so many eocene forms present, such as *Natica patula, N. sigaretina, Ostrea flabellula, Voluta jugosa,* etc., that it is somewhat difficult to decide to which subdivision the Nari beds should be assigned. They probably occupy an intermediate position, corresponding to the oligocene of continental geologists, a conclusion which is borne out by the detailed examination of the corals and echinoderms.

The lower Nari limestone is found in its typical form as far north as Bibi Náni in the Bolan pass, where it is overlaid by some grey sandstones and mottled beds, which probably represent the upper Nari of Sind. The lower Nari limestone, with *Nummulites garansensis, N. sublævigata* and *Orbitoides papyracea,* has not been found north of this, but to the north of the Gandahári hill, 20 miles east of Dera Bugti, and along the eastern slopes of the Suláimán range, a series of sandstones, with subordinate bands of conglomerate and clay, occupy a position intermediate between the upper eocene and the overlying Siwáliks. They are described as apparently conformable to both, and were regarded by Dr. Blanford as probably the equivalents of the upper Nari of Sind.

Resting upon the Nari group, almost throughout Sind, and forming the base of the upper tertiary series, there is found a mass of highly fossiliferous limestones and calcareous beds, usually more or less shaly, always distinctly stratified, and easily distinguished from the limestones of the older tertiary formations by the absence of nummulites. A superb section of the strata forming this group is exposed on the banks of the Gáj river, from which its name is derived.[1]

On the eastern flanks of the Kirthar range in upper Sind, the Gáj group forms a conspicuous ridge, the hard dark brown limestone bands near the base of the formation resisting the action of denudation far more than the soft sandstones of the Nari beds, and rising every here and there into peaks of 1,000 and 1,500 feet, or even more, scarped to the westward and sloping to the east, Amru, the highest summit of the Gáj ridge, being 2,700 feet above the sea. Still, the limestone bands, although so conspicuous, are subordinate, the greater part of the group consisting of sandy shales, clays with gypsum, and sandstones towards the base. Many of the bands of limestone appear very constant in position and may be traced for a long distance; they are dark brown in colour as a rule, but one bed is white and abounds in corals and small *Foraminifera* (*Orbitoides*), whilst some of the darker bands contain *Echinodermata* in large quantities.

The uppermost portion of the group is usually argillaceous, being chiefly composed of red and olive clays with white gypsum, and these beds pass gradually into precisely similar strata belonging to the overlying Manchhar group. The passage beds contain *Corbula trigonalis, Lucina* (*Diplodonta*) *incerta Tellina subdonacialis, Arca larkanensis*, amongst other fossils, such as *Turritella angulata*, and forms of *Ostrea* and *Placuna*.

All of these have allies living in estuaries at the present day, *Arca granosa*, a recent representative of *A. larkanensis*, being one of the commonest and most typical of Indian estuarine mollusca. To these estuarine passage beds further reference will be made presently when the relations of the Manchhar to the Gáj beds are discussed.

The Gáj beds at the Gáj river are very nearly 1,500 feet thick, but they appear to be less developed to the northward in the Kirthar range, and not to be much more than half the thickness named west of Lárkhána, where, however, they are nearly vertical and have probably suffered from pressure. In lower Sind, the Gáj group, like the Nari, disappears to the eastward of the Lakhi range, where it is either entirely wanting, or else represented by a thin band at the base of the Manchhar group, containing one of the characteristic fossils, *Ostrea multicostata*. There is, however, a very large area of Gáj beds north and north-east of Karáchi, and the appearance of the formation here is somewhat different from what it is in the Kirthar range, for the greater portion of the group

[1] *Memoirs*, XVII, 53, (1879).

consists of pale coloured limestones, almost horizontal or dipping at very low angles, and forming plateaux 400 or 500 feet high, bounded by steep scarps, which rise from the low ground of the soft Nari sandstones east of the Hab valley. A low range of hills, formed of Gáj beds, extends to the south-west, past the hot spring at Pir Mangho (Mugger or Manga Pir) to the end of the promontory known as Cape Monze, and the same beds form the low hills east and north-east of Karáchi, and furnish the materials of which the houses of the town are mostly built. A small island called Churna, in the sea west of Cape Monze, also consists of Gáj rocks. To the northward the Gáj area of lower Sind extends with very irregular outline to the neighbourhood of Tong and Karchat, almost due west of Hála, and there are several outliers farther north, connecting the southern portion of the group with the typical outcrop in the Kirthar range. East of Karáchi, Gáj beds extend in the direction of Tatta, until they disappear with the other tertiary rocks beneath the alluvium of the Indus. It is quite possible that the present group, as well as the Nari, never was deposited in the neighbourhood of Kotri and Jerruck.

It has been already stated that the Gáj beds, throughout the greater portion of the Kirthar range, rest conformably upon the Nari group, although there is a change in mineral character, and that, in lower Sind, the passage from one group into the other is gradual, calcareous bands, with Gáj fossils such as *Ostrea multicostata* and *Pecten subcorneus*, being found interstratified with the uppermost Nari sandstones. At one place, however, near Tandra Ráhim Khán, west by north of Sehwan, the outcrop of the Gáj beds, here dipping at a high angle to the westward, runs nearly in a straight line across the mouth of a valley, composed of a deep synclinal of the Nari group between two anticlinal ridges of Kirthar limestone. As the Gáj beds do not share the synclinal curve of the Naris, it is difficult to see how the two can be conformable, but an examination of the boundary between the two groups failed to show any clear evidence of unconformity. There are, however, some places south of Sehwan where the Gáj group overlaps the Nari beds and rests upon the Kirthar limestone, but it must be recollected that the Gáj group is itself overlapped by Manchhar beds in the immediate neighbourhood.

The commonest and most characteristic fossils of this group are *Ostrea multicostata*[1] and *Breynia carinata*. There cannot be any question that the Gáj fauna is newer than eocene; some of the species are recent (for instance, *Dosinia pseudoargus* is identical with the recent *D. exasperata*), and it is probable that many others, when they are compared with recent forms more carefully than has hitherto been done, will prove to be the

[1] It is not quite certain whether this species is identical with the European form, but it is certainly the shell figured by Messrs. D'Archiac and Haime. There is another species known by the same name and found in triassic beds in Europe.

same as living species. Several genera, too, as *Maretia*, *Breynia Meoma*, *Echinodiscus Clypeaster*, *Cladocora*, and *Mycedium*, are rare or unknown in the older tertiaries, and there is almost a complete diappearance of eocene forms, very few species being common even to the Nari beds.

The only mammal yet obtained from the Gáj beds is *Rhinoceros sivalensis*—a species found also in the Siwáliks.

The highest subdivision of the tertiary series in Sind was originally named[1] from a large lake, the Manchhar, a few miles west of Sehwan, but there can no longer be any doubt that in age, as well as mode of origin, they are part of that system of pliocene fresh water deposits which, under the name of Siwálik, ranges round the extra-peninsular area from Sind to Burma. It will be well, however, to retain the local name in the description of the Sind Siwáliks.

The Manchhar series of Sind consists of clays, sandstones, and conglomerates, and attains a thickness of but little, if at all, less than 10.000 fect in places on the flanks of the Kirthar range. Although it is difficult to draw an absolute line between the subdivisions, the whole group may be divided, wherever it is well exposed, into two portions. The lower consists mainly of a characteristic grey sandstone, rather soft, moderately fine grained, and composed of quartz, with some felspar and hornblende, together with red sandstones, conglomeratic beds, and, towards the base, red, brown, and grey clays, the latter, however, being much less largely developed than in the upper subdivision. The conglomeratic beds chiefly contain nodules of clay and of soft sandstone, apparently derived from beds precisely similar to those of the typical Siwáliks, but, so far as has been observed, do not contain any fragments derived from the older tertiary rocks, no pebbles either of the characteristic Gáj limestones or of the still more easily recognised nummulitic limestone of the Kirthars having been noticed in the beds of the lower Manchhars, although both abound in the upper strata of the group. These conglomeratic beds of the lower Manchhars are frequently ossiferous, the bones and teeth contained in them being, however, usually isolated and fragmentary.

The upper Manchhar subdivision, where it is best seen on the flanks of the Kirthar range west of Lárkhána, is thicker than the lower, and consists towards the base of a great thickness of orange or brown clays, with subordinate bands of sandstone and conglomerate. The sandstones are usually light brown, but occasionally grey, like the characteristic beds of the lower subdivision. The higher portion of this upper subgroup contains more sandstone and conglomerate, and the whole is

[1] *Memoirs*, XVII, 57, (1879).

capped by a thick band of massive, coarse conglomerate, which forms a conspicuous ridge along the edge of the Indus alluvium throughout upper Sind. This conglomerate contains numerous large pebbles of nummulitic and Gáj limestone, together with fragments of quartzite and other rocks of unknown origin. Throughout the conglomeratic beds of the upper Manchhars, pebbles of nummulitic limestone and of the brown Gáj limestone occur, showing that these older tertiary beds must have been upheaved and denuded in the latter Siwálik period, although there is a complete passage between the Gáj beds and the lower Manchhars.

There appears, however, good reason for supposing that some disturbance of the older rocks took place before the deposition of the lower portion of the Manchhar group. To the east of the Lakhi range the Manchhar beds, themselves disturbed, rest unconformably on the beds of the Kirthar group, which are vertical in many places, so that it is manifest that the Kirthars had, in this locality, been upheaved before the deposition of the Manchhars. The presence of the lower portion of the latter series appearing to be proved by the occurrence of teeth and bones of the same mammals as are found in the lower Manchhars elsewhere.

In one case a few estuarine fossils were found, near the Nari stream, in a Manchhar bed 300 or 400 feet above the base of the group. The only form recognised was *Corbula trigonalis*, already mentioned as characteristic of the estuarine passage beds between Gáj and Manchhar. With this exception, and that of some rolled oyster shells, possibly derived from a lower formation, no marine or estuarine fossils have been observed in the Manchhar beds of upper Sind, above the passage beds at the base of the group, and there appears every reason to believe that the group is of fluviatile origin.

In lower Sind, however, there is a very considerable intercalation of marine or estuarine beds with the Manchhars, and this evidence of deposition in salt water increases in the neighbourhood of the present coast. Around Karáchi beds of oysters, and sometimes of other marine or estuarine shells, are not unfrequently found interstratified with the Manchhar group. There is also some change in mineral character, the sandstones becoming more argillaceous and associated in places with pale grey, sandy clays and shales. The passage from the Gáj beds is very gradual, calcareous bands with Gáj fossils, such as *Ostrea multicostata* and *Pecten subcorneus*, being found some distance above the base of the Manchhar group.

Although there is no difficulty in drawing a line between Manchhar and Gáj beds, except in the neighbourhood of the coast, everything tends to show that there is no break in time between the two, the lower portion of the upper group being an estuarine or fluviatile continuation of the underlying marine beds. But the great thickness of the Manchhar group in upper Sind

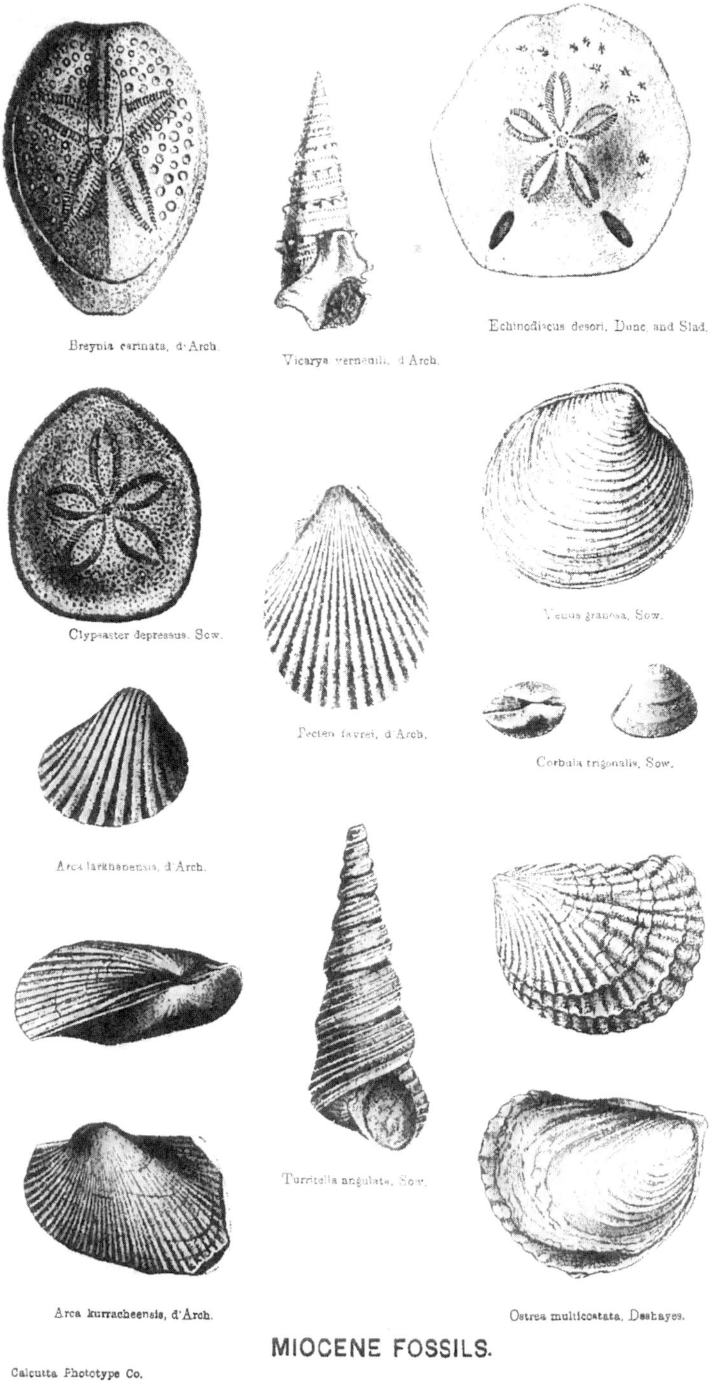

MIOCENE FOSSILS.

Calcutta Phototype Co.

alone would suffice to prove that a considerable period of time must have elapsed during the deposition of this formation, and it is far from improbable that the lower Manchhars may be upper miocene, whilst the upper Manchhars are pliocene.

The Manchhar beds extend along the age of the alluvium, and form a broad fringe to the Kirthar range, throughout upper Sind from west of Shikárpur to the Manchhar lake, but the breadth of the outcrop varies greatly, being as much as 14 miles where broadest west of Lárkhána, and diminishing both to the north and south. As already noticed, the Manchhars are thickest just where their outcrop is widest; the breadth of the area occupied by them is not, however, due simply to their vertical development, but chiefly to their forming a synclinal and anticlinal roll before disappearing beneath the alluvial plain, whereas in other parts of the range the same beds are exposed in a simple section, all the strata dipping to the eastward. To the north the section is complicated by faults, but to the south the thickness of the Manchhar group diminishes greatly, and near Tandra Ráhim Khán, west of Sehwan, although both upper and lower subdivisions of the group are developed and the uppermost conglomerate is exposed, the whole thickness of the Manchhar strata cannot be much more than about 3,000 feet. The Manchhar beds are seen west, south, and east of the Manchhar lake; they are well developed and occupy a large plain to the east of the Lakhi range, and west of the nummulitic limestone tract, near Kotri and Jerruck, they re-appear in many places in the different synclinal valleys to the west of the Lakhi range, and they occupy a considerable tract of country east and north-east of Karáchi. Throughout these areas in lower Sind the rocks are not nearly so well seen as to the northward, the soft sandstones and clays of the Manchhar group having been denuded into undulating plains, covered and concealed in general by the pebbles and sands derived from the comparatively hard older tertiary rocks of the neighbouring hills, and it is far more difficult than in upper Sind to distinguish the different portions of the group, or to form a correct idea of the thickness of strata exposed.

The Manchhar beds extend along the edge of the sea west of Karáchi, almost to the end of cape Monze, but no representative of this formation is seen for a considerable distance to the westward of the cape. The greater part of the country near the shores of Sonmiani bay consists of alluvium, and the few exposures of rocks are older tertiary, or perhaps cretaceous, a low cliff near the coast, north of Gadáni, being apparently composed of subrecent deposits. But west of Sonmiáni bay, in the neighbourhood of Hingláj, a well known place of Hindu pilgrimage, there are high hills of hard greyish white marls or clays, usually sandy often highly calcareous, and occasionally intersected by veins of gypsum. With this clay or marl, bands of shaly limestone, dark calcareous grit, and sandstone are interstratified, usually forming but a small portion of the mass, although their greater hardness renders them conspicuous. This marl formation extends for many hundreds of miles along the coasts of Balúchistán and of the Persian Gulf, forming great horizontal plateaux, surrounded by cliffs of whitish marl or clay and capped by dark coloured calcareous grit, at the headlands of Rás Malán, Ormára, and Gwádar, Rás Malán especially being a table land rising abruptly to a height of 2,000 feet from the sea. These remarkable rocks have been called the Mekrán group[1] from the name usually applied to the littoral tracts of Báluchistán.

The Mekrán group is of shallow water marine origin, and abounds in mollusca, echinoderms, etc., many of the species being apparently the same as living forms. The echinoderms alone haveas yet been examined in detail, they belong without exception to living genera, while most of the species are very closely allied to recent forms, and one species alone is doubtfully identical with a Gáj form. The general facies of the fauna is distinctly pliocene.[2]

Although there is no resemblance between the typical Manchhar beds and the characteristic rocks of the Mekrán group, nor, from the widely different conditions under which the two formations must have been deposited, would any similarity in mineral character be probable, some of the soft argillaceous shaly sands in the Manchhar beds near Karáchi closely resemble some similar beds in the Mekrán group near Gwádar. As all that is known of the geology of western Balúchistán is the result of brief visits to a few widely separated points, it is uncertain to what extent the rocks of Sind extend to the westward, and whether any representatives, of the Gáj group especially, exist in that direction, but it appears probable that the marine Mekrán group in western Bálúchistán represents the fresh water Manchhar series of Sind.

The only fossil remains of any importance hitherto detected in the Manchhar series are bones of mammalia, and all that have been recognised belong to the lower Manchhars, the upper subdivision of the group having hitherto furnished only a few bones, in too poor and fragmentary a state of preservation for the species, or even the genera, to be determined. The few estuarine shells found in the lowest Manchhar beds in upper Sind, and a portion at least of the marine fossils procured from a similar horizon near Karáchi, appear to be Gáj forms, and to indicate a close connection between the lower Manchhars and the underlying group

[1] W. T. Blanford, *Records*, V, 43, (1872); Eastern Persia, London, 1886, II, p. 462.

[2] Duncan and Sladen, *Pal. Indica*, series xiv, I, pt. iii, p. 370, (1886).

In places, and especially in the neighbourhood of the Lakhi range, silicified fossil wood is found in abundance in the Manchhar beds, stems of large trees being of common occurrence. The majority are dicotyledonous, but some fragments of monocotyledons are also found.

The vertebrate remains are extremely fragmentary, and chiefly consist of single teeth and broken portions of bones,[1] and the fauna is chiefly remarkable for the prevalence of artiodactyle ungulates, allied to pigs, or intermediate between pigs and ruminants. The majority of the genera are extinct; *Rhinoceros*, *Sus*, and *Manis* being the only living types, and the last named has only been recognised from a single digital phalange, so that the generic identification is far from sufficient. Both *Rhinoceros* and *Sus* existed in miocene times, whilst *Amphicyon*, *Anthracotherium*, *Hyopotamus*, and *Dinotherium*, which are also found in the Manchhar, are not known to occur in Europe in beds of later date than miocene. The genera *Hemimeryx* and *Sivameryx* are peculiar, both being allied to the Siwálik *Merycopotamus*.

The species found also in the pliocene Siwáliks are *Rhinoceros palæindicus*, *Acerotherium perimense*, *Chalicotherium sivalense*, *Sus hysudricus*, *Dorcatherium majus*, *D. minus*, *Mastodon latidens*, and *M. falconeri*; but as the presence of these forms in the Manchhars is inferred for the most part from fragments, the identifications are by no means quite certain, whilst the general facies of the fauna, the absence of characteristic living forms like *Equus*, *Bos*, *Antilope*, *Cervus*, and *Elephas*, and the presence of several extinct genera not hitherto detected in the Siwáliks, show that the mammaliferous beds of Sind are of older age than the typical Siwálik strata. It should be recollected, moreover, that the precise horizon at which the Siwálik forms are found is but rarely known with accuracy, that some of the Siwálik strata are as old as the lower Manchhars, if not older, and that a portion at least of the older types of mammals are from beds low down in the Siwálik series. None of the remarkable series of types allied to the giraffes and *Sivatherium*, nor of the peculiar bovine and antilopine forms so characteristic of the Siwálik fauna, have as yet been found in Sind. The only ruminant detected in the Manchhar beds is the miocene *Dorcatherium*, and the place of the more specialised *Pecora* appears to have been occupied by the less specialised even toed ungulates allied to the pig. While, therefore, it is probable that some extinct types, such as *Anthracotherium* afnd *Hyopotamus*, which are not known in Europe above the lower miocene, existed in India at a somewhat later period, together with species which survived till pliocene times, it is evident that the lower portion of the Manchhar group can scarcely be considered of later date than upper miocene. The palæontological

[1] Details will be found in Mr. Lydekker's papers in *Records*, IX, 91, 93, 106; X, 76, 83, 225; XI, 64, 71, 77, 79, (1876–78) and *Pal. Indica*, series x, I—IV, *passim*.

evidence is in accordance with the geological, and both show the close connection between the lower Manchhar beds and the Gáj group.

Further north, in the country between Sibi and Quetta the Siwáliks rest with a slight, but distinct, unconformity on the Spintangi limestone. There is everywhere a perfect parallelism of stratification between the two and a superficial appearance of conformity, but on a close examination it is seen that there are slight, though distinct, traces of erosion at the contact, that pebbles of nummulitic limestone are found near the base of the Siwáliks, and that there is a progressive thinning out of the nummulitic limestone from east to west by removal of the upper beds. The lower Siwáliks, which are conspicuous on the section in the Bugti hills further east, are represented by greenish and grey sandstones which do not attain a thickness of much over 100 or 200 feet. Above them the upper Siwaliks consist principally of red earthy clays, with interbedded sandstones, which become more and more frequent till it becomes a sandstone formation with subsidiary bands of clay and conglomerate, the whole capped by a great thickness of strong conglomerates.

These Siwáliks extend westwards, in the country intervening between the Bolan and Hurnai routes to Quetta, almost to the Quetta plain, but their original western extension has been obscured. In the valleys of Quetta and Pishín there are some deposits of conglomerates and red clays which have in places undergone considerable disturbance, and have been referred to a Siwálik age.[1] On the accompanying map they have been coloured as upper tertiary, but, as they appear to be distinct in age and origin from the Siwáliks referred to above, and much more closely connected with the recent deposits, their description will be deferred to a subsequent chapter.[2]

In the Bolan pass, in the Gandak valley, north-east of Quetta, and doubtless in many other valleys of these hills, there are sandstone and conglomerate deposits which have undergone considerable disturbance, and dip at high angles. In accordance with the custom, which separates the deposits which have undergone considerable disturbance from the undisturbed recent deposits that unconformably overlie them, these have been called Siwálik, but they are in a manner distinct from the older Siwáliks for the former were deposited after the valleys, in which they lie, had been excavated, while the latter date, in this neighbourhood, from a time when the disturbance of the strata and consequent elevation of the hills had not commenced.

In the Bugti hills and southern portion of the Suláimán range, the lower Siwáliks, which attain a maximum thickness of 5,000 feet, have the appearance of lying conformably on the nummulitic limestone. They

[1] W. T. Blanford, *Memoirs*, XX, 117, (1883). | [2] *Infra*, p. 416.

consist of moderately soft, fine grained, pepper and salt grey, sandstone, interstratified with conglomeratic beds, composed of fragments of clay and soft sandstone, apparently derived from contemporaneous deposits, imbedded in an argillaceous matrix and unaccompanied by any pebble of harder rock. Clay beds also occur and are usually of a red colour.

In the lower beds of this group vertebrate remains occur in considerable abundance near Dera Bugti, among which are *Mastodon*, *Rhinoceros*, *Dinotherium*, *Anthracotherium*, and *Hyopotamus*. The vertebrate fauna has a distinctly miocene facies and is associated with seven species of fluviatite mollusca, of which four belong to the genus *Unio*, two to *Melania* and one to *Paludina*. All seven are extinct, and none are nearly related to forms now living in western India, though two are allied to species still existing in Burma. Three of the species of *Unio* are aberrant forms with ribbed shells, exhibiting a superficial resemblance to the marine genus *Cardium*.[1]

East of the Suláimán range the upper Siwálik conglomerate is on some sections overlaid by a more recent conglomerate deposit, which has been disturbed and dips towards the plain at moderate angles. The newer conglomerate has the appearance of passing upwards into the recent deposits, and though it has been regarded as uppermost Siwálik[2] should probably, like the disturbed river gravels of the Bolan valley, be more correctly classed as recent than as tertiary.

The various localities referred to in the previous passages form part of one geological province throughout which the lower tertiaries maintain a certain constancy of character, allowing the rock groups on one section to be recognised on another, but when we pass northwards to the Punjab, or north-westwards to Afghánistán, it is no longer possible to apply the subdivisions adopted in Sind, and a fresh classification has to be adopted. Before passing on to these areas it will be well to notice the exposures to the east, in which the Sind subdivisions can be more or less recognised.

The tertiary rocks in the Cutch peninsula occupy a belt, varying in breadth from about four miles to twenty, between the alluvium near the coast and the older rocks in the interior of the country. Tertiary formations also fringe the Deccan traps and jurassic beds, on the borders of the two openings by which the Rann of Cutch communicates with the sea east and west of the province, and patches of the same tertiaries are found here and there on the shores of the Rann, not only in the main region of Cutch itself, but also around the detached hilly tracts or islands, Patcham, Kharir, etc., and in Wágad. The evidence of unconformity

[1] W. T. Blanford, *Memoirs*, XX, 162, 233, (1883).

[2] W. T. Blanford, *Memoirs*, XX, 219, (1883).

between the eocene rocks and the Deccan traps is very strong in Cutch. The lava flows which appear to have covered the greater part, if not the whole, of the jurassic region had been completely swept away from the surface of the country, and the underlying jurassic rocks exposed and largely croded in places before the eocene marine beds were deposited.[1] Despite this evidence of unconformity, there is every appearance, along the southern border of the trap area, of the tertiary beds resting conformably on the lava flows of the Deccan period.

The tertiaries of Cutch are far better known than those of Gujárát and Káthiáwár, the materials for the first descriptions of marine fossils from the later Indian deposits having been furnished by the rocks of the first named province. Attention was first directed to the Cutch tertiaries through the labours of Captain Grant, who carried with him to England a considerable collection of tertiary organic remains, together with the jurassic fossils mentioned in a former chapter. In accordance with the ideas prevailing amongst geologists at the time, he separated nummulitic rocks from the true tertiaries on his map,[2] and the same distinction was preserved in the description of the fossils, but subsequently all the forms described were classed as eocene by D'Archiac and Haime.[3] When the rocks of Cutch were mapped in 1867-69 by Wynne and Fedden, and described by the former, it was found that several distinct groups could be recognised, and that the fossils of these groups differed, and it was afterwards discovered that the groups corresponded very closely to those determined in Sind. The succession of the rocks in Cutch, according to Mr. Wynne, is the following, the probable Sind representative being appended in each case. The supposed European equivalents differ somewhat from those originally suggested before the corresponding beds in Sind had been examined[4]:—

	Cutch.		Sind.	European equivalents.
Alluvium, blown sand, etc.			Alluvium, etc.	*Peistocene and recent*
Tertiary	Upper tertiary (unconformity).	200 to 500 ft.	Manchhar	*Pliocene and upper miocene.*
	Argillaceous group	800 to 1,200 ,,	Gáj	*Miocene.*
	Arenaceous group	130 ,,	Nari (?)	*Lower miocene and upper eocene.*
	Nummulitic group	700 ,,	Kirthar	*Eocene*
	Gypseous shales	100 ,,	Ránikot	*Lower eocene.*
	Subnummulitic	100 ,,		
Stratified traps			Trap	*Uppermost cretaceous.*

[1] This view is opposed to Mr. Wynne's opinion. He considered that the lower eocene beds are conformable to the traps, and that the traps never existed in northern Cutch. —*Memoirs*, IX, 72, (1872).

[2] *Geol. Soc. Trans.* 2nd series., V, 303, Pl. xx, (1840).

[3] Description des Animaux Fossiles du groupe nummulitique de l'Inde, Paris, 1853.

[4] *Memoirs*, IX, 48, (1872).

The subnummulitic group consists chiefly of peculiar purple and red, mottled with white, soft argillaceous beds, laterite of various kinds, and coarse sandstones distinguished by brilliancy of colouring, white, red, lavender, purple, and orange tints prevailing. There are also some shales with impressions of leaves and carbonaceous layers, and occasionally with gypsum.

Some of the peculiar argillaceous beds have a distinctly volcanic aspect, but as they are much decomposed it is impossible to say that they are really of eruptive origin. The occurrence of these peculiar beds away from the traps, in places where there is good reason to suppose that the traps were removed by denudation in pretertiary times, and the fact that beds reconsolidated from trap fragments must, when decomposed, frequently be undistinguishable from a disintegrated eruptive rock, render it probable that these soft mottled beds are of sedimentary origin and composed of the detritus of volcanic rocks. Fossils are rare in the subnummulitic group which extends along the southern edge of the traps in Cutch, overlapping the volcanic rocks to the westward, and resting upon jurassic rocks near Lakhpat. The same group is represented in several small patches, deposited upon jurassic beds on the borders of the Rann, both on the mainland of Cutch and on some of the detached hills or islands, especially south of the hills in Patcham, Kharir, Bela, and Chorar, and intervening in the hollow between two ranges on the first named. The group is nowhere more than about 200 feet thick, and it frequently does not exceed 20 feet.

Above the subnummulitic beds there are in places from 50 to 100 feet of fine laminated shales, bituminous and often pyritous, with fragments of wood and leaf impressions. All the above rocks are classed by Mr. Wynne apart from the true tertiaries, and with the bedded traps. It appears, however, more probably correct, and more in accordance with the sequence in Sind, to consider the main break in the series as taking place between the traps and the next formation in ascending order.

The gypseous shales form a local and unimportant subdivision, not more than from 50 to 150 feet in thickness, occurring in western Cutch, round the Gaira hills and in a few other places. They consist of shales, with calcareous nodular bands and much gypsum, and with some beds of laterite. Some of the marly beds abound in *Nummulites* and other *Foraminifera*, oysters, etc.

The next group is of more importance, being the representative of the massive nummulitic (Kirthar) limestone of Sind. In Cutch these beds consist of pale yellow and white impure limestones, in bands of no great thickness, interstratified with marls and sandy beds. The upper portion consists chiefly of marls, limestones being more abundant below; *Nummulites*, *Alveolinæ*, and echinoderms of several kinds abound and corals and mollusca are locally common. The nummulitics of Cutch are, however, almost confined to the western part of the province, and occupy a band extending from Lakhpat round the western termination of the Deccan trap range in the Gaira hills.

Upon the nummulitic limestones and their associates there is usually found a thin and unimportant band of light coloured or white sand and sandy shales, having at the base some finer dun, or blue coloured, silty shales. These sandy beds are soft, friable, and obliquely laminated. In the lower portion of the group the carapace of a small crab and casts of bivalve shells have been found, in the upper part impressions of dicotyledonous leaves occur. This group, originally described as the arenaceous group, corresponds in mineral character and position to the upper Nari of western Sind.

The Gáj group of Sind is represented in Cutch by what was originally described as the argillaceous group, the best developed and most fossiliferous of the tertiary rock groups of that district, and it is this group which yielded the bulk of the fossils described as tertiary in the appendix to Captain Grant's paper, although it appears probable that there were among these fossils some admixture of species from a lower horizon. Until the whole of the Cutch and Sind fossils are thoroughly compared and determined, some doubt must remain as to the original horizon of a few Cutch species, but when the forms are common to the Gáj beds of Sind, and are not known to occur in older group of that area, it may fairly be inferred that they are probably restricted to the same horizon in Cutch.

The Gáj, or miocene, rocks of Cutch consist of sandstones at the base, with a few nodular, marly and ferruginous bands often containing *Turritella*, *Venus granosa*, and *Corbula*. Above the sandy beds are marly limestones and shales, next calcareous grits, and then a considerable thickness of shales, clays, and marls. The most fossiliferous beds are the marly limestones and shales. Only the echinoderms from these beds have as yet been critically examined, and of 16 species no less than 8 are also known from the Gáj group of Sind, one being *Breynia carinata*, one of the most characteristic Gáj forms.

The miocene beds are more extensively developed in Cutch than are the nummulitics. They are found not only in the west of the province around the extremity of the jurassic and trap area, but eastwards, resting upon the subnummulitic group, as far as about half way across the province. To the westward, however, the present group is overlapped by the next in ascending order.

The representative of the Siwálik rocks in the sub-Himaláyan tract, and of the Manchhar beds in Sind, appears to be widely developed in Cutch, and covers a large area, but it is very ill seen, being greatly concealed by alluvial deposits. The principal beds are more or less ferruginous conglomerate, at the base, followed in ascending order by thick brown sands and obliquely laminated, nodular, calcareous, and sandy clays. Marine beds, with large oysters, are intercalated, as in southern Sind. It will probably be found on further examination that this uppermost tertiary group in Cutch, as in Sind, passes down into the underlying subdivision in places, although to the eastward the latter appears to have been denuded before the deposition of the former. The upper tertiary group extends throughout southern Cutch from east to west, resting on the older tertiaries to the westward, but gradually overlapping them and the traps to the eastward, and resting upon jurassic rocks in the extreme east of the province.

In Káthiáwár eocene beds have been recorded from Beyt island, off the north-west extremity of the peninsula, by Dr. Carter,[1] and though the Geological Survey of Káthiáwár failed to show the existence of any lower tertiary beds on the mainland, it is impossible to accept Mr. Fedden's suggestion[2] that the record of the presence of nummulitic beds is due to a confusion between *Patellina* and *Nummulina*. Dr. Carter's statements are too specific to allow of such an explanation, and unless he was misinformed as to the locality from which his specimens came, we may accept the presence of lower tertiary beds at the north-west extremity of the Káthiáwár peninsula.

With this exception the only known tertiary deposits are miocene or newer. They are found along the southern edge of the trap area, the principal exposures being at the eastern and western extremities, with some narrow strips in the intervening area. They lie almost horizontal, are much obscured by recent deposits and cultivation, and in the absence of any deep cut sections no good general succession has been made out. In the south-eastern area they consist of shales and marls, which contain many marine fossils, with interbedded bands of a rusty conglomerate of clay pellets, and agates derived from the trap.

With the exception of the echinoderms, the marine fossils collected in Káthiáwár have not been examined in detail, but many of them are identical with Gáj species, and of the 13 species of *Echinodermata*, 6 are also found in the Gáj groups of Sind, the whole distinctly indicating a miocene age.[3]

The uppermost beds of the Káthiáwár tertiaries appear to be sandstones and conglomerates, of which the best known exposure is that in the small island of Perim or Piram—not to be confounded with the island of similar name at the entrance to the Red Sea. Perim island is a small reef of sandstones and conglomerates, only 1,800 yards long by 300 to 500 yards broad, which first achieved a geological celebrity through the

[1] Geological Papers on Western India. Bombay, 1857, p. 743.
[2] *Memoirs*, XXI, 122, (1884).
[3] Duncan and Sladen, *Pal. Indica*, series xiv, I, pt. 4, p. 80, (1880).

discovery of fossil mammalian bones by the Baron von Hügel in 1836. According to the most recent examination these bones are found in the conglomerate bands, but principally in a conglomerate bed which lies considerably below high water level and is obscured by a thick covering of mud for the greater part of the year. During the months of April, May and June, however, the south-east end of the reef becomes scoured and free from mud, and specimens of fossil bones can be found. As in all other Indian localities for fossil bones, the first collectors found a rich harvest in the accumulation of ages, while their successors have to be content with but occasional and fragmentary specimens.[1]

The known mammals from Perim island comprise ten species, of which four, *Mastodon latidens*, *M. perimensis*, *Rhinoceros perimensis*, and *Sus hysudricus*, are common to the Siwálik beds. All these forms are, however, found in other fossil faunas; *Mastodon perimensis* and *Sus hysudricus* being met with in the lower Manchhar beds of Sind, *Rhinoceros perimensis* in the Irawadi deposits, and *Mastodon latidens* in both, so that all the forms common to Perim island and the Siwáliks are clearly species of wide range. The absence of *Elephas* and its subgenera, and of bovines, and the presence of *Dinotherium*, tend strongly to make the Perim island fauna appear of greater age than the Siwálik generally, but, on the other hand, the presence of so highly specialised a genus as *Capra*, if the generic determination be accepted,[2] the occurrence of *Camelopardalis* and *Antilope*, and, above all, the absence, so far as is known, of *Anthracotherium*, *Hyopotamus* and other older ungulate types so abundant in the miocene beds of Sind and the Punjab, are opposed to the idea that the Perim island rocks can be of higher antiquity than pliocene. They possibly occupy an intermediate position between the Siwáliks proper and the Manchhars of Sind, but they are more nearly allied to the former.

In the north-western portion of the Káthiáwár peninsula the fossiliterous Gáj beds are overlaid by what have been described as the Dwárká beds. They consist of soft yellow earthy or marly clays, gypseous in part, overlaid by more or less marly or arenaceous limestones, generally soft and porous, largely composed of *Foraminifera* cemented by calcite, or of comminuted shells and corals. No recognisable fossils have been found in these beds, and their relation to the underlying fossiliferous miocene beds was not determinable; it is probably one of conformity in spite of the sudden change of lithological character.

[1] F. Fedden, *Memoirs*, XXI, 39, (1884); see also W. T. Blanford, *Memoirs*, VI, 374, (1869).

[2] It should not be forgotten that *Capra perimensis* is founded solely on a frontlet with the horn-cases, and that nothing is known of the greater part of the cranium, the teeth or the limb bones. See Lydekker, *Pal. Indica*, series x, I, 83, 170, Pl. xxviii, fig. 4, (1884).

Turning to Afghán-Túrkistán, there appears to be, as in Sind, a perfect conformity between the cretaceous and tertiary beds. In the synclinal of Mathar, 100 miles south-east of Balkh, the upper cretaceous limestones are overlaid by sandstones and greenish shales in which no fossils were found, except an *Exogyra* and *Cerithium*, and some fucoid-markings and badly preserved remains of fishes and crustaceans. These beds; may be regarded as eocene; they are overlaid by red sandstone, with a few clay bands and, towards the top, conglomerates composed principally of pebbles of upper cretaceous limestone, covered by a great thickness of sandstones and shales, containing a few fresh water shells and passing upwards into soft gypseous clays.[1]

Besides the exposures in synclinal basins within the hills, a zone of tertiary beds is found all along the edge of the alluvium of the Oxus valley. Here, however, the eocene clays are wanting and the red beds with conglomerates rest directly, in apparent conformity, on the upper cretaceous limestone. They pass upwards with a gradual transition into the recent deposits, and in the upper portion of the section, there occur beds which are undistinguishable from the recent wind blown loess of the Oxus valley.[2]

The northern extension of the Suláimán range has not been examined geologically, but there is good reason to suppose that the fringe of Siwálik rocks is continuous with great area of tertiary deposits, extending from the north-west corner of the Punjab along the outer edge of the Himálayas to the border of Nepál. The general description of the Himálayan tertiaries, with which must be included those of the north-west corner of the Punjab, will be deferred to the following chapter, but it will be well to notice in this place a portion of the Kohát district west of the Indus, where the lower tertiaries exhibit some peculiarities not noticed elsewhere.

The eocene rocks are well developed in this region, and the section exposed, although only a few miles distant from parts of the Salt range, differs in some important points. The following is abridged from Mr. Wynne's summary of the rocks exposed :—[3]

		Thickness in feet.
Pliocene and Miocene .	*Upper sandstones.*—Soft, grey sandstones, clays and conglomerates	500 to 1,500
	Lower sandstones.—Harder grey and purple sandstones, bright red and purple clays, slightly calcareous and pseudo-conglomeratic bands . .	3,000 to 3,500

[1] C. L. Griesbach, *Records* XIX, 255, (1886).

[2] C. L. Griesbach, *Records*, XIX, 257, 259, (1886).

[3] Wynne, *Memoirs*, XI, 101, (1875).

		Thickness in feet.
EOCENE	*Upper nummulitic.*—Nummulitic limestone and some shaly bands	60 to 100
	Red clay zone, or lower nummulitic.—Red clay, lavender coloured near the top, occasionally with *Nummulites.* The lower portion of the red clays in places is partly or wholly replaced by fossiliferous sandstones, thick greenish clays and bands of limestones, all containing *Nummulities* . .	150 to 400
EOCENE?	*Gypsum.*—White, grey or black gypsum with bands of clay or shale	50 to 300
	Rock salt.—Thickbeds of salt, almost pure. The base not seen	300 to 700 (1,200).

The region examined is the hilly tract north of the Bannu plain and of the Chicháli hills, and extending from the Indus, on the east, to the British frontier. The ground is traversed by a series of east and west ranges, chiefly formed of crushed and broken anticlinals of the nummulitic limestone and the associated rocks.

The rock salt and gypsum at the base of the tertiary series in the Kohát region are very important and remarkable. The salt consists of a more or less crystalline mass, usually grey in colour, with transparent patches, and never reddish like the salt of the Salt range. A few earthy bands occur, but the portion of the whole mass too impure to be worked

Fig. 18.—Hill of rock salt, 200 ft. high, at Bahádur Khel, after Wynne.

for commercial purposes is but small, although there is no attempt at refining the salt, which is exported for sale in the form in which it is mined. In some places the uppermost layer is dark coloured, almost black, and bituminous. The quantity of salt is something marvellous; in the anticlinal near Bahádur Khel alone, rock salt is seen for a distance of about eight miles, and the thickness exposed exceeds 1,000 feet, the width of the outcrop being sometimes more than a quarter of a mile. As a rule,

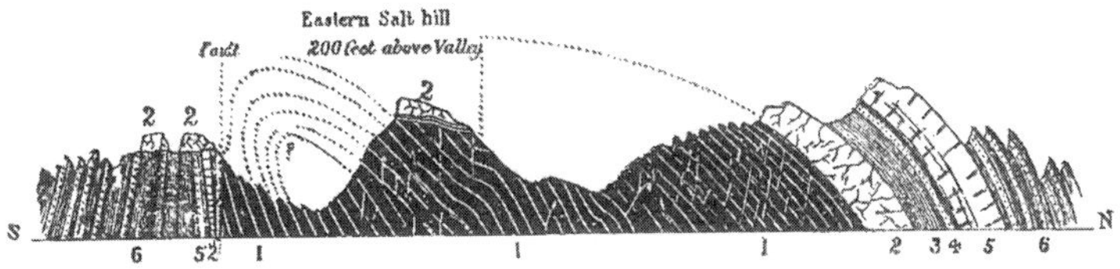

Fig. 19.—Section through the Bahádur Khel salt locality, after Wynne.

the salt contains sulphate of lime (gypsum), but none of the potassium and magnesium salts of the Salt range beds.

Above the salt come gypsum and clays, as in the Salt range, but the colours, white and grey, are very different, and the whole appearance, of both salt and gypsum, is so distinct from the Salt range marl that, although there is no indication of salt beds at a higher level in the Salt range itself, and although the outcrop of the salt marl close to Kálabágh on the Indus is only eighteen miles from one of the Kohát rock salt regions, it appears probable that the salt bearing rocks in the Kohát district may belong to a very different horizon from that occupied by the same minerals in the Salt range series, a supposition supported by the fact that a great series of mesozoic and palæozoic beds intervenes, throughout the Salt range, between the nummulitic group and the salt marl, whilst in Kohát the former rests with apparent conformity upon the gypsum and salt. It is by no means certain that the Kohát salt and gypsum are eocene, but, in the absence of any evidence to the contrary, it appears best to class them with the nummulitic beds immediately overlying them.[1]

Overlying the gypsum, there is usually found a thick bed of deep red clay, whose eocene age is proved by the occasional occurrence of nummulites in the upper portion. Sometimes the red clay is wanting, and apparently replaced by grey or olive coloured clays, marls, and limestones containing nummulites, but the replacement is not clearly proved. Above the red clay zone come earthy limestones, clays, and shales with nummulites. The main band of limestone is very much thinner than in the Salt range, but is, as usual, massive, pale coloured, and full of *Nummulites, Alveolina,* etc. The overlying formation, consisting of sandstones and clays, in which dark red and purple colours predominate, exceeds all the eocene beds in thickness, and is probably, like the sand-

[1]These conclusions would have to be modified if the hypothesis of hypogene origin of the salt and gypsum is adopted,—see *supra*, p. 111.

stones and clays overlying the nummulitic limestone of the Salt range, really of much later age than the limestone on which it rests.

Pebbles of nummulitic limestone are said[1] to be found in the lowest beds of the sandstone, and some reptilian bones (not determined), silicified fossil wood, and a few ill preserved, ribbed, bivalve mollusca have been found, but no characteristic organic remains. The beds resemble those of the Murree group and pass upwards into undoubted newer tertiary (Siwálik or Manchhar) strata of the usual character, the red colours becoming rarer, and the usual drab grey sandstones and orange or drab clays being the prevailing rocks.

When traced in a westerly direction towards Thal, the eocene beds are said to change in character, the limestones being replaced by hard sandstones and greenish grey or purple shales and, on the right bank of the Kuram river opposite Thal, volcanic beds occur among them. The beds are said to be penetrated and indurated by intrusions of a dark coarse crystalline trap, and besides the intrusions agglomerates and fine grained tufaceous trappean beds are found in alternating layers.[2] No other occurrence of volcanic beds in the eocene of northern India is known, and as the Thal locality was visited under circumstances extremely unfavourable for geological investigation, too much importance must not be attached to the observations.

No lower tertiary rocks are known south of the great snowy range of the Himálayas, between the small outliers folded among the pretertiary slates in western Garhwál and their re-appearance on the southern side of the Assam range. Nearly the whole of this tract of the Himálayas is inaccessible to geological observation and it cannot consequently be stated with confidence that the lower tertiaries are altogether wanting, but the general run of their northern boundary, in the north-west portion of the Himálayas, appears to indicate that the eocene coast line trended south of the present limit of the Himálayas, and makes it improbable that any lower tertiary marine beds would be found east of Garhwál, even if the country now inaccessible were open to observation.

At the western extremity of the Gáro hills the nummulitic deposits contain only one thin band of limestone about 40 feet thick, resting conformably upon the cretaceous sandstone. Even this limestone is often earthy, nodular. and ochreous, with shaly partings, the purer portion being generally formed of a mass of *Nummulites granulosa*, in various stages of growth. The series generally exhibits a shallow water type and an approach to the original margin of the eocene sea. Throughout the formation there are no clear

[1] A. B. Wynne, *Memoirs*, XI, 170, (1875). | [2] A. B. Wynne, *Records*, XII, 111, (1879).

sandstones; clays and soft earthy sandstones overlie the limestone and are with difficulty distinguished from the succeeding upper tertiary deposits. This character is most pronounced at the west end of the hills, where the formation strongly resembles the most characteristic beds of the Subáthu group in the north-west Himálayas. No coaly layer has been found in the series in the Gáro area. The lower tertiary beds are not confined to the southern margin of the Garo hills, outliers being found in some of the basins of the cretaceous rocks north of the Turá range. The country has not, however, been sufficiently closely surveyed to show whether these indicate the same thinning out and approach to the original shore line as is seen in the Khási hills further east.

A number of fossils obtained from the lower tertiary beds of the Gáro hills appear to belong, so far as they can be identified, to eocene species,[1] and the presence of bands of limestone full of nummulites is in itself sufficient to fix the age of these beds as eocene.

To the east, the nummulitics show a change to deposits of more pelagic character, in which there is a great development of clear nummulitic limestones. The transition between the prevailing types of deposits in both the cretaceous and nummulitic strata is in the ground, described by Colonel Godwin-Austen, on the confines of the Gáro and Khási hills, where the upper tertiaries have been almost denuded away from the base of the range, and here a partial unconformity was noticed between these systems.

The general character of the nummulitic series at the foot of the Khási hills is shown by the section in the Tharia river, where the following beds are exposed, all with a high southerly dip:—[2]

	Feet.
7. Limestone, coarse, massive, blue	200
6. Sandstone, clear, yellowish, coarsish	100
5. Limestone, fine, compact, blue or pink	200
4. Sandstone, earthy, greenish, ochrey	50
3. Limestone	50
2. Sandstone, yellowish	100
1. Limestone	200
TOTAL	900

On the highlands immediately to the south-west of the station of Cherra Punji, there is a small plateau of nummulitic strata, very much reduced in thickness. The bottom 80 feet are of limestone, covered by about an equal thickness of sandstone, not markedly different from the underlying cretaceous rock. At about ten feet above the limestone there is a thick seam of bright coal, the well known Cherra coal.[3] The limestone rests

[1] *Records*, XX, 42, (1887).
[2] *Memoirs*, VII, 164, (1869).
[3] *Memoirs*, I, 140, 185, (1858); *Records*, XXII, 167, (1889).

directly upon the surface of the Cherra sandstone, without any sign of denudation, and the bedding is parallel, having a southerly slope of three degrees.[1] The fossils from this limestone were determined by Dr. Stoliczka; portions of the rock consist almost entirely of small specimens of *Operculina canalifera*, *Nummulites lucasana*, and *N. ramondi*, both the last species very small, species of echinoderms, fragments of oysters, *Pecten*, *Cardium salteri* and fragments of *Natica Cerithium*, *Turritella*, etc.[2]

Close to the north of the coal hill, the nummulitic beds occur again in equal thickness, under the native town of Cherra Punji. The limestone is not seen here, but this may be partly due to concealment. The nummulitic sandstone forms the highest ground of the plateau from Cherra Punji to beyond Surarim. Carbonaceous markings are frequent in it, and at Lairangao, four miles north of Cherra village, there is a workable seam of coal. It is at about the same height over the cretaceous sandstone as the seam at Cherra, but the underlying beds are all sandstone and shale, except one bottom bed of limestone, resting on the Cherra sandstone. In this limestone bed *Operculina canalifera* and *Nummulites lamarcki* are very common; with these occur a *Trochocyathus*, *Stylocœnia vicaryi*, *Echinolampas spheroidalis*, a small *Cardita*, *Pecten*, *Natica roualti*, *Keilostoma marginatum*, a *Ziziphinus*, the small *Cerithium hookeri*, casts of a large *Natica*, *Cerithium* and *Terebellum*.[3]

Under Surarim, only a mile from Lairangao, this bottom bed of limestone is wanting, being apparently overlapped by the carbonaceous sandstones, which themselves come to an end in a low bluff, two miles further north, near where the road bifurcates.

In the corresponding section on the Bogapáni, below Chela, there is a coaly layer in the sandstone just over the bottom limestone, which may sufficiently establish the identity of these two bottom bands with those on the plateau above, the marine bed having increased in thickness, and this marine character is here strongly stamped upon the whole series.

In the Khási hills, at least on the Cherra section, the northern thinning

[1] This little plateau at Cherra offers a remarkable instance of a form of denudation that is nót, perhaps, taken sufficient account of in geological explanations. The scarp is very regular and well defined at many points; but the upper surface of the area, about a square mile in extent, is a chaos of tilted masses of the upper sandstone. This is clearly due to the more or less complete removal by solution of the supporting limestone. Colonel Godwin-Austen has described a much larger instance of this form of denudation in the Gáro hills, where a considerable enclosed catchment basin is drained underground.—*Jour. As. Soc. Beng.*, XXXVIII, pt. ii, 21, (1869).

[2] *Memoirs*, VII, 167, (1869). The fossils described from eastern Bengal by MM. D'Archiac and Haime in their "Groupe Nummulitique de l'Inde" were all fróm the Khási hills; but their specimens were so mixed, probably even including fossils from the cretaceous beds, that the value of their identifications is doubtful. As the authors themselves remark—"On voit qu'il y a un certain vague dans les rapports de plusieurs des assises que nous venons d'indiquer."—*l. c.*, p. 177.

[3] These fossils were determined by Dr. Stoliczka,—*Memoirs*, VII, 167, (1869).

out of the nummulitic series upon the gneissic plateau is not so clearly seen as in the case of the cretaceous rocks, because only remnants of the bottom bands are preserved there. It will be seen however, from what has gone before, that there is a distinct indication of a northerly thinning out of the lower tertiary beds and a disappearance of the marine type of deposits. The nummulitics have not been observed anywhere to rest upon the gneiss, so there is no proof of their having overlapped the cretaceous deposits, but it is presumable that they did so, for the sequence is conformable, or at least parallel and undisturbed, and the nummulitics extend close up to the northern boundary, where the cretaceous beds are very thin.

The great thickening of the series in the sections at the southern foot of the hills points to the same conclusion, and if the occurrence of coaly beds in the Bogapáni below Chela at a horizon corresponding to bed No. 2 of the section may be taken to indicate an equivalence with the coal bearing sandstone of the plateau, there is not only a great thickening of this and the underlying group, but there are some 600 feet of overlying strata which do not appear to be represented on the plateau to the north.

The nummulitic limestones form the most prominent features of the low hills at the foot of the scarp along the Khási area, where they are in much greater force than elsewhere, to the east or west. This may be owing to a greater elevation in this position, whereby a deeper zone of the basin of deposition has been exposed to view. Pure limestone is still the chief rock of the group on the eastern confines of the Gáro hills, west of the Umblai, as described by Colonel Godwin-Austen, although the total thickness must be much less than in the Tharia section.

To the east the nummulitic limestone is known to be well developed in the North Cachar hills, where it has a thickness of 300 feet, resting on the cretaceous, and overlaid by the upper tertiary sandstones. In a northeasterly direction it has been traced as far as the hot springs of the Kapili valley.[1]

No beds known to be of nummulitic age have yet been discovered in upper Assam, but a series of coal and petroleum bearing rocks, conformably overlaid by sandstones of the upper tertiary type, which appear to represent them, is exposed near the edge of the hills north of the Brahmaputra valley in the Sibságar and Dibrugarh districts.

For detailed information regarding the distribution of the rocks Mr. Mallet's Memoir must be referred to.[2] The measures are very much alike in all the exposures, seams of less than a yard in thickness are very numerous in some sections, and not unfrequently the coal beds attain much greater dimensions. In the Námdáng, south of Rongreng in the Mákum field, there is a seam 100 feet thick, containing at least

[1] T. D. LaTouche, *Records*, XVI, 201, (1883). | [2] *Memoirs*, XII, pt. 2, (1876).

75 feet of solid coal, and some very thick seams have been traced for more than a mile without diminution. The sandstones and shales often contain nodules and layers of clay ironstone. Earthy and ferruginous limestone occurs sparingly in thin concretionary bands, also some layers of hard tough magnesian limestone. The coal measure shales decompose into a very tenacious blue clay, differing in this respect from the Disang shales, which are more clunchy.

On the interesting question of the age of these very important and extensive coal measures, there is little evidence for opinion and that little suggests a middle tertiary horizon. The coal itself is a true coal of superior quality, not lignite, as is attested by its composition—

	Fixed carbon.	Volatile matter.	Ash.
Average composition of 27 Assam coals .	60·0	36·2	3·8
Ditto of 17 Ránîganj coals .	51·1	32·6	16·3

It is not unlike the nummulitic coal of the Khási hills, and quite unlike the cretaceous coal, which maintains its peculiar characteristics into close proximity to the Assam fields. The only fossils found in the seams are bad impressions of dicotyledonous leaves and no trace of animal life has been seen in any of the associated rocks. The strongest point in the argument is the closely transitional relation between the measures and the Tipam sandstone, which is a very typical representative of the Siwálik rock, and almost undoubtedly belongs to the upper tertiary period. In the Khási hills, as in the Punjab, the nummulitic coal occurs near the very base of the formation. There are about 1,000 feet of marine nummulitics above the coal bed in the Tharia section, but the upper Assam coal measures, if they are nummulitic at all, would seem to belong to the upper limits of the series.

The nummulitic beds in the Gáro hills pass into the upper tertiaries, which may here be divided into a lower group of fine grained sand stones with marine fossils, and an upper unfossiliferous group of rusty sandstones, variable in grain, with grey shales. These uppermost tertiary beds have not yielded any fossils to recent observers, but in 1821 Mr. Colebrooke read to the Geological Society of London [1] some observations on these rocks by Mr. David Scott, then Commissioner of Kúch Behar. In those days the Brahmaputra flowed at the foot of the hills, and at one spot on the left bank Mr. Scott found some fossils. The exact locality is not given, but it was somewhere between the Kálu and Mahendraganj (or Karibári), probably nearer the latter place. The position in the series is not known, but it seems certain that the bed belonged to the rocks of the hills, not to the older alluvium, for it is described as at the foot of a

[1] *Geol. Trans.*, series ii, I, 132, (1824).

small hill, rising about twenty feet over the general elevation of the plain. The fossils were a strange mixture of marine with land and fresh water forms, and amongst them Mr. Pentland described[1] the teeth of *Anthracotherium silistrense*, a species that has since been found in the Manchhar (upper miocene) beds of Sind.[2]

The change of character that is so marked in the cretaceous and nummulitic deposits from west to east has not been observed in the upper tertiaries; there certainly can be no striking feature of this kind. A massive soft greenish sandstone is the most prominent rock. It is somewhat like the common Siwálik rock, but more earthy and of darker hue. The associated beds are mostly grey shales, unlike the brown and ochrey clays of the sub-Himálayan series, and the fossils show that even at the western end of the range the deposits are, in part at least, marine. In Mr. Scott's notes of the section on the Brahmaputra, local contortions are noticed, and this condition increases eastwards. On the Sumesari, nearly sixty miles further east, where the tertiary zone is fourteen miles wide, the state of disturbance is still only partial.[3] At the southern edge of the hills the dip is 40° to the south, in very new looking strata, there is then a broad band in which the beds are flatly undulating, after which they rise again with a steady and increasing southerly dip.

Nothing like a general unconformity in the tertiary series was noted in the section of the Sumesari.

Further east lies the ground described by Colonel Godwin-Austen on the confines of the Gáro and Khási areas, where the upper tertiary rocks have been almost denuded away from the base of the range, the little that is left of them being nearly vertical. Here, too, is the transition ground of the prevailing marine type of deposits in both the cretaceous and nummulitic strata. Here a partial unconformity was noticed between these formations and this is the only position in the western part of the range where the upper tertiaries have been found on the plateau inside the line of disturbance. The summit of Nongkulang hill (2,070 feet) is formed of rusty sandstones and shales, resting horizontally upon the undisturbed nummulitic limestone. In a collection of fossils from these beds, sent by Colonel Godwin-Austen, Dr. Stoliczka found the genera *Conus*, *Dolium*, *Dentalium*, *Cardita*, *Cardium*, *Tellina*, *Nucula*, *Leda*, *Cucullæa*, and several others, and he remarked that none of the species, so far as recognisable, appeared to be identical with those known from the nummulitic beds of the same district. This fact suggests that these detached beds on the heights may be an overlap of some beds of the series higher than those that seem to be in transitional sequence with the

[1] *Geol. Trans.*, series ii, II, 393, (1829).
[2] *Records*, X, 77, (187;).
[3] *Memoirs*, VII, 193, (1869).

nummulitics in the sections to the west. This view is strengthened by the fact that Colonel Godwin-Austen observed some cases of local denudation unconformity between these fossiliferous sandstones and the nummulitic limestone, the strata being still quite parallel.[1]

The exposure of upper tertiaries south of the Gáro hills narrows in an easterly direction, and from about twenty miles east of the Sumesari river the disturbed upper tertiaries have been removed by denudation, along the foot of the Khási and Jaintia hills. They come in again where the strike of the disturbance, which marks the southern margin of the plateau of the Assam range, turns north-easterly, and occupy a large area in the hilly country between Assam and Burma, and to the south of the Cachar valley.

The observations in this country are scanty and scattered. In the Kasom range, on the eastern limit of Manipur territory, they consist of sandstones with but few argillaceous bands, containing numerous carbonised fossil tree trunks, and nests of coaly matter, but no coal seams were observed. Further south, however, well marked coal seams have been found in what appear to be the same beds on the west side of the Chindwin river. Fossil resin was found in the Manipur hills, and it is probable that the amber mines of Upper Burma are excavated in beds of this age.[2]

Further north marine fossils, among which is an undetermined species of *Venus*, were found near Sámaguting in beds which probably represent the lower group of the Gáro hills.

In upper Assam, the upper tertiary beds were described by Mr. Mallet as the Tipam and Dihing groups. The greenish grey, pepper and salt, sandstone of the Tipam range undoubtedly alternates with the top beds of the coal measures, but the shaly beds rapidly cease and the sandstone becomes very massive. Some coaly partings were also observed well up in the Tipam group, and fossil wood, whether silicified or semicarbonised, sometimes in very large blocks, is common throughout this great sandstone formation, as well as in the upper conglomeratic beds.

The upper (Dihing) group is less like the corresponding group of the Siwáliks than is the Tipam sandstone, and two points are noteworthy in it. Some of the bottom beds are coal conglomerates, made up almost exclusively of rolled fragments of coal, presumably of the coal seams underlying the Tipam group. This peculiar conglomerate has been observed, at great distances apart, at the top of the Tipam sandstone along the main fault. The ordinary conglomerates of the group are composed of well rolled pebbles of fine hard sandstone, identical in appearance with the Disang sandstone. The beds associated with these

[1] *Jour. As. Soc. Beng.*, XXXVIII. pt. ii, 14. 16, (1869).

[2] *Memoirs*, XIX, 226, (1883).

conglomerates are blue sandy clays, not like the brown and red clays of the upper Siwáliks, at least of north-western India.

The distribution of these upper tertiary sandstones presents some peculiarities which require notice. At the eastern end of the Assam range they come in where the zone of disturbance which marks its southern face diminishes in intensity and at the same time turns to the north-east. From here they extend eastwards into the Patkoi range, but do not extend across the hill country of Manipur. In the parallel of Manipur city there is a stretch of about 100 miles broad of pretertiary beds, before the upper tertiaries are again found capping the range which looks down on the Chindwin valley. To the north of the valley of Manipur the upper tertiaries are found capping the higher ranges, and further north, in the Angámi Nágá hills, there is only some ten or twelve miles between the boundary of the sandstones forming the high peak of Japvo and the margin of the eastern area of upper tertiaries. No details are known of the geology of the hills to the north-east of this, but the pre-tertiary slates and sandstones of Manipur and the Angámi Nágá country, appear to be represented by the Disang group in the coal fields of upper Assam. Whether there is a continuous outcrop of pretertiary slates in the intervening ground is not known, but in the Singpho country further east, the pretertiary beds are completely cut out by the tertiaries, which extend continuously from the plain to the crest of the range.

In a southerly direction from Manipur much the same appears to take place, for east of Chittagong Mr. La Touche found nothing but tertiary rocks on the route across the hills. It would seem, therefore, that the Manipur hills are an area of special elevation from which the newer beds have been removed by denudation.

In the last paragraph it has been assumed that these deposits originally extended over the area where they are now wanting. There is no direct proof that they ever did so, and the general resemblance in type to the Siwáliks along the foot of the Himálayas might suggest that they were deposited under similar circumstances, and that their original was not very different from their present extension, but in spite of a certain lithological similarity to the Himálayan Siwáliks, there is a much greater uniformity on different sections than they exhibit. The present boundaries, too, are in all cases evidently due to denudation, and there can be little doubt that they once extended right across the hills which separate Assam from Burma, at any rate north of Manipur. The presence of marine fossils in these sandstones shows that they were deposited under different conditions to the Himálayan Siwáliks, which we will see were formed subaerially by streams, under circumstances closely resembling those that now prevail along the foot of the Himálayan range.

Upper tertiary beds occupy a large area in Upper Burma, both in the valleys of the Irawadi and Chindwin rivers, and in the hills to the west of them. Little is known of this area, as no detailed geological survey has yet been practicable. Coal seams are found interbedded with the sandstones west of the Chindwin river, and of the Irawadi above Mandalay, as also in the small outlying patches of upper tertiaries in the Shan hills. Further south the petroleum of Burma is derived from upper tertiary beds, though there are also some occurrences of very minor importance in the older tertiary rocks of Pegu.

It is not till we reach the province of Pegu that anything approaching a detailed account of the tertiary deposits is available, where they were divided by Mr. Theobald into three main groups, supposed to correspond more or less to the eocene, miocene, and pliocene of the European classification, but it will be best to retain the local nomenclature, as the fossil evidence is by no means sufficient to establish their complete equivalence. The three main tertiary groups are as follows:[1]—

3. Fossil wood group	Sand gravels, etc., with silicified wood and mammalian bones.
2. Pegu group	Shales and sandstones, occasionally calcareous; fossils numerous.
1. Nummulitic	Shales and sandstone, with some limestone bands containing nummulites, etc.

There is but little to distinguish the nummulitic beds of Pegu from the Ma-í and Negrais rocks, beyond the much smaller amount of alteration that they have undergone, the more frequent appearance of fossils, and the occasional occurrence of limestone containing nummulites, especially in the higher part of the group. The ordinary beds are sandstones and shales, unaltered but frequently hard and compact. The distinction from the Negrais rocks is far from absolute. The tendency to a passage between the two, at the foot of the hills, has already been noticed, and there are in places, within the nummulitic area, hills formed of hardened masses, perhaps older than the rocks around, but which have much the appearance of being the same beds, slightly altered.

The main outcrop of the nummulitic rocks extends from north to south throughout the province of Pegu, between the Arakan hills and the Irawadi river. The beds have a general dip to the eastward, but to the southward it is difficult, if not impossible, to define the base of the formation, on account of the apparent passage from the nummulitic into the Negrais rocks. To the northward, west of Thayetmyo, near the former boundary of British territory, the section is better defined.[2] In the Hlwa (Lhowa) stream,

[1] *Memoirs*, X, 227, (1873).

[2] The beds formerly classed as triassic, on the strength of a mistaken identification of a fossil (*supra*, p. 145) must be included with the nummulitics. The information necessitating this change was not received in time to be incorporated in the text or in the geological map.

sixteen miles west by south of Thayetmyo, upwards of 4,000 feet of hard sandstones, mostly grey, and of blue, grey, or yellow shales, are exposed, but throughout all this thickness of beds no fossil remains have been detected, except a few carbonaceous markings. Apparently, at a somewhat higher horizon on the Ma-tun stream, which joins the Hlwa from the north, there is a great thickness of massive blue shales, of rather a dark indigo blue in general, but sometimes of lighter colour. These shales cannot be much less than 3,000 feet in thickness, but they are almost as unfossiliferous as the sandstones and shales on the Hlwa, the only organic remains found being some cycloid fish scales. Above these there is again a great thickness of sandstones and shales, mostly unfossiliferous, but containing a few layers with nummulites, and at the top of the whole group is a band of nummulitic limestone, from 10 to 100 feet thick. This limestone, however, is by no means continuous. Where it occurs, it seems to be the uppermost band of the group, but it frequently appears to thin out, and in fact to consist of irregular lenticular bands in shale, rather than of an unbroken bed. Denudation may, perhaps, have removed the limestone in places before the deposition of the next group. Other bands of limestone occur at a lower horizon, but they are more irregular than that at the top of the group.

The whole thickness of the formation must be considerable—probably not less than 10,000 feet, but no estimate of any value can be made, on account of the imperfect manner in which the rocks are seen. In northern Pegu, west of Thayetmyo, the breadth of the eocene outcrop from east to west is seventeen miles, but, a few miles to the south, the width diminishes, till, west of Prome, it is not more than six. The belt again expands in breadth near Akauktaung, on the Irawadi above Myanaung, but the beds are very poorly exposed in general, being covered with gravel and other later deposits. Farther to the southward, west of Myanaung and Henzada, the nummulitic rocks are much concealed by posttertiary gravels and from Henzada to Bassein the only rocks seen west of the Irawadi plain are the altered Negrais beds. The nummulitic strata re-appear west of Bassein, and continue thence to Cape Negrais, but still the rocks are much concealed by gravel. Throughout the area, however, limestone with nummulites occasionally appears amongst the higher beds of the group, and a peculiar, very fine, white or greenish, argillaceous sandstone, with *Foraminifera*, seen at Puriam point east of the Bassein river, and in Long island of that river, is also probably one of the uppermost eocene beds. This rock, known as *Andagu-kyauk*, or image stone, is employed by the Burmese for carving into images of Buddha, and is quarried to some extent for that purpose.

It is possible that nummulitic beds may crop out in places amongst the miocene rocks of the Prome district, but the only known exposure of the former in Pegu, apart from the belt just noticed as extending along the eastern side of the Arakan Yoma, is in a small ridge, known as Thon-doung, or lime hill, about five miles south of Thayetmyo. This ridge consists in great part of nummulitic limestone, resting upon shales and sandstones. In 1855 a promising bed of coal, 4 feet thick, was discovered in the latter, but it proved so irregular as to be of no value, the coal thinning out, and passing into a clay with mere laminæ and patches of coaly matter, in the course of a few feet.[1]

To the west of the Arakan range, limestone with nummulites has been noticed near Keantali, and there can be but little doubt that eocene beds extend along the coast for a considerable distance. The islands of Ramrí and Cheduba consist of sandstones and shales closely resembling those of Arakan, and doubtless belonging to the same series.[2] These beds are also very similar to the nummulitic rocks of Pegu. A few seams of coal have been found, resembling in character the nummulitic coal of Assam, and petroleum is obtained in several places. The limestone on the eastern side of Ramrí island, as already mentioned, resembles that of the Arakan coast near Ma-í and Taung-gup, and may, therefore, be cretaceous, but there is no marked character by which the rocks of the island can be divided into two series.

Above the nummulitic formation of Pegu there is an immense thickness of soft shales and sandstones, often fossiliferous, but almost destitute of any horizon distinguished either by mineralogical characters or by organic remains. The base of this group is assumed to coincide with the band of nummulitic limestone already mentioned, but there is no clear evidence that this bed is the uppermost rock of the eocene group, and no unconformity has been detected between the nummulitic rocks and the next strata in ascending order. The upper limit of the middle tertiary rocks of Pegu is equally ill defined, there being a gradual passage from clays and sandstones with marine fossils into the gravels and sands with silicified fossil wood and mammalian bones.

The fact is that without a thorough knowledge of the fossils the classification of rocks so obscure and so ill seen as those of Pegu is a simple impossibility, and until the tertiary molluscs, echinoderms, and corals of southern Asia are better known, it is hopeless to attempt more than a general rough arrangement of the Burmese tertiaries. In the absence of sufficient fossil data for the proper determination of different beds, all that has been attempted at present is to class together all the marine beds of Pegu above the nummulitic limestone, and without nummulites. The group thus constituted has been named the Pegu group from its forming the greater part of the Pegu Yoma between the Irawadi and Sittaung.

[1] T. Oldham, *Sel. Rec. Govt. India*, X, 99, (1856). | [2] F. R. Mallet *Records*, XI, 191, (1878).

There can be no doubt that a portion of this group is of miocene age, and corresponds generally to the Gáj group of Sind but it is probable that representatives of other groups are included.

The only approach to a subdivision of the Pegu group that has been suggested is the separation of a considerable thickness of soft unfossiliferous blue shales, which rest upon the upper nummulitic strata near Prome, and underlie the typical fossiliferous middle tertiary beds. These shales have been called the Sitsyahn shales, from a village on the Irawadi, eight and a half miles above Prome, whilst the overlying sandstones and shales with fossils are distinguished as Prome beds, from their occurrence in the neighbourhood of Prome. The Sitsyahn shales consist of blue clunchy clay with indistinct bedding, and greatly resemble some of the nummulitic shales, except that they are somewhat paler in colour. The thickness of the subdivision is about 800 feet, and the beds have been traced for a considerable distance along the upper limit of the nummulitic rocks in the Prome district.

The Prome beds succeed the Sitsyahn shales conformably, and are composed of grey sandstones, occasionally hard, but frequently argillaceous or shaly, hard yellow sandstones, and shales or clays of various colours. A section of about 2,500 feet of these beds is seen opposite Prome on the right bank of the Irawadi, and probably a much greater thickness exists east of the river. One of the most fossiliferous beds is a band of blue clay exposed at Ka-ma on the Irawadi, eighteen miles above Prome. The position of this band is high, and, above it, a bed, abounding in *Turritellæ*, and a hard sandstone containing corals belonging to the genus *Cladocera*, are the highest rocks of the group, and mark the passage into the fossil wood beds.

It is almost useless to give any palæontological details. *Foraminifera* and *Echinodermata* are rare, and the mollusca are not, as a rule, very characteristic forms. A sessile cirriped, very common in some beds, closely resembles *Balanus sublævis* of the miocene in Sind. A few small crabs occur, and small corals and sharks' teeth are common.

In one locality, Minet-taung (Myay-net-toung), twenty-four miles east-south-east of Thayetmyo, a bedded volcanic rock, consisting of greyish trap, occurs interstratified with the rocks of the Pegu group and, to all appearance, contemporaneous. Nothing has been ascertained as to the source of this igneous formation.

The Pegu group forms nearly the whole of the great range of hills, known as the Pegu Yoma, between the Irawadi and Sittaung, no older rocks being known, with any certainty, to occur in the country between the two rivers. The area occupied by the middle tertiary beds is very broad to the northward, where it extends from considerably west of the Irawadi to the base of the metamorphic hills east of the Sittaung, and contracts gradually between the alluvial plains of the two rivers to the southward, till it terminates in a long, narrow spur at Rangoon. West of the Irawadi, the Pegu group extends to a little below Prome, and some hills on the opposite side of the river below Prome are formed of the same beds. It is, however, not quite certain that no older rocks appear between the Irawadi and Sittaung for a species of *Pseudodiadema*, a genus of echinoderms with cretaceous affinities, has been found in some beds in eastern Prome and a *Terebratula* with a very cretaceous aspect was obtained near the town of Pegu. In the former case the beds appear to be high in the Pegu group, but owing to the great extent to which the surface of the country is concealed, both by gravel and other alluvial deposits, and by forest, it is most difficult to make out the geology satisfactorily, so that lower beds may have been brought up to the surface by faults or otherwise. In the case near Pegu the position of the beds is uncertain.

A small island, known as Kau-ran-gyi on the Arakan coast, is composed of a very pale brown or cream coloured, calcareous sandstone or earthy limestone, containing echinoderms, molluscs, sharks' teeth, and other fossils. The same rock occurs also at Nga-tha-mu on the mainland opposite Kau-ran-gyi island, but has not been detected elsewhere. The most abundant amongst the fossils are a species of *Lobophora* (*Echinodiscus*) and an *Echinolampas*, apparently *E. jacquemontii*, one of the commonest fossils of the Gáj group in Sind. The *Echinodiscus* also closely resembles a Gáj species. The bed is somewhat similar to the miliolite of Káthiáwár, and may represent a portion of the Pegu group, but it is more probably of later date. One of the sharks' teeth, however, closely resembles one found in the Pegu group south of Thayetmyo.

The highest member of the tertiary series in Pegu is distinguished by the abundance of silicified dicotyledonous wood, and is the source whence all the fragments of that substance, so abundant in the older and newer alluvial gravels of the Irawadi, are derived. The fossil-wood group is much coarser than the underlying formations, and consists of sands, gravels, and a few beds of clay or shale, all, as a rule, being soft and incoherent, although occasionally hard sandstone or conglomerate bands occur. The group is thus subdivided :—

a. Fossil-wood sands.—Sand, in part gravelly and conglomeratic, characterised by a profusion of conncretions of iron peroxide.

b. Fine silty clay, with a few small pebbles.

c. Sands, shales, and a few conglomerate beds, with a little concretionary iron peroxide.

The lowest beds, which pass downwards into the marine bands of the Pegu group, contain occasional silicified rolled fragments of wood, and a few mammalian bones. Some sharks' teeth also occur. The thickness of none of the subdivisions has been clearly ascertained, but the lower sands must comprise beds some hundreds of feet thick. The fine silty clay does not exceed about 40 feet in thickness. This bed is quite unfossiliferous, neither fossil wood nor bones having been found in it, and pebbles are rare, though a few occur. It thus forms a marked band in the group, and contrasts with the beds above and below it.

The upper fossil wood sands and gravels are by far the most important members of the formation, and it is from them that the greater portion of the silicified wood is derived. This wood occurs in the form of large and small masses, some being trunks of trees 40 or 50 feet long; usually, however, such masses display marks of attrition, as if the tree stems had been transported to a distance and rolled, before being silicified. The wood is always, or nearly always, exogenous, a few rolled fragments of endogenous wood, found in newer formations, being, nevertheless, probably derived from the present group. The wood is not coniferous, but owing to the very considerable amount of decomposition it had undergone previous to silicification, its nature is difficult to determine. Besides the fossil wood another characteristic of this portion of the group is the abundance of concretionary nodules of hydrated iron peroxide, which are in places so numerous as to have furnished a supply of iron ore for the native furnaces. Mammalian bones are of only local occurrence.

The following is a list of the *Vertebrata*, exclusive of sharks' teeth hitherto obtained in the Irawadi valley from the beds of the fossil wood group. Those marked with an asterisk being also found in the Siwaliks of the sub-Himálayas:—

MAMMALIA.

Ursus, sp.	*Tapyrus*, sp.
* *Elephas* (*Stegodon*) *cliftii*.	*Equus*, sp.
* *Mastodon latidens*.	*Hippopotamus* (*Hexaprotodon*) *iravadicus*.
* „ *sivalensis*.	* *Merycopotamus dissimilis*.
Rhinoceros iravadicus.	*Cervus*, sp.
* „ *perimensis*.	*Vishnutherium iravadicum*.
R., sp.	*Bos*, sp.

REPTILIA.[1]

Crocodilus, sp.	*Emys*, sp.
Gharialis, sp.	*Trionyx*, sp.
Testudo, sp.	*Emyda*, sp.
* *Colossochelys atlas*.	

The proportion of species identified with Siwálik forms is rather less than in the case of Perim Island, only five out of fourteen species of mammals being regarded as identical, yet the general facies of the two faunas is

[1] These genera are recorded amongst the Ava specimens in the collection of the Asiatic Society (Falconer, Catalogue of the Fossil Remains of Vertebrata, in the Museum of the Asiatic Society of Bengal, Calcutta, 1859, p. 30); but as all the specimens were unlabelled, there is some doubt about the locality.

very similar. Both contain a considerable proportion of living genera unknown in the middle tertiaries of Europe, together with some older forms, and the Irawadi fauna may be regarded as approximately of the same age as the upper Siwáliks, or pliocene if the views as to the relations of the Siwálik fauna advocated in the present work be accepted. Silicified wood abounds in places in some of the Siwálik beds of the Punjab and in the Manchhar beds of Sind, and is occasionally found in the sub-Himálayan Siwáliks though the remains of trees are for the most part carbonised.

Independently of the fact that the rocks supplying the materials from which the beds have been derived east and west of the Bay of Bengal are probably very distinct, there is some doubt as to the conditions under which the Burmese beds were deposited, owing to the frequent occurrence of sharks' teeth, and it has been suggested that the fossil wood group may have been, in part at least marine or estuarine. The silicified wood itself is never bored by xylophagous mollusca (*Teredinidæ* or *Pholadidæ*) and as not only all wood floating on the sea, but all found anywhere in tidal creeks in India at the present day, and even the dead trunks and branches of trees in places flooded by the tide, are riddled by boring molluscs, it is extremely improbable that the wood found in the Burmese pliocene beds can have been immersed for any length of time in salt water, whilst the tree stems can scarcely have been silicified before being imbedded, as they would have been in that case too heavy to be transported. It is true that the beds containing sharks' teeth are not those in which the fossil wood is most abundant, but still some fragments of wood occur with the teeth, and mammalian bones are common. The beds generally are much too coarse for estuarine deposits, and if they are marine it is difficult to understand why no molluscs, echinoderms, or corals occur. On the whole, it is most probable that the fossil wood beds, like the Siwáliks and Manchhars, are fluviatile or subaerial, deposited by streams and rainwash, and that the sharks inhabited rivers, as some species do at the present day.

The fossil wood beds in Pegu are evidently the mere remnants of a formation which once occupied a far more extensive area, the former existence of the beds being shown by the occurrence in abundance of fragments of silicified wood far beyond the present limits of the group. Judging from the occurrence of the larger blocks of fossil wood alone, and neglecting the small fragments in the alluvial gravels, the beds of the present group formerly extended far to the southward of their present limits, probably along the whole eastern side of the Arakan Yoma, and almost certainly as far as Rangoon along the Pegu range. A considerable area in the Sittaung valley, north of Taung-ngu (Tonghoo), is also occupied by the fossil wood beds, but no traces of the former existence of this group is found south of the Kabaung stream, which joins the Sittaung from the westward a little below Taung-ngu. There is rather more clay associated with the pliocene beds in the Sittaung than in the Irawadi valley whilst in two small outlying patches, east and north of Taung-ngu, the group is represented by a form of laterite containing numerous pebbles.

In Upper Burma the same beds are very extensively developed. They occupy large areas both east and west of the Irawadi, and between the Irawadi and the Chindwin above their confluence, as well as west of the latter river. The details of their stratigraphy and distribution have not yet been worked out, and there is no published information sufficiently detailed or connected to be incorporated in this work. The petroleum of Upper Burma is derived from beds which appear to lie at the base of the fossil wood group, or the upper limit of the Pegu group.

North of where the Irawadi crosses from the eastern to the western side of the depression between the hills of western China and the Shan states on the one hand, and of Manipur and the Arakan Yoma on the other, there are a number of intrusions and volcanic outbursts in the tertiary beds where exact age has not been determined, but appears to range from upper tertiary to pleistocene.

Although nothing definite is known as to the age of the beds associated with coal in Tenasserim, except that they are in all probability tertiary, there is more likelihood that they belong to the newer tertiaries than to the older. They consist of conglomerates, sandstones, soft shales, and beds of coal. The conglomerates are never coarse, the pebbles seldom exceeding a few inches in diameter; the sandstones are fine, gritty, and pebbly, clean white quartzose sands, or earthy and of a yellowish tint; and the shale beds are of a bluish green or blackish tint, very regularly disposed in thin laminæ. The coal is also in thin laminæ, with earthy bands.

These coal bearing deposits, whose total thickness nowhere exceeds 900 to 1,000 feet, are never traceable continuously over any extended area. They are found occupying isolated and detached basins in the great north and south valley of the Tenasserim river, between the main dividing range separating British Burma from Siam to the eastward, and the outer ridges near the sea coast to the westward. The small tracts of tertiary rocks are in all probability of fresh water origin, and have much the appearance of having been deposited in the small basins they now occupy. The only organic remains found are dicotyledonous leaves and scales and bones of fish.[1]

In the Andaman and Nicobar islands the upper tertiaries are represented by soft limestones formed of coral and shell sand, soft calcareous

[1] T. Oldham, *Sel. Rec. Govt. India*, X, 34, (1856).

sandstones and white clays with some bands containing pebbles of coral.[1] They form the whole of the archipelago east of the great Andaman, whence they have been called the Archipelago series; they are also developed in the great Andaman, more especially in the northern islands and on the west coast, as well as in the Nicobar islands. The only fossils that have been found are *Polycystinæ*, which were described by Ehrenberg from the clays of Nancowry harbour.[2]

No contact section has yet been observed, showing their relations to the rocks, sandstones and shales, which are regarded as the equivalents of the Negrais rocks of the Arakan Yoma, but the generally newer appearance of the rocks, their less disturbed condition, and the fact that pebbles of serpentine have been found in beds which there seems good reason for referring to this series, leaves little room for doubt that they are newer, and not, as has been suggested, merely a lithological variety indicating different conditions of deposition.

[1] For details, see H. Rink, Die Nikobarischen Inseln., Copenhagen, 1847, and *Sel. Rec. Govt. India*, LXXVII, 109—154, (1870); F. von Hochstetter, Reise der Novara, II, 83—112, 1864), and *Records*, II, 59—73, (1869); V. Ball, *Jour. As. Soc. Beng.*, XXXIX, pt. 2, 25, 231, (1870); R. D. Oldham, *Records*, XVIII, 135, (1885).

[2] *Abhandl. K. Akad. Wiss, Berlin*, 1875, p. 116.

CHAPTER XIV.

TERTIARIES OF THE HIMÁLAYAS.

(*Including the N.-W. Punjab.*)

Tertiaries of the central Himálayas—Tertiaries of the outer Himálayas—Sirmur series—Siwálik series—Homotaxis of the Siwáliks—Relations of Siwálik and recent faunas.

The description of the tertiary deposits of the Himálayan range was excluded from the last chapter, as their bearing on the question of the age of the Himálayas as a mountain chain, and certain peculiarities which they exhibit, resulting from their mode of origin, render it more convenient to consider them separately. They are exposed in two distinct areas, and may be distinguished, according to their geographical position, as the tertiaries of the central, and of the outer Himálayas. The latter of these are much the most important, whether on account of the superficial area they cover, or the interest of the sections they exhibit, but it will be most convenient to take up the description of the central Himálayan tertiaries first and then pass on to the tertiaries of the outer or southern edge of the range.

The existence of tertiary rocks in the central Himálayas has long been known, but even now the only information available is derived from observations made on rapid journeys through an elevated and inhospitable region, where the rarefaction of the atmosphere offers a serious impediment to physical exertion.

The best known area is that of the upper Indus valley, where the tertiaries extend, for a distance of two hundred miles, in a south-easterly direction from Kargil in Kashmír territory. Along the whole of the north-eastern boundary, from Kargil to beyond Leh, if not to the extreme limit of Kashmír territory, they rest in unconformable contact with the metamorphic rocks. Between Khalsi and Leh the lowest beds consist of coarse grained, sharp, felspathic sandstones, containing a large proportion of grains of undecomposed felspar, and including numerous large boulders of syenite and angular blocks of an intensely hard hornstone porphyry, whose original source is unknown. These beds have been looked upon, with some degree of reason, as probably of glacial origin.[1] They occur at the base of the nummulitics and no break or unconformity has been detected; they are however overlaid by black carbonaceous shaly beds very like those of the carboniferous of Kashmír, and as no similar rock has been detected in the western portion of the outcrop, where the original contact with the pretertiary rocks is exhibited, it is possible that the glacial beds belong to the carboniferous period, and that there is an undetected break between them and the overlying unmistakeable nummulitics.

The conglomeratic beds are succeeded by orange and brown sandstones, often calcareous, which form the lowest member of the series further to the north-west. *Melania*, and a bivalve shell, which is probably a *Unio*, though it has been referred to *Pholadomya* or *Panopæa*, have been found in the neighbourhood of Kargil, marking the beds as fresh water or estuarine in origin.

The sandstones are succeeded by green and purple or dark red shales, and these are overlaid, between Khalsi and Nurla, by a thick band of coarse, blue, shelly limestone containing numerous discs, which are probably ill preserved nummulites. Above this comes a coarse limestone conglomerate containing pebbles of the same limestone succeeded by shales and slates, generally of a grey colour.

In the sections eastwards of Leh conglomerates are said to occur near the upper limit of the series, and these conglomerates contain pebbles of the volcanic beds, which will presently be described, and of nummulitic limestone. The occurrence of these last shows that the beds had locally been elevated and exposed to denudation, while elsewhere the process of deposition had gone on continuously.

In the central portion of the exposure the sedimentary beds are in direct contact with the older rocks along their south-western margin, but at either extremity they are separated by a great series of volcanic rocks of a very basic type. There can be no doubt that these rocks, which form the upper limit of the tertiary system of this region, are in the main contemporaneous eruptive products, as they include beds of volcanic ash and agglomerate,[2] but there are also numerous intrusive masses associated with the bedded traps. Basic trappean intrusions are also found in the pretertiary rocks south-west of the boundary, which are evidently connected with these same eruptive rocks. These intrusions are interesting as, at Pugha and in the Markha valley south of Leh, they are composed of peridotite, until lately the only recorded instances of ultrabasic rocks having been found in India.[3]

On the north-eastern and north-western boundaries the tertiaries rest

[1] R. Lydekker, *Memoirs*, XXII, 104, (1883); R. D. Oldham, *Records*, XXI, 155, (1889).

[2] C. A. McMahon, *Records*, XIX, 118, (1886); R. D. Oldham, *Records*, XXI, 154, (1888).

[3] C. A. McMahon, *Records*, XIX, 115, (1886).

on an eroded surface of metamorphics, showing that the present boundary marks an original limit of deposition in these directions. The south-western boundary, on the other hand, is marked by great disturbance and the tertiaries certainly extended some distance beyond their present limit.[1] Clear evidence of this original extension is to be found not merely in the nature of the boundary but in the existence of an outlier of nummulitic limestone, originally discovered by Dr. Thomson in 1852, on the Singhe Lá. The correctness of the observation was at one time questioned, but was completely confirmed in 1888 by Mr. La Touche,[2] who described the rock as a black fœtid limestone, full of nummulites, resting directly on the palæozoic quartzites, without any intervening deposits of littoral type. The occurrence of an open sea formation in this outlier points to a southerly extension of the eocene sea, and it would be interesting to know whether there was direct communication with that in which the nummulitics of the Punjab were deposited. No outliers of tertiary beds are known in the country intervening between the two principal exposures, but this country has not been examined in any detail, and has undergone such extensive denudation that, even if the nummulitics once extended over it, they may well have been completely removed or only represented by small patches, folded up with the older rocks, which would escape notice in any but the most detailed examination. The considerable lithological resemblance between some of the beds of the tertiaries of the upper Indus valley and those of the Subáthu and Dagshái groups has been noticed by more than one observer,[3] but this would not in itself be proof of former continuity, while the general shallow water type of the Subáthu group, and its complete overlap by the Dagshái group on the inlier of pretertiary limestone which lies north of that at Riási, indicate a northern limit of deposition of the tertiaries on the southern face of the Himálayas.

The fossils do not help us in any way, for, with the exception of those already mentioned, the only recorded fossils are *Nummulites ramondi* and *N. exponens*, from the Markha valley;[4] the former species has also been found in the outlier of the Singhe Lá,[5] and is known from the Kirthar and Ránîkot groups of Sind. Besides these, some obscure remains were obtained near Khalsi, which have been supposed to be *Hippurites* and a cephalopod allied to *Hamites*, but the fossils are altogether too ill preserved for determination.

Besides the outlier already mentioned, there is one composed of basic traps forming the peaks known as D 24 and D 25 in Zanskar. It is not

[1] R. D. Oldham, *Records*, XXI, 156, (1888).

[2] *Records*, XXI, 160, (1888).

[3] F. Stoliczka, *Memoirs*, V, 343, (1865); R. Lydekker, *Memoirs*, XXII, 118, (1883).

[4] F. Stoliczka, *Memoirs*, V, 344, (1865).

[5] D'Archiac et Haime: Groupe Nummulitique de l'Inde, Paris, 1853, p. 176.

known whether these are bedded or intrusive. If the former, they are a true outlier of the Indus valley eocene volcanics; if the latter they are doubtless an old volcanic core, and indicate a former extension of the volcanic beds beyond their present limits. Some small patches of sandstone and conglomerate, which have been regarded as tertiary,[1] are also found in the Chang-cheng-mo valley and near Drás, associated in the latter locality with basic traps. The information regarding them is scanty, and they do not need detailed notice here.

Further to the east, in Hundes, the tertiaries are described by Mr. Griesbach[2] as composed of highly altered rocks, schists, phyllites and crystalline limestones, in which some distorted sections of *Nummulites* were observed associated with intrusive diorite. Above these rocks come pepper and salt grey sandstones, very like some of the lower Siwálik sandstones, whose relation to the nummuliferous beds was not observed, but is said to be probably one of unconformity. Both these are highly disturbed, dipping to the north-east, and are unconformably covered by the horizontal deposits of the Hundes plain, once regarded as upper tertiary, but now shown to be of pleistocene age.

The only igneous rocks in this section appear to be intrusive diorites, but a large development of basic traps was observed by Gen. R. Strachey further to the north, in the neighbourhood of the great lakes of Tibet, which probably represents the volcanics of the Indus valley tertiaries.

Nothing is known of the geology of the whole northern face of the Himálayas east of the Hundes plain except for one spot north of Sikkim. In the neighbourhood of the Cholamo lakes Sir J. Hooker observed conglomerates, slates and earthly red clays, and a compact blue limestone "full of encrinitic fossils and probably nummulites" which may be a continuation of the central Himálayan tertiaries.[3]

The tertiaries of the outer Himálayas are found in a narrow zone of upper tertiary rocks, extending the whole length of the Himálayas, and continuous at the surface, so far as is known, except for a stretch of about fifty miles at the foot of the Bhután hills, where they are covered by a great accumulation of recent deposits. West of the Ganges the tertiary area begins to widen out, and lower beds come in, till in the extreme north-west, beyond the Jehlam, there is a very complete representation of the tertiary sequence. It is only that portion of the tertiary area which lies beyond

[1] F. Stoliczka, *Records*, VII, 15, (1874); R. Lydekker, *Memoirs*, XXII, 113, 115, (1883).

[2] *Memoirs*, XXIII, 83, (1891).

[3] Himalayan Journals, London, 1855, II, 156, 177.

the western frontier of Nepál that has been examined in any detail, and here the best known areas are the hills below Kumáun and Garhwál, the section south of Simla, and the area beyond the Jehlam river. The classifications of the rocks that were adopted by the surveyors of the north-western area and of that south of Simla differ from each other, but the intermediate country has since been examined sufficiently to show the probable equivalence of the rock groups as indicated in the following table :—

North-West Area.		Simla Area.
Upper Siwálik . . . Lower (red and grey) Siwálik	Upper Tertiary . Siwálik series .	Upper Siwálik. Middle Siwálik. Lower (Náhan) Siwálik.
Murree beds . . . . Upper Nummulitic . . .	Lower Tertiary . Sirmur series .	Kasauli group. Dagshái group. Subáthu group.
Lower Nummulitic.		

According to the order of description adopted in this work the lower nummulitic strata of the hills beyond the Jehlam, which do not, strictly speaking, form part of the Himálayas, would stand first for description, but as the reasons for considering them older than the Subáthu group can only be understood after the description of the Sirmur series, the strict chronological order will not be adhered to, and the description of these beds will be left till after that of the series which overlies them. It will be best to begin with the area which, from priority of description, must remain the standard to which the sections of other districts must be referred.

In the hills near Simla, the marine nummulitics, with the upper groups of the lower tertiaries, appear from below more recent deposits a short way west of the Sutlej river, and rising into the high ground of what are known as the lower Himálayas, extend for some 80 miles till they disappear 12 miles east of Náhan, re-appearing as outliers east of the Ganges. Throughout part of this region the Sirmur series, as it has been named,[1] is separated from the upper portion of the tertiary system by a narrow strip of pretertiary rocks, and by the great dislocation, or main boundary, which forms one of the leading features in Himálayan geology. The rocks are everywhere highly disturbed, and structurally belong rather to the Himálayan area proper than to the tertiary area of the sub-Himálayas, but at the north-western extremity of the exposure they run down into the sub-Himálayas, and the marine group becomes covered by more recent deposits in a manner that has not been worked out in detail.

[1] *Manual*, 1st ed., p. 524. The series was first described as the Subáthu group (*Memoirs*, III, pt. ii, p. 74), a name which was subsequently restricted to its lowest member. The description in the text is partly based on Mr. Medlicott's descriptions (*loc. cit.*), and partly on subsequent unpublished observations by the present writer.

The lowest of the three groups, into which the Sirmur series has been divided, is named after the military station of Subáthu, near which it is well exposed. It consists principally of greenish grey and red gypseous shales, with some subordinate lenticular bands of impure limestone and sandstone, the latter principally found near the top of the group. The beds are everywhere highly disturbed, and the boundary with the pretertiary slates and limestones is almost always faulted, but wherever an original contact section is found, in the Jammu hills, at Subáthu, at the termination of the main Sirmur area east of Náhan, and again in the outliers east of the Ganges, there is always a parallelism of stratification between the beds below and above the junction, and the bottom bed of the Subáthu group is a peculiar ferruginous rock, containing pisolitic grains of iron oxide and closely resembling the laterite of the Peninsula, whose occurrence at this horizon in the extra-peninsular area is interesting in connection with the occurrence of laterite in the nummulites of Sind, Jaisalmer, Cutch and Surat. This rock is very well seen at Subáthu itself, where it was first observed and described, and in the shaly beds immediately overlying it there is a seam of impure coal. The coal is too impure and too crushed to be of any economic value, but its occurrence is of interest, as will appear in the sequel.

The Subáthu group is overlaid, with perfect conformity, by a great thickness of hard grey sandstones, interbedded with bright red nodular clays, known as the Dagshái group. The transition from the Subáthu to the Dagshái group, though perfectly conformable, is somewhat abrupt, and marked by the presence of a group of passage beds, comprising a peculiar pisolitic marl with small calcareous concretions scattered through a matrix of red clay, a white sandstone full of irregular shaped highly ferruginous concretions of some inches in diameter, and pure white sandstones associated with dark purple or liver coloured shales, differing markedly in appearance from the general run of those above or below them.

The beds of the Dagshái group proper consist almost exclusively of two distinct types of rock. One is a bright red or purple, homogeneous clay, weathering into small rounded nodular lumps; the other a fine grained hard sandstone of grey or purplish colour. The clays prevail in the lower part of the group and the sandstones, in beds of 10 to 50 feet thick, form but a small proportion of the total thickness, but in the upper portion of the group they increase, at the expense of the clay beds, till at the top there is about 200 or 300 feet of sandstones, with a few thin bands of red clay, which it is impossible to class definitely either with this group or the succeeding one.

As will have appeared from the preceding paragraph, the passage from the Dagshái to the Kasauli group is perfectly transitional, indeed the distinction of the two merely depends on the absence of the bright red nodular clays of the Dagshái group. The Kasauli group is essentially a sandstone formation in which the argillaceous beds are quite subordinate in amount. The sandstones are mostly of grey or greenish colour, and though some of the beds are as hard as anything in the Dagshái group, they are, as a rule, softer, coarser, more micaceous, and at times distinctly felspathic. The clay bands are gritty, micaceous, and but seldom shaly; in the lower part of the group they often have a remarkably trappoid appearance, owing to their dull green colour and mode of weathering, first into rounded masses and afterwards into small angular fragments.

At the upper limit of the Kasauli group some reddish clay bands are seen on the cart road to Simla. These clay bands are softer and paler than those of the Dagshái group, and resemble the clays of the lower portion of the upper tertiaries near Kálka. This, the only trace of a connection between the Sirmur series and the upper tertiaries of this region, will be referred to further on.

The Subáthu group is most palpably of marine origin and of nummulitic age, as is shown by the numerous fossils it contains. The Dagshái group has yielded no fossils, except some fucoid markings and annelid tracks, which are of no use for determining either the age or mode of origin of the beds, but the great contrast of lithological character suggests a corresponding change of conditions of formation, and it is probable that they were deposited either in lagoons or salt water lakes cut off from the sea, or were of subaerial origin. The Kasauli group has so far yielded no fossils but plant remains, and this, taken in conjunction with its general similarity to the upper tertiary deposits, renders it probable that it is composed of fresh water, if not subaerial deposits.

A short distance west of the Sutlej river the Subáthu group become covered up by the newer beds, and is not again seen till the Jammu hills are reached. Here there are some inliers of marine nummulitics, but the most interesting exposures are those in which they rest on the pretertiary limestones of Riási and the Punch valley.[1] In these the Subáthu beds rest, with perfect parallelism of stratification and every appearance of conformity, on the older rocks, and at their base is found the same peculiar pisolitic ferruginous bed as is seen at Subáthu. Separated from this by about 70 feet of shales, there is a coal seam of 2 to 5 feet thick, overlaid by some 350 feet of shales, with a couple of thin bands of nummulitic limestone,[2] the group being conformably overlaid by the sandstones and red clays similar to those of the Dagshái group.

On the Riási inlier the pisolitic bottom bed of the nummulitics is

[1] H. B. Medlicott, *Records*, IX, 53, (1876); R. Lydekker, *Memoirs*, XXII, 90, (1883).

[2] T. D. La Touche, *Records*, XXI, 62, (1888).

underlaid everywhere by a silicious breccia of variable thickness, composed of perfectly angular silicious fragments, cemented in places by cellular limonite. The true age or nature of this rock is uncertain. Mr. Medlicott thought it was a shattered condition of a sandstone band which often occurs at the top of the pretertiary limestone series, and in any case the angular nature of the fragments forbids the supposition that they have been transported for any great distance.

So far there has been no difficulty in recognising the equivalent of the Subáthu group, but in the sections west of the Jehlam there is a very great thickness of marine nummulitics, which doubtless in part represent the Subáthu group, but probably cover a larger period of time.

In the Salt range the principal member of the marine nummulitics is a band of fine, compact, grey or white, limestone, frequently cherty, of some 400 or 500 feet in thickness, which is unconformably overlaid by upper tertiary beds. Below the limestone there is from 50 to 100 feet of soft variegated shales or clays, with one or more coal seams. The clays are pyritous and decompose readily on exposure, the decomposed shales being burnt and employed in the manufacture of alum. These shaly beds contain a number of fossils, but the collections have not yet been examined in detail. *Cardita beaumonti* is, however, known to occur, and marks the age of the beds as lower eocene at latest.

North of the Salt range, in the hills of the Hazára district,[1] the oldest member of the tertiary system is a great thickness of dark bluish grey, or blackish, limestone, with brownish olive shales. The rock is generally fœtid and massive, sometimes distinctly, and sometimes obscurely, stratified. The distribution of this hill type of nummulitic limestone, as it has been called, is peculiar. It forms a broad belt throughout the Hazára and Murree hills, from the neighbourhood of Abbottábád past Murree, and along the spurs traversed by the Grand Trunk Road north-west of Ráwalpindi. The same rock forms the greater part of the Chittapahár range, and is continued west of the Indus in the Afrídí hills, which are principally composed of this formation.

The tract so defined lies immediately to the north of a great line of disturbance, accompanied by much faulting and displacement of the strata, which runs westwards from the neighbourhood of Murree. Though less sharply defined, this zone of disturbance corresponds to the great faulted boundary which separates the pretertiary rocks of the Himálayas from the tertiary formations of the sub-Himálayas, and is in the main a line of separation between the pretertiary deposits and the hill type of nummulitic

[1] The account of the tertiary beds of the North-West Punjab is derived from Mr. A. B. Wynne's papers, of which the principal are in *Records*, X, 107, (1879); XII, 114, 208, (1881).

limestone on the one hand, and the newer tertiaries on the other. The demarcation is not absolute, for outliers of the newer beds are found to the north, and inliers of the hill nummulitic limestone to the south of the line.

The most important of these inliers is that of the Khaire Múrut ridge, south-west of Ráwalpindi. Here, as in the hills north and west of Murree, the hill nummulitic limestone is overlaid by an upper nummulitic group, composed of grey, red, and deep purple clays or shales, associated with masses of gypsum, and alternating with thin bands of limestone. The composition of the group varies; sometimes the limestones and at others shales are most developed in the lower part, and at times there are but few calcareous beds. Strong zones of yellowish grey sandstones are found in the western localities, while eastwards hard grey sandstones and purple clays, resembling those of the overlying group, are said to be included. Except in the sandstones, which are unfossiliferous but for a few fucoid markings, marine fossils are abundant, among which *Nummulites*, *Oporculina*, etc., are common. None of the fossils have been critically examined, but the age of the beds is evidently eocene, and probably upper eocene.

There is still some doubt as to the equivalence of the marine nummulitics west of the Jehlam with those to the east, that is to say, it is uncertain whether the thicker series to the west merely represents the same period of time as the thinner series to the east, or a more extended one. Although the fossils of the Jammu inliers have not been examined, the presence of the ferruginous bottom bed and associated coal seam, together with the general similarity of the type of deposit and relations to the Dagshái beds above, leave little room for doubt that the Jammu nummulitics are the equivalent of the typical Subáthu group.

It would be natural also to regard the shaly marine beds, immediately underlying the sandstones and red clays of Dagshai type west of the Jehlam, as of contemporaneous origin with the typical Subáthus, but as the distinction between the two groups is merely due to the cessation of marine conditions of deposit, there is an uncertainty as to whether the change took simultaneously on all the sections, and it is possible that the upper portion of the nummulitics west of the Jehlam were formed contemporaneously with part of the typical Dagshái group, while the hill type of nummulitic limestone may be merely a deep water formation of the same age as the typical Subáthus.

In this connection the Salt range section is of some importance. The presence of coal seams in the shaly beds at the base of the tertiary rocks naturally suggests their equivalence to the Subáthu group, and it is even stated that the pisolitic lateritic bottom bed of the Subáthus is found in the Salt range. There is, however, some doubt as to this last. It is not possible to recognise this bed with certainty in Mr. Wynne's description, and it is not evident whether the only specific statement of identity is based on actual observation.[1]

The mere presence of coal seams is no proof of contemporaneity, and what palæontological evidence is available points to an older age for the Salt range nummulitics than that of the Subáthu group. The presence of *Cardita beaumonti*, a species characteristic of the passage beds between the cretaceous and nummulitic in Sind, in the shales below the main limestone of the Salt range, shows that they cannot well be later than oldest eocene in age. Of the 46 species described by Messrs. D'Archiac and Haime from the nummulitics of the Salt range, 13 are found in the Ránîkot, group of Sind, 14 in the Kirthar, of which 6 are also found in the Ránîkot, while 3 are found in Gáj and Nari beds. From the Subáthu bed 49 species were described, of which not one is also found in the Salt range, a difference of fauna which must be almost entirely due to the Subáthu bed having been deposited in shallow muddy water, while the Salt range species inhabited a deeper and clearer sea. The same cause that led to the distinction between the fauna of the Subáthu and Salt range nummulitics restricted the number of Sind species found in the former, and we find that there are only 10 species in all, of which 2 are Ránîkot, 7 Kirthar and 1 Gáj. The general facies of the fauna is consequently, so far as the more limited evidence allows us to judge, newer than that of the Salt range nummulitics.

The upper portion of the nummulitic series is wanting in the Salt range, where the clear nummulitic limestones are unconformably overlaid by the upper tertiaries. It is not clear how far the hill type of nummulitic limestone to the north may represent the nummulitic limestone of the Salt range. Palæontological evidence is wanting, and though there is a considerable lithological diversity, there is none that could not be accounted for by local variations in the conditions of deposition and by the greater disturbance that the northern beds have undergone.

Whether there is an exact equivalence in time of the two types of nummulitic deposits or no, the existence of marine conditions in the Salt range, at a period anterior to the formation of the typical Subáthu group, deprives us of any compulsion regard to the nummulitic series west of the Jehlam as coeval with that further east, and it is on the whole more natural to adopt Mr. Wynne's original correlation of the upper shaly beds in the western area with the typical Subáthus,[2] and to look on the underlying limestones and associated beds as older, and unrepresented on the Subáthu section.

In the north-west Punjab and in the Jammu hills the eocene beds are conformably overlaid, and pass by interstratification into a great thickness of red and purple clays, with interbedded grey or purplish sandstones,

[1] *Records*, IX, 54, (1876).
[2] *Records*, X, 109, (1877).

precisely similar in lithological composition as in stratigraphical position to the Dagshái and Kasauli groups of the Sirmur area. The ridge on which the hill station of Murree stands, as well as the hills to the south-east of it, are composed of these beds, which have consequently been known as the Murree beds.[1] They have, however, been traced to the Rávi, where they are directly continuous with a band of rocks, originally regarded as representing the Náhan or lower portion of the upper tertiaries, but now recognised as belonging to the Sirmur series. There is thus a direct continuity of outcrop, as well as a similarity of stratigraphical position with regard to the marine nummulitics, which leaves little room for doubting that the Murree beds represent the Dagshái and Kasauli groups of the Simla region. What little possibility of doubt there might remain is removed by the discovery of the palm *Sabal major* at Kasauli, and in the Murree beds in the Jehlam valley,[2] and by the recognition of the Kasauli plant bed in the Rávi valley.[3]

The plant in question ranges from lowest to middle miocene in Europe, and as it is found in India near the top of a series of beds intermediate between the eocene and the lower Siwálik beds, which are regarded as upper miocene on independent grounds, it will be seen that the stratigraphical position of the species is similar in the two areas, and that the Dagshái and Kasauli groups may be regarded as covering the oligocene and lower miocene periods of European geology.

Apart from these fossil plants no organic remains are known with certainty to occur in the Murree beds, any more than in the Dagshái group in its typical area. Bone fragments, crocodilian scutes and exogenous fossil timber are found in certain beds, resting on the nummulitic limestone of the Salt range, which Mr. Wynne regarded as forming part of his Murree group.[4] It is, however, very doubtful whether these beds can be regarded as the equivalents of any portion of the Dagshái or Kasauli groups, or of the Murree group as originally defined, in spite of the very indefinite nature of its upper limit. There is a well marked unconformity between the nummulitic limestone and the beds immediately overlying it, whose importance Mr. Wynne was inclined to minimise, but it certainly represents a considerable lapse of time, as is shown by the occurrence of conglomerates composed of pebbles of the underlying nummulitic limestone at the base of the upper tertiaries, by the complete cutting out of the eocene beds at either end of the Salt range, and by the occurrence, as determined by Mr. Theobald, of *Mastodon latidens* and *Rhinoceros palæindicus* in a fossiliferous zone, about 100 feet above the nummulitic limestone. These two animals mark the age of the beds in which they are found as upper miocene at oldest, and show that a great

[1] A. B. Wynne, *Records*, VII, 66, (1874).
[2] O. Feistmantel, *Records*, XV, 51, (1882).
[3] H. B. Medlicott, *Records*, IX, 52, (1876).
[4] *Records*, X, 119, (1877).

interval of time must have elapsed after the deposition of the middle eocene nummulitic limestone, an interval which would be filled by the oligocene and lower miocene Dagshái and Kasauli groups.

How far the lithological distinction between the Dagshái and Kasauli groups is maintained in the Jammu and Hazára hills is not clearly determinable from the published descriptions, but as the Murree beds are said to pass with perfect transition into that great series of upper tertiary deposits known as the Siwálik series, it is probable that there is a distinction between the lower and upper portion, analogous to that between the two groups in the Simla region.

The name Siwálik, originally applied to the range of hills separating the Dehra Dun from the plains, has been extended by geographers to the fringing hills of the southern foot of the Himálayan range, and applied by geologists to that great system of subaerial river deposits which contains remains of the "*Fauna antiqua sivalensis.*" In spite of local variations of texture, inevitable from their mode of formation, these upper tertiary beds of the Siwálik series maintain a great uniformity of type along the whole length of the Himálayan range.

Lithologically the lower portion of the system is characterised by a great thickness of fine grained grey, micaceous, pepper and salt sandstone, interbedded with clay bands near its lower portion, while the upper part of the system is composed of soft earthy clays, undistinguishable from the alluvium of the plains except by the disturbance they have undergone, and coarse conglomerates of well rounded pebbles and boulders of crystalline and metamorphic rocks derived from the Himálayan ranges.

In the neighbourhood of Náhan this system was originally divided into two members,[1] a lower, to which the name of Náhan was applied, and an upper, to which the name Siwálik was restricted. In this area the boundary between the two groups is a great fault, but there must be a real, if local, unconformity, for the upper Siwálik conglomerates contain numerous pebbles[2] of the Náhan sandstones they are faulted into contact with. The distinction between the Náhan and Siwálik zones appears to be well maintained in a south-easterly direction as far as the borders of Nepál, but to the north-west it disappears, and there appears to have been a continuous series of deposits, ranging from the bottom to the top of the upper tertiary formations. No fossils have yet been found in the typical Náhans, though it would appear that they do occur,[3] but to the north-west representatives of the Siwálik fauna

[1] H. B. Medlicott, *Memoirs*, III, pt. i, pp. 17, 101, (1864).

[2] H. B Medlicott, *Records*, XIV, 172, (1881).

[3] See H. B. Medlicott, *Memoirs*, III, pt. ii, p. 16, (1864); *Records*, XIV, 71, footnote (1864).

occur low down in the series, in beds which very possibly represent the Náhan group as originally defined. Under these circumstances it has been found inadvisable to retain the separation between Náhan and Siwálik, and the former are now classed as lower Siwálik, though the term may be retained as a useful local designation for a particular type of formation.

The Náhan group is composed of alternating beds of a fine grained, usually grey, firm sandstone, and of clays, usually bright red in colour and almost always some shade of red or purple, which weather in a nodular manner. The clays usually prevail in the lower part of the group and the sandstones in the upper.

The lithology of this group resembles very closely that of the Dagshái group, and one might be tempted to regard them as equivalent to each other. The equivalence cannot be absolutely disproved till the area west of the termination of the typical lower Himálayas, in the Kángra valley and the Jammu hills, has been examined in greater detail than has yet been done, but in the meanwhile there are good reasons for supposing that the lithological similarity between the two groups is due to a similarity in their condition of deposition and does not mean contemporancity of origin. In the first place the two groups are found in distinct areas, separated by a marked structural feature, exhibiting itself at the present day as a fault of many thousand feet throw. As will be shown in a subsequent chapter, this fault—commonly known as the main boundary—is connected in a peculiar manner with the elevation of the Himálayas, and it is highly improbable that the beds exposed south of it are of the same age as those found to the north. Another argument depends on the fact that no exposure of the Subáthu group has been found even in the deepest cut sections of the typical Náhan group, and a third may be derived from the smaller degree of induration, indicating, though not proving, a younger age. In the country north of Náhan town, where the Náhan and Dagshai groups are brought into contact with each other, on opposite sides of the main boundary fault, the sandstones of the former always weather into soft rounded lumps, while the Dagshái sandstones weather into angular fragments, which have lost the sharpness of their angles, but exhibit a much less degree of weathering than that to which the Náhan beds have undergone. Finally the red clay beds which have already been mentioned as occurring at the top of the Kasauli group, though they differ somewhat from the typical Náhan clays, resemble them sufficiently to point to a return of the conditions of deposition which prevailed in the Dagshái and Náhan periods, and suggest that on an unbroken section the Náhan would be found to overlie the Kasauli group.

No fossils have been described from the typical Náhans. It is possible that some of the lower Siwálik fossils found in the north-west Punjab may have been derived from beds of the same age, but the supposition lacks proof. There seems however to be little room for doubt that Sir Proby Cautley did find fossils on the northern side of the hill on which the town of Náhan stands, and consequently in the beds of the Náhan group, but the specimens were lost before they had been examined by a palæontologist.

The Náhan group is succeeded, on those sections where the sequence is complete, by an immense thickness of soft sandstones, generally coarser in grain and more micaceous, mostly of a pepper and salt grey colour, with some interbedded bands of earthy clay, occasionally slightly tinged with red, but never assuming the bright red colours of the Náhan clays. The argillaceous beds are, for the most part, confined to the lower part of the group, the middle part being usually composed of some thousands of feet of sandstones, without any intercalation of shale on the one hand, or any included pebble on the other. In the upper part of the group strings of pebbles occur among the sandstones, which become more numerous till bands of conglomerate appear and increase in abundance and coarseness.

The uppermost group of all varies very much in character. Near the large rivers draining from the central Himálayas, it consists principally of coarse conglomerate, composed of rounded boulders of the harder rocks of the Himálayan chain. In the intermediate stretches of ground it is composed largely of soft earthy beds precisely similar to those of the modern alluvium of the plains.

The details of this variation in lithology of the upper Siwálik beds, as well as the structural features of the Siwálik zone, are of great interest, but their principal interest lies in their bearing on the age and elevation of the Himálayan chain, and they will consequently be considered in the chapter devoted to that question. At present it will be sufficient to point out the conditions under which the Siwálik series was deposited. The earlier observers regarded this great series of beds as having been deposited in a sea, a supposition which is sufficiently disproved by the complete absence of any marine organisms, and by the occurrence of the remains of fresh water molluscs, fishes and tortoises. It is hardly possible that they could have been deposited in a fresh water lake, for it is not conceivable that a fresh water lake extending the whole length of the Himálayas could have existed. Moreover, the fresh water organisms whose remains have been found are all such as inhabit streams, and not lakes. But the most conclusive proof of all lies in the evident unity of the whole Siwálik series, pointing to the whole of it, with the possible exception of the Náhan group, having been formed under very similar conditions, while the very close resemblance between the upper Siwálik beds and the recent deposits of the Gangetic plain leaves little room for doubt that the Siwálik beds were deposited subaerially by streams and rivers.

The thickness attained by the Siwálik series is immense. Mr. Wynne estimated it at 14,000 feet in the north-west Punjab. In the Siwálik hills there are at least 15,000 feet of beds, and the series is by no means complete, and similar vast thicknesses may be measured on any section.

The few mollusca which have been found in the upper Siwáliks belong solely to fresh water or terrestrial forms, and the first comparison that was made[1] was carried out under circumstances so unfavourable, with so poor a collection of recent species from India, and at a time when the latter were so imperfectly known, that but little weight can be attached to the conclusions formed. The majority of the specimens obtained are in poor preservation, but all the forms collected from upper or middle Siwálik beds, since the recent fresh water shells have been better known, have proved to be either identical with living species, or closely allied to them. Amongst those hitherto identified, the only land shell is *Bulimus insularis*,[2] a species which ranges at the present day from Africa to Burma, whilst amongst fresh water molluscs, the two common Indian river snails *Paludina bengalensis* and *P. dissimilis* have been recognised, and forms of *Melania*, *Ampullaria* and *Unio* also occur.[3]

So far as the evidence extends, therefore, the few mollusca of the Siwáliks tend to show that the beds must be of upper and middle tertiary date. But the evidence afforded by the mollusca is imperfect, and both closer comparison and a larger series of fossil specimens are desirable before any very positive assertions can be made as to the antiquity of the Siwálik series, on the data afforded by the invertebrata. In investigating the question of age, we are consequently forced to depend, first upon the vertebrata, and especially the mammalia, and secondly upon such geological evidence of connection with other formations of known age as the rocks afford.

The first question, then, is the homotaxial relation of the mammalian

[1] E. Forbes, in Falconer's Palæontological Memoirs, London, 1886, I, p. 389.

[2] Theobald, MS.; Geoffrey Nevill, *Records*, XV, 106, (1882). As in this and the followng paragraph Mr. Theobald's view of the affinities of the Siwálik mollusca has been accepted in preference to the high authority of Prof. E. Forbes, it is only just to say that Mr. Theobald has a far more extensive knowledge of living Indian fresh water shells than it was possible for any naturalist in Europe to acquire at the period when Prof. E. Forbes' note was written. Indeed, it is evident from Prof. Forbes' remarks that the collections of recent Indian shells examined by him were too imperfect to enable him to form a competent opinion. Mr. Benson, a better authority on this particular subject than Forbes, considered the most, if not the whole, of the Siwálik mollusca identical with existing species (Falconer: Palæontologica Memoirs. I, pp. 26, 181). Of three species in the Survey collections, two were identified by Mr. Geoffrey Nevill with living forms.

[3] Some extinct fluviatile mollusca have been found in the lower Siwálik strata of the Bugti hills associated with a vertebrate fauna of miocene facies. See *supra*, p. 319.

fauna which has been obtained from the two upper groups of the Siwáliks series. The true age of this fauna, whether miocene or pliocene, was at one time disputed and, though there is happily no further controversy regarding this point, the question presents points of sufficient interest to claim somewhat extended notice.

The following list of genera, with the number of species of each that are known, comprises all that have been found in the typical Siwálik area, excluding the fauna of the Manchhar beds in Sind and of Perim island which appear to belong to an older period than the fossiliferous beds of the sub-Himálayas; genera still living are distinguished by an asterisk :—

PRIMATES—

* *Troglodytes*, 1; * *Simia*, 1; **Semnopithecus*, 1; **Macacus*, 1; **Cynocephalus*, 2.

CARNIVORA—

* *Mustela*, 1; **Mellivora*, 2; *Mellivorodon*, 1; **Lutra*, 3; *Hyænodon*, 1; *Ursus*, 1; *Hyænarctus*, 3; *Amphicyon*, 1; **Canis*, 2; **Viverra*, 2; **Hyæna*, 4; *Lepthyæna*, 1; *Hyænictis*, 1; *Œluropsis*, 1; *Œlurogale*, 1; **Felis*, 5; *Machaerodus*, 2.

PROBOSCIDEA—

**Elephas*, 6; *(*Euelephas*, 1, **Loxodon*, 1, *Stegodon*, 4); *Mastodon*, 5.

UNGULATA—

Chalicotherium, 1; **Rhinoceros*, 3; **Equus*, 1; *Hipparion*, 2; **Hippopotamus*, 1; *Tetraconodon*, 1; **Sus*, 5; *Hippohyus*, 1; *Scnitherium*, 1; *Merycopotamus*, 3; **Cervus*, 3; *Dorcatherium*, 2; **Tragulus*, 1; **Moschus*, 1; *Propalæomeryx*, 1; **Camelopardalis*, 1; *Helladotherium*, 1; *Hydaspitherium*, 2; *Sivatherium*, 1; **Alcelaphus*, 1; **Gazella*, 1; **Cobus*, 2; **Antilope*, 1; *Hippotragus*, 1; **Oreas*(?) 1; **Strepsiceros*, (?) 1; *Boselaphus*, 1; *Palaeoryx*, (?) 1; *Hemibos*, 1; *Leptobos*, 1; **Bubalus*, 2; **Bison*, 1; *Bos*, 3; *Bucapra*, 1; **Capra*, 2; **Ovis*, 1; **Camelus*, 2,

RODENTIA—

* *Mus* (*Nesokia*), 1; **Rhyzomys*, 1; **Hystrix*, 1; **Lepus*, 1.

AVES—

**Phalacrocorax*, 1; **Pelecanus*, 2; **Leptoptilus*, 1; **Mergus*, 1; **Struthio*, 1; **Dromaeas*, 1.

REPTILIA—

Crocodilia—**Crocodilus*, 1; **Gharialis*, 3; *Rhamprosuchus*, 1.
Lacertilia—**Varanus*, 1.
Chelonia—*Colossochelys*, 1; **Damonia*, 1; **Bellia*, 2; **Kachuga*, 3; **Hardella*, 1; **Emyda*, 4; **Trionyx*, 1; **Chitra*, 1.

PISCES—

**Carcharias*, 1; **Ophiocephalus*, 1; **Clarias*, 1; **Heterobranchus*, 1; **Chrysichthys*, 1; **Macrones*, 1; **Rita*, 1; **Arius*, 1; **Eagarius*, 1.

Only very imperfect information exists as to the exact horizon in the Siwálik series at which the bones of a large proportion of the species have been found, but the great majority are from the upper and middle

Siwáliks, none in the typical area being known to occur in the lower or Náhan subdivision. It is, however, by no means certain that some of the specimens from the north-western Punjab are not derived from beds of the same age as the Náhan group, and it is highly probable that some other forms with middle tertiary affinities would be found to be confined to lower Siwálik beds, if the precise horizon of all the bones collected were known.[1]

On the other hand, one pleistocene form, *Bos* (*Bubalus*) *palæindicus*, has been found in the highest Siwálik strata, associated with *Camelus sivalensis*, *Colossochelys*, etc.; and two species of elephant, belonging to the subgenus *Stegodon*, viz. *E. insignis* and *E. ganesa*, range throughout the upper Siwáliks, and recur in the pleistocene deposits. The species of proboscidians generally appear to have had a more extensive range, both in space and time than most of the forms belonging to other mammalian orders, but *Bos palæindicus* is an animal of exceptionally recent aspect, even in the pleistocene mammalian fauna, since it is only distinguishable from the living *Bos bubalus* (*Bubalus buffelus* v. *B. arni*, auct.) by comparatively trifling and unimportant osteological details. It must evidently have been a very near ally, and in all probability the not very distant progenitor, of the buffaloes which now inhabit the Ganges valley, Assam, and parts of the Central Provinces of India.

The Siwálik forms, however, which might be excluded on account of belonging to an older or a newer fauna, and of being supposed, on more or less strong evidence, to be confined to either the lowest or the uppermost portions of the series, are too few in number to affect the general facies, and there are unquestionably several miocene types and some pleistocene species found in the highly fossiliferous upper Siwálik beds. It is best therefore, for the present, to include all the forms enumerated.

Proceeding then to classify the genera of mammalia above given, it will be found that thirty-nine, comprising seventy-one species, still exist (the living species being, however, different in nearly every case), whilst twenty-five with thirty-seven species, are extinct.

Of the extinct genera, excluding those that are purely Indian, *Œlurogale* and *Hyænodon* are found in oligocene beds, the latter ranging into the miocene; of the miocene genera, *Dorcatherium* and *Amphicyon* are not known from newer beds, while *Hyænarctus*, *Chalicotherium*, and *Hipparion* range into the pliocene, *Machærodus* and *Mastodon* into the pleistocene; *Helladotherium* and *Palæoryx* are purely pliocene forms, while the distinction of *Hemibos* from the living *Bubalus* is very doubtful.

Of the other extinct forms, not known out of India, two (*Hippohyus* and

[1] A very large proportion of the Siwálik remains have been obtained by native collectors, and of course the precise locality of the bones is in most of these cases doubtful.

Merycopotamus) belong to the less specialised types characteristic in general of the older and middle tertiaries. Several others, such as *Tetraconodon*, with its enormously developed premolar teeth, and the huge four-horned *Sivatherium*, differ widely from anything now existing, but, being highly specialised forms there is nothing in their organisation to indicate that they are of earlier age than newer tertiary.

Amongst the recent genera represented in the Siwáliks, ten, viz. *Mustela Felis*, *Canis*, *Viverra*, *Lutra*, *Rhinoceros*, *Sus*, *Cervus*, and *Hystrix* are known to range as far back as upper miocene, and in one or two cases even further; twelve, viz. *Macacus*, *Semnopithecus*, *Ursus*, *Hyæna*, *Elephas*, *Equus*, *Hippopotamus*, *Camelopardalis*, *Gazella*, *Bos*, *Capra*, and *Mus*, are known from the European pliocene beds, but not earlier; whilst *Troglodytes*, *Simia*, *Cynocephalus*, *Mellivora*, *Cobus*, *Antilope*, *Oreas*, *Strepriceros*, *Capra*, *Ovis*, *Camelus*, and *Rhizomys*, have hitherto only been found in recent or pleistocene deposits, outside of India.

This examination of the relations between the Siwálik genera and the distribution of similar forms in European tertiaries leads, as might be anticipated, to a somewhat uncertain result. The proportion of living to extinct genera is greater than is found in most miocene deposits, but not more than appears to exist in the characteristically middle tertiary ossiferous beds of Sansan in France.[1] The presence of four extinct genera not known to range above the miocene period elsewhere, is contrasted with the occurrence of twenty-five genera not found elsewhere at a lower horizon than pliocene or pleistocene. There is perhaps rather more probability that early forms, like *Dorcatherium* and *Hyænodon*, should have survived longer in India than they did in Europe, just as rhinoceroses, tapirs, and elephants still exist in the tropics, associated with a fauna amongst which they appear antiquated and out of place, than that such eminently specialised types as *Macacus*, *Bos*, *Capra*, or *Equus*, should have lived in miocene times, but the argument is of small value, for the miocene *Cervus* and *Antilope* were in all probability as highly specialised, or nearly so, as the Siwálik genera. The fact, however, that the recent genera contain more species than the extinct forms is of some importance, since it is probable that types which were dying out would be represented by fewer species than those which were supplanting them, and which might fairly be credited with the vitally important power of producing distinct specific stocks by variation. A stronger argument for the newer age of the Siwálik beds is to be found in the close approximation between some of the mammals and the living species of the same genera, the most remarkable of all being the connection already noticed

[1] Gervais, Zoologie et Paléontologie Françaises, 2nd ed., Paris, 1859, p. 338.

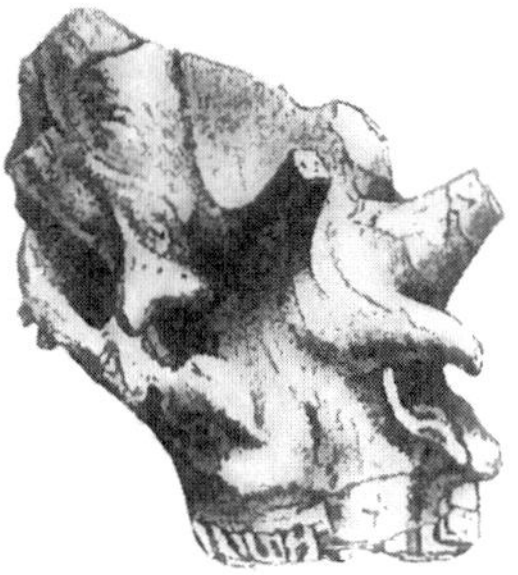

Sivatherium giganteum, F. & C. × $\frac{1}{12}$

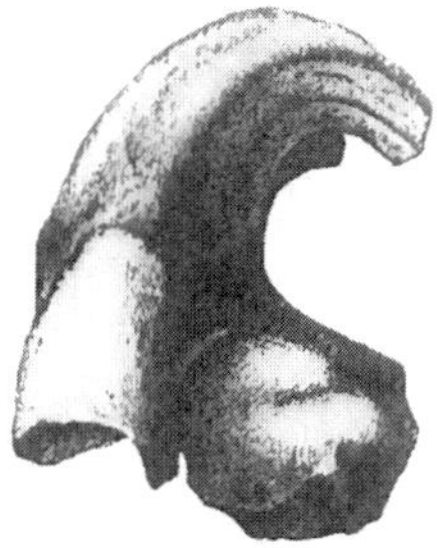

Capra sivalensis, Lyd. × ¼.

Rhinoceros sivalensis, F. & C. right upper molar × $\frac{2}{3}$.

Camelopardalis sivalensis, F. & C. left upper molar

Cynocephalus subhimalayanus, Theger, right upper molar.

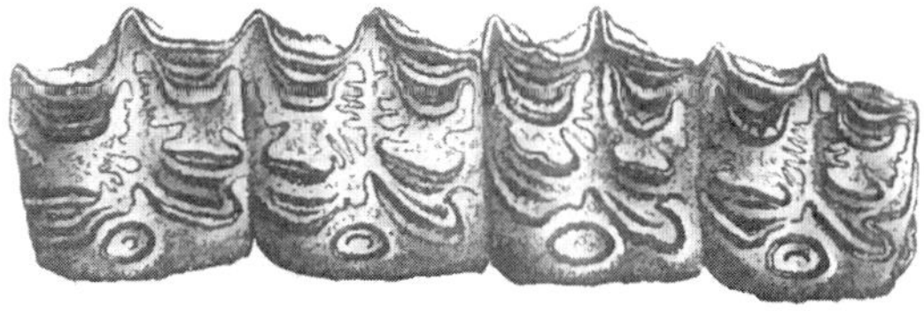

Hippotherium antilopium, F. & C. left lower molars and premolars.

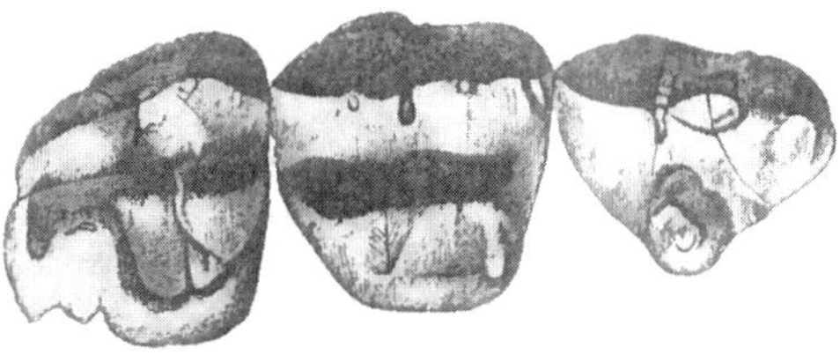

Hyœnarctos palœindicus, Lyd, right upper molars and premolars

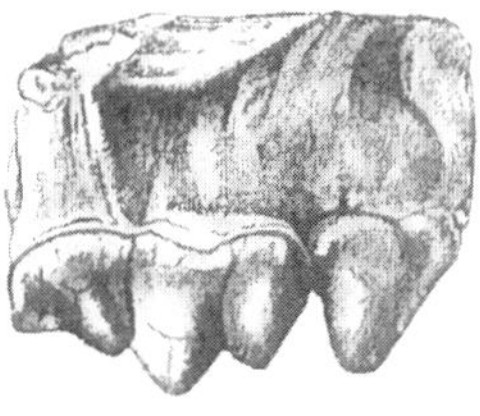

Hyœna colvini Lyd right upper premolars.

SIWALIK FOSSILS.

Calcutta Phototype Co.

Stegodon ganesa, F. & C. Cranium × $\frac{1}{24}$.

Stegodon ganesa, F. & C. × $\frac{1}{4}$

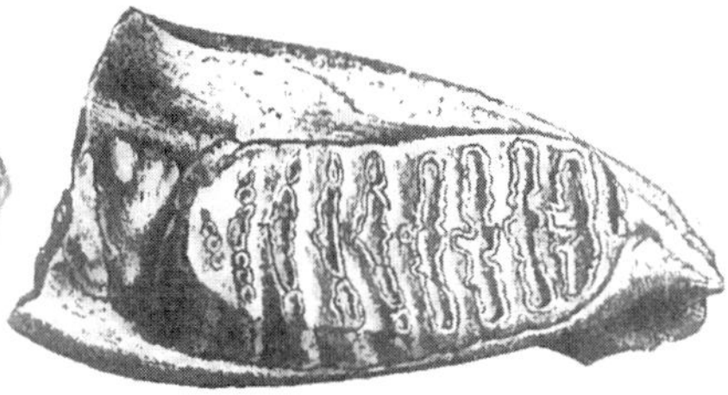

Elephas hysudricus, F. & C. left lower molar × $\frac{1}{5}$

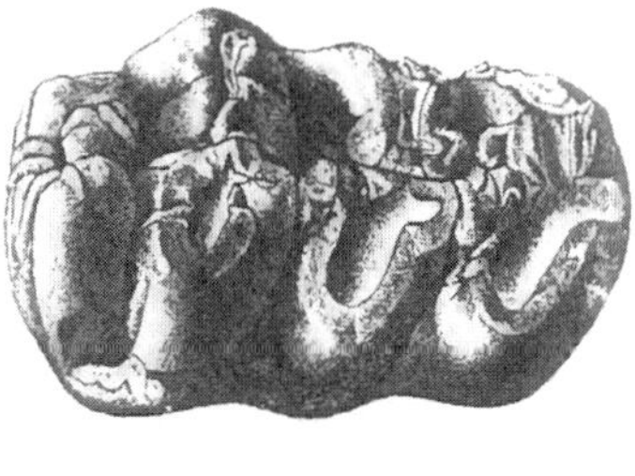

Mastodon falconeri, 2nd right lower molar × $\frac{1}{2}$

Mastodon latidens, 2nd right upper molar × $\frac{1}{2}$

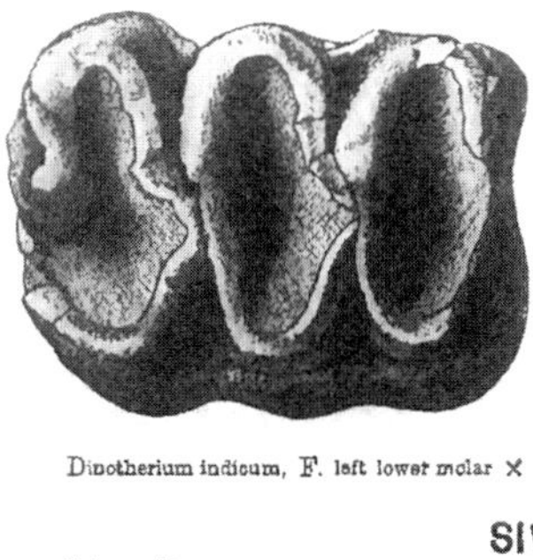

Dinotherium indicum, F. left lower molar × $\frac{1}{2}$.

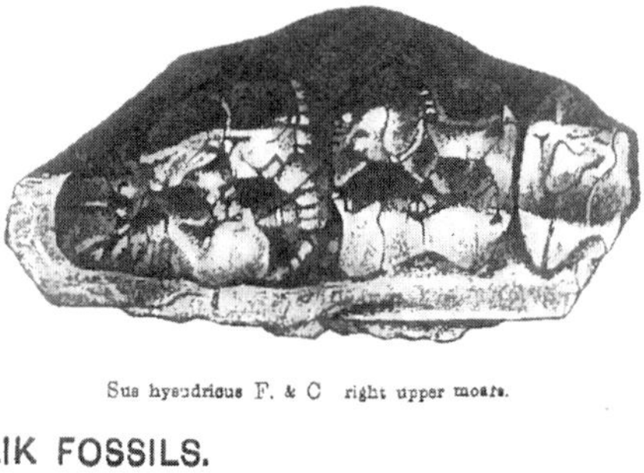

Sus hysudricus F. & C right upper moars.

SIWALIK FOSSILS.

Calcutta Phototype Co.

between the fossil buffalo of the uppermost Siwálik strata, that of the pleistocene Jumna and Narbadá beds, and the common Indian species now existing.

The anomalies exhibited by the mammalian fauna taken as a whole would doubtless disappear to a considerable extent if we knew the exact horizons from which the fossils were obtained, those forms with miocene affinities being more characteristic of the lower beds and those closely allied to recent forms of the upper. Yet, after allowing for this, when we consider that the great bulk of the fauna was obtained from what is practically a single group of beds, and that the life of a species is probably in every case, and certainly in most, more extensive than the period represented by the particular beds in which its remains are found, there remains a strong probability that there was an admixture of types not found in beds of later than miocene age in northern Europe with the more typically pliocene forms which prevail in the Siwálik fauna.

The evidence afforded by the reptiles, birds, and fishes is even more decidedly in favour of attributing a later data than miocene to the Siwálik beds. Among the reptiles only two out of twelve genera, and only eleven out of twenty species are extinct, the remainder, *Hardella thurgi*, *Damonia hamiltoni*, *Kachuga lineata*, *K. tectum*, *K. dhongoka*, *Chitra indica*, *Emyda vittata*, *Gharialis palustris*, are all found living at the present day,[1] while *Crocodilus sivalensis* seems undistinguishable from the recent *C. palustris*.

The six genera of birds are all represented in the living fauna of the world, and the same holds good of the fishes, no extinct genus of either of these orders having been determined in the Siwálik fauna. Moreover, two fishes (*Machrones aor* and *Bagarius yarrelli*) are still living.

Putting together the data derived from the fauna as a whole, it is impossible to deny that the balance of evidence is in favour of a pliocene age[2] and this view is supported by some stratigraphical data. As the approximate age of the Siwálik rocks is a necessary element in any argument founded upon their fauna, it will be best to show how the

[1] R. Lydekker, *Records*, XXII. 58, (1889).

[2] This appears at first sight to be in direct opposition to Dr. Falconer's conclusions, but a study of his writings leaves it doubtful whether he ever expressed any decided conviction on the subject. He repeatedly noticed the close connection between some Siwálik forms and those now found in India, and appeared for a long time rather disposed to consider that the tertiary mammalia of India "lasted through a period corresponding to more than one of the tertiary periods of Europe" (Palæontological Memoirs, I, p. 28) than to class the Siwálik fauna with the miocene of Europe. In his later writings he certainly spoke of the Siwálik fauna as miocene, but only incidentally, whilst in some of his latest papers he argued in favour of man having been a probable contemporary of *Colossochelys* and the Siwálik mammalia—an idea which it is difficult to reconcile with the miocene age of the fauna.

stratigraphical evidence supports the view that these beds are of pliocene, and not of miocene age, before proceeding to notice some other interesting points of connection between the Siwálik and other faunas, recent and extinct.

The stratigraphical, as distinguished from the purely homotaxial relations just discussed, depend entirely upon the connection between the typical Siwáliks and the Manchhar beds of Sind. The position of the latter has been already described, and it was shown that the Manchhar beds, comprising where thickest but little less than 10,000 feet of strata, rest upon the Gáj group, whose age is determined by the far more satisfactory data afforded by marine organisms, and is shown to be more probably upper than lower miocene. The lower Manchhar beds pass down into the Gáj rocks, so that it is reasonable to believe that no important difference of age exists between the two. It has also been shown that the mammalian fauna of the lower Manchhars, although containing several species in common with the Siwáliks, is altogether older in aspect, and that the majority of the forms hitherto recognised belong to the peculiar types of even toed ungulates allied to *Merycopotamus* and *Anthracotherium*, intermediate in character between pigs and ruminants, and characteristic of the miocene epoch. In these lower Manchhar beds there is also found a form of *Dinotherium*, miocene type unknown in the Siwáliks proper. though found (the species being distinct) in the beds of Perim island. Now, there can be no reasonable doubt that the Manchhar beds of Sind, as a whole, correspond with the Siwálik formation of northern India, for the two are portions of one continuous band of upper tertiary rocks, and viewed in this light, the relations of the faunas are very striking, the fossiliferous lower beds of the Manchhar group corresponding to the unfossiliferous Náhans, and the almost unfossiliferous upper Manchhar beds to the ossiferous strata of the Siwáliks. If, therefore, the lower Manchhars of Sind are upper miocene, so is the Náhan group of the Punjab, and it is impossible, either on stratigraphical or palæontological grounds, to class the fossiliferous middle Siwáliks lower than pliocene, the upper Siwáliks, which contain *Bos* (*Bubalus*) *palæindicus*, being probably upper pliocene. Briefly stated, the evidence is that the Siwálik fauna is newer than the Manchhar fauna, and found in higher beds, and that the Manchhar fauna is not older than upper miocene.

It is impossible to enter at length into the detailed relations of the Siwálik fauna with the forms found in various newer tertiary strata. There is however one European fossil fauna which is of singular interest from its resemblance to that of the Siwálik beds. In this collection of extinct mammalia, which was first discovered at Pikermi in Attica,[1] and has since been found at Samos and Mytilene on the coast of Asia Minor, at Baltavar in Hungary and at Maragha east of lake Urmia in northern Servia, not only is there a remarkable admixture of typically miocene forms with other species which have a later aspect, but there is the same remarkable abundance of true ruminants, amongst them several species of *Giraffidæ* and of *Antilope*, as in the Siwáliks. In miocene strata, although ruminants occur, they are in general but little, if at all, superior in number to the other artiodactyle ungulates, but in the Pikermi beds (including only the forms found at Pikermi) there are fifteen ruminants to one pig and one *Chalicotherium;* in the Siwálik fauna thirty-seven ruminants and but twelve other artiodactyle ungulates. Another point of similarity in the two faunas is the absence of small mammals.

The following is a list of the genera found in the beds of Greece, with the number of species belonging to each genus:—

PRIMATES—
Mesopithecus, 1.

CARNIVORA—
Simocyon, 1; Mustela, 1; Promephitis, 1; Ictitherium, 3; Hyæna, 1; Hyænarctus, 1; Hyænictis, 1; Machærodus, 1; Felis, 4.

PROBOSCIDEA—
Mastodon, 2; Dinotherium, 1.

UNGULATA PERISSODACTYLA—
Rhinoceros, 3; Aceratherium, 1; Leptodon, 1; Hipparion, 1.

UNGULATA ARTIODACTYLA—
Sus, 1; Chalicotherium (Ancylotherium), 1; Dremotherium 2; Antilope, 3; Palæotragus, 1; Palæoryx, 2; Tragoceros, 2; Palæoreas, 1; Antidorcas, (?) 1; Gazella, 1; Camelopardalis, 1; Helladotherium, 1.

RODENTIA—
Mus (Acomys), 1; Hystrix, 1.

Of birds, a *Phasianus*, a *Gallus*, and a *Grus* have been identified; of reptiles, bones of *Testudo* and *Varanus.*

Of the above twenty-nine genera of mammals eighteen, including *Helladotherium*,[2] are found in the Siwáliks of India, besides which the fauna bears in many respects the same similarity to that of Africa at the present day as the Siwálik mammals bear to their living Indian representatives. Now, this Pikermi fauna has been freqently quoted as upper miocene, and its connection with the miocene beds in other parts of Europe is unmistakable,

[1] Gaudry: Animaux fossiles et Géologie de l'Attique, 4° Paris, 1862.

[2] See Forsyth, Major, *Proc. Zool. Soc.*, 1891, p. 323.

no less than fifteen species being undistinguishable from those found in various miocene deposits. Several of these species are doubtfully identified, but amongst the number are such characteristic forms as *Machærodus cultridens*, *Mastodon turicensis*, and *Hipparion gracile*. But, as M. Gaudry points out in the clearest manner, the ossiferous beds of Pikermi contain at their base, and below the horizon whence the bones have been obtained a layer with pliocene marine fossils, and all the beds containing the bones, together with the pliocene marine beds, rest unconformably on lacustrine miocene rocks. There can be, therefore, no reasonable doubt that the Pikermi fossils, like the middle Siwáliks, are of pliocene age, and that the quotation of them as miocene is an error.[1] At the same time the absence of some characteristic living genera, such as *Elephas* and *Bos*, gives a somewhat older facies to the Pikermi than to the Siwálik fauna.

The points of similarity between the European miocene faunas and the animals now inhabiting either tropical Asia or Africa south of the Sahara may be due either to migration, and survival in a more favourable climate, or to the fauna having been formerly more uniform over large areas, and to the modified descendants continuing to live in one region, whereas they have died out and been replaced by distinct types in other parts of their old province.[2] On the latter hypothesis we may suppose that the fauna of central Europe and Malayasia was more or less uniform in the lower miocene period, and that Greece and Africa formed a single zoological province in pliocene days, but that the gibbon-like apes, *Tupaiadæ* and other Malay types, died out in central Europe, and the giraffes, antelopes, etc., in Greece, whilst the descendants of their relatives survived in the Malay countries and Africa respectively. The theory of migration presents, on the whole, fewer difficulties, and is in accordance with the little we know of the Indian miocene (Manchhar) fauna, in which living tropical forms appear to be less represented than in the deposits of that age in Europe. It is not unreasonable to suppose that some of the forms named, and especially the ruminants, migrated into southern Asia at the close of the miocene period.

[1] For the theory adopted by M. Gaudry to account for the survival of these miocene animals in pliocene times, see "Animaux fossiles et Geologie de l'Attique," p. 343. It appears simpler to believe that the miocene fauna of Europe migrated to the southward, and that many species survived in Greece after they had died out north of the Alps. Hence the admixture of pliocene and miocene types. A further contribution to the question, by M. Gaudry, was published in 1886, *Bull. Soc. Geol. Franc.*, 3rd series, XIV, 288.

[2] It is assumed in the present and in other arguments employed in this work that similarity of organisation implies relationship by descent, *i.e.* that animals having similar structure are descended more or less remotely from the same ancestors. The theories of evolution, and of origin of species by descent with modification, are now so widely accepted amongst naturalists that it is unnecessary to explain or defend them.

It is true that amongst the marine invertebrates there is a well marked resemblance between the miocene genera of Europe and living tropical forms. The Indian and African land faunas of the early and middle tertiaries are as yet too imperfectly known for any comparison to be made between them and those of the same epoch in extratropical regions. It is not improbable that there may prove to have been a greater similarity than exists amongst the terrestrial forms living at present, and it is also probable that, if such similarity existed, it will be found to have consisted mainly in the greater richness of the extratropical fauna in middle tertiary times, and in a number of types now extinct or confined to the tropics having been represented in both tropical and extratropical zones of climate. This last probability is founded on the fact that the temperature of Europe in the miocene epoch was in all probability nearer to that of the present tropics than to the temperate climate of recent times, and that, consequently, whole families of animals, and of plants intolerant of cold, then ranged to much higher latitudes than they now do.

It is by no means an improbable inference that the representation of European miocene genera in the Indian Siwáliks is due to changes of climate in later tertiary times and to a migration of the fauna towards the tropics. There is good reason for believing that Europe and south-eastern Asia were connected by land after the eocene period, and as it is certain that, a great portion of the disturbances affecting the Himálayan strata are of pliocene or postpliocene date, it is reasonable to conclude that, at the close of the miocene epoch, the mountain barrier which now separates the Indian peninsula from Central Asia did not exist, or was so much lower that it afforded little or no obstacle to migration.

But the immigration of the European miocene forms may not be the only way in which only the Siwálik fauna was affected by the secular refrigeration of the earth's surface, culminating in the glacial epoch. It is true that there is a considerable amount of similarity between the Siwálik fauna and that of India at the present day, but, nevertheless, there is a very striking distinction—a distinction due less to change and replacement than to disappearance. Even after making allowance for the fact that the whole assemblage did not exist contemporaneously, there is nothing so striking in the fauna of the Siwálik epoch as the wonderful wealth and variety of forms. It must be recollected that we know little or nothing of the smaller mammals, and that animals of size inferior to a pig or a sheep are scarcely represented. It would be premature to infer that, as at the present day, the more minute forms exceeded the larger types in abundance, for the conditions of intermediate ages may have affected the more bulky

[1] This will be found to agree with the conclusion regarding the probable date of origin of the Himálayas derived from other considerations. See Chap. XVII.

animals far more than the minute *Rodentia*, *Insectivora*, *Chiroptera*, etc. Still it is only reasonable to suppose that the ancestors of the present *Micro-mammalia* lived in the same profusion as they do now, and it is incredible that the living rodents and insectivores can play the parts on the modern stages, and fulfil the functions, of the great ungulates and carnivores of past times. Comparing like with like, and especially passing in review the *Carnivora*, *Proboscidea*, and *Ungulata*, all represented, and all, except the *Proboscidea*, well represented in the living fauna of India, indeed better than in most other parts of the world at the present day, it is impossible not to be struck with the comparative poverty in variety of the existing mammalian types. We have of course but an imperfect knowledge even of the larger Siwálik animals, and remains of *Carnivora* are rare, so much so that probably many species remain undiscovered, but even at present the known Siwálik carnivores are more numerous than the living forms of similar size in the same area, and the ungulates exceeded their living representatives in number in the proportion of more than three to one, there being fity-six known Siwálik species and only eighteen recent. The superior wealth of the older fauna is both generic and specific; not only are the types more varied, but there is a greater variety of forms in many of the genera, and no less than eleven extinct elephants and mastodons are represented by a solitary living form. Even such modern types as *Bos* have dwindled in numbers from six to two.

This great impoverishment of the recent mammalian fauna is not peculiar to India. It is found in other parts of the Old World and in America, wherever remains of animals have been preserved in sufficient quantities amongst the deposits of the later tertiary epochs for a good idea of the fauna to be presented. In the words of Mr. Wallace, "We live in a zoologically impoverished world, from which all the largest and fiercest and strongest forms have recently disappeared;" and he makes the happy suggestion,[1] that this enormous reduction in the numbers of the greater mammals is due to the glacial epoch. Thus, we have an addition to the arguments urged in the first chapter,[2] in favour of India having been affected by a cold period in the geologically recent past.

[1] Geographical Distribution of Animals, I, p. 150.

[2] *Supra*, p. 14.

CHAPTER XV.

LATERITE.

General characters and composition.—Distinction of high level and low level laterite—Distribution and mode of occurrence—Theories of the origin Resumé.

All who have paid any attention to the geology of India must be familiar with the term 'laterite,' and no one can have travelled far in India without meeting with the substance itself, which is still one of the stumbling blocks of Indian geology. Although it is difficult to conceive that a rock, so widely spread in India and Ceylon, and said to be extensively developed in Malacca and Sumatra, while some occurs in Burma, can be peculiar to these regions,[1] it is uncertain if anything precisely similar has hitherto been detected elsewhere. It is almost invariably a surface formation, and according to some observers, nothing but a form of soil; yet it becomes an important formation from the very large area in India, which it superficially covers, and a treatise on Indian geology would be imperfect without a full description of the rock.

The order in which its description comes is governed by the fact that while there can be but little doubt that some forms of laterite are in process of formation at the present day, others date from tertiary, and perhaps from eocene, times, and as the rock is usually unfossiliferous it appears best to describe all the varieties together before proceeding to the description of the posttertiary rocks.

The description of laterite, given in many geological works, is far from accurate, although the rock has been well described by several Indian geologists.[2] In its normal form it is a porous argillaceous rock much

[1] Voysey states that it is found at the Cape of Good Hope. It is a noteworthy fact that no laterite has been detected in Abyssinia, where the rocks throughout a large area of country are precisely similar to those of the Bombay Deccan. In map No. 4 of Berghaus' Physical Atlas, laterite is represented as covering nearly a quarter of the dry land of the earth. The term is, however, used in a different sense to that here applied.

[2] It would be difficult to give a description of any rock more clear and accurate than Newbold's account of the laterite of Bídar, *Jour. As. Soc. Beng.*, XIII, 989, (1844); *Jour. As. Soc. Beng.*, XIV, 299, (1845) and *Jour. Roy. As. Soc.* VIII, 227 (1846). The descriptions of laterite scattered through the writings of various Indian geologists are too numerous to quote. Amongst the more important are the following—Buchanan, Journey from Madras through Mysore, Canara, and Malabar, London, 1807, II, p. 440. Stirling, *As. Res.*, XV, 177, (1825); Christie, *Edin. New Phil. Jour.*, VI, 117, (1829);

impregnated with iron peroxide irregularly distributed throughout the mass, whose composition may be gathered from the following analysis of a very richly ferruginous variety from Rangoon.[1]—

Soluble in Acids.	
Peroxide of iron	46·279
Alumina	5·783
Lime	·742
Magnesia	·090
Silica	·120
Insoluble in Acids.	
Silica (dissolved by potash)	6·728
Silica (by fusion)	30·728
Lime, iron, and alumina	2·728
Combined water alkalies and lose	6·802
	100·000

The iron exists either entirely in the state of hydrated peroxide (limonite) or else partly as hydrated and partly as anhydrous peroxide. The surface of laterite after exposure is usually covered with a brown or blackish brown crust of limonite, but when freshly broken, the rock is

and *Mad. Jour. Lit.. Sci.*, IV, 468, (1836); Calder, *As. Res.*, XVIII, 4, (1833); Cole, *Mad. Jour. Lit. Sci.*, IV, 100, (1836).; Voysey, *Jour. As. Soc. Beng.*, XIX, 273; (1850); Kelaart, *Edin. New Phil. Jour.*, LIV, 28, (1853); Carter, *Jour. Bo. Br. Roy. As. Soc.*, IV, 199, (1852); V, 264, (1857); Aytoun, *Edin. New Phil. Jour.*, 2nd series, IV, 67, (1856); Buist., *Trans. Bo. Geog. Soc.*, XV, p. xxii, (1859).

The subject has also been frequently treated in the publications of the Geological Survey, especially *Memoirs* I, 69, (1856), 265-280, (1859); II, 78, (1860); IV, 260, (1864); X, 27, (1873); XII, 200, 224, (1876); XIII, 222, (1877); XVIII, 122, (1881); XXIV, 217 and 239, (1890); where an account of all previous notices is given. See also *Records*, XV, 93, (1882); XXII, 220, (1889).

[1] *Jour. As. Soc. Beng.*, XXII, 198, (1853). The result given is the mean of three analyses made in the laboratory of the School of Mines, London. The following are assays of the quantity of iron contained in the portion of laterite soluble in acids. The analysis of first five and No. 8 were made by Mr. Mallet for the first edition of the present work; the other three are from the paper on the laterite of Orissa (*Memoirs*, I, p. 288) :—

			Percentage of metallic iron.	Percentage of iron peroxide.
1.	High-level laterite	overlying Deccan trap, Amarkantak	35·6	50·8
2.	Ditto	from Máin Pát, Sargúja	16·6	23·7
3.	Ditto	from Baplaimali plateau, Káláhandi, south of Sambalpur	15·	21·4
4.	Ditto	from top of Moira hill in the Kharakpur range, south of Monghyr	28·3	40·4
5.	Ditto	from Mahuágarhí hill, Rájmahál hills	15·8	22·5
6.	Laterite (high-level)	from Káthiáwár, Western India	22·8	32·5
7.	Low-level laterite	from Daltola, Cuttack, Orissa	24·5	34·9
8.	Ditto	from near Cuttack	25·6	36·5
9.	Ditto	from Tanjore	23·4	33·4

mottled with various tints of brown, red, and yellow, and a considerable proportion sometimes consists of white clay. The difference of tint is evidently due to the segregation of the iron in the harder portions, the pale yellow and white portions of the rock, which contain little or no iron, being very much softer, and liable to be washed away on exposure. Occasionally the white portions have a brecciated appearance, and consist of angular fragments in a ferruginous matrix. In this case the rock has not unfrequently a compact texture resembling jasper, but it is never so hard as a purely silicious mineral.

The iron peroxide not unfrequently occurs in the form of small pisolitic nodules, which are sometimes employed as iron ore. Veins and nests of black manganese have been observed by Newbold in some laterites of the Deccan.[1]

In many forms of laterite the rock is traversed by small irregular tortuous tubes, from a quarter of an inch to upwards of an inch in diameter. The tubes are most commonly vertical, or nearly vertical, but their direction is quite irregular, and sometimes they are horizontal. They are usually lined throughout with a crust of limonite, and except near the surface are often filled with clay. Besides these, there are sometimes horizontal cracks, occasionally expanding into small cavities, and giving an appearance of irregular stratification to the formation. In the more massive forms of laterite some horizontal banding is usually present, the cavities beneath the surface being mostly filled by more or less sandy clay. When first quarried, the rock is so soft that it can easily be cut out with a pick, and sometimes with a spade, but it hardens greatly on exposure.

The exposed surface, whether vertical or horizontal, is characteristic and peculiar. It is extremely irregular, being pitted over with small hollows, caused by the washing away of the softer portions, and generally, though not always, traversed by the tubes and cavities just described. At times it is so much broken up by small holes as to appear vesicular, whilst the crust of limonite forms a brown glaze, often mammillated or botryoidal, so that the rock has a remarkably scoriaceous appearance and bears a very curious resemblance to an igneous product. It is not surprising that many observers should have looked upon laterite as volcanic, for not only does it often present this remarkable superficial resemblance to a scoriaceous lava flow, but it is found, in several parts of India, associated with basalt and other igneous rocks. Laterite, however, as will be shown presently, is never an original form of igneous rock. It is in all cases either produced by the alteration of other rocks, sometimes igneous sometimes sedimentary or metamorphic, or else it is of detrital origin.

The laterite frequently appears to pass into the underlying rock,

[1] *Jour. As. Soc. Beng.*, XIII, 992, (1844).

whether this be igneous, metamorphic, or sedimentary. In the case of basalt or gneiss underlying laterite the upper part is decomposed, forming a clay, which becomes a kind of lithomarge passing by insensible gradations into laterite itself, through its impregnation with iron by the water trickling through the laterite above. In fresh sections, where a detrital form of laterite is the overlying rock, the limit of the two can usually be traced without difficulty, but surfaces which have been exposed for a length of time are generally covered with more or less of the limonite glaze and the lithomarge can no longer be distinguished from laterite.[1] This lithomarge is always more ferruginous above than below; it varies in colour from red through yellow to white, being usually mottled, not frequently coloured purple or lilac in patches, and a few pipes often occur, apparently produced by the percolation of water.

Another form of lithomarge, found beneath the laterite in many places, and especially to the northward, consists of hardened clay, sometimes sandy and generally highly ferruginous, which shows no tendency to pass into the underlying rock, although it usually exhibits unmistakable transition into the laterite above. In these cases, the laterite and lithomarge together form a group of beds superposed, as a rule unconformably, upon older rocks of various kinds. In some instances, as in Bundelkhand, this infra-lateritic formation contains pebbles,[2] and there is every reason for believing that it is a rock of sedimentary origin. In some cases the present form of lithomarge contains hæmatite or limonite in quantities sufficient to enable the mineral to be collected for iron ore, as in Bundelkhand, near Jabalpur, and on the eastern flanks of the Rájmahál hills.[3]

One peculiarity possessed to an eminent degree by all forms of laterite is the property of broken or detrital fragments being recemented into a mass, closely resembling the original rock. Laterite itself has great powers of resisting atmospheric disintegration, being produced by long action of the atmosphere upon various ferruginous clays, but the underlying formation decomposes, is slowly washed away, and the originally horizontal cap of laterite, falling down, becomes reconsolidated on the irregular surface, which it still covers. Another form of reconsolidated laterite is composed of broken fragments, washed down by rain and streams to a lower level, at which they become recemented.

The surface of the country composed of the more solid forms of laterite is usually very barren, the trees and shrubs growing upon it being thinly scattered and of small size. This infertility is due, in great part, to the rock being so porous that all water sinks into it, and sufficient moisture is not retained to support vegetation. The result is that laterite plateaux are usually bare of soil, and frequently almost bare of vegetation.

[1] *Memoirs*, I, 283, (1859).
[2] *Memoirs*, II, 84, (1860).
[3] *Memoirs*, II, 81, (1860); XIII, 241, (1877).

Of course, this barrenness is not universal, soil sometimes accumulates on laterite caps, and some of the more gravelly or more argillaceous varieties support a moderate amount of vegetation. Still the general effect of the rock is to produce barrenness.

Several writers have divided the laterite into two forms, high level and low level laterite, the former of which was supposed not to be, the other to be, of detrital origin, and by some it has been urged that the term should only be applied to the latter of these.

The high level form, which is found capping the summits of hills and plateaux on the highlands of central and western India, is a rock of fine grain, and apart from the irregular distribution of the iron it contains, fairly homogeneous in structure; it is not sandy, and only exceptionally shows any indication of a detrital origin. This type Mr. Foote [1] has proposed to distinguish as iron clay, a term used by Voysey, one of the earliest observers, but also used by him in describing the low level detrital laterite of Nellore.

The low level laterite, which covers large tracts in the neighbourhood of both coasts, on the other hand, frequently contains grains of sand and pebbles, imbedded in the ferruginous matrix. It is, as a rule, less homogeneous than the high level form, and passes by insensible gradations into sandy clay or gravel with a very small proportion of iron, especially in the exposures that have been classed as laterite on the east coast, many of which have little or no claim to the name if it is to be used in any lithological sense. On the west coast the exposures are more truly lateritic in their nature, and there are large areas of rock which do not appear to be detrital in their origin and are undistinguishable, except by position, from the high level laterite of the Deccan.

This fact shows that no hard and fast distinction can be drawn between the high and low level laterite, but there is undoubtedly on the whole a difference in age and origin between the two types, and those geologists who consider that the name of a rock should distinguish not only its composition and structure, but also its mode of origin, are justified in refusing to use the same word to designate both. At the same time there is nothing in the description of the original propounder of the name to indicate that he restricted the word to one form rather than the other. There is good reason to suppose that some of the laterite in the district he examined is, and some is not, of detrital origin; moreover, the word has by convention come to be used so generally as an ill defined but convenient term, descriptive of the constitution and nature of the rock, irrespective of its mode of origin, that it would be inconvenient, if not impossible, to attempt a restriction of its meaning. But, though it is impossible to distinguish between the high level and low level laterites, if by

[1] *Memoirs*, XII, 201, (1870).

those terms there is implied any theory involving the supposition of an essential difference in age or mode of origin, yet the terms are convenient when they merely refer to the position occupied by the rock, whether on the highlands of the interior, or on the lowlands near the coast.

The fact is, that it is frequently impossible to determine whether the laterite is detrital or not, a difficulty well shown by the beds which occur interstratified with lower tertiaries near Surat. These beds are sedimentary, for marine fossils were found in one of them, yet they are not in the least sandy, and they resemble the laterite or iron clay of the Bombay Deccan so closely that, when these laterite beds were first examined by the Geological Survey, they were supposed to be volcanic rocks altered by surface action. The mistake, it should be added, was partly due to an apparent intercalation of basalt and nummulitic limestone, subsequently found to be caused by faulting.

The high level laterite is chiefly developed on the Deccan plateau, especially on the highest portions of the Sahyádri range, and of the spurs running from the Gháts. It forms a cap on the uppermost traps exposed on the plateau, being also found at lower elevations, but the lower beds are, as a rule, of small extent or thickness. The summit bed, as it is termed by Mr. Foote,[1] is not more than 50 to 90 feet in thickness in the southern Maráthá country. It is about the same at Mahábaleshwar, but from 100 to 200 feet at Bídar north-west of Haidarábád. It occurs at varying heights above the sea, 4,700 feet at Mahábaleshwar being probably the highest point, whilst at several places in the southern Maráthá country it is found capping ridges and isolated hills from about 2,000 to nearly 3,500 feet above the sea, always overlying the highest lava flow in the country. At Mátherán, near Bombay, and on some neighbouring plateaux, there are caps of laterite at about 2,000 feet above the sea, but these do not belong to the summit bed, as the traps on the tops of these hills are not high in the series. Some of the laterite caps are very extensive. The bed at Bídar is said to be twenty-eight miles long from west-north-west to east south-east and twenty-two miles broad, and the area of laterite at Kaliani, forty miles west of Bídar, is of even greater extent.[2]

The greater portion of the trap area in the Deccan has not been closely examined, but so far as is known, laterite is of rare occurrence, except near the Western Gháts and in the southern Maráthá country. A few very small caps are found south-west of Nágpur, in south-east Berar, and probably similar small outliers occur here and there along the south-eastern margin of the volcanic region. Farther to the north-east, laterite occurs at Amarkantak and on the eastern outliers of the Deccan traps, at Máin Pát and Jamíra Pát in Sárgúja, being from 100 to 200 feet thick on the

[1] *Memoirs*, XII, 203, (1876).

[2] Newbold, *Jour. Roy. As. Soc.*, VIII, 228, (1846).

former. North of the Narbadá also, in Rewá, Bundelkhand, and in other states as far west as Gujerát, laterite is found, sometimes as much as 200 feet in thickness, capping outliers of the trap series.

In all the localities hitherto mentioned laterite occurs resting upon the Deccan traps, but the high level laterite overlaps the traps, rests upon older rocks, and is found in places some hundreds of miles beyond any existing outlier of the volcanic series. Instances of this kind have been noticed by various observers in the southern Maráthá country,[1] the same laterite bed being apparently sometimes continued from the trap surface on to the transition or metamorphic rocks, whilst numerous outliers on the older formations are known to exist. Caps are said also to occur at high elevations on the Dambal or Kappatgod hills, east of Dhárwár, and on hills in the neighbourhood of Bellary and Cuddapah.[2] More to the north-east, in the high grounds of Patná, Káláhandi, Bastár, Jaipur, etc., between the Mahánadí and Godávari, caps of laterite, 50 to 100 feet thick, occur on many of the higher hills[3] at elevations of between 2,000 and 4,000 feet above the sea. The most eastern exposure known to occur in this neighbourhood is on the Kopilas hill, about 2,050 feet above the sea, and 12 miles nearly due north of Cuttack.[4] To the northward a great expanse of laterite is found on the Chutiá Nágpur plateau at elevations varying from 2,000 to 3,000 feet above the sea in several places, and especially to the north-west of Jashpur;[5] it caps ridges and peaks in the usual manner, but differs from the usual high level laterite in covering hills and valleys alike, and is probably in part a reconsolidated formation. Leaving, for the moment, the Rájmahál hills, which require separate notice, a thick mass of laterite occurs at an elevation of 1,500 feet on Moira hill, the highest peak of the Kharakpur range. Turning thence westward, caps of the same rock are found, outside of the trap area, at several places in Bundelkhand,[6] and at two near Gwalior,[7] all on the highest ground of the country.

Besides the above mentioned localities laterite has been reported to occur on some of the hills of Southern India, but ferruginous clays have possibly been described under the name of laterite, which have little of its true character. Such is the case with the Nílgiris, one of the localities mentioned by several geologists. No well authenticated occurrence of laterite is known at an elevation exceeding 5,000 feet above the sea.

There is, however, a very important bed of this rock on the Ráj-

[1] Newbold, *Jour. As. Soc. Beng.*, XIII, 996, (1844); *Jour. Roy. As. Soc.*, VIII, 228, (1846); Foote, *Memoirs*, XII, 205, 217, (1876).

[2] Newbold, *Jour. Roy. As. Soc.*, VIII, 228, (1846).

[3] Ball, *Records*, X, 169, (1877).

[4] The information of the occurrence of laterite on Kopilas hill was obtained by Mr. Ball from Dr. Stewart, of Cuttack.

[5] Ball, *Records*, X, 170, footnote, (1877).

[6] H. B. Medlicott, *Memoirs*, II, 82, (1860).

[7] Hacket, *Records*, III, 41, (1870).

mahál hills in Bengal.[1] These hills, like the highlands of the Bombay Deccan, are composed of bedded basaltic traps, and, as in the Deccan, the very highest bed consists of laterite, Mahuágarhí, the highest plateau in the range, 1,655 feet above the sea, being capped by this formation. The laterite in the Rájmahál hills is, in places, as much as 200 feet thick and it slopes gradually from the western scarp of the hills, where it attains its highest elevation, to the Gangetic plain on the east. Here, too, there is sometimes, as in the Deccan, an apparent passage from basalt into laterite, but the latter rock to the eastward is distinctly identical with the low level laterite of Bengal, and is clearly of detrital origin, whilst, even at considerable elevations in the hills, fragments, derived from the shales interstratified with the basaltic flows, are found imbedded in the laterite, so that, no distinct line having ever been drawn between the beds at different elevations, we appear in this case to have a passage from the high level into the low level laterite, and reasons for supposing that both were originally of sedimentary origin. The case, it should be remembered, is not clearly proved, the laterite of the Rájmahál country not having been specially examined with a view to test the connection between the beds to the eastward and those to the westward, but the two appear to be parts of the same formation, and it is certain that both are in this instance detrital.[2]

The evidence hitherto collected is insufficient to justify the conclusion that the high level laterite once formed a continuous bed, occupying the whole surface of the Indian peninsula from the Ganges valley to the neighbourhood of Madras, but the manner in which caps now occur upon isolated peaks and ridges clearly shows that they were once much more extensive, and that only the remnants have been left undenuded. It is difficult, in presence of the great amount of denudation which has taken place since the laterite caps were part of a more extensive formation to escape the conviction that the high level laterite must be of considerable geological antiquity.

Before proceeding to discuss the very difficult subject of the origin of laterite, it will be best to point out the general distribution of the low

[1] The laterite has been but briefly noticed; see T. Oldham, *Jour. As. Soc. Beng.*, XXIII, 273, (1854); Ball, *Memoirs*, XIII, 222, (1877).

[2] There is a possibility that the connection between the high level laterite and the low level laterite of the eastern coast is not confined to the solitary instance of the Rájmahál hills, although no other equally well marked case of passage can be traced, and in some cases, as at Kopilas near Cuttack, the difference in level is very great; but the low level laterite of the eastern coast rises gradually from the neighbourhood of the sea, at a slope which, if continued inland, would connect the bed with the high level formation. The latter is of greater antiquity than the low level bed, but the process of formation may have been continuous, the rock now found at a higher level being first formed, that at a lower elevation being gradually consolidated as the lower portion of the country was raised above the sea. It should be remembered that the higher part of the country was, in all probability, never depressed below the sea level.

level laterite, especially in the neighbourhood of the coast. On the west coast of the Peninsula laterite has not been observed in the Konkan, or lowlands, north of Bombay,[1] it appears, a little farther to the southward, between Bombay and Ratnágiri, and extends thence throughout large tracts of the low country, intervening between the Sahyádri range and the sea, as far as Cape Comorin. It does not, of course, cover the whole surface. In many places it has been cut away by streams, so that the lower formations are exposed, and in parts of the country it appears to be wanting. The greater part of the region, however, has never been geologically mapped, and very few details of the distribution of laterite are available.

In the country between Ratnágiri and Goa the rock appears to form a plateau, having a general elevation of 200 to 300 feet above the sea. On the coast it terminates in cliffs, the trap being exposed beneath it. The plateau extends for from 15 to 20 miles inland, and is cut through by numerous rivers and streams, in all of which the trap is exposed, the lignite and clays, which were mentioned in the last chapter, being found between the laterite and the traps at Ratnágiri. Farther inland the laterite is found at a higher elevation than near the coast, so that the rock appears to have a low slope towards the sea. The laterite is distinctly of detrital origin, and even conglomeratic in places, the thickness is considerable, but no exact measurements have been recorded, except at Ratnágiri, where it amounts to 35 feet, probably less than the average. It is evident that the plateau formerly extended much farther to the eastward, and it probably covered the whole of the country as far as the base of the Sahyádri range, for caps of laterite are found in places on the trap hills, and masses, reconsolidated from the detritus of the principal beds, are found at lower levels.

South of Malwán the underlying rock is no longer trap but gneiss, or some other metamorphic formation. The laterite, which is extensively developed, appears to be similar to that of the Bombay Konkan. In Travancore it overlies the fossiliferous tertiary beds.

On the east coast of India laterite occurs almost everywhere, rising from beneath the alluvium which fringes the coast, and sloping gradually upwards towards the interior, but this laterite is, as a rule, a much less massive formation than the rock of the western coast. It is seldom more than 20 feet in thickness, and is often represented by a mere sandy or gravelly deposit, not more than four or five feet thick. Where it is thicker the lower portion usually consists of lithomarge, produced by the alteration of the underlying rock. The laterite is frequently conglomeratic, and

This idea of the whole laterite being one continuous formation appears to have occurred to Newbold.—*Jour. Roy. As. Soc.*, VIII, 240, (1846).

[1] Except near Surat, where the outcrops are of nummulitic age. The rock differs from all superficial laterite, in being distinctly intercalated between other beds.

includes large rounded, or subangular, fragments of gneiss and other rocks, good instances being found at Trichinopoli, at many places near Madras, amongst which are the Red Hills, seven miles to the north-west of the city, and around the detached hills north-west of Cuttack, in Orissa. In the Madras area quartzite implements of human construction have been found in the laterite in considerable numbers.[1]

The fringe of laterite is of very unequal width. In places it forms a broad, low slope, stretching for many miles from the edge of the alluvium; in others it only remains as caps upon the older rocks. In one form or another it appears to be traced, at short intervals, from Cape Comorin to Orissa, and thence northward through Midnapur, Bardwán, and Bírbhúm, to the flanks of the Rájmahál hills, where it is well developed and, as already noticed, it appears to pass into the high level laterite.

The low level laterite is not confined to the neighbourhood of the coast. It is frequently found in patches over many parts of the country, but these patches are rarely of large size and they often appear to be due to local conditions, such as abundance of iron in the rocks, or reconsolidation of fragments derived from a bed of high level laterite. Many such lateritic deposits are rather of the nature of ferruginous gravel than of true lateritic. The small pisolitic nodules, so characteristic of some forms of laterite, are found abundantly in the older alluvium of the Ganges valley, and in many other superficial deposits in the plains of India, and whenever they are sufficiently abundant, appear to become cemented, with the accompanying sand and clay, into a rock closely resembling laterite in many of its peculiarities.

In Burma, laterite of the detrital low level type is found in places on the edge of the alluvial tracts of the Irawadi and Sittaung rivers in Pegu and Martaban, forming as usual a cap to other rocks, and having a very low dip towards the river from the sides of the valleys. The laterite appears to form the basement bed of the post-tertiary gravels and sands, and laterite gravels, apparently derived from the denudation of the massive laterite are largely dispersed through the older alluvial deposits.

West of the Irawadi only a few patches of laterite occur in the Myanaung district, but the rock is more common along the western foot of the Pegu Yoma. To the east of that range laterite is generally wanting, but there is a well marked belt of this formation along the base of the metamorphic hills east of the Sittaung river, forming a plateau which rises 50 or 60 feet above the alluvium of the Sittaung valley. Some laterite is also found in Tenasserim, whence it extends into the Malay peninsula.

Having thus stated, as briefly as is consistent with the object of afford-

[1] R. B. Foote, *Memoirs*, X, 27-58, (1873).

ing a tolerably complete account of the rock, the distribution and mode of occurrence of the different varieties of laterite, the question of the manner in which this rock has been formed, must next be considered. The subject has already been noticed as difficult, the difficulty arising from the fact that the rock has evidently undergone a considerable amount of change, both chemical and structural. The difference between laterite, when first cut from the quarry and the same rock after exposure, is well marked. The rock becomes harder, and the hardening appears not merely due to the desiccation of the argillaceous constituents, but also to a change taking place in the distribution of the peroxide of iron, the change being shown by the colour becoming darker, and by the surface being covered with a glaze of limonite. Whether the anhydrous iron peroxide, which occurs in some forms of high level laterite, becomes converted by exposure into hydrated peroxide, has not been ascertained, but it is quite clear that the process of segregation of the iron has tended greatly to obscure any structure which may have existed originally in the rock, and that this segregative action is constantly in progress. It has already been stated that iron has been dissolved out of the laterite and redeposited in the underlying lithomarge, where the latter is merely an altered form of the rock beneath, and it is a common circumstance to find pisolitic nodules of hydrated iron peroxide, evidently due to segregation, in some forms of laterite. These facts, and the process by which the surfaces of the rock, and of the tubes by which it is traversed, become coated with a glaze of limonite, render it evident that a transfer of iron oxide from one part of the rock to another is continually going on.

One view, which has been held by several good observers and has been strongly supported by Mr. Foote's examination of the laterite or iron clay in the southern Maráthá country, is that the high level laterite is simply the result of the alteration *in situ* of various forms of rock, and especially of basalt, by the action of atmospheric changes. Many of the dolerites of the Deccan contain iron in the form of magnetite, and large quantities of magnetic iron sand are found in the beds of streams which flow over the traps, whilst bands, both of magnetite and hæmatite, are locally common in the metamorphic rocks. The gradual change from doleritic trap into laterite has been noticed by several observers,[1] and so far as the Deccan alone is concerned, the evidence in favour of laterite being merely the result of atmospheric change acting upon very ferruginous volcanic rocks, appears so strong that, if there were no conflicting phenomena, it might be accepted as a satisfactory explanation. At the same time there are some difficulties, to which attention was first called by Captain

[1] Voysey, *Jour. As. Soc. Beng.*, XIX, 274, (1850); Foote, *Memoirs*, XII, 202, (1876).

Newbold,[1] and although Mr. Foote[2] has shown that they are not insuperable, they must not be overlooked, because the apparently sedimentary origin of the rock, in Bundelkhand and elsewhere, tends to invalidate the conclusion that the high level laterite is merely the result of surface change.

The main argument in favour of supposing the high level laterite of the Deccan to be merely altered basalt, is that the two rocks are seen to pass into each other. This fact, which is unquestionably established, may be considered proof that laterite *may* result from the alteration of basalt or a similar rock, but it is, of course, insufficient evidence to show that such is the origin in all cases. It is always possible that the upper portion of the laterite is, in each case, of extraneous origin, and that the surface of the basalt beneath has been affected by the infiltration of iron, in the manner already described when explaining the origin of lithomarge. Numerous instances are found, on the other hand, in which the laterite rests upon the surface of basalt, which is either hard and unaltered, or soft and decomposed, without any appearance of a passage from one rock to the other. But this, again, is no proof that the laterite above the unaltered trap is not itself the result of alteration of a different lava flow, the rock beneath not being susceptible of the same change. It is clear that the evidence afforded by the circumstance that basaltic trap sometimes passes into laterite, and sometimes does not, is insufficient to decide the question as to whether the latter is derived from the former by a process of chemical alteration.

It has been stated that magnetite occurs in many of the Deccan basalts, but until far more analyses have been made, it is impossible to say whether any of these rocks contain as large a proportion of iron as the laterite. It is probable that some may, but it is certain that so large a proportion of iron as 15 or 20 per cent.[3] in any basalt is exceptional, yet this is not above the average amount in the Deccan laterite. At the same time the larger percentage may perhaps be explained by a process of concentration, some of the other constituents of the rock having been removed, in the manner explained further on, but not the iron.

One difficulty, to which especial attention was drawn by Captain Newbold, is the complete absence in the laterite of those nodules, large or small, of various forms of silica, such as agate, jasper, and crystalline, quartz, so frequently found in the different forms of trap. It is difficult to understand, if laterite simply results from the alteration *in situ* of the Deccan basalts, why no amygdaloidal structure, especially where the amygdules contain so indestructible a mineral as agate, should be detected

[1] *Jour. As. Soc Beng.*, XIII, 995, (1844); *Jour. Roy. As. Soc.*, VIII, 238, (1846).

[2] *Memoirs*, XII, 203, (1876).

[3] That is of metallic iron; 15 per cent. of iron corresponds to 19·3 per cent. of protoxide, and 21·4 per cent. of sesquioxide.

in the altered rock. Mr. Foote suggests[1] that, in the case of the summit bed, which appears to rest upon the highest traps, the absence of amygdaloidal structure may be due, in the first place to the lava flow, having been of a peculiarly dense nature,[2] and secondly to the fact that, being the uppermost flow, the water which percolated it did not contain silica in sufficient quantity to form silicious nodules in the vesicular hollows. He also points out that the underlying bed into which the summit bed laterite is seen to graduate in several sections, is a very argillaceous rock without vesicular cavities or enclosed minerals.

One conclusion is clear. If the high level laterite of the Deccan has been produced by the alteration *in situ* of volcanic rocks, only particular varieties of such rocks are capable of undergoing the alteration. If all were similarly liable to be converted into laterite at the surface, the occurrence of that rock would be more general, and less restricted to particular elevations. The great difficulty in the way of explaining the origin of the high level laterite, so widely spread in Málwá and the Deccan, by a simple process of atmospheric alteration is, in brief, that the hypothesis demands the occurrence, over an enormous area of country, of a volcanic rock, whether a tuff or a true lava flow is immaterial, of peculiar and unusual composition, containing a much larger proportion of iron than usual, and wanting the amygdaloidal structure, so common in the Deccan traps. This difficulty, it must be remembered, is, so far, only a reason for caution in coming to a conclusion, and does not show that the hypothesis of alteration *in situ* is impossible.

The great extension of the laterite beyond the trap area might be explained by supposing that the highest volcanic stratum covered a wider surface than any of the inferior lava flows, but this theory is untenable in some cases, for instance in that of the Gauli plateau, south of Belgáum,[3] where a bed of laterite at a lower level than the summit bed was traced by Mr. Foote on both sides of the Mahádáyi ravine, passing into the underlying trap to the northward and into metamorphic rocks to the south, as is represented in the sketch section fig. 20, given on the following page. In this case, the southern portion must have been formed from gneissic rocks if the laterite be the result of alteration alone, and it is difficult to understand how two rocks, so totally dissimilar in constitution as basalt and gneiss, can have produced precisely the same rock, by a simple process of disintegration *in situ*.

On the other hand, the difficulties in the way of supposing the high level laterite to be sedimentary are considerable. The idea of its being a marine deposit can scarcely be entertained, as there is not a shadow of

[1] *Memoirs*, XII, 203, (1876).

[2] Such dense beds do certainly occur in the Deccan traps, indeed they cannot be said to be rare, although they do not, as a rule, preserve their non-vesicular character over large areas.

[3] *Memoirs*, XII, 217, (1876).

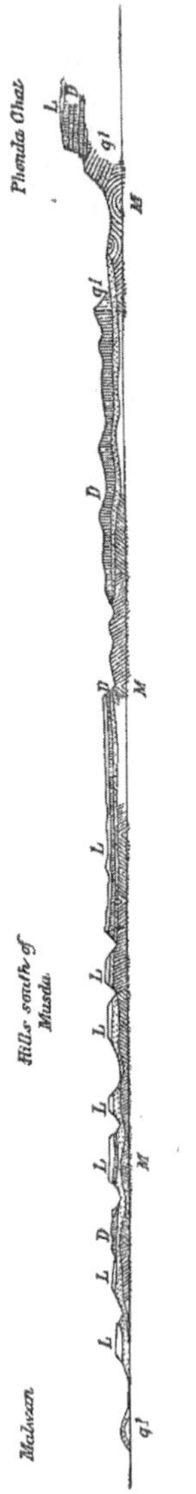

Fig. 20,—Section across the Gauli plateau, after Foote. L=laterite; D=Deccan trap; M=gneiss; Q=Kaládgi quartzites.

evidence in any part of India to render it probable that the whole of the great trap plateau has been beneath the sea in tertiary times. It is inconceivable that fluviatile deposits should be so enormously extended, yet so thin. One objection, which at the first glance appears important, is apparent rather than real. It is that a sedimentary deposit could not be formed on the highest portions of the country, because there could be no higher land in the neighbourhood from which the sediment might be derived, whilst the singularly small amount of disturbance which the Deccan rocks have undergone renders it improbable that any great relative change of elevation has taken place. But it must be remembered that laterite is a rock which resists atmospheric action far more than most forms of doleritic trap, as is shown by the manner in which hard unaltered caps of laterite rest upon softened and decomposing basaltic rocks. Consequently those portions of the plateau which were originally highest may, if not capped by laterite, have disintegrated more rapidly than those protected by the lateritic formation, until the latter remained, forming the highest ridges, long after the unprotected portions had been swept away.

The evidence afforded by the laterite outliers in Bundelkhand[1] is distinctly opposed to the theory of alteration *in situ*. The whole group, laterite above underlaid by ferruginous clay, frequently containing sand and pebbles, is found indifferently capping the trap and Vindhyan sandstones. Now, whatever may be the case with dolerite and gneissic rocks, no conceivable process of alteration could convert a purely quartzose rock, containing a mere trace of iron, like the Vindhyan sandstone, into an argillaceous one with 20 per cent. of iron entering into its composition, and the circumstance that the lower portion of the lateritic group is clearly detrital, proves that the laterite is not an altered outlier of the Deccan trap.

It appears almost impossible to separate the Bundelkhand laterite from the high level laterite of the Deccan. Lithologically and stratigraphically

[1] *Memoirs*, II, 79—86, (1860).

the two rocks are identical. There can be no reasonable doubt that the trap once occupied the surface of the ground now cut out into valleys by the feeders and main streams of the Son, Narbadá, and Mahánadí, and that Bundelkhand and Málwá were united with Mandlá and Sargúja into one plateau of horizontal trap rocks. If this be conceded, and it appears impossible to doubt it, the caps of laterite near Ságar (Saugor), occupying precisely the same relative position as those at Amarkantak and the Máin Pát, may fairly be considered part of the same bed, and the Rewá outliers, which are probably either beyond the original range of the trap, or else on ground which was above the general trap level, must be referred to the same origin. Now the Amarkantak and Sargúja laterites are not merely similar in every respect to the other Deccan high level outcrops of the rock, but they appear to be connected, by a series of small caps at intervals, with the typical formation of the southern Maráthá country. There may be a break in the chain, as the distances are too great for any safe conclusions to be formed, and all that can be done is to indicate the probabilities, but it appears to be a fair inference that, if the Bundelkhand laterite is of detrital origin, the rock of Amarkantak and the Deccan is the same.

The laterite of the Rájmahál hills is separated by so great a break from that of Sargúja, and the Rájmahál traps are in all probability so much older than those of the Deccan, that it is impossible to say whether the Rájmahál laterite is of the same age as that of central and western India. Lithologically it is identical, and like the Deccan laterite it occurs, in part at least, at a considerable elevation, whilst its sedimentary origin has already been mentioned.

On the west coast of the Peninsula we have a careful study of the low-level laterite of south Malabar by Mr. P. Lake.[1] He describes two distinct varieties, the first, distinguished as vesicular laterite, is characterised by numerous vermicular branching and anastomising tubes which, in the portion of the rock not exposed to the air, are filled with a white or yellow clay, containing a much smaller proportion of iron and a larger percentage of potash than the walls of the tubes. Away from the surface, as the rock becomes less affected by the weather, the distinction between the tube walls and their contents gradually disappears, till what was laterite above passes into clay below. The laterite is thus seen to be the result of a sort of superficial concretionary action, the iron tending to segregate in the form of tubes from which the clayey non-ferruginous parts are washed out, leaving the resulting laterite with a higher proportion of iron than the clay from which it was formed.

[1] *Memoirs*, XXIV, 217, (1890).

The other variety of laterite, distinguished as pellety, is derived from the vesicular laterite. After the tube contents have been washed out the tube walls slowly break up into little subangular pieces, which are washed away and deposited at a lower level, where they are recemented into a rock, usually much more solid than the first variety. It consists of small irregular ferruginous pellets cemented by a similar material, is devoid of vermicular tubes, and has a higher percentage of iron than the unexposed parts of the vesicular variety, the percentage being probably much the same as in the tube walls.

The origin of the pellety form is sufficiently obvious: it is a detrital rock formed of the debris of the vesicular form, which is regarded by Mr. Lake as a product of decomposition *in situ* of the gneiss. The gneiss of Malabar, consisting principally of quartz, felspar, hornblende or mica, and garnets, weathers readily into a ferruginous clay. If exposed to the weather, the iron segregates and hardens the clay, where it collects, while the rest, deprived of ferruginous cement, is loose and easily washed away. As this process goes on the gneiss becomes covered with a cap of laterite, gradually increasing in thickness till it acts as a protection to the underlying gneiss from further decomposition.

There can be no doubt that the vesicular laterite is, in part at least, formed by a laterisation of the decomposed gneiss, for the lamination of the gneiss can often be traced into the laterite, and the more quartzose beds stand up as ribs several feet into the laterite which has replaced the more felspathic portions on either side.

One more hypothesis of the origin of laterite requires notice. Mr. F. R. Mallet, in noticing the resemblance of certain ferruginous beds with underlying bole, interstratified with the volcanic rocks of Ulster,[1] to the laterite and lithomarge of India, suggests that the laterite is of lacustrine origin. During the decay of vegetable matter in the presence of the higher oxides of iron oxygen is absorbed, reducing them to protoxide, which unites with carbonic acid, a product of the decomposition of vegetation, to form ferrous carbonate, soluble in water containing an excess of carbonic acid. When the water carrying this ferrous carbonate in solution is exposed to the air in streams or lakes it absorbs oxygen, the ferrous carbonate is decomposed with the escape of carbonic acid and a re-oxidation of the ferrous oxide into insoluble ferric oxide, which is precipitated wherever the water comes to rest in a lake or marsh.

On this hypothesis the high level laterite would have been formed in the shallow depressions left between the lava flows and ash heaps of the surface of the land at the close of the Deccan trap period, while the low

[1] *Records*, XIV, 139, (1881).

level laterite, excluding the ferruginous gravels, would have been formed in the depressions on the plain of marine denudation whose surface it caps. As already explained, the resistance which laterite offers to denudation would be sufficient to account for its being now found capping the hills, and if it were originally formed in the depressions of the surface, for the complete reversal of contour is an indication of the time that has elapsed.

From what has gone before it will be seen that the subject of the origin of laterite is still wrapt in obscurity. None of the various hypotheses that have been propounded is completely satisfactory, nor is it possible to come to any final conclusion till an agreement is come to as to the meaning of the word laterite. It must be used either as purely a lithological, or as a chronological, term, not indifferently as either, but as the word was originally intended to describe a peculiar variety of rock, irrespective of its age or origin, for which too a fresh name would have to be adopted if the familiar one were abandoned, it is the lithological sense of the word which it is most desirable to retain.

Using the term as defined by its proposer for a vesicular, highly ferruginous, clay, soft in the mass but readily hardening on exposure to the weather, it seems that there is nothing essentially volcanic in the rock. The high level laterites are doubtless derived, directly or indirectly, from the debris of volcanic rocks, but it is to the gneiss and granite that we must look for the ultimate source of the laterite of the south Konkan, Travancore, and Ceylon.

According to some geologists this laterite is in reality a soil and formed by the direct decomposition *in situ* of the underlying rock, but some doubt attaches to the observations on which this conclusion is based, as the apparent transition may be due to an infiltration of iron from the overlying laterite and a conversion of the decomposed portion of the underlying rock, where it is suitably argillaceous, into laterite. It is certain that, though laerite appears to a certain extent to be still in course of formation, the bulk of the rock is of ancient date, for it is now found on the summits of steeply scarped trap hills, or a deeply eroded plain of marine denudation, the exceptions being for the most part those cases where it is palpably or probably of detrital origin, and derived from pre-existing laterite.

There are two difficulties in the way of accepting this explanation without modification. The first is the large proportion of iron present in laterite, a proportion which is only exceptionally found in any volcanic rock, and much exceeds that present in any of the gneissose rocks, apart from certain highly ferruginous bands. Such local and exceptional accumulations of iron are not sufficient to account for the amount present in the laterite caps far removed from them, nor can the concentration of iron caused by the washing away of the less ferruginous and consequently less coherent portions, account for its presence, any more than for the argillaceous nature of the rock, where it rests on sandstones almost devoid of any trace of clay or iron. The supposition that laterite is derived from the decomposition of a specially ferriferous eruption, marking the close of the Deccan trap period, is insufficient to account for the facts even within the Deccan trap area, and quite fails when applied to the remote exposures, far beyond the limits of the area within which the influence of these eruptions could have been felt.

One objection which might be urged to the hypothesis of the origin of laterite by the decomposition *in situ* of volcanic rock, that it should in that case be found interbedded with the trap, may be dismissed. However laterite originated, time was certainly a factor in its formation, and the occurrence of beds of bole suggests that the same causes which subsequently led to its formation, were at work during the Deccan trap period, but that the rapid succession of lava flows did not leave them time to produce the full effect which resulted when the eruptions had ceased.

The second difficulty is the great thickness of some of the patches of high level laterite, which seems to preclude their having been formed as a soil, by the segregation of the oxide of iron and removal of the less coherent non-ferruginous portions, and the more probable explanation is that they were formed in lakes or marshy hollows by the deposition of oxide of iron from the stagnant waters, mixed more or less with fine grained ash and decomposed volcanic debris. The laterite of the lowlands may have been similarly formed in marshy hollows, left on the surface of a plain of marine denudation after its elevation above sea level. It is described as a thin, fairly uniform layer covering the undulations of this plain, but in Mr. Lake's description of south Malabar there are said to be gneiss hills, or islands, rising from this plain, on which no laterite is found, and if the laterite were a direct product of decomposition of gneiss, it is difficult to see why it should be restricted to the lowland near the coast, which was evidently once covered by the sea.

As to the conditions necessary for the formation of laterite little can be said. Those countries where it has been supposed to be still in process of formation are characterised for the most part by a warm, moist, climate, and an abundant vegetation. But there is one characteristic of all the laterite regions that appears to be important; the laterite is without exception only found on level or gently undulating surfaces, if we ignore the irregularities produced by subsequent denudation. It is found on the terrace bordering the sea coast, and on the plateaux capping the hills further inland, but whenever a rock which could pass for laterite is found on the intermediate slopes, it is clearly of derivative origin. The rounded surfaces of the gneiss hills of Ceylon and Southern India are often covered to a great depth with a more or less ferruginous subsoil, which never passes into laterite, except in such localities as the summit plateau of the Shevaroy hills or the plain, now intersected with valleys of denudation, which borders the sea coast. It is also said to be found in the bottoms of the valleys of south Malabar, but wherever laterite is found on the sloping ground, it is clearly derived from the disintegration of some bed at a higher level.

The geological age of the high level laterite must, of course, remain undetermined, until the mode of formation has been more definitely ascertained. If the rock be merely the result of surface alteration, it may be of any date subsequent to the termination of the volcanic outbursts. Indeed, it must still be in process of formation, as has been justly pointed out by several observers. But, as its occurrence in the form of a few isolated caps shows that it was once a much more extensive formation, it must have existed before the denudation of the area had much advanced, and must, therefore, have been formed, in part at least, soon after the termination of the volcanic eruptions. The great similarity between the high level laterite and the beds of the same rock interstratified with the nummulitic limestones and gravels of Gujarát and Cutch, tends to suggest the possibility that the two are contemporaneous, and also that they may have been produced in the same manner, with this important distinction, however, that the Gujarát beds are marine, whilst there does not appear to be any evidence in favour of supposing that the highlands of the Deccan were submerged during any portion of the tertiary period. Had they been submerged, the amount of denudation which the traps must have undergone would, in all probability, have caused the high level laterite to be more distinctly unconformable. At the same time, it is far from clear that the laterite is truly conformable to the highest trap flows. It has been hitherto assumed, rather than proved, that all the beds of laterite, at lower elevations than the summit bed, are of later age. The occurrence of laterite at various elevations presents no difficulty on the theory of the laterite being an altered form of the traps, but if this rock be of any definite date, it is clear that extensive denudation must have reduced the level of such hills as Mátherán, the uppermost beds of which are at least 2,000 or 3,000 feet below the highest volcanic flows, before the laterite was deposited. Nevertheless, the laterite of Mátherán, although apparently non-detrital, may be a secondary product. This question of the conformity of the high-level laterite to the highest traps requires, in fact, further investigation.

Whether the true laterite of the low grounds near the sea is to be regarded as newer than that of the high level plateaux depends on the hypothesis of origin adopted. If they are both products of decomposition *in situ* of the underlying rock, they may both be of the same age, but if, as seems probable, the formation of the laterite was anterior to the excavation of the valleys which now limit its extent, then the low level laterite is shown, by the less degree of denudation that has taken place since its formation, to be newer than the high level. On any hypothesis, except the impossible one of direct volcanic origin, there is no reason why the production of laterite should be restricted to any particular geological age, and Mr. Lake's observation of the apparent passage of gneiss into laterite in the bottoms of some of the valleys of south Malabar,[1] would bring the date of origin of some part of the laterite down to a, geologically, very recent period.

The foregoing remarks regarding the origin of laterite refer only to what may be called the non-detrital form, in the sense that it is not palpably formed of the debris of pre-existing rocks of a similar nature; but the bulk of what has generally been described as low level laterite, principally along the east coast, is evidently of clastic origin, and is often merely a sub-recent gravel with a ferruginous matrix. Although, as has been explained, it seems advisable to use the term laterite in a purely lithological sense and not to apply it to such rocks, even though there is every degree of transition between the two, it is impossible to ignore these so called laterites, seeing that the term has been so generally used in the past.

Like the true laterite of south Malabar, they are found resting on what appears to be a plain of marine denudation, and the enquiry naturally arises as to whether the low level laterite is a marine formation. *A priori* it would appear improbable that a marine formation should be deposited during the process by which a plain of marine denudation is elevated above the sea. On the other hand, the frequent occurrence of pebbles, often of large size, in the laterites of the east coast appears due to the action of the waves, especially where, as around the isolated hills in Orissa, which may originally have been islands, a mass of well rounded shingle, in every way resembling a beach, is found cemented together by laterite. The absence of marine fossils may be due to their having been obliterated by the forces which produced the peculiar concretionary structure of the rock.

There are, however, two circumstances which appear to militate strongly against considering the laterite a marine formation. One of these is its remarkable thinness, which, so far as is known, rarely exceeds 20 feet along the east coast, and the other is the very frequent occurrence, in the Madras country, of palæolithic implements imbedded in the rock. Some of these might have been dropped into the sea from canoes, but

[1] *Memoirs*, XXIV, 226, (1890).

it is incredible that the men who used the stones should have lost them in the sea in such numbers as would account for their present abundance.

On the whole, it appears most probable that the low level laterite is a subaerial deposit, due, however, in many cases, to the rearrangement of marine gravels and sands by rain and streams. All rain and stream action would tend to carry away the lighter sand and clay, and to leave behind the heavy iron sand, and to this may be due the concentration of the ferruginous element.

The presence of palæolithic human implements in the Madras laterite proves that the rock is of post-tertiary origin. The implements[1] found are chiefly of quartzite, and have evidently been fashioned from pebbles, derived originally from the rocks of the Cuddapah system.

Despite the geologically recent origin of the low level laterite, the considerable amount of denudation which it has undergone shows that it is, in part at least, a formation of ancient date, counting by years. It has already been mentioned that, on the west coast the plateau near the sea has been cut through by streams to a great depth, and the underlying trap exposed, and that farther inland, at a higher level, only a few caps of the low level laterite remain. On the eastern coast, which, owing to the large amount of deposits brought down by rivers, is protected from the action of the sea, the laterite has undergone less denudation, in consequence of its being frequently covered by later alluvial deposits, but still it has been removed by atmospheric action over large areas away from the coast. It is probable that the land rose very slowly from the sea, the laterite forming on the raised slope *pari passu* with the elevation, and that, consequently, the farther inland the rock the older its date, and the longer the period during which it has undergone denudation from atmospheric agencies. But the deep ravines cut by the streams close to the western coast, near Ratnágiri, mark the lapse of a considerable period of time since the low level laterite was first consolidated, and a curious piece of evidence of the same kind has been recorded by Mr. Foote[2] in the neighbourhood of Madras.

Between two villages called Amerumbode and Maderapaucum, east of Sattavedu, and about 30 miles north-west by north of Madras, are some stone circles, made of blocks of the laterite, in which palæolithic implements are found in abundance in the immediate neighbourhood. The stone circles are known in the country as Karambar rings, and precisely similar rings of stone are found in many parts of India, associated with various other rude stone buildings such, as kistvaens and cromlechs. That these stone circles are of much later date than the palæolithic

[1] R. B. Foote, *Mad. Jour. Lit. Sci.*, Oct. 1866, 3rd series, Pt. 2, p. 1; also *Quart. Jour. Geol. Soc.*, XXIV, 484, (1868).

[2] *Memoirs*, X, 47, (1873).

implements is evident, first because in the particular case near Madras the circles are constructed of rock in which the implements are imbedded, and secondly, because iron implements, which mark a far more advanced stage of human progress, have been repeatedly found within the enclosures. Nevertheless, the stone circles themselves must be the work of a very ancient period, for all record of their construction, or even of the people who built them, has passed away.

CHAPTER XVI.

PLEISTOCENE AND RECENT DEPOSITS.

(Exclusive of the Indo-Gangetic alluvium.)

PENINSULAR AREA—Extent, and distinction from tertiary beds—Various forms of posttertiary deposits—Cuddalore sandstone—Warkalli beds—Cave deposits—Older alluvium of the Narbadá, Tápti, Godávari and Kistna—Newer alluvium of the east coast—Smooth water anchorages and recent deposits of the west coast—Lake deposits—*Regur*, or cotton soil—Blonw sand—EXTRA-PENINSULAR AREA—Hills west of the Indus—North-West Punjab—the Himálayas—Eastern frontier—Alluvium of the Irawadi.

The posttertiary (postpliocene, pleistocene or quaternary) and recent formations of India occupy an immense area. They form the wide plains of the Indus, Ganges, and Brahmaputra, and cover large tracts of country south of the Gangetic and east of the Indus plain. No older formation is exposed throughout the greater portion of the belt of alluvial lowland fringing the eastern coast of the Peninsula, and subrecent accumulations occupy a large area in Gujarát and in some other districts near the western coast. Large deposits in the valleys of the peninsular rivers and upon the fertile plains of the interior are also of recent or subrecent origin. The most important and extensive of these forms the great Indo-Gangetic plain, and, as the extent and variety of the recent and subrecent deposits render it impossible to treat of them all in a single chapter, its description, with all the important and interesting questions it raises, will be deferred for the present.

It is very difficult to draw a clear and distinct line between tertiary and posttertiary formations in India. The limit of the two in Europe coincides with the glacial epoch, but as no physical trace of this cold period has been detected in peninsular India, the distinction between the pliocene tertiary formations and the postpliocene beds is there less easily defined. Practically, no difficulty has hitherto arisen, because the tertiary beds of the Peninsula are comparatively unimportant, and those which occur belong to the older or middle tertiaries, and not to the newer beds, so that there is a marked break between the tertiary and posttertiary deposits; but in the extra-peninsular area, where the uppermost tertiary deposits are largely developed, it is often extremely difficult to say where the line should be drawn.

In dealing with the recent and subrecent deposits of India it is impossible to observe a strictly chronological order, and it is necessary to classify them more or less according to their nature. In the case of those in peninsular India the following classification will be followed, the oder being roughly, though not strictly, that of their date of origii. :—

1. Subrecent or doubtfully tertiary deposits of the coastal region.
2. Cave deposits.
3. Older alluvial deposits.
4. Newer alluvial deposits of the river valleys and deltas.
5. Raised littoral accumulations of sand, shells, etc.
6. Soils.
7. Blown sand.

Along the eastern coast of the Peninsula, from the neighbourhood of Rájámahendri to the Tinevelli district, a peculiar formation, consisting chiefly of sandstones and grits, is found underlying the laterite and associated gravels which form a low slope on the edge of the east coast alluvium. This sandstone formation has received several local names, but is now generally known as the Cuddalore sandstones,[1] from being well developed in the neighbourhood of the civil station of Cuddalore, about 100 miles south of Madras.

The greater portion of the Cuddalore group, throughout the area in which it is found, consists of gritty and sandy beds, sometimes highly ferruginous, and coloured of various tints of yellow, brown, red, and purple, sometimes white or pale coloured, and not infrequently mottled. In some cases the rock is argillaceous, and occasionally thin bands of clays or shales are interstratified. The beds are soft, loose textured and, as a rule, ill consolidated, being rarely sufficiently compact to be used as building stone. Bands of conglomerate have been found.

As already stated, these beds have been traced throughout a large portion of the east coast. Their most northerly extension known is between Vizagapatam and Rájámahendri. The coast north of Vizagapatam, as far as the Chilká lake, has not been examined geologically, and throughout Orissa no outcrops of the Cuddalore beds have been detected, but there is a possibility that they may be represented by some clays and sandy beds associated with the laterite of Midnapur.[2] There is rather more probability that certain sandstones, grits and clays, which occur east of Ráníganj, and extend northwards as far as Surí in Bírbhúm, belong to the same group as the Cuddalore sandstones of Madras.

From the neighbourhood of Rájámahendri the Cuddalore beds have been mapped at intervals along the coast for fully 600 miles to the

[1] For further information see more particularly, H. F. Blanford, *Memoirs*, IV, 165, (1863); King and Foote, *Memoirs*, IV, 256, (1864); Foote, *Memoirs*, X, 59, (1873) and XX, 35, (1883).

[2] *Memoirs*, I, 268, (1859).

southward. They usually form a low slope, dipping at a very slight angle to the eastward, in the direction of the sea, and are, as a rule, much covered and concealed by the deposits associated with the low level laterite of the east coast. To the westward they rest indifferently, but always unconformably, upon rocks of various ages,—metamorphic, jurassic or cretaceous,—and they often terminate in this direction in a low scarp. To the eastward they disappear in places, with their capping of laterite, beneath the alluvium of the coast, but they quite as often, especially to the southward, terminate in a small cliff. Their outcrop is repeatedly interrupted by the broad alluvial valleys of rivers, and in some places, as for nearly 100 miles south of Madras, they appear to be wanting altogether, whilst in other parts of the country they form a broad tract, usually sandy and infertile, raised above the general level, occasionally no less than 25 miles wide from east to west, as near Cuddalore, but generally much less.

From the paucity of sections and the extent to which the Cuddalore sandstones are concealed by laterite and sandy soil, their absolute thickness can nowhere be estimated with accuracy. The scarp in which they terminate to the westward is sometimes as much as 100 feet high and they must be somewhat thicker than this, but it is doubtful if they attain any considerable thickness. They are perfectly undisturbed, and have all the appearance of being a comparatively late formation.

The only fossils found in the Cuddalore beds consist of exogenous silicified fossil wood, some of which is coniferous and has been described under the name of *Peuce schmidiana*.[1] The genus *Peuce* is not acknowledged by all palæobotanists, and it appears too ill defined to justify any conclusions as to the age of the rocks being founded upon its occurrence.

This silicified wood is especially abundant at Tiruvakarai (Trivicary), about fourteen miles west north-west of Pondicherri, whence the name of Trivicary grits has been applied by some writers to the local development of the Cuddalore sandstones. The trunks of trees occurring at this place are of large size, one having been found as much as 100 feet in length, while stems 15 to 20 feet long and 5 to 6 feet in girth are not uncommon. They occur prostrate, imbedded in ferruginous grit.

The age and mode of origin of the Cuddalore sandstones are obscure, as but little importance can be attached to the identification of one generic form of coniferous wood. They are quite unconformable to the cretaceous beds, which they overlap in a most irregular manner, as near Pondicherri, where they rest on beds of the Ariyalúr group, forming the plateau near the town, known as the Red hills; six miles further westward, and west of the belt of cretaceous rocks, they are seen resting on Utatúr beds near Tiruvakarai, whilst a few miles further west they completely overlap

[1] Schmid u. Schleiden: Ueber die Natur der Kieselhölzer. Jena, 1855, pp. 4, 36.

the cretaceous beds and rest on gneiss. Fragments derived from the cretaceous beds and containing cretaceous fossils have been found near Tanjore. Near Rájámahendri the Cuddalore sandstones overlie the Deccan trap, the jurassic rocks and the gneiss. It is safe, therefore, to conclude that the sandstones cannot be older than upper tertiary. And the discovery by Mr. Foote, in the Tinnevelli district, of subrecent marine beds, containing only living species of mollusca, associated with grits which he believed to represent the Cuddalore beds,[1] appears to show that an even later date should be ascribed to them.

The origin of these rocks is as obscure as their date. Occurring as they do, parallel with the coast, it is natural to suppose that they are of marine origin and have been formed near the shore, when the general contour of the coast was the same as it now is, though the level of the land was somewhat lower. But the complete absence, so far as is known, of all marine remains is not easy to explain. Coarse sandstones and grits are usually unfossiliferous, but in beds which have undergone so little change some casts of shells, at least, would probably be found in the more argillaceous strata, if they were of marine origin. At the same time it is difficult to suppose that the western coast of the Bay of Bengal can have formed part of a river valley in tertiary times, and it is equally improbable that stratified grits, sandstones and conglomerates, like those of the Cuddalore beds, can be a form of subaerial wash.

On the west coast of the Peninsula a series of clays and sandy clays with lignite beds near their base, known as the Warkalli beds, are found for about twenty miles along the Travancore coast, from about three miles north of Quilon to the same distance south of Warkalli, and have been supposed to represent the Cuddalore sandstones of the east coast. Some doubt may be expressed as to the correctness of this correlation, for the Warkalli beds are said to attain a thickness of 200 feet, double the greatest recorded thickness of their supposed representatives on the east coast, and their upper surface is said to be formed by the same plain of marine denudation which cuts the gneiss further inland,[2] pointing to an older date than that of the Cuddalore beds. The Warkalli beds are said to be underlaid by the limestones containing eocene fossils, which have been referred to in a previous chapter, but though there is no reason to doubt the presence of these eocene limestones, they have not been examined *in situ* by a competent geologist, and we have no information at present as to whether or not they are conformable to the overlying beds. If so, it would point to a greater antiquity for the Warkalli beds than there seems to be any good ground for ascribing to the Cuddalore sandstones.

[1] *Memoirs*, XX, 41, (1883).

[2] W. King, *Records*, XV, 92, (1882).

On the coast of Káthiáwár a subrecent marine limestone, largely used as a building stone in Bombay, is found. It is commonly known as Porebandar stone from the name of the port whence it is shipped or, using the name proposed by Dr. Carter, miliolite. The typical miliolite is a finely oolitic freestone, largely composed of foraminifera, which form the nuclei of the oolitic grains, but near the sea coast the limestone is not infrequently mixed with a large proportion of sand. In the eastern part of Káthiáwár these beds are only found near the coast; further west, however, they form the whole surface mapped as recent, and extend on to the tertiary rocks and the Deccan trap. They attain a maximum thickness of about 100 feet, are extensively false-bedded in thin layers, and, though clearly a marine deposit land shells, which were doubtless washed down by small streams or floods, have been found in some of the more impure beds.[1]

There is only one locality in the Indian Peninsula where mammaliferous cave deposits have been detected. This is at a place called Billa Surgam, a few miles north of Banaganpalli in the Karnúl district. The caves are in the limestone belonging to the Karnúl series, and situated at a higher level than the beds of the present drainage, their floor is incrusted with stalagmite, beneath which red marl, full of bones of animals, large and small, is found. These caves were first discovered by Captain Newbold, whose collections were, however, never described, nor can the specimens now be found. More recently they have been explored by the Geological Survey, and the collection of bones, some of which exhibit traces of having been shaped by man, described by Mr. Lydekker. The fauna, besides many living forms, contains five species, *Viverra karnuliensis*, *Hystrix crassidens*, *Atherura karnuliensis*, *Rhinoceros karnuliensis*, and *Sus karnuliensis*, which are extinct, though closely allied to living forms. But the most interesting feature is the occurrence of four types identical with, or closely allied to, living African forms; these are *Cynocephalus*, sp., *Equus asinus*, *Hyæna crocuta*, and *Manis* cf. *gigantea*.[2]

The older alluvial deposits are well represented in the valleys of the Peninsula. Leaving the deposits of the Indo-Gangetic plain for separate description, the various older alluvial deposits of the peninsular rivers

[1] *Jour. Bo. Br. Roy. As. Soc.* V. 313, (1857); Geological papers on Western India, Bombay, 1857, p. 756; *Memoirs*, XXI, 126, (1884).

[2] For further details, see T. J. Newbold: *Jour. As. Soc. Beng.*, XIII, 610, (1844); R. B. Foote, *Records*, XVII, 27, 200, (1884); XVIII, 227, (1885); R. Lydekker, *Records*, XIX, 120, (1886); *Pal. Indica*, series x, IV, pt. 2, (1886).

deserve notice, both on account of the area they occupy, and of the organic remains they have yielded. The rivers of the Peninsula may be divided into two main groups; the first comprises the Narbadá and Tápti, which flow westward and drain the central portion of the Peninsula; the second includes the Mahánadí, Godávari, Kistna, Penner, Cauvery, and several minor streams which flow eastwards into the Bay of Bengal.

In the first named of the two groups one striking peculiarity is noticeable. Extensive alluvial plains, composed of clays and gravels, exist in the valleys of the Narbadá and Tápti. In the Narbadá valley the principal plain extends from a little east of Jabalpur to Hardá, a distance of more than 200 miles, and varies in breadth from 12 miles to 35. There is a smaller plain further down the river, extending for about 80 miles from Bárwai to the Harin Pal south of Bágh, but it is comparatively ill marked, the alluvial deposits are, so far as is known, much less deep, and no mammalian remains have been found. In the Tápti valley there is a large plain in Khándesh, extending east and west for about 150 miles, and terminating to the eastward close to Burhánpur. This plain lies chiefly to the north of the river, and is probably in places as much as 30 miles wide, but its limits have not been accurately determined. It appears to be connected by a narrow alluvial belt to the south-east with the plain of the Purna,[1] a tributary of the Tápti, draining a great portion of Berar. The Purna plain is at a higher level than Khándesh and is about 100 miles long, and in places 40 miles broad, its eastern extremity being near Amráoti, so that the whole length of the combined Tápti and Purna plains is about 240 miles. The Tápti and both the Narbadá plains are closed on the west by rocky and hilly country, through which the river has cut a channel with a rapid descent, and in the case of the Narbadá, as will be explained presently, it is ascertained that the alluvial deposits of the upper basin extend to a considerable depth beneath the level of the river bed at the point of exit, so that the plain lies in a great rock basin.

In the valleys of the eastward flowing rivers, such as the Godávari, Kistna and Cauvery, there are no such broad and well defined alluvial plains as in the drainage areas of the Tápti and Narbadá. There are numerous extensive alluvial flats in many places, but they are far inferior in extent to the Narbadá and Tápti plains, and they appear to be chiefly due to the river having worn a broad valley through soft, or easily disintegrated rocks. This is especially the case on the Godávari and its tributaries, the alluvial portions of the river valley being in the Gondwána rocks, or else in the Deccan traps, whilst the river traverses rocky gorges,

[1] This is not quite certain however, the ground not having been properly surveyed. There is a considerable amount of rock exposed in the rivers between the two plains, but the fall from one to the other cannot be much more than 100 feet, to judge by the railway levels.

through the metamorphic rocks forming the various barriers, at the places where the valley leaves the softer formations. On the Narbadá and Tápti it is otherwise, the rocks underlying the alluvial areas, so far as is known, are of the same kind as those cut through by the rivers at their exit from the plains. It is not improbable that the formation of these well defined plains in the Narbadá and Tápti valleys, and the absence of similar flats on the Godávari and Kistna, may be due to the rise of the Indian Peninsula in posttertiary times having been, as already suggested, greater or more rapid to the westward than to the eastward.

Partly in consequence of mammalian bones having been discovered in considerable quantities, and partly because the geology of the neighbouring country is of so much interest and variety as to have attracted the notice of many geologists, the alluvial deposits of the Narbadá valley have received far more attention than similar formations on the banks of the other Indian rivers.[1] The great plain already mentioned as extending from Jabalpur to Hardá is chiefly composed of a stiff, reddish, brownish or yellowish clay, with numerous bands of sand and gravel intercalated. Kankar abounds throughout the deposit, and pisolitic iron granules are of frequent occurrence in the argillaceous beds. Occasionally pebbles and sand are found cemented together by carbonate of lime, so as to form a hard compact conglomerate. This rock is especially developed at the base of the alluvial deposits, and is often found forming a coating to the underlying rock, not only in the Narbadá but in many other river valleys. The clay is frequently quite devoid of stratification, but it appears never to attain any great thickness without sandy layers intervening. The river, in many places, cuts through the clays, sands, and gravels to the underlying rock, usually belonging to the transition series, and the section of old alluvial deposits on the banks of the stream never greatly exceeds 100 feet in depth, this being about the usual difference in elevation between the bed of the Narbadá and the general surface of the alluvial plain in the neighbourhood of the river. But in a boring which was made at Súkakheri, north of Mohpáni and south of the Gádawárá station on the Great Indian Peninsula Railway, a depth of 491 feet was attained, without the base of the alluvial deposits being reached; another bore-hole was made through alluvial beds close to Gádawárá station to a depth of 251 feet. Throughout the thickness of nearly 500 feet, no change of importance was detected in the alluvial formations. By far the greater portion of the beds traversed consisted of clay with calcareous and ferruginous grains, sand and pebbles being found occasionally throughout. The bottom

[1] For description of the Narbadá alluvial deposits see *Memoirs*, II, 279, (1860); VI, 227, (1869); *Records*, VI, 49, (1873); VIII, 66, (1875).

of the bore-hole was in lateritic gravel, and it is possible that rock was not far distant.

The evidence thus obtained of the depth to which the alluvial deposits of the Narbadá valley extend proves that they fill a rock basin, for the bed of the Narbadá river, at the point where it leaves the alluvial plain near Handiá and commences to run through the rocky channel which extends to Bárwai, is not more than 200 feet below the level of the surface at Gádawárá and Súkakheri, and the valley is surrounded by higher rocky ground in every other direction. A slight prolongation of the alluvial basin to the south-west in the direction of Hardá, the prevalence of alluvium in parts of Nimár, and the circumstance that there is a great break by which the railway traverses the Sátpura range, immediately east of Asírgarh, may indicate that the upper Narbadá formerly joined the Tápti in Khándesh,[1] and that the lower part of the valley of the former river, as it now exists, is due to changes of level in the later posttertiary period.

The surface of the Narbadá alluvium is undulating, and evidently denuded by the action of rain and streams. There is a slight slope of the surface to the westward throughout the plain, the elevation of the railway station at Hardá, at the western extremity of the alluvial tract, being 947 feet above the sea, whilst Sohágpur station is 1,103 feet, Narsinghpur 1,185, and Jabalpur, which is, however, on rock a little above the plain 1,351. The fall of the surface in 200 miles is probably about 300 feet.

Different views have been put forward as to the marine, lacustrine, or fluviatile origin of the Narbadá alluvial deposits, but, before considering these, it will be well to give a list of the organic remains hitherto identified. They consist of bones and shells, and the following species have been determined :—

VERTEBRATA.[2]

MAMMALIA—

Ursus namadicus.
Bubalus palæindicus.
Leptobos fraseri.
Bos namadicus.
Cervus (? *duvaucelli*).
Sus, sp.
Hippopotamus palæindicus.
 „ *namadicus.*
† *Equus namadicus.*

MAMMALIA,—*contd.*

Rhinoceros unicornis.
Elephas namadicus.
† „ *insignis.*
† „ *ganesa.*

REPTILIA—

Pangshura flaviventris.
Batagur, cf. *dhongoka.*
Trionyx, cf. *gangetius.*
Crocodilus, sp.

[1] The greatest elevation on the G. I. P. Railway between the Narbadá and Tápti valley, is 1,245 feet above the sea, or only 300 feet above Hardá in the alluvial plain of the Narbadá.

[2] Lydekker, *Pal. Indica*, series x, III, (1884-86). The species marked with a dagger are found also in the Siwáliks.

MOLLUSCA.[1]

GASTEROPODA—
Melania tuberculata.
Paludina bengalensis.
„ dissimilis.
* Bythinia cerameopoma.
* „ pulchella.
Bulimus insularis.
* Lymnea acuminata.
Planorbis exustus.
* „ compressus?

LAMELLIBRANCHIATA—
Unio corrugatus? var.
„ indicus.
„ sp. near U. shurtleffianus.
„ marginalis.
Corbicula, sp. near C. striatella.

The only trace of man hitherto found in these deposits consists of a chipped stone scraper or hatchet discovered by Mr. Hacket *in situ* near the village of Bhutra, eight miles north of Gádawárá.[2] The material is Vindhyan quartzite, and the form similar to that of some of the implements fo und inthe lateritic deposits of Southern India, and in the postpliocene formations of Europe.

The only form identical with existing Indian species is *Rhinoceros unicornis*, originally described under the name *R. namadicus*, but according to Mr. Lydekker the bones are not distinguishable from those of the living species.[3] *Elephas namadicus* is allied to the existing Indian elephant, *Bubalus palæindicus* is very close to the living Indian wild buffalo, and the deer is a near relation to, if not identical with, the bárasingha (*Cervus duvaucelli*). On the other hand, *Elephas insignis* and *Hippopotamus namadicus* belong to extinct subgenera, the first being found, and the latter represented by a nearly allied species, in the pliocene Siwálik rocks. *Hippopotamus palæindicus* and *Bos namadicus* are not nearly allied to any Indian living species, the first belongs to a genus now only found in Africa, whilst the second, although having some characters in common with the living wild cattle of India, *Bos* (*Bibos*) *gaurus*, differs from the latter in many important particulars, and appears to be quite as closely connected with true taurine oxen belonging to the type of *Bos taurus*. *Bos namadicus*, indeed, cannot be classed in the subdivision *Bibos*. The relations of the remaining mammals are less distinctly made out, the specimens on which the species are founded being for the most part fragmentary.

The only reptile clearly identified is *Emys tectum*, which is considered identical with a living Indian form. It is very singular that only fragmentary remains of crocodiles occur, for they abound in the Siwálik rocks and a species is common in the Narbadá at the present day. The mollusca appear to be the same as species now living in the area, and all the

[1] *Memoirs*, II, 284, (1860); *Records*, VI, 54, (1873). The nomenclature in this list is that adopted in the first edition of the Manual. The species marked with an asterisk are not determined with certainty, no specimens having been preserved in the Geological Museum.

[2] *Records*, VI, 49, (1873); two figures of the implement are given.

[3] *Pal Indica*, series x, I, p. viii, (1880).

commonest forms now known to occur in the river valley are represented,[1] except some minute species of land shells. Their absence is not surprising, because land shells for the most part float, when washed away, and are left on the surface, where they decompose, instead of being preserved in alluvial deposits.

The examination of the molluscan remains in the Narbadá clays and gravels completely disproves the idea of a marine origin, but it has been considered by some observers that the deposits are lacustrine.[2] This view was principally based upon the uniform appearance of the clay and the absence of stratification. But this very uniformity and want of stratification are common characters of undoubted river deposits, and may be observed on the banks of most large streams, whilst the frequent deposition of pebble beds throughout the clays could not have taken place in the still waters of a large lake. The bones too are isolated and broken, sometimes even being rolled, whereas, if deposited in a lake, different bones would in all probability be found together, because away from the margin there could be no current in the lake of sufficient force to transport bones divested of flesh, and any mammalian remains deposited in the bottom of the lake must be derived from floating carcases or portions of carcases. Moreover, the *Chelonia* and fresh water mollusca are all forms which inhabit either rivers or shallow marshes in river valleys, and it is improbable, if so great a change took place in the area as would be involved in the replacement of lakes by a river valley, that a greater difference would not be produced between the tortoises and fresh water shells formerly inhabiting the waters and those still living.

The fact of the alluvial formation occupying a rock basin shows, however, that a considerable upheaval of land must have taken place to the westward, and it is possible that this upheaval may for a time have given rise to a lake, and the lower beds may consequently be lacustrine even though those from which the fossils were obtained are alluvial. If the Narbadá has really been diverted from its original course as suggested above this could only have happened through a movement of elevation sufficiently rapid to pond back the drainage and produce a lake.

The alluvial plains of the Tápti valley require but brief notice.[3] In

[1] The only exception of any importance is *Melania spinulosa*, but that is not by any means so generally distributed a form as *M. tuberculata*. The absence in the Narbadá deposits of *Melania variabilis* and *M. spinulosa*, the latter of which is included in Mr. Theobald's lists of living Narbadá species (*Memoirs*, II, 287), was noticed by Dr. Falconer, *Quart. Jour. Geol. Soc.*, XXI, 382, (1865); but it is extremely doubtful whether *M. variabilis* does exist in the Narbadá valley or its neighbourhood. The occurrence of *M. lyrata* included in Mr. Theobald's list, *loc. cit.*, is also very doubtful.

[2] *Memoirs*, II, 283, (1860).

[3] For a few additional details, see *Memoirs*, VI, 276, 286, (1869); and Wynne, *Records*, II, 1, (1869).

their principal characters they resemble the Narbadá plain, but the depth of the deposits is unknown, no deep borings having been made. As in the Narbadá valley, the river now runs at a considerable depth below the alluvial plain and is evidently cutting its channel deeper. The whole basin is composed of the Deccan trap, and the Tápti cuts its way out to the westward through the same formation. No remains of mammalia have hitherto been detected in the alluvium, but they will probably be found if sought after; the few mollusca found, as in the Narbadá plain, belong to recent fresh water species inhabiting rivers.

The difference in elevation between the Tápti and Purna plains is not accurately known, nor are the levels of different parts of the plains well determined, the only data available being those furnished by the railway. The height above the sea at Bhusáwal, just south of the alluvial flat, near the eastern extremity of Khándesh, is 677 feet. This cannot be much above the flood level of the Tápti river, for the rail level at the bridge over the Tápti, only about six miles distant, is 685 feet. At Malkápur, close to the western extremity of the Purna alluvial plain, the level is 816 feet, at Akola 917, at Murtazápur 986, and at Badnera south of Amráoti, 1,093. The last locality, however, is some miles distant from the south-eastern edge of the alluvium, and none of the railway stations are out in the alluvial plain, as in the Narbadá valley.

The only peculiarity of the Purna alluvial deposits, which deserves notice, is the occurrence of salt in some of the beds at a little depth below the surface. Throughout an area more than 30 miles in length, extending from the neighbourhood of Dahihánda (Dhyanda), north of Akola, to within a few miles of Amráoti, wells are sunk for the purpose of obtaining brine in several places on both sides of the Purna river. The deepest wells are about 120 feet deep. They traverse clay, sand and gravel, and finally, it is said, a band of gravelly clay, from which brine is obtained. No fossils have been found in the clay and sand dug from the wells. The occurrence of salt in the alluvial deposits of India is not uncommon, and it is impossible to say, without further evidence, whether it indicates the presence of marine beds. The absence of marine fossils in all known cases is opposed to any such conclusion, but still it is not impossible that the land may have been 1,000 feet lower than it now is in late tertiary, or early posttertiary times, and this difference in elevation would depress the Purna alluvial area beneath the sea level.

It has already been mentioned that the alluvial deposits of the Godávari do not occur in distinct basins, like those of the Narbadá and Tápti. This river in general has but a slight fall, and forms a broad alluvial plain where it traverses softer beds, whilst it cuts a steeper slope through harder rocks. There is an exception to the latter rule in the gorge above Rájámahendri.

Extensive alluvial areas occur along the upper part of the Godávari in the Bombay presidency and the adjoining portion of the Nizam's dominions, and similar tracts are found on the Pengangá, Wardhá, and Waingangá, tributaries of the Godávari, in Berar, Nágpur and Chándá.

The composition of these deposits differs in no important particular from that of the Narbadá and Tápti alluvium. The gravels are chiefly composed of rolled agates and fragments of basalt derived from the Deccan traps, which are the prevailing rocks in the upper part of the valley. Silicified fossil wood in all sizes from small fragments up to trunks 10 and 15 feet long [1] is abundant along the west margin of the Chikiála sandstones, from near the Wardhá to the Godávari near Enchapalli, and is found less abundantly from here on to Albaka on the Godávari. The greater portion of the alluvium in all cases consist of brown clay with kankar. In the Wardhá valley beneath the clay and calcareous conglomerates some fine sandy silt, light brown or grey in colour, occurs west of Chándá, and contains salt, with a considerable proportion of sulphate of magnesia [2] (Epsom salts).

Mammalian bones have been found, sometimes it is said in large numbers, in the Godávari valley, but very few appear to have been preserved, and the only species identified is *Elephas namadicus*.[3] Bones of *Bos* and other animals occur, and it appears probable that the fauna is similar to that of the Narbadá valley. From the gravels near Múngi and Paitan (Pytun) on the road from Ahmednagar to Jálna, Mr. Wynne obtained an agate flake, [4] apparently of human manufacture, thus affording a second instance of traces of man occurring in the pleistocene river gravels of the Peninsula.

The most important localities at which bones have been observed are the neighbourhood of Múngi and Paitan already mentioned, and one or more places on the Pengangá or its tributaries in the neighbourhood of Hingoli.[5] At one spot near Hingoli bones are said to have been found in immense quantities, but unfortunately they were not preserved.

The valley of the Kistna resembles that of the Godávari in many respects. There are similar plains of alluvial clay with beds of sand, gravel and calcareous conglomerate, but none of these plains appear to be of great extent. Beds of gravel have been observed in many places at a height of 60 to 80 feet above the present course of the river and its tributaries.[6]

[1] W. King, *Memoirs*, XVIII, 298, (1881).
[2] Hughes *Memoirs*, XIII, 92, (1877).
[3] Falconer, *Quart. Jour. Geol. Soc.*, XXI, 381, (1865), *Memoirs*, VI, 232, (1869).
[4] For a description by Dr. T. Oldham and figures, see *Records*, I, 65, (1868).
[5] Capt. O. W. Gray, *Mad. Jour. Lit. Sci.*, VII, 477, (1838); Carter, on the authority of Dr. Bradley, *Jour. Bo. Br. Roy. As. Soc.*, V, 304, (1854); Newbold, *Jour. Roy. As. Soc.* VIII, 246, (1846). See also *Memoirs*, VI, 232, (1869).
[6] Newbold, *Jour. Roy. As. Soc.*, VIII, 247, (1846); Foote, *Memoirs*, XII, 237, (1876).

The only important mammalian remains hitherto found in the alluvial deposits of the Kistna and its tributaries consist of portions of the cranium and mandible of a *Rhinoceros*, and some bovine teeth and jaws, found on the Gatparba near the town of Gokák.[1] The bovine remains have not been determined but the *Rhinoceros* has been described under the name of *R. deccanensis* by its discoverer, Mr. Foote;[2] the species differs widely from all living forms, and does not appear to be very nearly connected with any known fossil Indian species. Some fresh water shells of living species were found with the bones.

It was probably from some part of the upper drainage area of the Kistna, also, that Colonel Sykes obtained the teeth of a trilophodont *Mastodon* described by Falconer[3] under the name of *M. pandionis*.

Large numbers of chipped quartzite implements of human manufacture, and belonging to the same type as that discovered in the Narbadá alluvium, have been found in various gravels in the southern Maráthá country on the Malparba and other affluents of the Kistna.[4] The relations between the ossiferous gravels and those containing the implements are, however, somewhat obscure.

Nothing of importance is known concerning the older alluvial deposits of the remaining rivers in the Indian peninsula.

It is in the Mahánadí, Kistna, and Penner valleys that the principal diamond gravels are found, frequently at heights considerably above the present stream level.[5] The pebbles in the gravels are composed of various kinds of metamorphic and transition rocks.

Throughout the east coast of the Peninsula, from the delta of the Ganges to the neighbourhood of Cape Comorin, with the exception of a few miles near Vizagapatam, there is a belt of alluvial deposits, varying greatly in breadth, but nowhere exceeding about fifty miles. In places the hills approach the sea, leaving only a comparatively narrow belt of sandy foreshore, as south of the Chilká lake in Orissa and again near Pondicherri, whilst broad alluvial plains extend inland for many miles, near the mouths of the great rivers Mahánadí, Godávari, Kistna, Cauvery, etc., where there is actually a slight projection beyond the general coast line, owing to the quantity of sediment deposited, although the strong currents which sweep up and down the coast prevent any great seaward extension of the deltas.

To the northward the east coast alluvium joins the older alluvial deposits on the western side of the Ganges delta, and the two resemble each other closely in mineral characters. The coast alluvium consists chiefly of

[1] *Memoirs*, XII, 232, (1876).

[2] *Pal. Ind.*, series x, I, pt. i, (1874).

[3] Palæontological Memoirs, London, 1868, I, 124.

[4] Foote, *Memoirs*, XII, 241, (1876).

[5] Newbold, *Jour. Roy. As. Soc.*, VII, 226, (1853).

clays with kankar and, near the hills, pisolitic nodules of iron peroxide, the latter being in places sufficiently abundant to render the deposit a kind of laterite gravel. Gravels and sand also occur, frequently more or less mixed with ferruginous concretions, and there is, in many localities, an apparent passage between the ferruginous gravel of the alluvium and the low level form of laterite, but in other places this older alluvium rests unconformably upon the low level laterite, which has been shown, by the occurrence of palæolithic implements, to be itself of posttertiary age.

The surface of the coast alluvium is usually quite flat near the sea and in the river deltas, but towards the hills it is more uneven, and the surface has undergone a considerable amount of denudation, evidently from being at a higher level.

At Madras and Pondicherri, shells belonging to recent species have been found at depths of from 5 to 20 feet beneath the surface, or considerably above the present sea level. Farther south also, near Porto Novo in the lower valley of the Vellar,[1] a bed of estuarine shells is found above the present flood level of the river, and consequently at a considerable height above the sea Similar deposits of shells have also been noticed near Cuddalore and Tanjore.[2]

The shells, as a rule, are estuarine forms, such as now live in the creeks and backwaters of the coast,[3] but in several cases true marine species have been found. The subfossil shells near Madras are so abundant in places that they have been collected for burning into lime.

Another place where estuarine shells have been observed is close to the Chilká lake in southern Orissa. The forms found were *Cytherea casta* and *Arca granosa*, and the deposit containing the shells is now at elevations of from 20 to 30 feet above the level of the highest tides.

The thickness of the alluvium has been tested at Madras by a boring which went through it, and struck the crystalline rocks at 55 feet from the surface.[4] Further south, at Pondicherri, the thickness of the alluvium is much greater, one boring having been put down 550 feet without reaching its base. The alluvial deposits of Pondicherri are both interesting and important, in that they yield a supply of artesian water at various depths below the surface,[5] and in one boring, at Bahúr, a bed of lignite, 10·65 m. (35 feet) in thickness, was struck at a depth of 73·38 m. (240 feet) from the surface.[6] The lignite is too impure to be of commercial importance, but since it

[1] H. F. Blanford, *Memoirs*, IV, 192, (1863).

[2] King and Foote, *Memoirs*, IV, 254, (1864).

[3] The following are the most characteristic species. They are seldom, if ever, found in the open sea, but they are always met with in backwaters, and at the mouths of rivers, and many of them occur in creeks of deltas near the sea:—*Potamides telescopium*, *P. fluviatilis*, *Arca granosa*, *Cythera casta*, *C. meretrix*, *Ostrea*, a large species.

[4] Newbold, *Jour. Roy. As. Soc.*, VIII, 248, (1846).

[5] For details see W. King, *Records*, XIII, 113, 194, (1880).

[6] Geological Survey, MS. Records.

must have been formed at or near the surface, it is interesting as evidence of an amount of subsidence corresponding to the depth at which it was found beneath the surface.

Evidence of subsidence to a less degree is again found in a submerged forest at the western end of Válimukam bay in the Tinnevelli district. The forest, or rather so much of it as can be seen, is described as about half an acre in extent, lying at or just below high water mark; the stumps have a diameter of one to one and a half feet at the base of the hole, and are surrounded by black mud containing remains of twigs and detached branches. An incised bone pendent was found, which appeared to have been washed out of this mud, showing that the forest flourished since the advent of man.

The trees of this forest could hardly have flourished at sea level or on the coast, so that there has certainly been some subsidence in this neighbourhood, but indications of a contrary movement are found close by in the occurrence of *Potamides* and other littoral marine shells in clay above high water level, showing that this clay must have been elevated since it was formed.[1]

Before proceeding to the description of the recent accumulations on the west coast of the Peninsula it will be well to notice the remarkable smooth water anchorages of Aleppi and Narakal. These are mud banks of about four miles in length, whose position varies in the course of years within the extreme limits of about eleven miles. The sea bottom on these banks is composed of a very soft mud, which readily mixes with the sea water, and smooth water can always be found over the mud banks, though open to the full force of the south-west monsoon, however tempestuous the sea outside may be. It was this peculiarity which first attracted attention, and rendered them important to the navigators of a coast where there are no sheltered harbours, and the accounts which have been written from time to time constitute a tolerably extensive literature.[2] According to the most recent investigation of the subject, these smooth water anchorages owe their origin to a bed of very soft, fine grained, greenish clay, containing foraminiferæ and diatomaceæ, which underlies the soft recent sandstones of the surface of the narrow strips of land separating the sea from the backwaters of Travancore and Cochin. When the water level in these backwaters is raised by the monsoon rains, this mud is forced outwards, and rises in cones and ridges along the shore and under the sea, and once it has become

[1] Foote, *Memoirs*, XX, 83, (1883).

[2] A good account of these mud banks by Dr. W. King is published in *Records*, XVII, 14, (1885), where an account of the previous literature will also be found. A more recent investigation by Mr. Lake is printed in *Records*, XXIII, 41, (1890).

thoroughly mixed with the sea water the waves of the open sea are smoothed off and reduced in size over the mud banks.

This result appears to be due to two separate causes. In the first place the mud contains an appreciable proportion of oily matter, and the action of oil in stilling stormy waters is now well known, but the second cause appears to be much the more important. The large quantity of impalpable mud mixed with the water increases its density, and, consequently, the waves, on entering this denser water, decrease in size and are retarded. Moreover, as the proportion of mud is much less at the surface than lower down, the lower part of the wave is retarded more than the upper, and the wave may actually break if the increase in density be sufficiently rapid, or merely be obliterated if it is sufficiently gradual. This action is intensified by the large amount of fresh water falling on the sea as rain and poured out by the rivers, which floats on the surface in such quantities that ships may often replenish their stock of fresh water by dipping over the side of the vessel with a bucket. It is doubtless due to the greater density of the deeper layers of water, owing to the smaller proportion of salt and mud in the upper layers, and the consequent retardation of the lower portions of the larger waves that they are broken up, while the film of oil derived from the mud causes the smaller wavelets to be smoothed off.

There is no such continuous plain of alluvium along the western shore of the Peninsula as on the east coast. The ground between the Sahyádri range and the sea, where not hilly, consists generally of a gentle slope towards the coast, composed of rock, covered in many places by laterite. The coast itself is rocky in parts, and the alluvial deposits are chiefly confined to the neighbourhood of the small streams, which run from the Western Gháts to the sea, or of the backwaters, or lagoons, which have been cut off by banks of sand along the coast. The backwaters are of considerable extent in Travancore and Malabar, but they are wanting farther north and on the coast of the Bombay presidency. The alluvial valleys between the hills are unimportant south of Bombay itself, although they gradually increase in extent to the northward.

Alluvial plains, evidently of comparatively recent formation, connect the hills of Bombay and Salsette island, a few creeks alone remaining to show the position of the marine channels which formerly existed. Farther north these plains gradually increase in extent, until they merge into the alluvial flat of Gujarát.

At Bombay the alluvial deposits consist of blue and yellowish brown clay. The former varies in thickness from a few inches to several feet, its upper surface being at present about one or two feet below high water level. It is very salt, and contains small grains and nodules of kankar, and occasionally plates of gypsum; it is frequently penetrated by mangrove roots, which are usually riddled by *Teredo* borings, just as in the mud of tidal creeks, and at one spot large masses of oyster shells have been found in it. The yellowish brown clay appears to be the older of the two deposits. Its surface is frequently above the sea level, it abounds in larger masses of kankar, and it has occasionally yielded estuarine shells, *Placuna*, *Ostrea*, etc. That these alluvial deposits are estuarine, and precisely similar to the mud now deposited in the creeks and backwaters of the coast, or on the shores of Bombay harbour, is shown by the similarity of mineral character and by the organic remains, both vegetable and animal, found in the clay.[1]

Some very interesting indications of subsidence were found in the excavation of the Prince's dock at Bombay. A large number of tree stems and roots were found in the blue clay, many in the position in which they originally grew and some of the stumps were 30 feet below high water level. The evidence of subsidence here is unmistakeable, but the littoral concrete, seen on the west side of the island, must have been formed at a lower level than it now stands at. The elevation on one side of the island and depression on the other could not have been contemporaneous, so that we have clear proof of oscillations of level similar to, but of greater extent than, those Dr. Buist recognised many years ago.[2]

It is evident that Bombay harbour is the last remaining inlet out of many which formerly indented the Bombay coast, and that this harbour is gradually silting up and being converted into dry land. The process, however, is slow, and it may be ages before its progress is such as to affect the trade of Bombay, but, unless depression takes place in the area, or means are devised for checking the deposition of mud, there can be no question of the ultimate result. Except at Bombay, little has been recorded concerning the alluvium of the western coast south of Damán, and that little presents no features of interest.

In the neighbourhood of the rivers Tápti and Narbadá there is, however, a broad and fertile alluvial plain[3] near the sea, resembling in some of its features the alluvium of the east coast. Commencing to the southward near Damán, this plain covers the greater portion of the Surat, Broach, and Ahmaḍábád districts, and continues as far as the Rann, where it joins the area of recent deposits connected with the Indus valley. Near Surat this plain is about 30 miles in breadth, and near Baroda it is 60 miles wide.

The alluvium of eastern Gujarát consists of brown clays with kankar, resting upon sands and sandy clays with occasional gravels. The surface

[1] Buist, *Trans. Bom. Geog. Soc.*, X, 181, (1852); Carter, *Jour. Bom. Br. Roy. As. Soc.*, IV, 204, (1853).

[2] *Trans. Bo. Geog., Soc.*, X, 177, (1857).

[3] *Memoirs*, VI, 233, (1869); *Records*, I, 30, (1868); VIII 49, (1875).

is covered with black soil to the southward, though not in the district of Ahmadábád, and is frequently flat over considerable areas, but in parts of the country the ground is undulating, evidently in consequence of having been denuded by rain action. The deposits appear to have been chiefly estuarine or marine, and have probably been raised, as on the east coast, but no fossils have been found. The Gulf of Cambay is said to be gradually silting up, and there can be very little doubt that it was formerly part of a broad inlet leading from the Rann, then an inland sea, to the ocean, and that the remainder of the inlet has been converted into the alluvial plains of Ahmadábád, Broach, Surat, and north-eastern Káthiáwár.

In north-eastern Káthiáwár, on the borders of the Rann, there is a large alluvial tract,[1] continuous with the alluvium of Ahmedábád, and similar in character. Between Káthiáwár and Ahmedábád, in the line of depression between the head of the Gulf of Cambay and the Rann of Cutch, there still exists a large shallow lake of brackish water, called the Nal, about twenty miles in length by three or four broad. In the neighbourhood of this marsh shells of a form of *Cerithium* (probably *Potamides telescopium* or *P. fluviatilis*) are found, showing that estuarine conditions have prevailed at no distant period, and tending to confirm the probability that the depression between Káthiáwár and Ahmedábád is an old marine inlet, silted up in recent times. The distribution of black soil in the neighbourhood of the Nal will be noticed presently.

Along the south coast of Káthiáwár there is very little alluvium, its place being taken by a calcareous grit, with marine shells, which is evidently of late formation. A glance at the map will show that this coast is exposed to the full action of the currents which sweep along the shores of the Peninsula, so that it is unlikely that any accumulation of sediment would take place. A patch of recent deposits has been mapped at the western extremity of the Káthiáwár peninsula, but along its north-western coast the Deccan traps extend down the sea shore. The belt of alluvium reappears in Cutch,[2] where it is from three to ten miles broad, there being only one place where rocks come down to the shore. This is in the Gulf of Cutch. The alluvial plain of Cutch consists of a brown loam, resting upon mottled clay, with kankar and grains of quartz.

An agglutinated calcareous shelly grit is found, a little raised above the sea level, in several places on the west coast of India. This deposit, which was called littoral concrete by Dr. Buist,[3] consists of shells, corals, pebbles, and sand, cemented together more or less thoroughly by carbonate of lime, and sufficiently compact in places to be employed as an inferior kind

[1] Rogers, *Quart. Jour. Geol. Soc.*, XXVI, 118, (1870); F. Fedden, *Memoirs*, XXI, 130, (1884).

[2] Wynne, *Memoirs*, IX, 81, (1872).

[3] *Trans. Bo. Geog. Soc.*, X, 179, (1852); *Jour. Bom. Br. Roy. As. Soc.* IV, 206, (1853).

of building stone. The best known locality is in Bombay island, where the shelly grit forms the flat ground of the Esplanade and part of the surface on which the fort was built, the same deposit is also found at Mahim and other places in the island, resting sometimes upon rock, but more often upon the blue alluvial clay, described a few pages back. The same formation is found to the southward at Malwán,[1] and northward here and there as far as Damán, where it was observed by Mr. Wynne, apparently in process of formation.[2] Near Bulsár, a little north of Damán, the littoral concrete was observed to be stratified, the strata dipping at a low angle towards the sea.

In western Káthiáwár the same formation is much more widely developed. It here assumes the character of an earthy calcareous grit, is usually of a dark ashy colour, and contains marine shells and corals. Occasionally it attains a thickness of 60 feet, and it rests unconformably on the denuded surface of the miliolite. The fossils found in the calcareous grit, so far as is known, are all species now living on the neighbouring coast, but no thorough comparison has ever been made.

There can be very little doubt that the shelly calcareous grits of the Bombay and Káthiáwár coast are truly marine, not estuarine, and that they are the result of a littoral accumulation of the sand and pebbles found on the shore, together with marine shells and corals. The beds may have originally been sand spits or beach deposits, very little, if at all, above high-water mark, and consolidated by the cementing action of carbonate of lime after being raised. In any case there appears to be evidence of a rise in the land, trifling at Bombay, but greater in Káthiáwár.

Indications of local deposits, supposed to have been formed in lakes, have been noticed on the Nílgiri hills of Southern India[3] and in the southern Maráthá country,[4] and have been supposed to indicate changes of level. No fossils have been found in these deposits, nor does the evidence in either case amount to clear proof of the former existence of lacustrine conditions, although the probabilities are in favour of this view.

It would be beyond the scope of the present work to enter into the question of Indian soils. Consisting as they do of the surface of the ground altered by the action of the air and rain, by impregnation with organic matter, and by the results of agricultural processes, they necessarily vary with every difference in the underlying formation, whether it be one of the older rocks or of the more recent unconsolidated deposits. There are, however, two forms of superficial formations which, having received repeated

[1] *Memoirs*, XII, 243, (1876).
[2] *Records*, I, 32, (1868).
[3] H. F. Blanford, *Memoirs*, I, 243, (1858).
[4] Foote, *Memoirs*., XII, 228, (1876).

notice in Indian geological works, require a few remarks to be devoted to them, and one of the two, the *regur*, or black soil, is a very remarkable substance. The red soil also requires notice, because it has been so frequently mentioned in geological treatises.

The somewhat ferruginous soils common on the surface of many Indian rocks, and especially of the metamorphic formations, would probably never have attracted much attention but for the contrast they present in appearance to the black soil. They have only been noticed, as a rule, in papers relating to the black soil country in the western and southern portions of the Peninsula. The commonest form of red soil is a sandy clay, coloured red by iron peroxide, and either derived from the decomposition of rock *in situ* or from the same products of decomposition washed to a lower elevation by rain. The term is, however, frequently used in a very vague sense, apparently to distinguish such soils as are not black, and hence many alluvial soils may be comprehended under the general term. In very many cases, too, it appears to have been applied in Southern India to thick alluvial beds of sand or sandy clays, which are in fact ordinary river or rain-wash deposits.

The regur of Peninsular India, called black soil from its colour, and cotton soil from its suitability to the cultivation of cotton, occupies the surface of a very large portion of the country, and Newbold considers that at least one-third of Southern India is covered by it. The name is a corruption of the Telugu *regada*, or of cognate words in affined languages.[1]

Regur, in its most characteristic form, is a fine dark soil, which varies greatly in colour, in consistence, and in fertility, but preserves the constant characters of being highly argillaceous and somewhat calcareous, of becoming highly adhesive when wetted (a fact of which any one who has to traverse a black soil country after a shower of rain becomes fully aware) and of expanding and contracting to an unusual extent under the respective influences of moisture and dryness. Hence, in the dry season the surface is seamed with broad and deep cracks, often five or six inches across

[1] The following are some of the principal writers who have described regur:—

Christie, *Edin. Phil. Jour.*, VI, 119, (1829); VII, 50, (1829); *Mad. Jour. Lit. Sci.*, IV, 469, (1836).

Voysey, *Jour. As. Soc. Beng.*, II, 303, (1833).

Newbold, *Proc. Roy. Soc.*, IV, 54, (1838); *Jour. As. Soc. Beng.*, XIII, 987, (1844); XIV, 229, 270, (1845); *Jour. Roy. As. Soc.*, VIII, 252, (1846).

Hislop, *Jour. Bo. Br. Roy. As. Soc.*, V, 61, (1853).

Carter, *Jour. Bom. Br. Roy. As. Soc.*, V, 329, (1854).

Theobald, *Memoirs*, II, 298, (1860); X, 229, (1873).

H. F. Blanford, *Memoirs*, IV, 183, (1852).

King and Foote, *Memoirs*, IV, 352, (1864).

W. T. Blanford, *Memoirs*, VI, 235, (1869); *Records*, VIII, 50, (1875).

T. Oldham, *Records*, IV, 80, (1871).

Foote, *Memoirs*, XII, 251, (1876).

and several feet deep. Like all argillaceous soils, regur retains water, and consequently requires less irrigation than more sandy ground; indeed, as a rule, black soil is never irrigated at all in the western Deccan, Nágpur, and Haiderábád. When dry, it usually breaks up into small fragments; on being moistened with water it gives out an argillaceous odour. It is said to fuse, when strongly heated, into a glassy mass, but this is not invariably the case, and is probably dependent on the proportions of iron and lime present.

The chemical composition of regur has not received much attention. From the few and partial analyses[1] which have been made the proportions of iron, lime, and magnesia seem to vary, and there appears always to be a considerable quantity of organic matter combined. The black colour appears to be due either to the carbonaceous elements of the soil, or to organic salts of iron, but the tint varies much, being frequently brownish, and sometimes grey.

Christie made some experiments to determine the absorbent power of regur. He first dried a portion at a temperature nearly sufficient to char paper; he then exposed to the atmosphere of a moderately damp apartment 2615·6 grains of the dried soil, and found after a few days that it had

[1] The following are the analyses. In neither case is it stated how the analyses were made, nor which ingredients were determined by loss. In the first, by Dr. Macleod and published by Captain Newbold (*Jour. Roy. As. Soc.*, VIII, 254), a complete analysis of a dried sample appears to have been made, but the locality from which the specimen was derived is not stated:—

Silica	48·2
Alumina	20·3
Carbonate of lime	16·0
Carbonate of magnesia	10·2
Oxide of iron	1·0
Water and extractive	4·3
	100·0

In the other analysis by Mr. Tween (*Memoirs*, IV, 361), undried soil was used, and the component parts were only determined in the soluble portion. The residue in all consisted chiefly of magnesia and alkali; in A1, B1, B2, there were traces of sulphuric acid.

A and B were from near Seoni, C from Indore, D from Barwáni, and E from Burhánpur; Seoni and Barwáni are in the Narbadá valley, and Burhánpur in the Tápti.

A1, A2 represent the surface soil and subsoil taken from the same locality, A1 being the surface, A2 from 5 feet below surface. The two marked B1, B2, are, in like manner, the soil and subsoil (3 feet deep) from one locality, while C, D, and E are the soils taken from only a few inches below the surface. B1 is considered the best quality of soil:—

	A		B		C	D	E
	1	2	1	2			
Insoluble	62·7	47·61	62·8	63·7	68·61	57·91	61·80
Organic matter	9·2	8·4	9·	8·7	7·2	8·7	7·65
Water	8·4	7·6	8·2	6·5	9·4	9·9	7·35
Oxide of iron	11·	15·9	10·9	11·8	6·76	4 36	5·7
Alumina	7·5	8·6	7·6	8·4	5·81	8·75	7·67
Carbonate of lime	1·2	11·89	1·5	1·3	1·57	9·28	8·53
	100·	100·	100·	100·	99·35	98·90	98·70

gained 147·1 grains. He then exposed the same sample to an atmosphere saturated with moisture, and found that the weight increased daily, till the end of a few weeks, when it was found to be 2828·4 grains. The soil had, therefore, gained 212·8 grains, or about 8 per cent.

As a rule, the purest beds of regur contain no pebbles, although this soil usually abounds in kankar. Fragments of chalcedony or zeolite are, however, often found in the black soil, where it is derived from the decomposition of basalt, and in Southern India regur occasionally contains debris of the metamorphic rocks, sandstone or limestone, on which it rests.

Where uncultivated, black soil plains usually support but few trees, and those, as a rule, of no great size, but the principal product is grass, commonly growing to a height of three or four feet, but sometimes considerably higher. The growth of grass on the uncultivated plains of India is, however, greatly promoted, and the trees injured or killed, by the universal practice of burning the grass annually in the dry season, so that it is probable that the plains of black soil would support forest if left to themselves.

The fertility of this soil is so great that some of the black soil plains are said to have produced crops for 2,000 years without manure, without having been left fallow, and without irrigation. On the other hand, some varieties of black soil, occurring near the coast of Southern India, are comparatively infertile.

The typical appearance is only presented by this soil near the surface of the ground; if the regur is more than about 6 to 10 feet deep, it usually passes down into brown clay with kankar. It is never, except where it has been carried down and re-arranged as a stream deposit, met with at any depth beneath the surface.

The distribution of black soil in the Indian peninsula is of some importance, because it affords a clue to the origin of the formation. Regur is found everywhere on the plains of the Deccan trap country, except in the neighbourhood of the coast. A very similar soil is found locally in the basaltic Rájmahál hills, but with this exception nothing of the kind appears to be known in Bengal or the neighbouring provinces. In Southern India, however, tracts of black soil are found scattered throughout the valley of the Kistna, and occupying the lower plains and flats of Coimbatore, Madura, Salem, Tanjore, Rámnád, and Tinnevelli. There is but little on the Mysore plateau. Some occurs on portions of the coast plain on the eastern shore of the Peninsula, and the great alluvial flat of Surat and Broach in eastern Gújarát consists of this soil. The soils of Ahmadábád are light coloured, but regur occupies the surface of the depression lying between Ahmadábád and Káthiáwár, and connecting the head of the Gulf of Cambay with the Rann of Cutch.[1]

[1] Rogers, *Quart. Jour. Geol. Soc.*, XXIV, 118, (1870).

In many cases there cannot be a question that regur is simply derived from basalt by surface decomposition, and it is not surprising that numerous observers, from Christie and Voysey to Carter and Theobald, should have contended, and should still contend, that all cotton soil is derived from disintegrated trap rocks. Throughout the immense Deccan trap area, the passage from decomposed basalt into regur may be seen in thousands of sections, and all the alluvial valleys, most of which contain black soil, are filled with deposits derived from the disintegration of basaltic rocks. More than this, the boundary of the trap is approximately the boundary of the black soil over enormous areas; where the latter is found beyond the trap boundary, volcanic rocks may very probably have existed formerly, and have disappeared through disintegration, or the soil have been washed down from the neighbouring trap hills. This is admirably seen around Nágpur and Chándá in the Central Provinces, where regur occurs everywhere upon the trap, but is never seen upon the metamorphic rocks a few miles to the eastward, except where there is reason to suppose it has been transported, as in the alluvial flats of rivers which flow from the trap country. Again, whilst nothing resembling regur is found in the metamorphic region of Bengal, Behar, Orissa, Chutiá Nágpur, Chhatísgarh, and the neighbouring provinces, soils, undistinguishable from those of the Deccan traps, are found in the basaltic Rájmahál hills, and a similar formation has also been observed in Pegu,[1] derived from the decomposition of basalt. It has been urged that basalt may have been more widely spread in Southern India than is now the case, and that, where none is now found, its disappearance is due to its having been converted, by disintegration into, regur.

This view cannot, however, be accepted. In the first place, as was shown by Newbold, basalt generally disintegrates into a reddish soil, quite different from regur in character. This reddish soil may be seen in places passing into regur, but the black soil is, as a rule, confined to the flatter ground at the bottom of the valleys or on flat hill tops, the brown or red soil occupying the slopes. Again, the masses of black soil in the valleys of the Godávari and Kistna might be due to the alluvial deposits having been derived from the trap rocks, through which both rivers flow in the upper part of their course, but hundreds of square miles in the basins of the Penner, Pálár, Cauvery, and other rivers still farther to the south are composed of precisely similar regur to that of the trap area. There is no reason for supposing that the Deccan trap ever extended to the valleys of the rivers named, or can there be any reasonable doubt that the alluvial flats contained in these valleys are mainly formed from the detritus of metamorphic rocks.

Captain Newbold considered[2] all regur to be of subaqueous origin in

[1] Theobald, *Memoirs*, X, 229, (1873).

[2] *Jour. Roy. As. Soc.*, VIII, 256, (1846).

India, and compared it to the deposits in tanks, and to the mud of the Nile. Mr. H. F. Blanford suggested[1] that the cotton soil of Trichinopoli had accumulated in lagoons or backwaters near the sea, and he showed that in one place, near Pondicherri, regur was actually being formed in a nearly dry lagoon separated from the sea by a sand spit. Messrs. King and Foote, on the other hand, considered[2] it more probable that the Trichinopoli regur was a fresh water deposit accumulated in marshes. It has since been shown[3] that a complete passage takes place in the neighbourhood of Surat between the deposits formed in tidal estuaries and the regur of the surrounding country, and it appears probable that much of the black soil of eastern Gujarát may have been originally a marine or estuarine (brackish water) formation. On the other hand, Hislop[4] objected to the theory of formation by deposition in water, and he appears to have been the first to suggest that regur may really be of subaerial origin and due to the impregnation of certain argillaceous soils by organic matter. This appears to be the most probable theory; there can be no doubt that some forms of regur originate from the decomposition of basalt *in situ*, others from the disintegration of other argillaceous rocks, whilst other varieties again were originally alluvial clays formed in river valleys, or deposited in fresh water marshes, estuarine flats, or salt water lagoons. The essential character of a dark colour appears due in all cases to the admixture of organic matter, and perhaps the presence of a small quantity of iron. It is far from improbable that most of the black soil flats of India were covered with luxuriant forest, before the vegetation was annually exposed to the effects of fire. The increased dampness of the soil, the protection from denudation by rain, and the supply of decomposing vegetable matter may have contributed to the formation of the more fertile forms of regur. That the process of regur formation is purely superficial, and that it is due to surface action of a past time, is well seen in many of the regur plains with a slightly undulating contour. In such places the earth is black on the flats above, where the superficial layer has not been washed away, brown where the wash of rain has swept away the surface soil, and the black soil washed from the sides of the hollows has frequently accumulated towards the lower portion of them.

The abrupt termination of regur in places at the edge of the trap country is simply due to the change from an argillaceous soil to a sandy one. The basalt appears generally to decompose into a highly aluminous substance, the metamorphic rocks, on the other hand, produce sand to a large extent. At the same time it should be stated that it is not quite clear why argillaceous deposits should have become regur in Southern India, whilst nothing of the kind is known in Bengal, except in the basal-

[1] *Memoirs*, IV, 191, (1863).
[2] *Memoirs*, IV, 357, (1864).
[3] *Records*, VIII, 50, (1875).
[4] *Jour. Bom. Br. Roy. As. Soc.*, V, 61, (1857).

tic region of the Rájmahál hills. A dark coloured soil certainly forms in the marshes of eastern India, but it has not the character of regur, and no cotton soil has been noticed in the dense forests of Chutiá Nágpur and Bastár, nor, except on the surface of basalt, in the forest-clad plains of Burma. It is doubtful whether true regur occurs on the Malabar coast between Bombay and Cape Comorin, and the marshy soils on the top of the Sahyádri range do not form cotton soil. The black soil plains appear to be almost confined to those parts of India which have a moderate rainfall, not exceeding about 50 inches, but it is impossible to say whether this is a necessary condition.

It may then be stated that regur has been shown on fairly trustworthy evidence to result from the impregnation of certain argillaceous formations with organic matter, but that the process which has taken place is imperfectly understood, and that some peculiarities in distribution yet require explanation.

True peat forms in the hollows on the Nílgiris and some of the other mountains in Southern India, such as the Shevaroys,[1] at elevations above 4,000 feet, and its formation is due, as in temperate climates, to the growth and decomposition of a moss. In the marshes of the Gangetic delta an inferior kind of peat is also formed by the decomposition of various aquatic plants, and especially of wild rice.[2] The peat like beds found so widely distributed in the neighbourhood of Calcutta at a little depth below the surface appear to be derived from the decomposition of forest vegetation. A somewhat similar substance has been obtained from beneath a marsh in Oudh.[3]

Sand drifted by the wind forms low hillocks on many parts of the Indian coast. A series of parallel ridges of sand hills along the shore of Orissa has been supposed to mark successive positions of the shore line. A similar tract of blown sand is found north of Orissa in the Midnapur district, and southwards at intervals throughout the whole of the east coast. The sand is, of course, derived from the sea shore and blown up into ridges at right angles to the prevailing wind, with their longer slope to windward and a shorter and steeper surface to leeward. Smaller patches of sand are sometimes found on the banks of backwaters. The sand hills frequently extend for two or three miles inland from the coast, and in such cases the inner ridges are covered with a peculiar vegetation, amongst which the cashewnut tree (*Anacardium occidentale*) and a screw-pine (*Pandanus*) are conspicuous, and in some cases between the parallel ridges coinciding in direction with the coast the ground is flat, and even occasionally marshy,

[1] Foote, *Memoirs*, XII, 252, (1876).
[2] *Jour. As. Soc. Beng.*, XXIII, 400, (1854).
[3] *Proc. As. Soc. Beng.*, 1865, p. 85.

as in parts of Midnapur. In the latter case it is probable that a lagoon has existed, which has been gradually silted up, the origin of the lagoon being due to the formation of a sand spit outside it. As already noticed, the existence of several parallel sand ridges probably indicates a rise of land, each ridge coinciding with a former coast line.

On the Malabar coast, sand dunes are equally common, and contribute greatly to the formation of lagoons or backwaters[1] by accumulating on spits of sand. In the northern portion of the western coast about Bombay no sand hills have been noticed, probably because the detritus from the trap rocks does not form a suitable material, but further north again, in Surat and Broach,[2] in portions of Káthiáwár, and in Cutch, blown sand occupies more or less ground in many places in the neighbourhood of the shore.

Sand dunes in India are not confined to the sea coast, but are frequently found on the banks of rivers. And the accumulation of blown sand on river banks is of common occurrence on many of the peninsular rivers, such as the Godávari, Kistna, and Cauvery. In some instances noticed by Newbold,[3] villages have been buried by the sand blown from the river beds during the dry season.[4]

One peculiar form of sand hill, known as *teri*, is developed to a large extent along the Tinnevelli coast, and to a small extent in the north-western parts of Nellore and in the south of Travancore. The sand of which these hills are composed consists of rounded grains of colourless quartz, stained red, often bright red, by a thin film of ferruginous stain, which is easily dissolved by acids. In the Tinnevelli district they owe their origin to the dense clouds of sand and dust blown by the south-west monsoon off the bare red soil plains towards the coast, where the wind meets the sea breeze, is checked, and the sand dropped to form the téris.[5]

In the extrapeninsular area we find recent and subrecent river gravels in every valley, but the more extensive accumulations, if we except the alluvium of the Irawadi river, are all found in rock bound basins of closed or arrested drainage, which have been formed by differential movements of the surface during the elevation of the hill ranges among which they are found.

In the dry country west of the Indus there are extensive accumulations of recent deposits, of which only a small proportion can be regarded as alluvium in the true sense of the word. Beyond the frontier there are immense stretches of blown sand and loess in western Balúchistán and Afghánistán, of which very little is known, but it is probable that they

[1] Newbold, *Jour. Roy. As. Soc.*, VIII, 268, (1846).

[2] *Memoirs*, VI, 235, (1869) ; IX, 82, (1872).

[3] *Jour. Roy. As. Soc.*, VIII, 269, (1846).

[4] The principal accumulation of blown sand in India, that of the great Rájputána Desert, will be described in the next chapter.

[5] R. B. Foote, *Memoirs*, XX, 87, (1883).

are composed principally of the same types of accumulation as are seen in the smaller valley plains around Quetta.

First among these, as being the oldest, is a series of usually more or less bright red clays, sands and gravels which, in some of the publications of the Geological Survey,[1] have been regarded as tertiary, and have been coloured as such on the accompanying geological map, but although it is possible that they may belong to the newest portion of that period, they are so intimately connected with the recent deposits that, as has already been noticed,[2] it is more convenient to describe them in this place rather than in what might be regarded as their more proper place. These deposits are frequently undisturbed, especially towards the centre of the valley plains and are then difficult to distinguish from more recent deposits, except that the latter are seldom so deep a red in colour. More usually, however, they have undergone some slight disturbance, which has enabled the drainage to cut into them and form an irregular surface dotted with small hills, devoid of soil or vegetation, owing to the saline nature of the clays and the steepness of their slopes. Towards the margins of the valleys where these deposits abut against the hills, they are sometimes tilted up at high angles of dip, as in the Mashálak range west of Quetta.

Though they occur in close proximity to typical Siwáliks, no actual contact section has yet been found, but there are certain indications that the red clays of the valleys are considerably newer than those of the Siwálik system, and it is certain that they were deposited after the main features of the orography had been marked out by disturbance and erosion.[3]

The most important of the recent deposits of these plains are the extensive gravel slopes at the foot of the hills, and the loess.

The great gravel slopes, or *dháman*, which everywhere fringe the foot of the hills, and often reach a width of many miles in this comparatively rainless country, form one of the most conspicuous features in the scenery of the more open parts of the hill country west of the Indus.

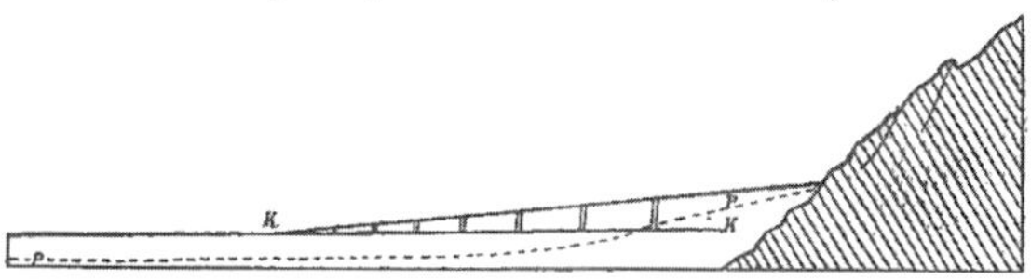

Fig. 21.—Diagram illustrating the theory of the karez. PP=limit of permanent saturation of subsoil; K.K=karez.

They occur as great inosculating fans, spreading with a slope of 300 to 600 feet per mile from the mouths of the stream valleys. It is into these

[1] C. L. Griesbach, *Memoirs*, XVIII, 18, (1881); W. T. Blanford, *Memoirs*, XX, 115, (1883).

[2] *Supra*, p. 319.

[3] These valley deposits have not yet received the detailed attention they deserve. See *Records*, XXV, 36, (1892); see also *Memoirs*, XVIII, 18 (1881); XX, 115, (1883).

fans that long underground tunnels, known as *karez*, are driven, with a slope less than that of the surface, till they pass below the level of permanent saturation, and, acting as a subsoil drain, carry the water out to the surface.

The loess deposits consist of a fine grained, usually grey coloured and unstratified accumulation of wind blown dust, precisely similar to the great loess deposits of China, which have been described by the Baron von Richthoven. They vary in size from small patches of a few yards across to great plains like that of Thal Chotiáli. In the Kachi, as the plain south of Sibi is called, the deposits of the plain appear to be principally wind blown loess, more or less mixed with true alluvium.

Closely connected with the true loess is a more or less finely stratified type of deposit, which is formed in the low lying parts of the loess plain. After every heavy shower the drainage from the higher parts of the plain, as well as from the surrounding hills, collects in these depressions, whence it gradually disappears by percolation and evaporation. The water, when it first collects in these depressions, always carries a large amount of solid matter in suspension, which is deposited when it comes to rest, the coarser particles sinking first of all and the finer afterwards. By a repetition of these floods, a finely bedded accumulation of alternately finer and coarser grained material is formed, which presents a great similarity to a lacustrine formation, though it was not deposited in a lake in the true sense of the word, but in mere temporary collections of flood water.

On the great plains of Ráwalpindi (known as the Potwár), Bannu, and Pesháwar, extensive deposits of gravel, sand, and silt exist. Little is known about the later deposits in the Pesháwar and Bannu plains, but those of the Potwár present some features of interest. The surface consists of a rather light brown alluvial clay, often containing kankar, and passing in places into fine silt. Beneath this alluvial deposit there is a mass of gravels and sand, sometimes enclosing boulders of large size. The boulders are not, however, confined to the pebble beds, many have been observed imbedded in fine silt, and this circumstance, together with the great size of many of the blocks found, and the distance to which they have been transported, has induced several observers to attribute the transport of the larger masses to ice, whether floating down a river or in a lake. It has been suggested that the Potwár may have been converted into a lacustrine basin in post-tertiary times by the elevation of the Salt range and the ridges west of the Indus. There is but little evidence in favour of this view, but still it is not impossible, for, although the pebble beds underlying the finer silt of the Soh́án valley

For additional details concerning these alluvial deposits of the Potwár, see *Records*, X, 122, 140, 223; (1877), XIII, 221, (1880).

appear too coarse for lacustrine[1] deposits, the silt may be, in part at least, a later deposit.

The posttertiary deposits are quite unconformable to the Siwálik rocks, which had been greatly disturbed and denuded before the later beds were formed. These later beds themselves, however, are occasionally found dipping at a considerable angle, due, it is said, to original deposition. The pebble beds are found around Ráwalpindi and in the neighbourhood of the Indus; they overlie the Rotás gorge near Jehlam, occur on some of the Salt range plateaux, and cap the mountain above Kálabágh on the Indus. They are found at a considerable elevation above the present river beds, some fragments of crystalline rocks in the neighbourhood of the Indus, apparently brought down by the stream, having been observed 2,000 feet above the river.

The large blocks attributed to ice flotation appear to have been derived from the Himálayas. They are abundant along the Indus as far up as Amb on the left bank of the river, in the gorge of the Siran and for some miles below Attock, around Jhand about twenty miles farther south, and farther still to the southward near the village of Trap on the lower course of the Soháan. Some of the blocks measure nearly 50 feet in girth and others are even larger. In places such blocks have been found 20 miles away from the banks of the Indus.

The Indus, as is well known, is subject to extraordinary floods, due to a portion of the upper valley becoming blocked by landslips (or according to some by glaciers) and to the sudden destruction of the barriers thus formed. Such floods occurred in 1841 and 1858, and have doubtless taken place in past ages.[1] In the flood of 1841 the waters of the Kábul river were checked and forced backwards for twenty miles by the rise of the Indus, and Drew has shown that the lake in Gilgit, formed by the landslip in 1840-41, must have been 35 miles long and upwards of 300 feet deep. Enormous quantities of detritus must be carried down by the violent floods produced by the bursting of such barriers, and if, as appears probable, the low temperature of the glacial epoch was felt in India, such lakes at an elevation of 5,000 or 6,000 feet above the sea would have been deeply frozen in winter, and large blocks from the river bed and dam might easily have been imbedded in the ice, glaciers also in the north-western Himálayas must have been more extensive than they now are, and the formation of lakes dammed up by glaciers was probably of more common occurrence than at the present day. Shaw[2] has called attention to the occurrence of heaps of stone and gravel of all sizes brought 80 miles down

[1] For accounts of these floods, see Cunningham's Ladak, London, 1854; Montgomerie, *Jour. As. Soc. Beng.*, XXIX, 128, (1860); Shaw, High Tartary, Yarkand, and Kashghar, London, 1871, p. 433, etc., and Appendix, p. 481; and especially Drew, Jummoo and Kashmir Territories, London, 1875, p. 44. Numerous references to other accounts are given by the last named writer.

[2] *l. c.*, p. 486.

the Shâyak, one of the tributaries of the upper Indus in Ladákh, by blocks of ice, and a similar action on a larger scale on the Indus may easily have supplied the erratics of the upper Punjab. If the Potwár was a lake, the dispersion of the erratic blocks is easily understood; if not, the area over which the masses of rock are found may be due to variations in the course of the Indus, and the reversed flow of its tributaries in great floods.

In one locality near Fatehjang a number of land and fresh water shells were found in silt, apparently the same as that in which boulders are elsewhere imbedded. The species found, including *Lymnea rufescens, Planorbis exustus, Paludina bengalensis, Bythinia pulchella, Melania tuberculata, Bulimus insularis, Opeas gracilis,* etc., are the same as are now common in the country, and it appears doubtful if they would have survived any very great diminution of temperature. At the same time it is possible that the beds containing shells may be of later date than those with boulders.[1]

The recent and subrecent deposits in the Himálayas are represented by lacustrine deposits, moraines, talus accumulations, and more conspicuously, by the river gravels abundantly developed in nearly every valley, as well as along the outer foot of the range. So far as these require notice in this work, they will be referred to in the chapter devoted to the Himálayan range, but there are three larger expanses of recent and subrecent deposits in Kashmír, Hundes and Nepál which, being extensive enough to be depicted on the accompanying geological map of India, require some description of their principal characteristics.

The alluvial basin of Kashmír has a length of about 84 miles with a breadth of some 20 to 25, and is in part occupied by low lying alluvial deposits, not much raised above the level of the Jehlam river, but principally by older deposits forming elevated plateaux on the borders of the alluvial plain and islands rising from it. These elevated plateaux of alluvial and lacustrine deposits are locally known as *Karewa*, a name which has been adopted for the deposits of which they are formed.

The Karewa deposits consist principally of sand and shingle, with some fine grained clayey silt towards the centre of the valley; the beds are for the most part horizontal or slope with a gentle dip, which is probably merely the original slope of deposition, but near the flanks of the Pír Panjal they dip away from the hills at angles rising to 20°.

The best published descriptions of these beds are those of Colonel Godwin-Austen[2] and Mr. Drew,[3] by both of whom they are regarded as of lacustrine origin, an opinion also adopted by Mr. Lydekker[4] but difficult to accept in its entirety. It is very probable that some of the finely bedded

[1] Theobald, *Records*, X, 141, (1877).
[2] *Quart. Jour. Geol. Soc.*, XX, 383, (1864).
[3] Jummoo and Kashmir Territories, p. 210.
[4] *Memoirs*, XXII, 72, (1883).

fine grained deposits described by Colonel Godwin-Austen were deposited in still water, but the frequent alternations of beds cf shingle with sand and the layers of lignite from one to three inches in thickness, point to subaerial conditions of formation. Even the presence of true lacustrine deposits does not prove that the whole of the Kashmír lake basin was ever occupied by a lake. This rock basin was probably gradually formed by a deformation of the earth's crust, and the hollow so produced was filled up almost, if not quite, as soon as formed. At the present day true lacustrine deposits are being formed in those places on the northern limit of the valley where, owing to a deficiency of deposition, hollows have been left in which water has accumulated, and it is probable that the conditions have been much the same as at present throughout the geological history of the Kashmír valley, and that a minor area of true lacustrine deposits has been accompanied by a greater area where subaerial accumulation of sediment has been in progress.

It is possible that some of the older beds of the *karewas* may be contemporaneous with part of the upper Siwáliks, but the only fossils yet found, besides undetermined fish scales and plant remains, have been land and fresh water shells, all apparently belonging to living species.

The only other valley at all comparable with that of Kashmír is Nepál.[1] The superficial differences correspond with those that mark the structural characters of the two regions; both are longitudinal valleys, lying in the general strike of the strata, but the clear open oval area of Kashmír approximately coincides with the elliptical synclinal depression of the calcareous upper palæozoic strata. Nepál, on the contrary, is rather a group of confluent valleys, with high dividing spurs in both directions. On the prolongation of the strike of the rocks there is a continuation of the special excavation of the mountain zone, and the rocks of this zone, being prevailingly calcareous, has suggested the conjecture that the feature is primarily due to erosion by solution, as may also be the case with Kashmír. Another cause, however, and the proximate one of the formation of a lake-basin in Nepál, was probably, in part, a relative rise of the hills on the south, for here also the bottom beds of the valley deposits have undergone local disturbance on this side.

These deposits correspond very closely with those of Kashmír. There is no remnant of a lake, but the other features are alike. An extensive upland area, known as *tánr* land, corresponds to the *karewa* of Kashmír, and to the *bhángar* of the Gangetic plains. It is the surface of the old deposits, no doubt considerably modified by waste in the central parts, and by rainwash accumulations near the hills. The streams flow at a depth of from 50 to 500 feet below this surface, according to position, but

[1] H. B. Medlicott, *Records*, VIII, 93, (1875)

here, as in Kashmír, they are now, for the most part, subject to overflow, and thus form the alluvial valleys, known as *kholas*, corresponding to the *khádir* land of the plains. Beds of serviceable peat, much used for brick and lime burning, occur at various levels in the valley deposits, and there is also a blue clay, extensively used for top dressing the fields, whose fertilising virtue seems to be due to the phosphate of iron (vivianite) freely scattered through it in blue specks. No fossil remains have as yet been found in any of these deposits.

The subrecent deposits of Hundes occupy an area of some 120 miles long by from 15 to 60 miles broad in the upper valley of the Sutlej, which now flows in a deep and narrow gorge, not much less than 3,000 feet deep, cut through the horizontal deposits it had formed at an earlier period of its history.

Our knowledge of these beds is almost entirely dependent on the description of the two brothers Richard and Henry Strachey. It seems possible, from an observation of the latter of these, that the deposits are in part of lacustrine origin, for he mentions that, in the central part of the valley, the cliffs exhibit throughout their height a fine homogeneous clay, with but little gravel in it. There is some inherent probability in the supposition that part of these deposits were formed in local accumulations of water, but there is nothing to show that the whole of them might not be of subaerial origin, as it is almost certain that the bulk of them might have been.

But the chief interest of these deposits attaches to the mammalian fossils they contain. These have long been known, though it was only within late years that their derivation from the horizontal gravels was definitely determined by Mr. C. L. Griesbach.[1] The earlier specimens, brought across the frontier as curiosities by the Tibetan traders, had been regarded as tertiary by the late Dr. Falconer, who considered that the beds had undergone considerable elevation since their formation. A more recent revision of the fauna by Mr. Lydekker[2] renders the retention of this opinion, itself improbable on account of the horizontality of the deposits, impossible. With the exception of *Hippotherium*, which was determined by Mr. Waterhouse from specimens brought by Sir R. Strachey though it is questionable whether the materials were sufficient for its determination, only living genera are known; of these *Bos*, *Ovis* (?), *Capra*, and *Equus* are genera still living in the highlands of Tibet. *Hyæna* is not at present known in Tibet, though there is no reason why it should not formerly have ranged into high altitudes, and, besides, the correctness of the determination is open to question. There remains the genus *Rhinoceros*, which points to a warmer climate and a lower altitude than that in which the remains are found. It is, however, not impossible

[1] *Records*, XIII, 91, (1880).

[2] *Records*, XIV, 178, (1881.)

for a Rhinoceros, especially one of small size, to have lived on the bushes which grow in the neighbourhood of many of the Tibetan rivers, while the doubtful evidence of this genus is more than outweighed by the fragment of a skull figured by Royle,[1] which agrees so closely with that of *Pantholops hodgsoni* that there can be little doubt of at least generic identity, and *Pantholops* is a genus peculiar to the most elevated and coldest portions of Tibet. It is consequently more probable, so far as the palæontological evidence goes, that the subrecent deposits of Hundes were formed at or near the elevation at which they are now found than that they were formed at a much lower level and subsequently elevated without undergoing any disturbance.

In the hills east of India there are some rock basins occupied by alluvial deposits, of which the best known is that of Manipur. About 50 miles long by 20 broad, of an irregular shape, with many small hills rising like islands from the alluvial plain, it is not surprising that it should be generally regarded as a filled in lake. There are, however, no real reasons for supposing that any large proportion of the valley was ever occupied by deep water. There are no terraces round it such as would have resulted from the lowering of the outlet of the lake during the long period occupied by its filling up. The deposits are all of ordinary alluvial type, and the courses of the streams show that the present surface is the result of the gradual subaerial formation of an alluvial plain. The elevation of the surrounding hills is, geologically speaking, of comparatively recent date, probably not dating further back than the latter portion of the tertiary period, and it is probable that the origin of the Manipur basin was gradual, and that the active erosion of the surrounding hills, due to abundant rainfall, caused it to be filled up as fast as it was formed, with the exception of insignificant areas that partially escaped sedimentation, and were occupied by shallow lakes.

The upper waters of the Chindwin (Kyin-dwin)] river drain a number of alluvium filled valleys, the largest of which, on the upper part of the Chindwin itself, is known as the Hukong valley, while, in the hills west of the Chindwin, there are the Kubo valley and those south of it. No details are at present known regarding these alluvial spreads of Upper Burma, as no detailed geological investigations have yet been possible, but in Lower Burma more extended investigations have been made, and it is possible to distinguish posttertiary deposits of two distinct periods.

Along the margin of the Irawadi and Sittaung alluvium, there is a broad, but interrupted, belt of undulating ground, clearly distinguished

[1] Illustrations of the Botany, etc., of the Himalaya Mountains. 4° London, 1839, pl. III. fig. 1. The original specimen cannot now be found.

from the flat alluvial plains near the river both by the greater inequality of its surface and by its more sandy character. This tract is locally known as Eng-dain, or the country of the Eng tree (*Dipterocarpus grandiflora*), but the same name is naturally applied to the very similar sandy tracts occupied by the pliocene fossil wood group, so that the popular distinction does not precisely coincide with the geological limits of the formation.

The Eng-dain tract is composed chiefly of gravel, derived in a large measure from the neighbouring hills, but partly from a distance. A portion of the deposits, like the *bhábar*, the edge of the Ganges valley, may simply be the detritus washed from the surface of the hills by rain and small streams to form a slope at the base of the range, but in Pegu, as in other countries with a heavy rainfall, this slope is inconsiderable, and a great portion of the alluvial gravels are simply stream and river deposits. Similar beds of sand and gravel are found in many places underlying the argillaceous delta deposits of the Irawadi, and are evidently of more ancient origin.

Besides the fringe, of variable width, formed by the gravels along the edge of the older rocks, large tracts of the same older alluvial deposits are found in places isolated in the delta, occasionally forming ground raised to a considerable height above the flat country around. One such tract, about 20 miles long from north-east to south-west, by 10 miles broad, occurs east of Nga-pu-tau and south of Bassein; another, of about the same dimensions, lies south-west of Rangoon. These areas may be ancient *bhángar* deposits, or they may be caused by local upheaval.

Except in the immediate vicinity of the river channel, there is no important expanse of alluvial deposits in the valleys of the Burmese rivers; the beds of all, immediately above their deltas, are formed in places by older rocks, and there is no such continuous alluvial plain as is found along the course of the Ganges and Indus. Small tracts of alluvium occur, as usual, every here and there, but the wide undulating plains in the neighbourhood of the river in Upper Burma are largely composed, not of river alluvium, but of the pliocene fossil wood deposits.

Compared with the Gangetic and Indus deltas, those of the Irawadi and other Burmese rivers convey an idea of imperfection and backwardness, as though the latter were of more modern growth than the former, and had made less progress towards the formation of a great fertile plain. The Salwín cannot be said to have any delta at all, and in the Irawadi delta, as has already been mentioned, elevated tracts, both of rock and of the older alluvial deposits, occur in the neighbourhood of the sea. Considering the size of the river, the Sittaung delta, if the alluvial plain extending to the northward beyond Taung-ngu (Tonghoo) be included, is proportionally more extensive than that of the Irawadi, but still the broad Gulf of Martaban extends into the very mouth of the Sittaung river.

The Irawadi delta extends from the Rangoon river to the Bassein river, and the head of the delta may be placed near Myanaung.[1] The first important distributary—that forming the head of the Bassein river—leaves the main river a little above Henzáda, but water overflows in floods some miles above Myanaung, and finds its way to the sea by the Myit-ma-kha Khyaung, the origin of the Rangoon river. The various rivers and creeks of the Irawadi delta are said to be far less liable to change than those of the Ganges and Indus, but it must be remembered that the authentic history of the latter rivers, and especially of the Indus, extends much farther back than does that of the Irawadi. The general surface of the delta near the sea, with the exception of the higher tracts already mentioned, differs but little in elevation from that of the great Indian rivers and Mr. Theobald considers that at least 2,000 square miles must be below the level of high spring tides. Large marshes, or jhils ("*eng*" in Burmese), are found occupying the depressions between the raised banks of the principal streams, and the whole region, especially in the neighbourhood of the sea, consists of a network of the tidal creeks. Little appears to be known as to the progress of the delta seaward. judging by the contour of the coast, it would appear that the Irawadi, owing to its far greater size, and perhaps to the larger proportion of silt transported by its waters, had pushed its delta seaward far beyond the Sittaung. The Salwín traverses for the most part an area of hard metamorphic rocks, and probably brings down but little detritus, so that the conversion of the Gulf of Martaban into land, if it is ever to be effected, must depend largely upon the deposits from the Irawadi.

The alluvial plain and delta of the lower Irawadi consist mainly of a clay[2] very similar to that found in the Gangetic plain, but containing much less lime, and consequently poor in kankar. The colour is generally yellowish brown, sometimes reddish, owing to the presence of peroxide of iron. The proportion of sand varies, and is greater on the whole than in the Gangetic alluvium. A few thin layers of sand occur interstratified with the clay, and a band of dark blue or carbonaceous clay, a few inches in thickness, has been noticed in several localities.

[1] Mr. Theobald considers Min-gyi, 13 miles below Myanaung, the apex of the delta; and taking Puriam point, east of the Bassein river, and Elephant point, west of the Rangoon river, as the two lateral angles, he estimates the distances from Min-gyi to Elephant and Puriam points as 129 and 176 miles, respectively, the two points being 137 miles apart, —*Records*, III, 21, (1870).

[2] Mr. Theobald considers this clay marine or estuarine, but no fossils have been found in it and his main arguments, founded on the similarity between the clays of the Irawadi and Gangetic deltas, are of course favourable to the fluviatile origin of the Irawadi clay, if that of the Ganges be also of fresh water origin,—*Records*, III, 17, (1870).

The clay, in many places towards the head of the delta, is seen to rest upon pebbly sand, and the latter is frequently found beneath the clay in the delta itself, wells being sunk through the argillaceous surface formation to the porous stratum beneath. In the absence of any borings, however, it is impossible to say what the nature of the beds at a depth below the surface may be, and it is not clear whether the sand is the underlying formation throughout, or whether it is merely intercalated between beds of clay.

On the surface of the clay, in the immediate neighbourhood of the river, deposits of silt and sand are found in some places, and resemble the *khádar* deposits of the Ganges valley. No extensive area, however, is covered by these sandy beds. They form a narrow belt along the river channel above the influence of the tide, and occupy a rather larger area around Pantanau. The deposits of the Sittaung alluvial plain closely resemble those of the Irawadi.

CHAPTER XVII.

THE INDO-GANGETIC PLAIN.

Area and elevation—Fluviatile origin of the Gangetic plain—Subrecent marine conditions in the Indus valley—Character of Indo-Gangetic alluvium—Fossils in the alluvium—General features of the Indo-Gangetic plain—The Brahmaputra valley—The delta of the Ganges and Brahmaputra—The plains of upper Bengal and the North-West Provinces—*Kalar* or *Reh*—The Punjab—The lost river of the Indian desert—The lower Indus valley and delta—The Rann of Cutch—The desert of western Rájpútána.

The immense alluvial plain of the Ganges, Indus, and Brahmaputra rivers and their tributaries, the richest and most populous portion of India, covers an area of about 300,000 square miles, and forms approximately one fourth of the whole surface of British India, exclusive of Burma. The greater part of the provinces known as Assam, Bengal (including Behar), the North-West Provinces, Oudh, the Punjab and Sind, are included in the great plain which, varying in width from 90 to nearly 300 miles, entirely separates the geological region of peninsular India from the Himálayas to the north, the Suláimán and Kirthar ranges to the west, and the hill regions of Assam, Tipperah, and Chittagong to the eastward. Owing to the varying extent to which the surface is raised on the margins of the area by the detritus brought by rivers from the hills, and the gradation between the finer deposits of the plain and the coarser gravels forming the slope at the base of the Himálayas, it is difficult to estimate exactly the greatest height of the plain above the sea. The highest level recorded by the Great Trigonometrical Survey between the Ganges and Indus, on the road from Saháranpur to Ludhiána, is 924 feet,[1] and this may be fairly taken as the

[1] The following elevations of localities in the Indo-Gangetic plain will afford some idea of the general height of the surface above the sea. The figures, except in the case of Rájmahál, are taken from the maps and published sections of the Great Trigonometrical Survey, with a few additions kindly furnished by the Surveyor General, Colonel Walker. At all the localities quoted the height is the approximate level of the plain:—

Locality	Feet
BRAHMAPUTRA VALLEY—	
Sadiyá	440
Dibrugarh	348
Sibságar	319
Burámukh, near Tezpur	256
Gauháti	163
Goálpárá	150
GANGES VALLEY—	
Bardwán	102
Rájmahál	68
Benares	258
Allahábád	319
Cawnpore	417
Agra	553
Delhi	715
Meerut	739
Saháranpur	907

summit level at the lowest part of the watershed between the Indus and the Ganges. There is no ridge of high ground between the Ganges and Indus drainage, and a very trifling change in the surface might at any time turn the affluents of one river into the other. It is reasonable to infer that such changes have taken place in past times, and that the occurrence of closely allied species of *Platanista* (a fresh water dolphin peculiar to the Indus, Ganges, and Brahmaputra) in the two rivers, and of many other animals common to both streams, may thus be explained.

An idea once prevalent amongst both geologists and naturalists was, that the great Indian plain had been an arm of the sea in late geological times.[1] It is possible that this may have been the case, but there is absolutely no evidence whatever in favour of such a view, and some facts are opposed to it. On the southern flank of the Himálayas, no marine formations have been discovered of later date than eocene, and even these are unknown, except in one place east of the Ganges, between the spot where the Jumna leaves the Himálayas, and the Gáro hills, or throughout thirteen degrees of longitude, whilst the later tertiary formations, belonging to the Siwálik system, contain fresh water *Reptilia* and *Mollusca,* and not a single marine shell has been found in them. In Sind marine beds of miocene date are found, which become replaced by fresh water beds as they are traced up the Indus Valley, and in the Salt range the fresh water Siwáliks rest upon the nummulitic limestone. It is true that it is impossible to tell what beds may be concealed beneath the Indo-Gangetic alluvium, and marine strata may exist to an enormous extent without appearing at the surface. It is also unquestionable that the amount of information hitherto derived from borings is very small indeed, but so far as that information extends, and so far as the lower strata of the alluvial plain have been exposed in the beds of rivers, not a single occurrence of a marine shell has ever been observed, nor is there such a change in the deposits as would render it probable that the underlying strata are marine. As will be shown presently, the lowest deposits known in the plain itself are of posttertiary age, and they are certainly fresh water, whilst the tertiary deposits

INDUS VALLEY—

Umballa	901
Ludhiána	806
Firozpur	645
Lahore	708
Dera Ismáil Khán	595
Multán	407
Baháwalpur	375
Kashmor	246
Shikárpur	198
Sehwán	110
Kotri	66

[1] Hooker, Himalayan Journals, 1st ed., London, 1854, I, p. 378; Theobald, *Records*, III, 19. (1870). Mr. Theobald's main argument, derived from the clay at Pattharghatta, near Rájmahál, has been shown by a re-examination of the locality to be untenable, the deposit in question being merely a surface wash, containing fragments of bricks amongst other things (*Memoirs*, XIII, 224, (1877). Dr. Falconer considered that the Indo-Gangetic area was formerly an arm of the sea, but that it had been converted into land before the Siwálik epoch, —*Palæontological Memoirs, I,* 29.

are chiefly known to occur on the northern margin of the plain. The older pliocene deposits of Perim island in the Gulf of Cambay lie, however, to the south of the alluvial area, and five species of mammals found in them are also met with in the Siwáliks at the base of the Himálayas, so that there was probably land communication between the two areas. The only evidence known in favour of marine conditions having prevailed during the deposition of any portion of the Gangetic alluvium is the occurrence of brine springs at considerable depths in a few localities. These springs, however, are not numerous, and without additional evidence it is impossible to look upon them as proofs of marine deposits. At the same time it is by no means impossible that the sea occupied portions of Sind and Bengal long after the plain of upper India was dry land. With reference to Bengal there is very little evidence. Mr. Fergusson, in a masterly essay on recent changes in the delta of the Ganges,[1] has brought forward a quantity of historical data tending to show that the whole Ganges valley was probably not habitable 5,000 years ago, and that the extension of human settlements to the eastward from the Punjab has been gradual. The latter may be conceded, with the reservation that additional evidence as to the previous want of population is desirable. The Ganges valley 5,000 years ago, like that of the Bráhmaputra valley at the present day, may have been so swampy as to be ill suited for cultivation, and yet there is no reason for supposing that the area had recently been covered by the sea, for the state of the surface may have been due to an amount of depression sufficient to render the area marshy, but not enough to cause it to be overflowed by the ocean. That depression has taken place in the delta is shown by the records of the Fort William (Calcutta) borehole, to be described presently, but the only known marine beds in the neighbourhood of the Ganges delta, those at the foot of the Gáro hills, are of tertiary age, and probably pliocene.

In the Indus valley some evidence has been obtained of the sea having occupied part of the area in posttertiary times.[2] East of the alluvial plain of the Indus near Umarkot is a tract of blown sand, the depressions in which are filled by salt lakes. These lakes are supplied by water trickling through the soil from large marshes and pools supplied by the flood waters of the rivers, and it is evident that the depressions amongst the sand hills are at a lower level than the alluvial plain, and that the salt is

[1] *Quart. Jour. Geol. Soc.*, XIX, 321, (1863). There is one ethnological fact which Mr. Fergusson has not noticed. The population of Bengal, as any one who has seen much of Indian races will probably admit, is shown by colour, physique, and habits of life to contain a large proportion of the non-Aryan races, the people of upper India, on the other hand, having a much larger Aryan element. This mixed race may have migrated into the country, but it is at least as probable that the non-Aryan tribes were indigenous, and that the present Bengali race is due to an admixture of Aryan blood. The point is, whether Mr. Fergusson has not taken the south-eastern migration of the more civilised population amongst uncivilised tribes for the original peopling of the Gangetic plain.

[2] *Jour. As. Soc. Beng.*, XLV, pt. ii, 93, (1876); *Records*, X, 10, 21, (1877).

derived from the soil beneath the sand. To the southward is a great flat salt tract known as the Rann of Cutch, marshy in parts, dry in others, throughout the greater part of the year, but covered by water when the level of the sea is raised by the south-west monsoon blowing into the Gulf of Cutch and the old mouth of the Indus, and all water which runs off the land is thus ponded back. The Lúni river, which flows into the Rann, is, except after rain, extremely salt, and salt is largely manufactured from the salt earth at Pachpadra, close to the Lúni, more than 100 miles from the edge of the Rann, and nearly 300 from the sea. Both the present condition of the Rann and tradition point to the area having been covered by the sea in recent times, and having been filled up by deposits from the streams running into it, while the occurrence in some of the salt lakes near Umarkot, 150 miles from the sea, of an estuarine mollusc *Potamides (Pirenella) layardi*, common in the salt lagoons and backwaters of the Indian coast, seems to indicate that these lakes were formerly in communication with the sea. The enormous quantity of blown sand, also, which covers the Indian desert, can only be satisfactorily explained by supposing that it was derived from a former coast line north of the Rann and east of the Indus valley.[1]

It appears probable that in posttertiary times an arm of the sea extended up the Indus valley at least as far as the salt lakes now exist, or to the neighbourhood of Rohri, and probably farther, and also up the Lúni valley to the neighbourhood of Jodhpur, the Rann of Cutch being of course an inland sea. The country to the westward has been raised by the deposits of the Indus, and the salt lakes have been isolated by ridges of blown sand.

It is true that along the western margin of the Indus alluvium later tertiary (Manchhar) rocks are found, containing remains of mammalia and precisely resembling the Siwálik formation, and as there is nevertheless a probability that the lower Indus valley was an arm of the sea in posttertiary times, it may fairly be argued that the sub-Himálayan Siwáliks are no proof that the Ganges valley was not an inland sea at the same epoch. But in the Indus region the representatives of the Siwáliks pass downwards into miocene marine beds. In lower Sind the Manchhar formation itself becomes interstratified with bands containing marine shells, and not very far to the westward, there is a very thick marine pliocene formation on the Balúchistán coast, so that there is evidence in abundance of the sea having occupied portions of the area in later tertiary times, whilst there is no proof of any such marine conditions in the Ganges plain.

The various deposits of the Indo-Gangetic plain[2] may be roughly classed

[1] A description of this area will be found at the end of the present chapter.

[2] The authorities for the following account are manuscript reports by Mr. Theobald on parts of the alluvial area in Bengal, Behar, and the North-Western Provinces, some extracts from which were published in *Records*, III, 17, (1870); Medlicott, Sketch of the Geology of the North-Western Provinces, *Records*, VI, 9, (1873), and various papers referred to.

under two subdivisions, older and newer; the former consisting of beds which are undergoing denudation, whilst the latter form the newer accumulations, the flood and delta deposits now in process of formation. It is difficult, if not impossible, to draw any distinct line of separation between these two subdivisions, unless, as but rarely occurs, they contain fossils characteristic of their age, but, generally speaking, all the higher ground is composed of older deposits, whilst the newer alluvium is chiefly confined to the neighbourhood of the river channels, except in the delta of the Ganges, and in the Brahmaputra plain. Still, there are large parts, both of the Indus and Ganges plains, which are flooded every season, and on these areas newer deposits are formed by the flood waters. Moreover, as the rivers constantly change their courses, they often sweep away deposits only a few years, or even a few months old.

The prevailing formation throughout the Indo-Gangetic alluvial area is some form of clay, more or less sandy. The older deposits generally contain kankar, the newer deposits do not as a rule, but there are numerous exceptions in both cases. In the Indus valley the alluvial deposits are much more sandy than in the Ganges valley, and the surface of the ground is paler in colour, except where marshy conditions prevail. The deposits of the Brahmaputra valley in Assam are also sandy. In both these valleys the greater part of the area is occupied by the newer alluvial deposits whilst the greater portion of the Ganges plain, except towards the delta is composed of an older alluvial formation.

The older alluvium is usually composed of massive clay beds of a rather pale, reddish brown colour, very often yellowish when recently exposed to the air, with more or less kankar disseminated throughout. In places, and especially in Bengal and Behar, pisolitic concretions of hydrated iron peroxide, from the size of a mustard seed to that of a pea, are disseminated through the clay; occasionally these nodules attain larger dimensions, some being found near Dinájpur of the size of pigeons' eggs. In places kankar forms compact beds of earthy limestone. Sand, gravels, and conglomerates occur, but are, as a rule, subordinate, except on the edges of the valley, the quantity of sand in the clay decreasing gradually as the distance from the hills increases. Pebbles are scarce at a greater distance than from 20 to 30 miles from the hills bordering the plain. Beds of sandstone, sufficiently compact for building, have occasionally been found, but are of rare occurrence. On the whole, there is no great difference between the alluvial formations of the Indo-Gangetic plain and those of the Narbadá and Tápti, except that the latter are rather darker in colour, and perhaps less sandy.

The newer alluvial deposits consist of coarse gravels near the hills, and especially at the base of the Himálayas, sandy clay and sand along the course of the rivers, and fine silt consolidating into clay in the delta and in the flatter parts of the river plain. In the Ganges delta beds of impure peat commonly occur. Fresh water shells are of more frequent occurrence in the newer forms of alluvium than in the older, the species being those now living in the rivers and marshes of the country.

The only information of importance hitherto procured as to the nature and depth of the alluvial deposits beneath the surface is derived from five borings: one, 481 feet deep, at Fort William, Calcutta, within the delta and close to a tidal river; the second at Umballa, 701 feet deep, at nearly the highest level of the plain away from the slope of detritus along the margin; the third, carried to a depth of 464 feet, at Sabzal-kot west of the Indus, about 21 miles west by north of Rájanpur and about 400 feet above sea level; the fourth, at Agra, carried to a depth of 481 feet from a surface level of 553 feet above the sea; and the fifth, and deepest, having a depth of 1,336 feet from a surface level of about 370 feet, at Lucknow. All these boreholes were made for the purpose of obtaining water.

The Calcutta borehole is, with the exception of that at Lucknow, the most important, because it was carried down to a depth of about 460 feet below the mean sea level. The following account of the deposits passed through in the borehole is taken from the "Abstract Report of the Proceedings of the Committee appointed to superintend the Bore Operations in Fort William from their commencement, December 1835, to their close in April 1840:[1]"—

"After penetrating through the surface soil to a depth of about ten feet, a stratum of stiff blue clay, fifteen feet in thickness, was met with. Underlaying this was a light coloured sandy clay, which became gradually darker in colour from the admixture of vegetable matter, till it passed into a bed of peat, at a distance of about thirty feet from the surface.[2] Beds of clay and variegated sand intermixed with kankar, mica, and small pebbles, alternated to a depth of 120 feet, when the sand became loose and almost semi-fluid in its texture. At 152 feet the quicksand became darker in colour and coarser in grain, intermixed with red water worn nodules of hydrated oxide of iron, resembling to a certain extent the laterite of South India. At 159 feet a stiff clay with yellow veins occurred, altering at 163

[1] *Jour. As. Soc. Beng.*, IX, 686, (1840). See also an excellent account by Lieutenant (afterwards Colonel) R. Baird Smith, *Calcutta Jour. Nat. Hist.*, I, 324, (1841) and *Proc. Geol. Soc.*, IV, 4, (1842). From the latter the account in Lyell's "Principles of Geology" appears to be chiefly taken. Some additional details will be found in the *Jour. As. Soc. Beng.*, II, 369, 649, (1833); IV, 235, (1835); V, 374, (1836); VI, 234, 321, 498, 897, (1837); VII, 168, 466, (1838).

[2] Eighty feet in the original, but this is almost certainly a misprint; first, because Lieutenant Baird Smith mentions in his description the occurrence of peat between 30 and 50 feet from the surface, whereas from 75 to 120 feet sandy clay is said to occur, and this agrees with his descriptive catalogue of the specimens extracted from the borehole, and with his figured section; secondly, because, as will be shown hereafter, a bed of peat is found everywhere around Calcutta at a depth of 20 to 30 feet.

feet remarkably in colour and substance, and becoming dark, friable, and apparently containing much vegetable and ferruginous matter. A fine sand succeeded at 170 feet, and this gradually became coarser and mixed with fragments of quartz and felspar to a depth of 180 feet. At 196 feet clay impregnated with iron was passed through, and at 221 feet sand recurred, containing fragments of limestone with nodules of kankar and pieces of quartz and felspar; the same stratum continued to 340 feet, and at 350 feet a fossil bone, conjectured to be the humerus of a dog, was extracted.[1] At 360 feet a piece of supposed tortoise shell[2] was found, and subsequently several pieces of the same substance were obtained. At 372 feet another fossil bone was discovered, but it could not be identified, from its being torn and broken by the borer. At 392 feet a few pieces of fine coal, such as are found in the beds of mountain streams, with some fragments of decayed wood, were picked out of the sand, and at 400 feet a piece of limestone was brought up. From 400 to 481 feet fine sand, like that of the seashore, intermixed largely with shingle composed of fragments of primary rocks, quartz, felspar, mica, slate, and limestone prevailed, and in this stratum the bore has been terminated."

The first and most important observation to be made on the foregoing facts is that no trace of marine deposits was detected, but on the contrary there appears every reason for believing that the beds traversed, from top to bottom of the borehole, had been deposited either by fresh water, or in the neighbourhood of an estuary. At a depth of 30 feet below the surface, or about 10 feet below mean tide level, and again at 382 feet, beds of peat with wood were found, and in both cases there can be but little doubt that the deposits prove the existence of ancient land surfaces. The wood in the upper peat beds was examined by Dr. Wallich and found to be of two kinds, one of which was recognised as belonging to the súndri tree (*Heritiera littoralis*), which grows in abundance on the muddy flats of the Ganges delta, the other probably as the root of a climbing plant resembling *Briedelia*. Moreover, at considerable depths, bones of terrestrial mammals and fluviatile reptiles were found, but the only fragments of shells noticed, at 380 feet, are said to have been of fresh water species.

The next noteworthy circumstance is the occurrence at a depth of 175 to 185 feet, again at 300 to 325, and again throughout the lower 85 feet of the borehole, of pebbles in considerable quantities. The pebbles in the lower portion are especially mentioned as large, and their size is shown by the circumstance that they impeded the progress of the bore, and that it was necessary in several cases to break them up before they

[1] A ruminant bone, according to Dr. Falconer; Lyell, Principles of Geology, London, ed. 1867, I, p. 479. The specimen cannot now be found. Figures of this bone are given, *Jour. As. Soc. Beng*, VI, 234, pl. xviii, (1837); and *Calcutta Jour. Nat. Hist.*, I., pl. ix, (1841).

[2] Figured *Jour. As. Soc. Beng.*, VI, 321, pl. xxi; and *Calc. Jour. Nat. Hist.*, I, pl. ix.

could be extracted, so that it may be fairly inferred that they were at least two or three inches across (the borehole was six inches in diameter). The greater part of the pebbles were clearly derived from gneissic rocks, but some fragments of coal and lignite which were obtained were perhaps from the Dàmuda series, though their composition indicates the probability that they were derived from the tertiary or cretaceous coal seams of the Assam hills.[1]

The peat bed, it may here be mentioned, is found in all excavations around Calcutta, at a depth varying from about 20 to about 30 feet, and the same stratum appears to extend over a large area in the neighbouring country.[2] A peaty layer has been noticed at Port Canning on the Mutlá (Mutlah), 35 miles to the south-east, and at Khulná, in Jessor, 80 miles east by north, always at such a depth below the present surface, as to be some feet beneath the present mean tide level. In many of the cases noticed, roots of the *sundri* tree were found in the peaty stratum. This tree grows a little above ordinary high water mark in ground liable to flooding, so that in every instance of the roots occurring below the mean tide level, there is conclusive evidence of depression. This evidence is confirmed by the occurrence of pebbles, for it is extremely improbable that coarse gravel should have been deposited in water 80 fathoms deep, and large fragments could not have been brought to their present position unless the streams, which now traverse the country, had a greater fall formerly, or unless, which is perhaps more probable, rocky hills existed which have now been covered up by alluvial deposits. The coarse gravels and sands which form so considerable a proportion of the beds traversed can scarcely be deltaic accumulations, and it is therefore probable that when they were formed, the present site of Calcutta was near the margin of the alluvial plain, and it is quite possible that a portion of the Bay of Bengal was dry land.[3]

At Lucknow the deepest of all the boreholes in the Gangetic alluvium was driven to a depth of 1,336 feet from the surface, or nearly 1,000 feet below sea level. The beds passed through from top to bottom were of the same character, alternations of sand and sandy silt, with occasional bands of kankar, and beyond the mention of coarse sand near the bottom of the borehole, there are no indications of an approach to the base of the alluvial

[1] In a boring recently sunk at Chandernagore subangular gravel of quartz and felspar was met with at about 150 feet. The felspar fragments were extremely abundant and cannot have travelled any great distance. This indication of the vicinity of rocks exposed at the surface is interesting in connection with the hypothesis of the recent origin of the outlet of the Ganges into the Bay of Bengal,—*infra*, p. 443.

[2] Baird Smith, *Jour. As. Soc. Beng.*, IX, 686, (1840); H. F. Blanford, *Jour. As. Soc. Beng.*, XXXIII, 154, (1864). See also notices of earlier borings, *Jour. As. Soc. Beng.*, II, 369, 649, (1833).

[3] But whilst the depression of nearly 500 feet, probably since tertiary times, is unmistakable in the neighbourhood of Calcutta, the signs of elevation within the same epoch in Orissa, only 100 to 200 miles distant to the south-west, are equally distinct.

deposits. Like all the other boreholes, it was sunk in search of a supply of artesian water, and apart from its interest in showing the great thickness of the alluvial deposits, it is important as proving that artesian conditions do prevail under the Indo-Gangetic plain. After the surface water was shut out, a water-bearing stratum was struck at 158 feet, whose water stood at 61 feet from the top of the borehole; at 190 feet another was met with, and the water stood at 42 feet, at 341 feet the water rose to 24 feet from the surface, at 750 feet to 13 feet, at 783 feet to 9 feet, at 975 feet to 2 feet from the top of the borehole. At 990 feet and 1,040 feet two water bearing strata were struck in which the pressure was less and the water sunk to 5 feet from the top of the boring, but at 1,141 feet it again rose to within 18 inches of the top and at 1,189 feet a bed of quicksand was struck from which water flowed at the rate of 10 gallons per minute over the top of the casing, itself 24 feet above the mean level of the surrounding plain.[1]

There is very little of interest in the other three boreholes that have been sunk in the Gangetic alluvium, except in so far as they bear on the theory of the origin of the Himálayas, as will be mentioned in the subsequent chapter.

The Agra borehole, sunk near the southern margin of the alluvium, is the only one which traversed its whole thickness to the supporting floor of rock. The total thickness of alluvial deposits passed through was only 481 feet, composed of sand and sandy clays with some kankar, the uppermost 150 feet being apparently composed to a considerable extent of blown sand, as opposed to true alluvial deposits.[2]

Umballa is on the watershed of the Indo-Gangetic plain, between the Jumna, which flows into the Ganges, and the Sutlej, a tributary of the Indus. The locality is about 905 feet above the sea, and 20 miles from the base of the Himálayas. There is very little of interest in this borehole. The depth to which it was carried was insufficient to test the thickness of the alluvial deposits, and it ceased 200 feet above the level of the sea. No mention is made of any organic remains being found, but their occurrence could not be anticipated, as they occur but rarely in the alluvial formations of the Gangetic plain.[3]

The borehole at Sabzal-ka-kot is only four miles from the base of the hills, and by far the greater portion of the beds traversed consist of sand and pebbles, clays being subordinate, although several beds were met with.

The rarity of organic remains, especially in the older alluvial deposits, has already been referred to, but shells are occasionally found, belonging to species now inhabiting the rivers and marshes of the country. An important discovery of mammalian remains was made about 1830 in some

[1] For a detailed account and section of this borehole, see *Records*, XXIII, 261, (1890).

[2] For detailed section see *Records*, XVIII. 121, (1885).

[3] For detailed accounts, see T. Login, *Quart. Jour. Geol. Soc.*, XXVIII, 198, (1872).

calcareous shoals of the Jumna.[1] The bones were chiefly found cemented together with substances of recent origin, such as fragments of weapons and boats, into a mass of concrete, chiefly formed of the kankar washed from the river's bank, but in two cases the skeleton of an elephant was found preserved in the clay. In one instance, in which the bones were clearly *in situ*, they were found 4½ feet above the highest flood mark, and 80 feet below the summit of the clay cliff formed by the river, and there appears no reason to doubt that all the specimens found were originally derived from the clay. The following species have been recognised[2]:—

Semnopithecus, sp.	*Sus*, sp.
Elephas namadicus.	*Bos (Bubalus) palæindicus*.
Mus, sp.	*Bos*, sp.
Hippopotamus (Tetraprotodon) palæindicus.	*Antilope*, sp.
	Cervus, sp.
Equus, sp.	Fish and crocodile bones.

Three of the species, all that have hitherto been specifically identified, are found in the Narbadá alluvium also, whilst the only genus not now found wild in India is *Hippopotamus*; the species belong, however, to the same subgenus as the living African animal. The evidence is not sufficient to justify any decided conclusions, except that the Jumna clays must have been deposited in the same posttertiary epoch as the Narbadá alluvium, but so far as the specific identifications go, they tend to indicate that the Jumna fossils are newer than the Narbadá remains, as the extinct type *Hexaprotodon* and the foreign form *Bos namadicus* have not been recognised amongst the former.

Some bones were also found in the Betwá river in Bundelhkand and the Bugáoti between Mírzápur and Chanár,[3] but they have not been identified.

Before proceeding further a few words are requisite in explanation of a word which it will be found necessary to use occasionally in the following pages and of four Hindi terms applied in the Ganges valley to particular kinds of alluvial surface which require notice, because they will be found freely used, and because, with perhaps one exception,[4] they have no precise equivalents in English.

To Anglo-Indians it is quite unnecessary to explain the meaning of the term *kankar*, but the explanation may be of some use to European

[1] Sergeant E. Dean, *Jour. As. Soc. Beng.*, IV, 261, (1835). See also Falconer, *Quart. Jour. Geol. Soc.*, XXI, 377, (1865); Palæontological Memoirs, II, p. 640.

[2] Several are figured, *Jour. As. Soc. Beng.*, II, pl. xxv, (1833); and IV, pl. xxxiii, (1835).

[3] *Jour. As. Soc. Beng.*, IV, 571, (1835).

[4] The exception is *khádar*, which corresponds to the English word *strath*. The English term is, however, local; its exact meaning is far from commonly known, and it is only used in hilly country.

students. The original signification of the word is gravel, the term being applied to any small fragments of rock, whether rounded or not. By Anglo-Indians, however, the name has been especially used for concretionary carbonate of lime, occurring, usually as nodules, in the alluvial deposits of the country, and especially in the older of these formations. The commonest form consists of small nodules of irregular shape, from half an inch to three or four inches in diameter, composed of tolerably compact carbonate of lime within and of a mixture of carbonate of lime and clay without.[1] The more massive forms are a variety of calcareous tufa, which sometimes forms thick beds in the alluvium, and frequently fills cracks in the alluvial deposits, or in older rocks.[2] In the beds of streams immense masses of calcareous tufa are often found forming the matrix of a conglomerate, of which the pebbles are derived from the rocks brought down by the stream. There can be no doubt that the kankar nodules, calcareous beds, and veins are all deposited from water containing in solution carbonate of lime, derived either from the decomposition of the debris of older rocks of various kinds, or else from fragments of limestone and other calcareous formations contained in the alluvium.

Bhábar is the slope of gravel along the foot of the Himálayas. Compared with the slopes in the dry regions of Central Asia, Tibet, Túrkistán, Persia, etc., the gravel deposits at the foot of the great Indian ranges are insignificant, the difference in height between the top and bottom of the slope nowhere exceeding 1,000 feet. This difference is probably partly due to the much greater rainfall in India, and to streams being consequently able to carry away a much larger proportion of the detritus washed from the surface of the hills, partly also to the circumstance that the rocks in the lower regions of the hills are not subjected to the loosening effects of frost.

Streams issuing from the Himálayan ranges lose a part, or the whole, of their water by percolation through the gravel in the *bhábar* region.

[1] The following analyses will give a fair idea of the usual composition of nodular kankar:—
[1, Gházípur, Prinsep, "*Glean. Sci.*" III, 278, (1831); 2, 3, 4, Ránígánj, Dejoux, *Records*, VII, 123, (1874);—5, Barmuri;—6, Rámnagar;—7, Sanktoria, all near Ráníganj, Tween, *ibid*;—8, 9, Sahárunpur, Thomson, Rurki Treatise on Civil Engineering, I, p. 115.]

	1		2	3	4		5	6	7	8	
Carbonate of lime .	72·	...	72·	56·94	78·5	......	54·	65·4	66·3	57·18	72·33
Carbonate of magnesia . . .	0·4	...	1·3	1·72	2·	......	...	...	...	...	Trace.
Oxide of iron and alumina . .	11·	Oxide of iron.	·7	1·67	2·	Oxide of iron and alumina.	2·7	1·9	2·	10·32	·73
Water . . .	1·4	Clay .	22·	30·	10·5	Water and organic matter.	2·7	2·3	4·5	...	...
Silica . . .	15·2	Sand .	2·	9·67	7·	Insoluble . .	40·6	30·4	27·2	32·50	13·94.

[2] See the account by Captain E. Smith of the kankar in the Jumna alluvium, *Jour. As. Soc. Beng.*, II, 622, (1833); also Newbold, *Jour. Roy. As. Soc.*, VIII, 258, (1846).

The whole tract in its original condition is covered with high forest, in which the sál (*Shorea robusta*) prevails. At the base of the slope, much of the water which has percolated the gravel re-issues in the form of springs; the ground is marshy, and high grass replaces the forest. This tract is the *tarái*, a term not unfrequently applied to the whole forest-clad slope at the base of the Himálayas, known also as *morung* in Nepál.

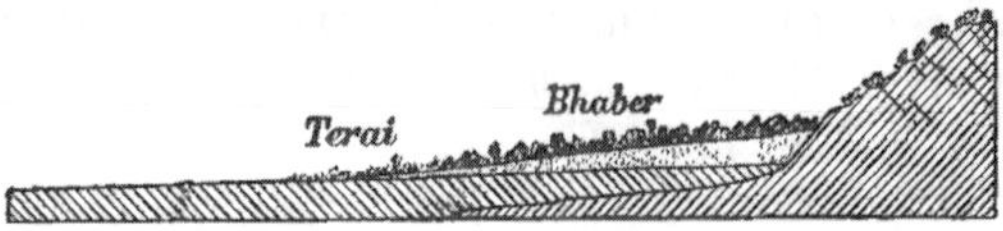

Fig. 22.—Diagram illustrating the relations of *Bhábar* and *Tarái*.

The alluvial plain itself, in the North-West Provinces especially, is composed of *bhángar* or high land, the flat of older alluvium, now at a considerable elevation above the rivers which traverse it, and *khádar* or low land, the low plain through which each river flows. The latter has evidently been cut out from the former by the streams. It is of variable width, and is annually flooded.

In the upper provinces the high banks of the rivers are frequently capped by the hills of blown sand, known in the North-West Provinces as *bhúr*. This is the extreme form of a rather important element in the formation of Indian river channels, and the same result in a less marked form may be traced in a rather sandy, raised bank, along the course of many large rivers down to the limits of tidal action in the deltas. In the lower parts of the river plains this bank, which is above the flood level and is usually selected for village sites, intervenes between the river channel proper and the marshy ground liable to annual floods on each side, the communication between the two latter being kept up by numerous creeks. The origin of the *bhúr* land, or raised bank, is the following. During many months of the year, and especially in the hot season, strong winds arise, frequently of a very local character, and sometimes apparently almost confined to the river channels. These, in the dry season, are plains of loose sand, often two or three miles across and sometimes wider, of which the river usually occupies not more than a fourth. The wind on the Indus and Ganges frequently blows in nearly the same direction as the river channel. Such winds are especially prevalent about midday and in the afternoon, and their effect in transporting the sands of the river bed is so great that the atmosphere becomes too thick for objects a few yards distant to be seen. All who have been in the habit of navigating Indian rivers must have noticed the prevalence of these sand storms. They are so marked that, where large sand banks exist to windward of the river, it is often impracticable for vessels to continue their course, except in the morning before the wind arises, or in the evening, when the motion of the air has diminished. Much of the sand raised by the wind falls again in the bed of the river, but quantities must fall upon the banks in the immediate neighbourhood, where the deposit is retained by vegetation and gradually consolidated into a firm bank. It is only where the quantity of sand is greater that blown sand hills are formed. The original raising of the river bank to the flood level is due to the deposition of silt in a manner which will be explained presently when treating of deltaic accumulations, but the elevation of the immediate neighbourhood of the river bed above the reach of the highest floods is due to the deposit of sand by the wind.

To enter at length into the various peculiarities of land surface[1] which are found in different parts of the great plain of northern India would be far beyond the scope of the present work. A brief account of the principal characters must suffice. The whole region may be roughly divided into five great tracts, each possessing marked peculiarities. These are, commencing to the eastward :—

1. The Brahmaputra valley in Assam.
2. The delta of the Ganges and Brahmaputra.
3. The plains of upper Bengal and the North-West Provinces.
4. The Punjab.
5. The lower Indus valley and delta.

The Assam valley is a gigantic *khádar*, or strath, the greater portion being liable to flooding and consequently not in a habitable state. There, are, however, higher tracts here and there, sometimes mere mounds, rising a little above the general level, and sometimes small plains,[2] which may be considered as representing the extensive *bhángar* of the Gangetic plain. Along the foot of the hills are gravel deposits, but they do not appear to be very extensive.

The quantity of silt carried down by the Brahmaputra is very great, far greater than in the Ganges. The comparative backwardness of the river valley, as shown by the small amount of habitable land, is surprising, since it is evident that the river is occupied in rapidly raising its plain by deposits of silt, and the necessary inference is that the alluvial

[1] The following papers may be consulted for fuller accounts:

For Assam— *Memoirs*, IV, 437, (1865); VII, 155, (1869).

For Lower Bengal and the delta—Mr. Fergusson's paper, *Quart. Jour. Geol. Soc.*, XIX, 321, (1863); also Colebrooke, *As. Res.*, VII, 1, (1801); and Rennell, *Phil. Trans.*, LXXI, 87, (1781).

For the plains of Upper India—*Records*, VI, 9, (1873); Sir P. Cautley, Ganges Canal, London, 1860, I, pt. ii; Falconer, *Quart. Jour. Geol. Soc.*, XXI, 372, (1865); Login, *Quart. Jour. Geol. Soc.*, XXVIII, 186, (1872); H. B. Medlicott, *Records*, XIV, 205, (1881); XVI, 205, (1883); XVIII, 112, (1885).

For the Punjab—Sketch of the Geology, *Punjab Gazetteer*, Provincial volume, 1889, p. 22.

[2] *Memoirs*, IV, 438, (1865). One of these plains is described by Major Godwin-Austen, *Jour. As. Soc. Beng.*, XLIV, pt. ii, 40, (1875).

plain of Assam, in its present form, is not only of later date than the Gangetic plain, but absolutely newer than many portions of the Ganges delta.[1] The difference may be due to a depression of the lower part of the Brahmaputra valley in Assam, to an elevation of the delta, or to a great increase in the supply of water. The second theory is distinctly disproved by the general evidence of subsidence in the delta, and the third is improbable; the evidence is therefore rather in favour of the Brahmaputra valley in Assam having been an area of subsidence in a relatively late geological period. As will be shown presently, there is some additional evidence in favour of this view within the delta itself.

The limits of the delta, or the places where the rivers first bifurcate and commence to give off disturbances, are between Rájmahál and Murshidábád on the Ganges, and on the Bráhmaputra opposite the south-west corner of the Gáro hills. But for a considerable distance above the actual delta the rivers flow through a broad plain of low ground, a large area of which is liable to flooding, and consequently to the deposition of silt. The delta is, in fact, the natural continuation of the *khádar,* or alluvial flat in the upper portion of the river's course, and this *khádar* becomes broader before it expands into the delta.

By far the best description of the Ganges delta, of the changes it is undergoing, and of the action of the rivers in raising the land by the deposition of silt, is that of Mr. Fergusson.[2] He has shown that rivers oscillate in curves, the extent of which is directly proportional to the quantity of water flowing down the channel. Thus, the oscillations of the Ganges where broadest (7,000 feet in the low season) between Monghyr and Rájmahál, average 9½ miles in length; where it contains less water, and is only half the breadth (3,500 feet), between Allahábád and Chanár, the oscillations are 3·7 miles long; in the Bhágírathí, where it averages 1,200 feet in breadth, the length of the oscillations is 1·5 miles; and in the Mátábhángá, where only 500 feet broad, the length of each oscillation becomes only half a mile.[3] The next point which he notices is well known, the tendency of rivers to raise their banks, but the explanation is partly novel. When the whole country is covered with water, moving rapidly towards the sea in the river channels, and stationary throughout the intervening marshes, the dead water of the marshes prevents the floods of the rivers from breaking out of the channels, and, by stopping the course of the silt charged water along the edges of the creeks and streams, forces it to deposit the

[1] For a full discussion see Fergusson, *Quart. Jour. Geol. Soc.*, XIX, 330, (1863). It should, however, be noticed that Mr. Fergusson was led by some published barometrical observations, now shown to have been insufficient, to suppose the level of the Brahmaputra valley at Gauháti to be only about 100 feet above the sea, instead of 163, the maximum flood-level since determined by the Great Trigonometrical Survey.

[2] *Quart. Jour. Geol. Soc.*, XIX, 321-54, (1863).

[3] *Quart. Jour. Geol. Soc.*, XIX, 324 (1863).

sediment it has in suspension. Hence gradually arises a system of river channels, traversing the country in many directions, between banks which are higher than the intervening flats, and these flats form persistent marshes, known in the Ganges delta as *jhils* or *bhils*.[1]

Each river frequently changes its precise course, the smallest alteration in its channel having an effect which is felt for many miles above and below. So that, just as the oscillations of a denuding stream produce a low alluvial flat between high banks, the curves of a depositing river gradually form a high alluvial flat, raised above the surrounding country. In course of time this raised tract is abandoned by the main river for the lower ground at the side, and the river bed is either filled up by silt, or, if near the sea, converted into a tidal creek.

The present Bengal delta, therefore, comprises a large area in which the ground has been raised above the general flood level, through having been traversed by the main branches of the Ganges in past times. Such is the case in the country north of Calcutta. The eastern part of the delta is more backward, the marshes, or *jhils*, are more extensive, and the banks of the streams less consolidated, and this is now the great depositing area. But large tracts of low country, such as the salt lake near Calcutta, are found in the western area also. The remarkable struggle which takes place between the Ganges and Brahmaputra, each tending, by raising the neighbourhood of its channel, to drive back the other, and to gain possession of a larger tract of delta, is most vividly told by Mr. Fergusson, but is too long for extract here. Mr. Fergusson refers many of the more modern changes in the delta to the upheaval of the elevated tract known as the Mádhupur jungle, which had the effect of diverting the Brahmaputra to the eastward into the Sylhet *jhils*, where the silt of the river was deposited. The result was that scarcely any sediment found its way to the sea by the Meghná, the great estuary of all the Sylhet rivers, and hence the sea face of the delta to the eastward curves back in the form of a gulf. The gap was much greater at the commencement of the present century, but about that time the Brahmaputra having, by the deposit of silt, greatly raised the portion of the Sylhet *jhils* into which it flowed changed its course completely in the course of a few years, and instead of flowing to the east of the Mádhupur jungle, cut out a new channel to the west of the raised tract. Since its change, of course, the Brahmaputra has been brought much nearer to the main stream of the Ganges, and the two rivers are now depositing silt so rapidly on the eastern sea face of the delta, that great changes are taking place, and new islands are rapidly forming, whilst the western portion of the deltaic coast line, through which but a small portion of the flood water of the great rivers finds its way to the sea, has undergone but little change since it was first surveyed in the last century.

[1] The former term is Hindi, the latter Bengali.

In the sea outside the middle of the delta there is a singularly deep area, known and marked on charts as the "Swatch of no ground," in which the soundings, which are from 5 to 10 fathoms all around, change almost suddenly to 200 and even 300 fathoms. This remarkable depression runs north and south, and has been referred to a local sinking, but it appears more probable, as has been shown by Mr. Fergusson, that the sediment is carried away from the spot, and deposition prevented, by the strong currents engendered by a meeting of the tides from the east and west coasts of the Bay of Bengal. Mr. Fergusson also shows that, so long as the Bay of Bengal has preserved its present form,[1] the meeting of the tides must have favoured the formation of a spit of sand along the present position of the Sundarbans, as the lower portion of the Ganges delta is called, and that any great deposit of silt to seaward of the present line is impeded by the fine sediment being washed away by the tidal currents, and deposited in the deeper parts of the Bay.

In spite of all that has been written on this subject the origin of the swatch of no ground has by no means been cleared up. A very similar depression has been shown to exist in the bed of the shallow sea off the Indus delta and the cause in both cases has probably been the same, a combination of an excess of subsidence with a deficiency of sedimentation, the latter due to the action of surface currents in sweeping away the silt-laden waters. It is not in accordance with what we know to suppose that at such depths as we are there dealing with, there can be any currents of sufficient velocity to account for the depression by actual erosion.

The chief point in the above theory to which exception might be taken is the question of whether the elevation of the Mádhhupur jungle is sufficiently recent to account for the changes in the course of the Brahmaputra river. This tract of country is composed of a red, iron stained, clayey soil, in which accummulations of pisolitic concretions of oxide of iron are found and worked in places as an iron ore. The clay is of the same type precisely as the older alluvium of the Brahmaputra and lower Ganges valleys. There seems little room for doubt that this is really a region of special elevation, for were we to suppose that it is part of the old surface of the delta, left standing at its former level while the surrounding area was depressed, the height of its ground level, which rises to 100 feet above the general level of the delta outside the Mádhupur jungle, would necessitate a much greater extension of the delta into the Bay of Bengal than there seems any ground for supposing to have ever been the case. The steeply scarped western face and the gentle fall to the level of the delta on the east show

[1] This is probably not so old as pliocene, because such gigantic disturbance has taken place throughout the Assam hills and Arakan, since the close of the Siwálik epoch, that the shape of the northern part of the Bay of Bengal must have changed greatly.

that the elevated area must once have been more extensive than it now is and that the western half has been washed away by the rivers that impinged upon it, while the deeply eroded undulating nature of its surface shows that it has been raised above the flood level of the rivers and so subject to denudation for too long a period to make it probable that the diversion of the Brahmaputra river to the east was entirely due its elevation.

Whether the elevation of the Mádhupur jungle was anterior to, or contemporaneous with, the depression, which it is difficult to suppose has not taken place in the area occupied by the Sylhet jhils, it is impossible to say, but the latter would in itself have been sufficient to account for the diversion of the Brahmaputra to the east of the Mádhupur jungle, and was probably its principal determining cause.

An interesting point to determine in this connection is the date of origin of the delta of the Ganges. Reference has already been made to the fact that the plateau of the Assam range forms structurally part of the Peninsula of India, and to the presence of rocks of peninsular type north of the Brahmaputra valley. Moreover, the upper tertiary deposits south of the Assam range differ from those at the foot of the Himálayas, so far as the latter are known, in being partly of marine origin, and as will be shown further on, the formation of the depression occupied by the Gangetic alluvium was most probably an integral part of the operations which resulted in the elevation of the Himálayas. In the same way it is probable that the transverse depression, through which the Gangetic drainage now finds its way to the sea, may have been formed at the close of the tertiary period *pari passu* with the elevation of the Tipperah hills.

There is some direct evidence in favour of the more recent origin of the Gangetic outlet in the presence of closely allied species of dolphins in the Ganges and Indus rivers, of a very different generic type from the cetacean inhabiting the Irawadi. These two species must be descended from a common ancestor which acquired a fresh water habitat, and the differentiation of the Indus and Gangetic species have arisen from a subsequent separation of the drainage areas.[1] The changes in the course of the drainage over what is now the watershed region, which will be referred

[1] The occurrence of allied forms of porpoise or dolphin in the Ganges and Indus, and the circumstance that the peculiar genus living in these rivers is unknown elsewhere (the cetacean inhabiting the Irawadi being of a very different generic type) have attracted the attention of naturalists already. The ova and young of fish are not difficult of transport, and a very trifling accident might place a pool of water to which the fish of one river have gained access in communication with the other stream. Crocodiles and river tortoises can live for a long time out of water, and have considerable powers of migration on land, but dolphins are confined to the rivers, and could neither live in a shallow pool, nor traverse dry land. The existence, therefore, of closely allied species, doubtless derived from a common ancestor, in two distinct rivers, is a very striking fact. Mr. Murray

to further on, though they opened water communication between the Indus and Ganges rivers, probably did so only in the torrential region, which is not frequented by the dolphins, and the difference existing between the two species indicates a more prolonged separation than could have been the case had there been migration from one drainage area to the other, when they were put into communication with each other by the wanderings of the rivers near the present limits of the two drainage areas. We are consequently driven to suppose, either that the two closely allied species originated independently of each other, which is extremely improbable to say the least, or that the great bulk of the Himálayan drainage once found its way to the sea by a single delta, instead of two, and this must have been either at the head of the Arabian sea, or of the Bay of Bengal. The indications of the sea having extended up the Indus valley within the recent period, and the absence of any similar indications in the delta of the Ganges, make it probable that the former was the original outlet of the drainage, and that the formation of the gap between the Rájmahál and Gáro hills, and of the Gangetic delta, is geologically of recent date.

On the western edge of the delta in Bengal there is a large area of older alluvium, whose surface is slightly undulating, evidently in consequence of denudation. This tract, which is continuous with the alluvial area of the east coast, comprises the greater portion of the country to the westward of the Bhágiráthi and Húghlí, and probably owes its comparative elevation to the deposits from the Mor, Adjai, and Dámodar rivers.

The great plain of Northern India is the area of an alluvial deposit older than that of the delta, and the greater portion of the area is composed of *bhángar* land, through which the rivers cut their *khádar* valleys at depths of from 50 to 200 feet below the general level. The *bhángar* surface, as a rule, is nearly flat, but is much cut up by ravines in the neighbourhood of the rivers.

The question as to whether the great rivers are on the whole, raising their

(Geographical Distribution of Mammals, p. 213) proposed an ingenious hypothesis to account for the phenomenon. He considered that the plain of upper India was once an arm of the sea, that it was cut off by the rise of the coast in Sind and Cutch, and gradually converted into a brackish, and then a fresh water lake, discharging itself by the Ganges, that meantime the marine dolphins inhabiting the sea had gradually become adapted to the changed conditions, and had in fact become *Platanistæ*. He then supposes that the Ganges was cut off from the lake, which overflowed again, and this time into the Arabian Sea, the dolphins of the Ganges and Indus being specialised during the change. It would be unnecessary to refer to this hypothesis, which of course is little more than a suggestion, but for the large amount of support the idea has received from naturalists. It is of course foreign to the purpose of the present work to discuss the genesis of *Platanista*, but, as will be shown, the geological phenomena of the Indo-Gangetic plain do not bear out Mr. Murray's hypothesis, which, it should be stated, was never proposed as a geological theory, but merely as illustrative of the possible mode of origin of allied species.

beds by a deposit of silt, or cutting their channels deeper, has been much discussed without leading to any definite conclusions. The abrupt scarps by which the *bhángar* is not unfrequently terminated, and the defined limits of the *khádar*, clearly prove that the latter has been at some time or other an area of denudation, but it is not easy to tell whether, at the present time in any given stream, the tendency is to raise or lower the general *khádar* level. It is also by no means so evident, as might at first sight be supposed, whether the *bhángar* land generally is an area of denudation or of deposition, although this can, as a rule, be easily seen in each particular area. Thus the minor hill streams from the lower ranges of the Himálayas between the Sutlej and Jumna must deposit sediment, for they cease within the area, whilst between the Jumna and the Ganges numerous streams rise in the *bhángar*, and they must be denuding agents. In the neighbourhood of the *khádar*, *bhángar* land is frequently cut into by ravines, which prove conclusively that the surface of the country is being washed away, but all such marks of rain action cease at no great distance from the low ground, and the principal secondary streams, instead of running from the upland *bhángar* by the nearest route, at right angles, or nearly at right angles to the main river, usually pursue a nearly parallel course down the middle of each *doáb*,[1] or triangular area between two principal streams.

As the velocity of the rivers where they leave the hills is much greater than in the alluvial plains, there must, so long as diminution takes place in velocity of the water when the river is carrying as much earthy matter as it can transport, be a continuous deposition of detritus, and a gradual raising of the area flooded by the stream. This is the case even in the larger rivers which carry a considerable body of water at all times, while the effect of the small streams, which dry up more or less for a great portion of the year, but are converted into muddy torrents charged with coarse sediment during the heavy rains of the summer monsoon, is necessarily to raise the surface of each *doáb*, especially in the neighbourhood of the hills, and to produce floods from which finer sediment is deposited on the surface of the *bhángar* land. Whether the addition thus produced is, on the whole, greater than the wasting of the surface from rain is a question which it is impossible to decide throughout a great part of the country.

One question, which presents itself, is the necessity of accounting for the rivers now cutting their channels at a level considerably below that of the alluvial *bhángar* flat, because this flat must, at all events in the neighbourhood of the *khádar*, have been deposited by streams from the same drainage area, at a period when the main river ran at a comparatively higher level. The change may be due to a general

[1] A Persian word, meaning 'two waters,' and applied to the confluence of two rivers, as well as to the land intervening between them.

elevation of the upper Gangetic plain, or to a depression in the deltaic region. Of the former there is no evidence, of the latter, as shown by the result of the Calcutta borehole, there is ample proof, and it is therefore quite possible that in early posttertiary times, when the animals lived, whose remains are found in the Jumna alluvium, the area of the Ganges delta had been raised to a considerably higher level than it occupies at the present time. Colonel Greenwood has shown[1] that the deposit of silt in river valleys must take place backward, that the lowest portion of the slope must be first raised, and that the check thus given to the flow of water will cause silt to be deposited, so as to raise the alluvial plain further up the course of the river, and if no change of level takes place, the gradual elevation of the Ganges delta by silt deposit will ultimately react on the higher portions of the valley until the rivers once more deposit alluvium on the high *bhángar* land, provided always that this has not been raised so much as to render the slope too great for the rivers to be depositing agents.

One point of interest has been explained by Mr. Fergusson in the paper so often mentioned. A glance at the map will show that the Ganges from Allahábád to Rájmahál, and the Jumna from Delhi to Allahábád, flow close to the southern margin of the great alluvial plain. This is due to the enormous quantity of silt brought down by the Himálayan rivers, and the comparatively small supply furnished by those streams which debouch into the Ganges valley from the southward. The northern portion of the plain has consequently been raised, and the main drainage of the whole forced to find its way as close to the hills of the southern margin as it can. During this process the courses of the tributary rivers running from the northward have been driven westward, and the confluence of these tributaries with the main stream of the Ganges has been shifted upwards along the course of the main river, owing to the tendency of the streams to deposit silt in the neighbourhood of the delta.

The *bhábar* slope of gravel along the foot of the Himálayas, although evidently of comparatively recent formation, has frequently, to the eastward, been cut into terraces by the streams from the hills.[2] This is a necessary consequence of the streams cutting deeper channels in the rocks of the hilly ground. It is curious to note, however, that to the westward the *bhábar* is being raised instead of being cut through by streams. It is

[1] Rain and Rivers, 2nd ed., London, 1866, pp. 173, etc.

[2] Hooker, Himalayan Journals, 1st ed., London, 1854, I, p. 378 (larger edition). Dr. Hooker very naturally, writing forty years ago, when the study of river action was in its infancy, and when nearly all great deposits and all extensive denudations were supposed to be marine, attributed the gravel to a beach deposit, and the valleys to marine denudation. There has been since a great revolution in those portions of geological dynamics which treat of the action of rivers and the sea, and especially in the views held, by English geologists at least, on the comparative amount of work done by the two agents.

not known how far this difference is due to the greater rainfall to the eastward, and to the streams being consequently able to carry away the gravel as they cut down their bed in the rock, whereas weaker streams are prevented from cutting back their channels by their inability to wash away the gravel they have already deposited. There have doubtless been alterations of the gradients of the stream beds through recent upheaval or depression of the surface, and that these would have as much influence on the present action of the streams where they cross the *bhábar* zone as the rainfall.

In connection with the surface of the upper provinces another peculiar local feature requires explanation. Many tracts of land in the Indo-Gangetic alluvial plain are rendered worthless for cultivation by an efflorescence of salt, known in the North-West Provinces as *reh*, and further west as *kalar* (kullar). The name *úsar*, meaning barren, is frequently applied to land thus affected. The salt varies in composition; it consists chiefly of sulphate of soda mixed with more or less common salt and carbonate of soda; it is only found in the drier parts of the country, being unknown in damper regions, such as Bengal.

The *úsar* plains have existed for an unknown time. Where the *reh* or *kalar* is abundant, the water in the upper stratum is impregnated to an extent that is productive of serious injury to the health of the population. To a greater or less extent this pollution of the water near the surface is general throughout Upper India, yet sweet water is obtainable, in the worst *reh* tracts, at depths below 60 to 80 feet.

It is consequently clear that the impregnation of the soil is superficial, and as the upper deposits are demonstrably of fresh water formation, they must originally have been comparatively free from impurities. Still all soils contain some salt, and all the water draining from soils is impregnated to a certain extent. The salts forming *reh* or *kalar* appear to be the refuse products, and to consist of such substances, resulting from the various processes involved in the decomposition of rock, or of detritus derived from rock, and the formation of soil, as are not assimilated by plants. Unless these salts are removed they must accumulate, and the natural process of removal is evidently by rain water, percolating through the soil and carrying off any injurious excess of the rejected salts. If the amount of water be sufficient, and through drainage exists, there will be a constant dilution and renewal of the subsoil water, but if the water reaching the subsoil can only be dissipated by evaporation during the dry season, salts will accumulate in such subsoil water, and as this water is brought to the surface by capillary action, and evaporated, the salts held in solution will be left as an efflorescence on the surface of the ground.

That the composition of *reh* does not differ greatly from that of the salts produced by the decomposition of such rocks as have contributed by their disintegration to the formation of the alluvial plains of Indlia, is shown by the composition of the river water [1] running from the Himálayas, the mountains from which the detritus, now forming the plains of India, was originally derived.

In the case of Upper India it is easy to understand how the destruction of the conditions necessary for cultivation has been established, and it is by no means improbable that a similar process has, in other parts of the world, changed countries, once fertile and populous, into barren deserts. The whole country is treeless. For a great part of the year a scorching sun and a parching wind dry up the moisture in the ground, rendering it hard and impervious to water. When the rains of the monsoon season fall, a large proportion of the water runs off the surface, and the earth is unable to absorb more than a portion of what remains. Thus a great part is evapcrated without penetrating the ground. The little that does percolate through cracks, and in a zig-zag way, through the more porous layers to the upper water stratum, is no more than sufficient to replace what has been dissipated by evaporation, fed by capillary action.

This more or less complete want of water circulation in the subsoil must have been gradually producing its effects in Upper India throughout many generations. The natural process is so slow that it would escape notice were it not that from time to time larger tracts of land become barren. A disturbing cause has, however, been introduced in the form of great irrigation canals. Their immediate effect is to raise the level of the *reh* polluted subsoil water, and thus to produce a great increase of evaporation, with the natural result of more *reh* being left on the surface, and more land being thrown out of cultivation. It is impossible to enter at length into the subject here, but it may be stated that, as all canal water contains salts in solution, whilst rain water contains none, the only change in conditions, so far as the concentration of salts in the soil is concerned, by the addition of canal irrigation, unless facilities for drainage of the subsoil water are also provided, must be the addition of all the refuse salts contained in the canal water to those which would be produced on the surface by the simple action of rain and evaporation.

South and west of Delhi and west of Agra, brine is obtained in places from wells in the alluvium. No particulars have been recorded which

[1] In several analyses of river and canal water from the Ganges and Jumna, the proportion of sulphate of soda varied from 0·0914 to 0·4325 part in 10,000; chloride of sodium from 0·0023 to 0·15 part. The proportion of the two to each other is similar to that found in *reh*. See *Sel. Rec. Govt., India, D. P. W.*, No. XLII, p. 47, (1864). An able and detailed account of the origin, composition and mode of concentration of the reh salts by Mr. W. Center, M.B., will be found in *Records*, XIII, 253—273, (1886).

explain the occurrence of salt in these localities. The case is similar to that already mentioned in the Purna valley in Berar. The distribution of the salt producing ground appears irregular, and this is in favour of the salt being derived from springs in the rock beneath the alluvium.

The plains intersected by the five great rivers which combine to form the lower Indus are not, as a rule, simply divided into *bhángar* and *khádar* like the plains of the North-West Provinces. Owing probably to the greater fall in the Punjab rivers, their deposits are very sandy, and this character tends to diminish the pluvial denudation of the surface by allowing the water to sink into the soil. The action of winds upon the sand of the river, the formation of *bhúr* land, and the elevation of the ground in the neighbourhood of the river banks above the intervening tracts, through the deposition of blown sand, are exhibited in the Punjab to a greater extent than in the Gangetic plain.

To the south-east the limits of the Punjab alluvium are difficult to trace, owing to the manner in which both alluvium and rock are concealed by blown sand. The same is the case throughout the eastern margin of the Indus alluvial plain in Sind.

The ancient geography of the Punjab is far better known than that of most parts of India, partly because the civilisation of north-western India is older than that of other parts of the country, but still more because of the accurate descriptions given by Greek writers of the Indian campaigns of Alexander the Great. It is consequently possible to form some idea of the principal alterations which have taken place in the course of the last 2,000 years, in the channels of the great Punjab rivers, but our best guide unfortunately fails us at the most critical point. Alexander never penetrated to the eastward beyond the land of the five rivers, and there is but little except vague tradition to tell whether the present tributaries of the Indus have ever flowed into the Ganges, or those of the Ganges into the Indus. Yet it is certain that in no part of the great Indo-Gangetic plain have more important changes taken place since the dawn of history than in the neighbourhood of the watershed between the Indus and Ganges.

An inspection of the map accompanying this chapter will show a dried up river channel, which can be traced from the neighbourhood of Sirsá into connection with the eastern Narra in Sind, and local tradition states that this was formerly occupied by a flowing river. At present this channel is dry, except in its upper part, where it periodically carries, for a greater or less distance, the flood waters of the minor streams which drain the outer Himálayas between the Sutlej and the Jumna. The origin of the channel is situated at the junction of the alluvial fans of the Sutlej and Jumna, as is shown by the course of the minor drainage channels, and there are abandoned river courses leading from it in the direction of the debouchures of these two rivers from the hills. There can be no room for doubt that, within the period known geologically as recent, this river channel carried a flowing stream to the sea, and there is some evidence, apart from oral tradition, to show that its drying up took place within the historical period.

The Muhammadan historians of the eleventh and twelfth centuries uniformly speak of the combined Sutlej and Biás rivers, now known as the Garrah or Sutlej, as the Biyah, a nomenclature which is also employed in the Hindu annals of Jaisalmer. This retention of the name of the smaller of the two rivers for the combined waters, where there is no superior sanctity to recommend it, shows that the rivers must have received their actual names at a period when the Sutlej did not join the Biás, but pursued an independent course, and the subsequent abandonment of the illogical nomenclature may be held to show that the alteration of the course of the Sutlej which took it into the Biás did not take place much before the eleventh century. Previous to this change it doubtless flowed down what is now the dry river bed known as the Hákra or Wandan, and there is some evidence, though far from conclusive, that it followed this course as late as the eleventh century of our era.[1]

The traditions of the Hindus point to a time when a large and sacred river, known as the Saraswatí and described as 'chief and purest of rivers flowing from the mountains to the sea,' pursued its course through the eastern Punjab. The modern Saraswatí is an insignificant stream fed by the drainage of the outer hills alone, becoming nearly dry in the hot season and losing itself in the sands of the Rájputána desert. It is absurd to suppose that the language of the Vedas could have been applied, or that any conceivable alteration of the rainfall could have made it applicable, to the Saraswatí of the present day, and the most reasonable explanation is, as suggested by Mr. Fergusson, that the Saraswatí was in fact the Jumna, which, in the Vedic period, pursued a westerly course to the sea, probably down the dry river channel just referred to.[2] It is certainly a suggestive fact in this connection, that when the Bráhmaputra changed its course through Bengal about the commencement of the present century and flowed west of the Mádhupur jungle to join the Ganges, the new channel was named Jamuna, a word etymologically identical with Jumna. On similar principles the old Saraswatí, when it broke eastwards to join the Ganges, may have assumed the name Jamuna or Jumna for its new course, and if this explanation be correct, the Hindu legend that the Saraswatí joins the Ganges at Prayág or Allahábád, is unwittingly a true statement of fact.

This bringing of the change in the course of the Jumna river, which

[1] See an interesting but anonymous article in the *Calcutta Review*, Vol. LIX, pp. 1 29, (1874) understood to be by Surgeon-Major C. F. Oldham; also *Jour. As. Soc. Beng.*, LV, pt. ii, 322—43, (1887).

[2] *Quart. Jour. Geol. Soc.*, XIX, 348, (1863).

has indubitably taken place, down to so recent a date is interesting, for the change must have occurred previous to the present distinction of *khâdar* and *bhângar*, and if this distinction has been produced since the Aryan invasion the question naturally arises whether it may be due, not to movements of elevation or depression, but to the clearing of the land from forest, and the extension of cultivation in the plains, and more especially in the hills, which, by allowing the rain to flow more quickly off the surface, would increase the erosive power of the rivers when in flood, and cause them to cut down their channels into the plains over which they formerly flowed.

The surface of the Indus alluvium in upper Sind differs but little from that of the Punjab, a considerable portion of the area is annually flooded, and the whole drainage of a great river being here, as in Assam, confined to a comparatively narrow tract, some permanent marshes of large size exist. The two most important marshy tracts are along the western edge of the valley from near Jacobábád to the Manchhar lake near Sehwán, and along the eastern edge from Khairpur to below Umarkot. The latter is the channel considered by some the ancient course of the Sutlej. In the neighbourhood of the Indus the ground is rather higher, having evidently been raised by the deposit of silt, aided doubtless by the action of the wind on the sands of the river bed.

Along the edge of the Kirthar range, west of Sind, there is a well marked *bhábar* slope of gravel, but the breadth seldom exceeds one to two miles except where rivers run out of the range. This gravel slope is absolutely barren, and, like other features in Sind geology, is more conspicuous on account of its barrenness.

There is one singular feature in the Indus valley to which nothing parallel is to be found in the Gangetic area. The river between Sukkur and Rohri has cut its way through a low range of limestone hills, surrounded on all sides by alluvial deposits. The eastern Narra, fed by the flood waters of the Indus, traverses an alluvial tract eastward of the hills. In fact, the circumstance that the flood waters of the Indus, both to the east and west, traverse plains at a lower level than the river bed, is shown by the course of the canals, and great fears have been entertained that the Indus may desert its present channel and break out to the westward, through the plain in which Jacobábád is built, into the line of marshes already mentioned. The curious features of the tract are not even confined to the present river course, for at Aror, four miles south-east of Rohri, there is another gap in the limestone range, said, on what is believed to be good historical evidence, to have been the bed of the river rather more than nine centuries ago.[1] At that time the main stream is supposed to have

[1] Cunningham, Ancient Geography of India, London, 1871, I, pp. 257, 264, etc.

traversed Sind considerably to the east of its present course; it passed by the old city of Bráhmanábád, and then probably ran southward by the Purán, an old river bed still existing, to the Kori creek, which was the principal mouth of the river. The Indus is said to have deserted its old bed at Aror for its present channel between Sukkur and Rohri, in consequence of an earthquake about A D. 962, and as Bráhmanábád was also, in all probability destroyed by an earthquake[1] at some period prior to A.D. 1020, it is not impossible that the two events were due to the same cause. The Indus is said to have deserted Bráhmanábád at the time when the city was destroyed. All the details preserved, however, are so much mixed up with mythical incidents that but little dependence can be placed upon them, and nearly all the circumstances mentioned are more or less open to dispute. It is questioned, for instance, whether Aror was ever situated on the Indus, and it is contended that Bukkur, a fortress on an island in the river opposite Rohri and consequently in the channel now cut through the limestone range, existed before the ninth century. Certainly, the channel through the hills at Aror is very narrow, and it is possible that it was never traversed by the main stream of the river, though the configuration of the ground supports the hypothesis that some stream has cut through the hills at the spot. Again, it is contended that Sehwán, the ancient Sindomána, was always on the Indus, and that consequently the main stream of the river must have run in ancient times where it flows now. But, on the other hand, Alexander is said to have left the river, and marched to the neighbourhood of Lárkhána, and thence to Sehwán, from which place he "marched back to the river."[2] It may be fairly concluded that important changes have taken place in the course of the river, without feeling certain that the precise nature of these changes has been correctly ascertained.

The accumulation of fluviatile deposits in the Indus plain, and the consequent elevation of the surface, is well seen in the neighbourhood of Umarkot, where, as has already been mentioned, the flood water from the Narra trickles through the sand hills forming the limit of the Indus alluvium, and fills large hollows between the ridges of sand. The level of the bottom of these hollows must have been, in all probability, at least as high as the general surface of the Indus plain at no distant date.

During the floods, water leaves the Indus, and its tributary the Sutlej, as far up as Baháwalpur, and flows southward by the eastern Narra, which must be regarded as a distributary, although its waters now seldom reach the sea. The true head of the delta, however, is generally considered to

[1] Bellasis, *Jour. Bombay Br. Roy. As. Soc.*, V, 413, 467, (1853). There is some doubt regarding the exact position of Bráhmanábád. According to General Haig the ruins usually known by that name are those of Mansurah, while he places the real Bráhmanábád some seven miles to the north-east at a place now known as Depar. *Jour. Roy. As. Soc.*, new series, XVI, 281, (1884).

[2] Arrian: "Anabasis," VI, 16.

be a little above Haidarábád, where the Phuleli stream leaves the river.[1] The channels of the delta frequently change, more frequently perhaps than in the case of the Ganges. The sea face is, in all probability, determined by marine currents, and it is improbable that any great change is likely to take place through the deposit of sediment.

The eastern part of the Indus delta now receives but little water from the river. It is said that a large area of country in the neighbourhood of the Kori mouth was depressed during the earthquake of 1819,[2] and that the great size of the Kori creek is due to the depression. A very large area north-west of the Kori creek is covered with salt, sometimes a foot or even more in thickness, deposited from sea water.

In the neighbourhood of the sea the soil is usually argillaceous and firm, but in the upper part of the delta the whole surface is composed of loose micaceous sand with but little clay, and the rivers consequently have unusual facilities for changing their channels. The littoral portion of the delta is so low that a broad tract of country is always overflowed at spring tides, whilst the bottom of the sea in the neighbourhood of the coast is so shallow, and the slope outwards so gradual, that large vessels cannot, in many places, come within sight of the land. A tract of country of variable width, but in places several miles broad, along the sea face of the delta, is annually flooded by the rise of the river, the water being kept higher than it would otherwise be by the influence of the south-west monsoon.

Reference has already been made to the Rann of Cutch, and it was pointed out that this tract of country is evidently an old marine gulf now silted up. A brief description of the area and its peculiarities may, however, be well added to the account of the Indus delta, which it adjoins to the eastward.

The Rann[3] consists of an immense marshy salt plain, scarcely above the sea level and stretching for 200 miles from east to west, and in places

[1] A very good description of the Indus delta has been given by Lieutenant T. G. Carless, *Jour. Roy. Geog. Soc.*, VIII, 328, (1838), reprinted in *Sel. Rec. Bombay Govt.*, XVII, 461-500, (1855). See also a memoir by Assistant Surgeon J. F. Heddle, (*ibid*, p. 403). For the ancient changes in the delta of the Indus see also Cunningham, Ancient Geography of India, p. 283, etc.

[2] It is stated by Carless, *Jour. Roy. Geog. Soc.*, VIII, 366, (1838), that the alluvial formations exposed on the bank of the Kori creek opposite Kotasir are, with the exception of the uppermost layers, broken up in confused masses, and inclined to the horizon at an angle of 30 or 40 degrees. The disturbance is attributed to the earthquake. It would be well however, that the spot should be examined by an experienced geologist, as the vagaries of false bedding (or oblique lamination) in sands and silts deposited by the strong currents of an estuary, are very likely to mislead any one unaccustomed to the peculiar appearance of these deposits.

[3] For a fuller description of the portion north of Cutch by Mr. Wynne see *Memoirs*, IX, 14, (1872). See also Burnes, Travels in Bokhara 2nd ed., London, 1835, I, p. 316; Grant, *Geol. Trans.*, 2nd series, V, 318, (1840); Frere, *Jour. Roy. Geog. Soc.*, XL 181, (1870); Rogers, *Quart. Jour. Geol. Soc* XXVI, 118, (1870).

nearly 100 from north to south. From the south-eastern extremity a low alluvial tract, dividing Ahmadábád from Káthiáwár, and including an extensive brackish water marsh called the Nal, connects the Rann with the head of the Gulf of Cambay. A very trifling depression, probably not amounting to 50 feet, would convert Káthiáwár into an island, and even a smaller amount of sinking would suffice to isolate Cutch completely; indeed, it is now an island during the prevalence of the south-west monsoon, when the sea, raised by the wind, dams back the water brought into the Rann by the various rivers which drain into the flat from Rájpútána, Gujarát, and Cutch, in the same manner as the level of the creeks is raised in the Indus delta. At this time portions of the Rann are seven feet under water, but the average depth does not exceed five feet. The inundation lasts from July to the end of November, and portions of the surface, especially a tract to the westward near Sindri, depressed by the earthquake of 1819, are constantly covered with water. Below this water there is, in places, a bed of salt, sometimes as much as three or four feet in thickness.

There can be little doubt that the Rann was a gulf of the sea within recent times. Not only do the traditions of the country all agree with this view,[1] but the present condition of the surface, an immense flat of sandy mud, can only be explained by supposing that the tract is the site of an inlet, now silted up. The barren condition of the surface is due to flooding by salt water at one season, and hot dry weather at other times; the soil is consequently too salt to support even the vegetation, such as mangroves, which will grow in ordinary sea water. Unless further depression takes place, the surface must be gradually raised by the silt brought in by rivers, and the tracts which support vegetation must extend.

The depression of an area of 2,000 square miles around the fort of Sindri in the western part of the Rann, at the time of great earthquake of 1819, has been described so often,[2] that it appears unnecessary to repeat the account here. In this case the circumstance which enabled the changes of level to be accurately estimated was the fact that the whole of the tract affected was very nearly at the sea level, and so close to the sea that it was

[2] There is some historical evidence also. When Alexander the Great sailed down the Indus he passed through the great eastern branch, then the main stream of the river, but now dry, to the Kori mouth. Near this mouth he came to a great lake (Arrian: "Anabasis," VI, 20). Mention is also made of a great lake-like expanse of water in this direction by some Mahomedan historians. Sir B. Frere also states on apparently good traditional evidence, that Viráwah, in Nagar Párkar, north-east of the Rann, was a seaport from 500 to 800 years ago, *Jour. Roy. Geol. Soc.*, XL, 195, (1876) No mention of any sea north of Cutch appears to have been made by the Chinese travellers of the seventh century; Cunningham, Ancient Geography of India, I, p. 302.

[1] An account is given in Lyell's Principles, ed. 1868, II, pp. 97-104, and has been copied into many text-books. For a very full description by Mr. Wynne see *Memoirs*, IX, 29-47, (1872). Mr. Wynne doubts whether the Allah Bund was really raised, and suggests, with much probability, that the appearance of elevation was due to the depression of the ground around Sindri, south of the Allah Bund.

flooded immediately. A further depression is said to have taken place in 1845 in the same neighbourhood.[1]

At first the effect of the depression in 1819 was to produce a great sheet of water, navigable by boats of some size, but this has gradually silted up, and Mr. Wynne, on visiting the ruins of Sindri in January 1869, found that the greater portion had been filled up to nearly the level of the Rann, and that but a small shallow pool remained around Sindri itself.

Though not, strictly speaking, part of the Indo-Gangetic alluvial plain, this will be the best place to notice that great accumulation of blown sand, in the tract between the Indus and the Arávallis, which is known as the Indian desert. The name implies a greater degree of barrenness and solitude than is actually the case. Shrubs and grass tufts are scattered thinly over nearly the whole area, small trees are not infrequently met with, and it supports large numbers of sheep and cattle, and a hardy population, civilised enough to build cities and palaces and wells of hundreds of feet in depth.

Over the whole of this area sand hills are scattered more or less thickly, but the great accumulation of blown sand forms a strip along the north of the Rann of Cutch, from which two arms run, one northwards by Umarkot and then turning north-east and running north of Jaisalmer to Bikaner; the other running north-eastwards between Bálmer and Jodhpur and coalescing with the first about Bikaner. The central area of Jaisalmer, Bálmer and Pokaran is rocky, with comparatively few and scattered sand hills.

The sand hills are of two types. One of these, admirably delineated in the Trigonometrical Survey maps, is of the ordinary type of sand dune. Its longer axis is at right angles to the prevailing direction of the wind, and it presents a long gently sloping face to windwards, up which the sand grains are driven, and a steep face to leeward, down which they roll, whose slope coincides with the angle of repose of the dry sand.

The other type is one which is not noticed in the text-books. It is very largely developed in the Thar district of Sind to the north of the Rann of Cutch, and appears, equally with the first type, to owe it sform to the

[1] Nelson, *Quart. Jour. Geol. Soc.*, II, 103, (1846). Before quitting the subject of the great alluvial region of Northern India, it may be as well to point out that by far the greater portion of the earthquakes, and especially of the more severe shocks felt in India, occur in the immediate neighbourhood of the Indo-Gangetic plain, and especially near the deltas of the great rivers. The earthquakes are, as a rule, felt much more severely on the rocky ground around the alluvial plain, than in the plain itself. When depression takes place, as in the case of Sindri in the Rann, the shock may be but slightly felt at the locality principally affected, although towns in Cutch, on rocky ground, at a distance of several miles, are thrown down; but this is in accordance with experience elsewhere.

prevailing winds. Throughout the area mentioned, the sand is heaped into

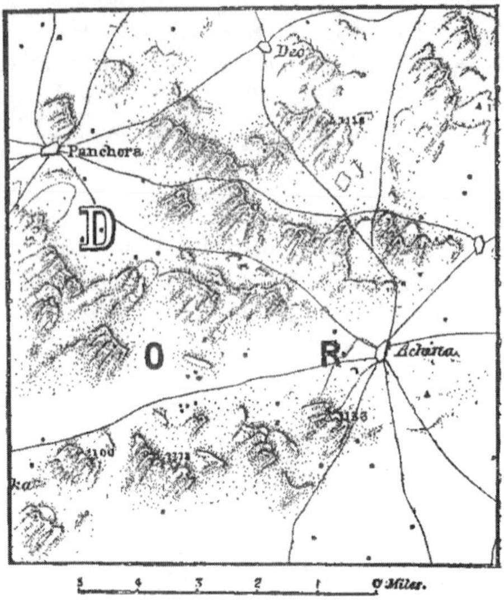

Fig. 23.—Sand hills of the transverse type; after the Topographical Survey of Rajputana.

long narrow ridges, running about north-east and south-west to north north-east and south south-west, with a steep slope on either side, the crest gradually rising in height to the north-eastern extremity, which is

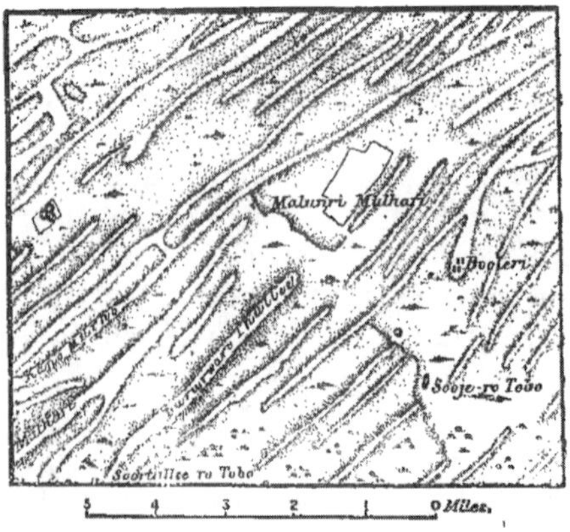

Fig. 24.—Sand hills of the longitudinal type; after the Sind Revenue Survey.

usually the highest point of the ridge, and beyond this is a steep slope downwards, coinciding with the angle of repose of the sand.

The mode of formation of this type of sand hill is not very easy to explain, but from the fact that, where the two types are found together, the longer axis of the one is at right angles to that of the other, we may conclude that in both cases the form is decided by the direction and force of the wind, the longer axis being parallel in one case, and in the other transverse, to its prevailing direction. The steep slope of repose at the northern end of these sand hills shows that they are formed of sand grains which are driven along the surface of the ground by the wind, and not of those light enough to be carried in suspension, so that no theory of accumulation under the lee of bushes, will account for the facts, and we must look elsewhere for an explanation. If one of the transverse type of sand hills be examined, it will be seen that the windward slope is by no means a uniform plane, but is composed of long narrow ridges, parallel to the direction of the wind, with intervening depressions, probably kept open by a concentration of the wind in them and a consequent increase of transporting power, if not an actual development of power of erosion. It seems probable that the longitudinal type of sand hill is due to an exaggeration of this effect, by which the depressions, instead of being comparatively shallow, and causing mere saddles in the general ridge, are carried almost, if not quite, to the base of the accumulation. However this may be, the restriction of the longitudinal type of sand hill to the seaward and western margins of the desert appears to show that they are connected with a greater wind force than the transverse type.

The height of these sand hills is considerable. They frequently exceed 100 feet, ranging to 200 feet, and, according to Sir Bartle Frere, 400 to 500 feet in the southern part of the desert. The size of these sand hills and the area they cover imply an accumulation of blown sand which it is not easy to account for.

It appears difficult to believe that all the sand found in the desert can have been derived from the Indus. The surface of the Rann at present is too muddy to furnish any large supply. The sand consists of well rounded quartz grains, mixed with smaller quantities of felspar and hornblende, and is undistinguishable from the sand of the sea coast except that the grains are better rounded, as is always the case with wind blown sand. That found in the bed of the Indus is also very similar in character. The most probable theory appears to be that the Rann of Cutch and the lower portion of the Indus valley were, as has already been shown to be probable on other grounds, occupied by the sea in posttertiary times, and that the sand of the desert was derived from its shore. The most sandy tracts are on the edge of the Indus valley, along the northern margin of the Rann, and along the depression of the Lúni valley, and these portions of the country were all probably situated on the coast. The form of the rocky hills around Bálmer and Jaisalmer shows that they have been shaped by subaerial, not by marine denudation, and it is probable that the more elevated central portion of the desert was land, whilst the Indus valley, the Rann, and the Lúni valley were occupied by sea.

The accumulation of sand in the desert region is evidently due to the low rainfall and to the consequent absence of streams, the effect being intensified by the accumulation of sand and the porous nature of the resulting surface. In other parts of India the sand blown from river channels or the sea coast is either driven by the wind into other river channels, or it is swept into them again by rain. There are sand hills in abundance in the alluvial plain of the Indus, but they attain no great size, because the sand is always swept sooner or later into some stream, by which it is carried away towards the sea.

Besides the occasional sand hills of the Indus valley in Sind, there are some much larger tracts in the Punjab, repeating, on a smaller scale, the phenomena of the Thar and the Rájputána desert. The most important of these is in the Sind-Ságar Doáb between the Indus and Jehlam, but there is a barren tract in the Rachna Doáb between the Chenáb and Rávi, and sand hills occur in places also in the Bári Doab between the Rávi and Sutlej.

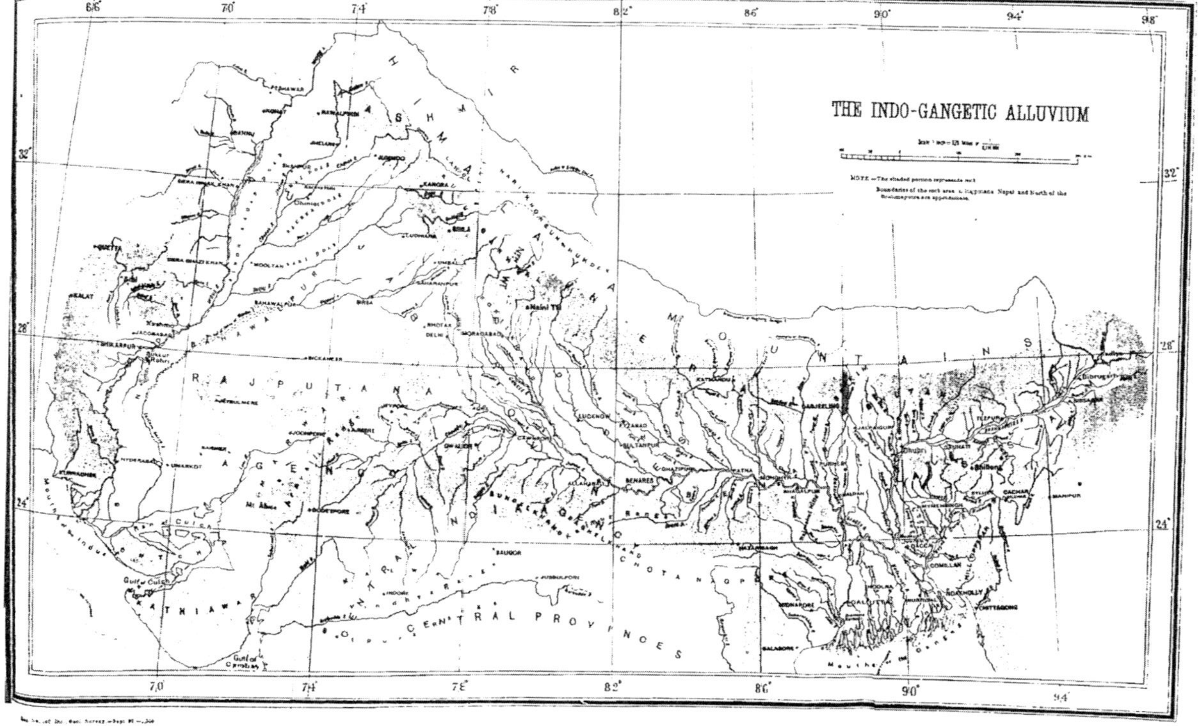

PHOTOLITHOGRAPHED FROM AN ORIGINAL MAP SUPPLIED BY THE GEOLOGICAL SURVEY OF INDIA,
at the Survey of India Offices, Calcutta Oct 1887.

CHAPTER XVIII.

THE AGE AND ORIGIN OF THE HIMÁLAYAS.

Geographical limitation of the Himálayas—Physical geography—Evidence of the tertiary deposits as to the age and elevation of the Himálayas—Revd. O. Fisher's theory of mountain formation—Age of the Himálayas.

In dealing with the great question of the date of commencement, cause, and history of the elevation of the Himálayas, the first point to determine is the exact signification in which this name shall be used. The Himálayas in a general sense are well understood to be the great system of mountains which rises to the north of the alluvial plains of upper India, and forms the southern margin of the highlands of Tibet, but the limits of the range at either end are difficult to define, for it becomes continuous with the mountain ranges between India and China on the one hand and those north of Afghánistán on the other, and though it is easy to regard these as distinct ranges, once the change of general direction is well established, the absolute continuity of each with the Himálayas, where the junction takes place, shows that the elevation of the whole was part of the same great series of movements of the earth's crust. It is, however, necessary to adopt some definite geographical limits to the Himálayas, and those used here will be the lines along which the strike of the chains of hills, and of the rocks they are composed of, takes a sudden bend. On the west this line may be taken to run through the hills west of the valley of Kashmír, from where the Karakoram range bends into the Hindu Kush to where the Jehlam leaves the hills. On the east neither the geology nor the geography of the hills is sufficiently well known to define the limit of the Himálayas, but it may be presumed to run in from the neighbourhood of Sadiyá in a north-easterly direction.

The descriptions of the orography of this great system of mountains vary very much according to the idiosyncracy of the writer, and the particular meaning he may have attached to the term 'mountain chain.' The earliest of the writers on the Himálayan mountains, Captain Herbert, regarded the spur on which Simla is situated as the natural termination of the main snowy range, giving as his reason that it was the watershed between the drainage of the Indus and the Ganges. But this system of classification would lead to endless confusion and completely obscure the true relations of the various chains of which the Himálaya mountains are composed, for the rivers, without exception, cut at some point of their course through a zone of special elevation, whether we look to present contour or to geological structure, and it is impossible on any rational ground to separate the portion of the range on one side of the river valley from its continuation on the other.

The most popular of the views regarding the physical geography of the Himálayas is probably that proposed by Clements Markham, which regards them as consisting of three more or less parallel ranges known as the northern, central, and southern, respectively. This view was most beautifully illustrated in the map attached to his edition of the travels of Bogle and Manning,[1] where three long ranges are depicted, stretching across the map from east to west. The view appears to have a certain resemblance to the truth, and cannot be absolutely disproved owing to our ignorance of the geography of the greater part of the Himálayas and to the indefiniteness of the term mountain range, but our knowledge of Himálayan geography is sufficient to show that the orography of the Himálayas is by no means so simple or well defined as it is represented on the map just referred to.

In the north-western portion of the Himálayas, where alone the geography is known with any degree of completeness, four principal ranges are commonly recognised. The most northerly and most elevated of these, which appears to bend round into the Hindu Kush at its north-westerly extremity, is the Mustagh or Karakoram range, whose culminating peak, 28,265 feet high and the second highest in the world, was formerly known as K_2, but is now often named after its discoverer Godwin-Austen.

South of, and more or less parallel with, this comes the Ladákh range, which may be regarded as commencing near the junction of the Sháyak (Shyok) and Indus rivers and running thence south-eastwards along the north side of the Indus valley. This range, which has a most marked individuality both geographically and geologically, is breached by the Indus at about 150 miles from its north-westerly termination in 79° of east longitude. The range continues as far as Hanle, forming there the south, instead of the north, side of the Indus valley, but its further continuation is imperfectly known.

The Zanskar range appears to owe its existence quite as much to the accident that it forms the watershed between the Indus and Chenáb drainage, and has consequently been less denuded than the regions on the north and south, as to any special elevation it has undergone.

The outermost of the principal ranges is that which is known as the

[1] "Narrative of the Mission of George Bogle to Thibet and of the Journey of Thomas Manning to Lhasa," 2nd edition, London, 1879.

Pír Panjál south of the valley of Kashmír, and as the Dháoladhár[1] south of Chamba. Though the unity of these two ranges is obscured by their being broken through by the Chenáb and Rávi rivers, their geological structure, so far as it is known, seems to show that they are in reality part of one and the same range. Much might, however, be said in favour of the view which would regard them as two separate ranges, écheloned along the northern margin of the plains of the Punjab.

Nothing definite can be said of the south-easterly continuation of the ranges. The Pír Panjál, Dháoladhár and Zanskar ranges may be regarded as coalescing and becoming continuous with the great range of snowy peaks, while the Ladákh and Karakoram ranges coalesce to continue as the range of mountains which runs north of the great longitudinal valley, of the upper Indus, Sutlej, and Sanpo rivers. It may, however, well be doubted whether either of these ranges has a real continuity along the whole length of the Himálayas, and it is altogether more probable that, whether we regard them structurally or according to the accidents of the existing contour of the ground, they consist of a series of comparatively short ranges overlapping each other at their extremities. The final classification of the minor ranges of the great Himálayan system of mountains must wait for a more detailed geological and geographical knowledge than is at present available.

Though it is impossible to give any definite idea of the detailed orography of the Himálayas it is possible to divide the mountains into orographical regions sufficiently distinct from each other, even if their exact boundaries are somewhat indefinite. The innermost of these is the upland of Tibet, characterised by great elevation and a dry climate with its concomitant of very extensive accumulations of detritus in the valleys.

The drainage of the southern portion of this region, except that of the comparatively small area which is drained by the upper Sutlej river, escapes into the Brahmaputra and Indus valleys at the extremities of the Himálayan range. But by far the greater portion has a closed drainage, or such as escapes finds its way into extra-Indian rivers.

South of the Tibetan region rises the great zone of snowy peaks whose drainage, from both northern and southern slopes, finds its way, in a more or less directly-transverse direction, on to the Indo-Gangetic plain. The watershed of this transverse drainage lies to the north of the zone of highest peaks, which is repeatedly interrupted by the deep valleys of the rivers traversing it.

South of the snowy peaks comes a zone of lower hills, seldom rising

[1] On the accompanying map the name Dháoladhár has been misplaced. It is the range which runs south-eastwards from Dalhousie, south of the Chamba valley.

much over 12,000 feet above the sea, which has been distinguished as the lower Himálayas. These lower Himálayas in many places graduate into the main snowy range, so that it is difficult to draw a definite distincton between the two; yet, they can be recognised as a fairly well marked feature of the range, forming a belt of hills some 50 or 60 miles broad, between the high mountains of the central range and the low hills of the sub-Himálayas. West of the Sutlej the lower Himálayas cannot be recognised as a distinct feature, the high ranges of the Dháoladhár and Pír Panjál rising directly from the sub-Himálayan zone, but the inner portion of what has generally been regarded as the sub-Himálayan zone in this region rises to greater altitudes than where the lower Himálayas are typically developed, and should possibly be regarded as the continuation of this feature.

The sub-Himálayas, which have been referred to in the last paragraph, form the outermost zone of the hills. They are usually marked by an abrupt drop in the average height of the hills, they are exclusively composed of tertiary and principally upper tertiary deposits, and except in the region west of the Sutlej, seldom rise over 4,000 feet.

There can be no doubt that this sudden drop in the average height of the peaks, between the lower and sub-Himálayan region, is principally due to the sub-Himálayan region having been subjected to a smaller amount of elevatory movement than the lower Himálayas, though it is doubtless also due in part to the greater softness of the rocks they are composed of, and their greater proximity to the lowlands of the plains, but it is not so easy to determine whether the distinction between the central ranges and the lower Himálayas is due principally to differences in the amount of upheaval they have undergone or to denudation. Doubtless both have co-operated. The bottoms of the river valleys near the plains being at a lower level than further into the heart of the mountains, and the average slopes at which the hillsides stand, which depend on the readiness of the rock to disintegrate and the amount and distribution of the rainfall, being probably less on the average in the lower than in the central Himálayas, the peaks could naturally not rise to the same altitude. This does not, however, seem to be a sufficient explanation of the facts, and it is only natural to suppose that the belt of mountains which contains the highest peaks in the world must have been an area of special upheaval, while there are some features in the profile of the main river valleys which support this conclusion. These valleys all penetrate the hills to within 10 miles of the line of highest peaks without rising more than 4,000 to 5,000 feet above sea level, but as they cross this line there is a sudden rise of the river bed which carries it up to 9,000 to 10,000 feet within a few miles. Above this the gradient falls again and, in the Tibetan region, the average slope does not seem to be more than a few feet in each mile of channel. This sudden rise in the river beds as they cross the line of highest peaks seems to show that this has been a region of greater and more rapid upheaval than those to the north or south, and that the rivers have not yet been able to cut down to the level they will ultimately reach.

It has already been mentioned that all the principal rivers draining from the Himáláyas have their sources to the north of the line of highest peaks, and that they cross this zone of special upheaval in deep valleys. The old explanation of this feature was that the valleys were great fractures in the range, through which the rivers found their way. This view has never been specifically disproved in the case of the Himáláyas, but it has been so frequently shown to be incorrect in other cases where it was maintained, it is so generally discredited, and moreover the shape of the valleys is so palpably due to subaerial erosion that it is unnecessary to devote further attention to it here, and we may accept these transverse valleys as having been entirely produced by the action of rain and rivers.

It will be shown further on that the sub-Himáláyan ranges are composed of the disturbed and upheaved deposits laid down by the same rivers which now traverse them. In this case it is evident that the rivers are older than the hills they traverse, and that the gorges have been gradually cut through the hills as they were slowly upheaved. In the same manner it might be supposed that the rivers, originally draining from the north of what is now the line of greatest elevation, were able to keep their valleys open by cutting them down, at a pace sufficient to prevent the upheaval producing an actual reversal of drainage, but in many cases the drainage area to the north of the line of highest peaks appears to be too small to have given sufficient erosive power to the stream to allow of this explanation. In the first edition of this Manual it was suggested [1] that these transverse river valleys may have formerly extended further to the north, draining a larger area of country beyond the snowy range than they now do, and that owing to the greater depth of the gorges of the Indus, Sutlej, and Dihing or Sanpo, their upper waters were cut off by a gradual encroachment of the longitudinal valleys of these rivers on the transverse drainage. The present writer is unable to accept this view. He regards it as altogether more probable that the first effect of the commencement of the upheaval of the Himáláyas, was to establish a pair of longitudinal valleys along its northern face, whose drainage escaped round the extremities of the upheaval, and that in the first instance the whole of the drainage north of what is now the line of highest peaks escaped by these rivers. As the mountains were upheaved the gradients of the rivers flowing directly to their southern margin became steeper than those of the longitudinal valleys north of the main range, the erosive power of the streams increased, and they were able to cut back through the line of maximum upheaval and rob part of the

[1] p. 677.

drainage which originally flowed east and west to the gorges of the Indus, Sutlej, and Sanpo

The few geological investigations which have been made along the southern margin of the Tibetan highlands have not been sufficiently detailed as regards the distribution of the recent deposits and forms of the valleys, to decide this question with certainty, but there is one specific observation, recorded by General Strachey,[1] which points to the conclusion that the explanation given here is the correct one. He records that the subrecent deposits of the Sutlej valley in Hundes extend right up to the crest of the Niti pass, and that a detached portion of it is to be seen two or three miles south of the crest. The mere fact of its extending up to the crest of the pass shows that there must originally have been higher ground to the south; in other words, that the original watershed of the Sutlej must have run further south than it now does, and the occurrence of an outlier in what is now the southern drainage area, if confirmed, gives a still further, though unnecessary, proof of the encroachment of the southern on the northern drainage areas.

The same is indicated by the shape of the valleys which drain in either direction from the watershed. So far as can be gathered from the admirable maps of northern Kumáun and Garhwál, and from the accounts of travellers, the slopes on the southern side of the passes are much steeper than on the northern; the erosion of these slopes would consequently be more rapid, and as it progressed the watershed would gradually be forced northwards.

The most conclusive evidence, however, seems to be that derived form the subrecent deposits of the Sutlej valley in Hundes. These show that the Sutlej was followed approximately its present course during a period sufficient, firstly, for the formation of a deep rock valley, secondly, for the accumulation in this of over 3,000 feet of subrecent deposits, and, thirdly, for the re-excavation of gorges, 3,000 feet deep, through these same accumulations. There can consequently have been no progressive cutting back of the head waters of the Sutlej during all this period.

From a stratigraphical point of view the Himálayan mountains may be divided into three zones, which correspond more or less with the orographical ones. The first of these is the Tibetan, in which marine fossiliferous rocks are largely developed, whose present distribution and limits are to a great extent due to the disturbance and denudation they have undergone. Except near the north-western extremity of the range they are not known to occur south of the snowy peaks. The second is the zone of the snowy

[1] *Jour. Roy. Geog. Soc.*, XXI, 63, (1851).

peaks and lower Himálayas, composed mainly of crystalline and metamorphic rocks and of unfossiliferous sedimentary beds, believed to be principally of palæozoic age. The third is the zone of the sub-Himálayas, composed entirely of tertiary, and principally of upper tertiary deposits, which forms the margin of the hills towards the Indo-Gangetic plains, and has so intimate a connection with, and so important a bearing on, the history of the elevation of the Himálayas that it will require a more detailed notice here than the others.

The stratigraphy and palæontology of the rocks composing this tertiary fringe have been referred to in a previous chapter, but it will be necessary to recapitulate part of what has been written, and to add some further details which are important from the present point of view.

The classification which will be adopted is the following:—

Upper tertiary or Siwálik series	Upper Siwálik. Middle Siwálik. Lower or Náhan Siwálik.	
Lower tertiary or Sirmur series	Kasauli group . . Dagshái group . . Subáthu group.	Murree beds.

The lowest of these groups consists everywhere of marine deposits, clays, shales with some limestone, and a few bands of sandstone. It passes upwards with perfect conformity into a series of interbedded sandstones and clays. The latter, almost always red in colour, prevailing in the lower part, the former in the upper, so that there is a gradual increase in the average coarseness of the débris from below upwards, a feature even more conspicuously displayed in the sections of the upper tertiaries.

The distribution of these rock groups is noteworthy. There is a long narrow outlier in western Garhwál just east of the Ganges, in which only the marine Subáthu beds are found. A larger area is found further west in the Simla hills, where all three groups are represented. For a part of its length this exposure is in direct contact with the Siwálik series along the great fault, which will be referred to further on, but along its western half it is separated by a narrow strip of pretertiary slates. At the western extremity of this outcrop of lower tertiary rocks, which belong by position to the lower Himálayas rather than the sub-Himálayas, they run down into a narrow strip, which, stretching along the south face of the Dháoladhár, connects them with the larger area of lower tertiaries in Jammu.

The upper tertiaries are, like the lower, divided into three groups. The lowest of these, known as Náhan consists of clays and sandstones. the former being mostly bright red in colour and weathering with a nodular structure, the latter firm or even hard, and throughout the whole not a pebble of hard rock is to be found.

The middle Siwáliks consist principally of clays, and soft sandstones, or sand rock, with occasional strings of small pebbles, which become more abundant towards the upper part, till they gradually merge into the coarse conglomerates of the upper Siwáliks. It must be understood that this classification, being dependent on lithological characters, not on the palæontology of the beds, is not strictly accurate, and it is certain that the different stages must more or less overlap each other on different sections. Any classification on palæontological grounds is unfortunately impossible at present, as most of the fossils have been obtained through native collectors, and their localities are not known with certainty. But this is unimportant for our present purpose, as it seems certain that the three successive lithological stages do represent successive periods of time, though part of the conglomerate stage on one section was certainly represented by a part of the sand rock stage on another.

In the north-west of the Punjab, beyond the Jehlam, the whole of the tertiary rock groups are said to form one conformable system from base to summit.[1] Further east their relations are less simple and at first sight somewhat perplexing. The true meaning of the anomalies was long ago pointed out by Mr. Medlicott,[2] but have been illustrated in so much greater detail by Mr. Middlemiss in his account of the sub-Himálayas of Kumáun and Garhwál[3] that it will be well to turn to this region for illustrative sections. Here there is normally a perfectly conformable transition from the Náhan group to the middle Siwálik sandstones, and again from these to the upper Siwálik conglomerates. This conformable succession, which is exhibited by many sections, is illustrated on two of the sections reproduced on the accompanying plate, but it is not invariable. Many sections, as No 3 on the plate, show the upper Siwálik conglomerates resting unconformably on the eroded edges of Náhan sandstones, and this peculiarity of unconformable contact between two members of a conformable system finds its most striking exemplification in the short section reproduced in figure 25.[4]

West of the Ganges the country has not been examined in the same detail, but it is certain that the same feature exists. In the neighbourhood of Náhan the Náhan and upper Siwalik groups are in contact along a line of fault, but the latter contain many boulders derived from the sandstones of the former, showing that they had been elevated and exposed to denudation at the time that the upper Siwálik conglomerates were being deposited. Beyond the Sutlej, on the other hand, it was found impossible to draw any boundary between the two groups, so gradual was the transition.[5]

[1] A. B. Wynne, *Records*, X, 112, (1877).

[2] *Memoirs*, III, pt. ii, (864).

[3] *Memoirs*, XXIV, pt. ii, (1890).

[4] Page 468.

[5] *Memoirs*, III, pt. ii, Chap. IV, (1864) ; see also *Records*, XIV, 169, (1881).

Throughout this eastern area just referred to, the upper and lower tertiaries are nowhere found in superposition. They occur on opposite sides of a great fracture, marking the limits of the sub-Himálayan region, and it is at present uncertain whether any beds whose age would place them with the Sirmur series conformably underlie the Náhans in this region.

The examination of the sub-Himálayas of Jammu has been even more cursory than that of the country further east, and it is at present uncertain whether the same relations, as exist further east between the different groups of the Siwálik series, may not be found to prevail between the lower and upper tertiaries. The unconformity between the two is proved by the presence of boulders of lower tertiary sandstone in the upper Siwálik conglomerates, but the conformity is not equally well proved. The map accompanying the only published account [1] of this region appears to indicate a conformity between the Siwáliks and the Murree beds of the small inliers at Naoshera, and between the Punch and Jehlam rivers, and on the whole it is probable, especially if we bear in mind the asserted conformity of the whole sequence on the further side of the Jehlam, that the relations of the lower and upper tertiaries are the same apparently contradictory ones, of conformity on one section and unconformity on another, as are exhibited by the groups of the Siwálik series.

There is but one explanation possible for the known facts, that this great thickness of deposits, whose unity of lithological type, no less than the special sections showing conformity between its subdivisions, prove that they belong to one rock series, must have been deposited during a period of disturbance, so that while a continuous sequence of conformable deposits was being laid down in one place, in another they were disturbed, elevated and exposed to denudation.

Along the whole length of the Himálayas, wherever the junction of the Siwáliks with the pretertiary rocks of the Himálayas has been seen, it is a great reversed fault. To the west of the Biás a similar reversed fault forms the boundary between the lower tertiaries and the secondary and palæozoic rocks of the Himálayas, and in the intermediate area, where the lower tertiaries rise up and form part of the lower Himálayan area between the Sutlej and the Jumna, this great fault forms, for part of its course, the boundary between the Sirmur and Siwálik series.

The fault is, however, not a mere boundary fault in the ordinary sense of the term, that is, the fault is not of a date subsequent to the deposition of the whole thickness of the series whose boundary it forms, nor did this ever extend, in its full development, far to the north of the line of fault.

[1] *Records*, IX, 55, (1876).

Mr. Middlemiss' section, reproduced in the woodcut below of itself proves this, for it is seen that the great boundary fault was fully developed previous

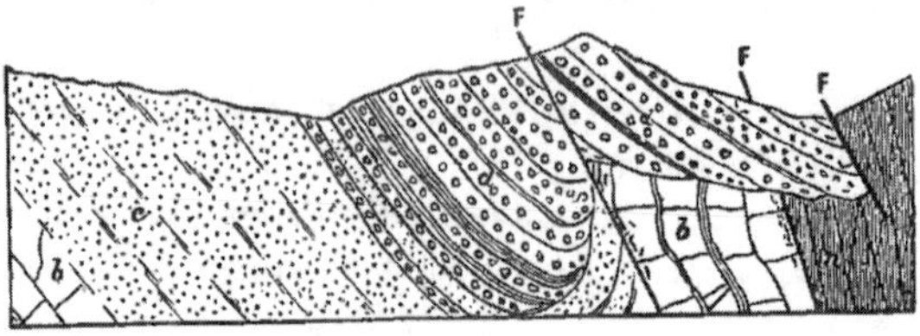

Fig. 25.—Section at the head of the Sara valley, eastern Kumáun, showing overstep of the main boundary fault by the upper Siwáliks.

to the deposition of the upper Siwálik conglomerates, which rest on the eroded Náhan sandstones and overstep the fault on to the pretertiary slates. A somewhat similar section appears to be found in the sub-Himálayas of the Dárjíling district,[1] but has not been worked out so completely.

Apart from these special sections there are some considerations of a general nature, which are in reality of greater importance. It will be seen that this line of faulting forms an absolute boundary for the Siwáliks in all the range east of the Sutlej, and in the hills west of the Rávi it similarly forms an absolute boundary for the tertiary rocks, no outlier being found to the north, and only a few small inliers to the south of it. Now, if the many thousand feet of tertiary strata found south of the fault had been laid down in a continuous sequence previous to its formation, they must have extended far to the north of it, and it is almost impossible to understand how they could have been so completely removed as to leave no trace of an outlier.

It was Mr. Medlicott who first, in 1859, pointed out the meaning of this, —that the fault is in effect an original limit of deposition, to the north of which the Siwáliks never extended. In order to lay special stress on this peculiarity, and to take the 'main boundary,' as he named it, out of the category of ordinary boundary faults, he described it as a cliff, against which the Náhan sandstones were deposited, and an original boundary of deposition, only slightly modified by subsequent faulting. Subsequent investigations have shown that this description requires some modification, but have fully established the conclusion that the main boundary is not merely a boundary of present distribution, but in effect an original limit of deposition. In order to understand the grounds on which this conclusion is based and was originally reached, it is necessary for a while to leave the tertiaries and consider the submontane deposits of the present day.

[1] P. N. Bose, *Records*, XXIII, 244, (1890).

The outer margin of the hills at the present day is everywhere fringed with a band of gravel deposits forming the "*bhábar,*" or gravel slope of the foot of the hills. The extent and constitution of this varies with its position. Opposite the debouchures of the great rivers draining the central portions of the Himálayas it reaches a great development, and is composed almost entirely of boulders of hard crystalline and metamorphic rocks, which have mostly been well rounded in their long journey down the river valley. Boulders of limestone are somewhat rare, while the softer varieties of slate and sandstone are almost absent, having been unable to withstand the severe treatment they received. In the stretches intermediate between the great rivers the nature of the gravel varies according to the rocks exposed within the drainage areas of the streams; where these drain only from the outer hills of Siwálik conglomerate, rounded boulders of hard rock will be found, elsewhere there are seen fragments of limestone, sandstone, or slate, which are often subangular, owing to the shorter distance they have travelled and the smaller degree of abrasion they have undergone, and are always less rounded than the hard boulders of the great rivers.

If we now turn from the submontane deposits of the present day to the upper Siwáliks we find a remarkable resemblance between them. Not only are the upper Siwáliks so similar to the recent deposits in general character that they have, not without reason, been compared to an elevated portion of the plains, but there is precisely the same connection between their composition and the existent lines of drainage. In the sub-Himálayas of Kumáun, there is a great development of the upper and middle Siwáliks, and especially of the conglomerates, where the Rámgangá and Kosi rivers issue from the hills. Further west, where there are no large streams draining from the interior of the hills, the whole Siwálik zone becomes constricted and only the Náhan group is seen. Between the Ganges and Jumna the upper Siwálik conglomerates again attain a great thickness, and are composed of well rounded boulders of hard rocks, precisely similar to the débris brought down by these rivers at the present day. West of the Jumna the conglomerates die out to a great extent, and those which are seen consist of fragments, to a large extent subangular, of the older tertiary sandstones, and of the formations found in the outer part of the Himálayas of this region. Where the Sutlej debouches from the hills there are at least 4,000 feet of coarse conglomerates, but in a parallel section, only seven miles off, there is only about 500 feet of them, in the middle of over 3,000 feet of brown sandy clays.[1] The same features have been noticed in the case of all the other great rivers, that the upper Siwálik conglomerates attain a great thickness in their neighbourhood and are composed of waterworn boulders of hard rocks, while in the intermediate country they are generally represented by brown clays undistinguishable

[1] H. B. Medlicott, *Records*, IX, 57, (1876).

from the recent alluvium, or if conglomeratic the pebbles are of local debris.

There is but one explanation possible of these features, that the Himálayan range already existed at the time when the upper Siwáliks were being deposited, with very much the same boundaries as at present, with the principal features of its drainage already established, and with an elevation comparable to that of the present day. The Siwáliks formed, therefore, the northern fringe of a series of alluvial deposits, whose southerly extension must be looked for beneath the undisturbed deposits of the Gangetic alluvium.

But the Siwáliks now form low hills, in which these once horizontal deposits have been disturbed, elevated, and exposed to denudation. There has consequently been a southerly advance of the margin of the hills since the upper Siwálik age.

The vast thickness of Siwálik deposits, whose upper division alone attains many thousands of feet in thickness, all of which were formed subaerially, and even now, after the elevation they have undergone, only reach a very few thousand feet above the sea, can only have been formed in an area which was gradually subsiding as the deposits were heaped up. We must conclude then that the plain country south of the hills, where the conditions are so similar to those under which the upper Siwáliks were formed, and where immense masses of debris have been heaped up without raising it very much above the alluvial plain to the south, is an area where considerable subsidence has taken place during the recent period.

There are of course no sections showing the actual nature of the boundary between this area of subsidence on the one hand, and the region of recent elevation occupied by the outer hills of the Siwálik zone on the other, but there are some considerations of a general nature which, apart from any reasoning from analogy, indicate its nature.

The steady sweep of the boundary along the length of the Himálayas, the absence of any deep re-entering angles or outlying patches, show that it is in the main a structural feature, and that only its details have been shaped by denudation and sedimentation. Nowhere are the upper Siwálik conglomerates found passing conformably beneath the recent deposits of the foot of the hills, and the section of the outermost ridge is always an anticlinal, whose southern half shows an increasing steepness of dip in a southerly direction. The beds actually in contact with the submontane gravels may be uppermost Siwáliks or belong to the lower part of the Náhan group, the dip may be moderate, vertical, or even inverted,[1] rarely the whole southern half of the anticlical may have been denuded

[1] Examples of all these cases may be found in the sheets of sections published by Mr. Middlemiss, *Memoirs*, XXIV, pt. ii, (1890). It must be borne in mind that the underground section south of the junction of the Siwáliks with the recent deposits is purely conjectural.

away and covered up by recent deposits, but where it is seen there is usually a rapid increase in the steepness of the southerly dip near the margin of the hills. These indications of a line of special bending of the strata close to the southern edge of the hills help out the supposition that the actual demarcation of the two contiguous areas—one of elevation, the other of subsidence—may be of the nature of a fault, one side of which has been raised and the other depressed, the depression being filled up as fast as made by the abundant debris brought down from the hills.

Though there are no sections showing the nature of the junction between the undisturbed recent deposits of the plains and the upraised Siwáliks the sections in the sub-Himálayas throw some light on its nature. Wherever the Siwálik zone attains any considerable width it is found to be traversed by one or more reversed faults of great throw, running more or less parallel to the outer boundary on the one hand and the main boundary of the Siwáliks on the other. These faults all show an ascending section on the outer (southern) side and the dip usually flattens towards the fault, where the uppermost beds are seen in contact with strata of a very much lower zone. Moreover, the older beds are invariably thrown into an anticlinal immediately north of the fault, while the southern half, when it is present, shows an increasing dip as the fault is neared, exactly as is the case in the anticlinal of the outermost ridge, and occasionally, as in that case, the southern half is cut out.[1]

The relations of the rocks on either side of these great faults are so similar to what we have inferred is probably the case between the Siwáliks of the outermost ridge and the deposits of the submontane region of the plains, that it is natural to regard each fault as marking a former limit of the disturbed tract, and the successive faults as indicating a step by step southerly advance of the outer margin of the hills. According to this hypothesis the great main boundary would mark approximately the southern limit of the Himálayas at the commencement of the Siwálik period, north of which the upper tertiary deposits did not extend to any great distance or in any great thickness.

We are not, however, confined to the conclusions that may be drawn from direct observations in dealing with this problem. It may also be attacked from the purely physical and mathematical side, as has been done by the Rev. O. Fisher in his great work on the physics of the earth's crust.[1] Mr. Fisher adopts the hypothesis that the solid crust of the earth is of limited thickness and rests on a magma of greater density, whose condition is actually or virtually that of a fluid. As the central core of the earth

[1] Physics of the Earth's Crust, 1st ed., 8° London, 1888, pp. 114-41.

cools down by the conduction of heat away from it, the outer crust is left partially unsupported by the consequent contraction, exposed to a greater strain than it is capable of bearing, and yielding along lines of weakness, is thickened both upwards and downwards from a zone somewhere in the thickness of the crust, above which the material will on the whole be forced upwards, and below it downwards. This zone, called the neutral zone, is placed, for reasons unnecessary to enter into here, as its exact position is not of great importance, at three-fifths of the thickness of the crust from its upper surface. If the subjacent magma had a density one and two-thirds as great as that of the overlying crust the upward protuberance would be supported by the buoyancy of that portion which had been thrust downwards into the magma. But such a great disparity cannot exist, and the extra weight of the elevated tract will, consequently, bear the crust downwards on either side as is indicated in the diagram, fig. 26, till sufficient of the lighter solid material is depressed into the denser magma to provide the requisite buoyancy.

As soon as an elevated tract is formed denudation will commence, and as it is extremely unlikely that the protuberance will be symmetrical, a larger amount of material will be deposited on one side of it than the other. In the diagram this is supposed to take place on the right hand side and its effect will be to depress that side of the range more than the other while the elevated tract will be lightened by the removal of material from its surface. As a consequence, the centre of gravity will be shifted

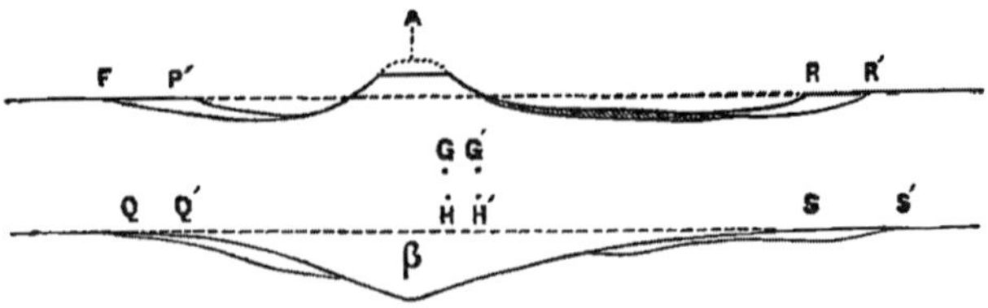

Fig. 26.—Diagram to illustrate Revd. O. Fisher's theory of mountain formation—

A.—Upward protuberance, or 'elevated tract.'
B.—Downward protuberance or 'root.'
P. R.—Original limits of depressions on surface.
P.′ R.′—Subsequent limits after the deposition of sediment represented by shading, and removal of the portion of the elevated tract represented by dotted lines.
Q. Q.′ S. S.′—Original and subsequent limits of downward protuberance.
G. G.′ H. H.′—Original and subsequent positions of centres of gravity of crust, and of displaced fluid respectively.

towards the side on which there is the greatest accumulation of debris, while the centre of flotation of the downward protuberance remains unchanged. The disturbed tract is consequently thrown into a state of unstable equilibrium, which can only be made stable by such a movement of rotation as will diminish the size of the depression on the left hand side, where there is least sediment, and extend it on the right hand side, where most has been deposited. As this action goes on, the depression on that side where least sediment is deposited may become obliterated, and as the crust is exposed to a tensional strain on this side it may be that fissures will open and volcanic outburst take place.

Such, briefly stated, is Mr. Fisher's theory of mountain formation, and there is, on the most superficial view, a considerable resemblance to what we know to have taken place in the case of the Himálayas. The great depression in which the Gangetic alluvium has been accumulated corresponds to that in which the greater bulk of debris derived from the denudation of the mountains is deposited, and the obliteration of the eocene sea of the central Himálayas, accompanied as it was by a great outburst of volcanic energy, would appear to correspond to what the theory points out as likely to happen to the depression on the other side of the range. But the greater part of the elevation has taken place since the obliteration of the eocene sea of the central Himálayas, and we are then met by the difficulty that we have to assume a rotation of the whole elevated tract of Central Asia. We have no reason to suppose that so large a mass of the earth's crust would have sufficient rigidity to allow of its rotation as a whole; rather, there is good reason to suppose that it would yield infinitely to a long continued stress. Moreover, Mr. Fisher's theory takes no notice of horizontal compression, which in the case of the Himálayas has certainly gone on up to a very recent period. The fact is this movement of rotation, with the degree of rigidity it demands, as well as the ignoring of the effects of lateral compression after the elevated tract is formed, are mere generalisations necessary to bring the complex conditions of nature within the powers of mathematical investigation, and are not intended to be taken literally, as an exact account of what actually takes place.

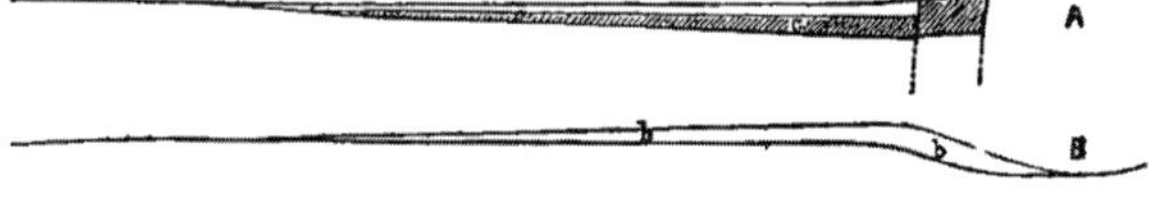

Fig. 27.—Diagram to illustrate the theory of the elevation of the Himálayas, corresponding to the right-hand half of fig. 26. Horizontal scale about 60 miles, vertical about 30 miles to 1 inch.

A.—Massif of the Himálayas.
B—Root of the same.
C.—Earlier marginal deposits, compressed and elevated.
c.—Continuation of the same, depressed and undisturbed.
D.—Subsequent deposits overlapping C.
b.—Sinking of lower surface of crust due to C and D.

Bearing this in mind, we have now to consider what modifications are required to fit the purely mathematical theory to the more complex conditions of actuality. In the first place the elements of rigidity and rotation being abandoned, we need not consider the left hand side of the diagram, and may redraw it in greater accordance with the conditions of the Himálayan region (see fig. 27). We have now an elevated region *A* subjected to denudation and adjoining it an area extending to *R*, on which deposition is taking place, the deposits being contributed by the elevated ground *A* to the north, and the waste of the rock area to the south. The tract being supposed to be in equilibrium, the surplus floating power of *B* will cause it to rise when *A* is lightened by denudation, and the load thrown on *D* will cause it to sink, especially in the neighbourhood of *A* where the load is greatest, till the magma displaced by the lower surface of the crust is sufficient to float the load. The result will be, firstly an extension of the depression in a direction away from the elevated tract *A*, and, secondly, a strong tendency to either fracture or flexure of the crust at the junction of *A* and *D*.

As we may take the crust to be infinitely yielding to long continued stresses, there is no reason why that produced by the lightening of the one area and the loading of the other should not be relieved simply by the sinking of the latter and the rising of the former on either side of a separating plane. But denudation and deposition are not the only forces at work, for to bring the case into connection with that of the Himálayas, we must suppose compression to be continually in progress. This will be relieved partly by an additional elevation of *A*, but also by the compression and consequent elevation of the marginal deposits D, which would not offer the same resistance as the already consolidated beds of *A*. In this way the deposits on the edge of the depression would gradually come to form part of the tract *A*, whose boundary would advance towards *R*, but not to the same extent as the shifting of the outer boundary of the depression towards *R*.

With this amplification,—for it is no modification of Mr. Fisher's theory, but merely a more detailed explanation of part of the process which does not lend itself to mathematic investigation,—we find it easy to explain the true nature of the great reversed faults which traverse and bound the Siwálik zone. They mark the successive limits between hill and plain, between the area of deposition and depression on the one hand and of denudation and upheaval on the other, and the small amount of disturbance which may generally be noticed in the beds immediately in contact with the fault plane is explained by the fact that their immense throw is not due to the effect of horizontal pressure acting on an inclined plane, but principally to a vertical pressure, downwards on the one side, and upwards on the other, of the fault.

It appears then, that so far as the tract along the foot of the hills is concerned, the conclusions drawn from observed facts and from theoretical deduction agree with each other, but we have another check on the theory, for, if it is true, the Gangetic plain must have originated at the same time as the great mountain range to the north, and gradually increased in width by the subsidence of the rock area along its southern margin. Now, when a stream issues from the rock area on to the alluvial plain, it is the coarsest debris which is first deposited, while the finer grained material is carried further and deposited at a greater distance from the margin of the alluvium. Consequently, if a boring be sunk near the southern limit of the plain, where it has been encroaching on the rock area, the beds passed through near the surface should be finer grained on the average than those passed through lower down, for these last belong to an earlier period, when the edge of the alluvium was less distant than it is at the present day. Near the northern margin of the plains the conditions should be reversed, for there the rock area has been encroaching on the plains and the upper bed should be composed of coarser debris on the average than those lower down.

The deep borings that have been put down in the Gangetic plain are four in number. Of these, two, at Umballa and Fort William respectively, are well situated for testing the hypothesis; a third, at Agra, is less suitable for reasons which will shortly appear, while the fourth, at Lucknow, being well out in the middle of the plain, does not appear to have gone deep enough to give any evidence of importance.

In dealing with the records of these boreholes we are harassed by the vagaries of nomenclature indulged in by the men, never trained geologists, to whom the conduct of the operations was necessarily entrusted. But by adopting two classes only, sand and clay, and by including in the former 'sand,' 'coarse sand,' 'clayey sand,' etc., and in the latter 'clay,' 'silt,' 'sandy silt,' 'limey silt,' etc., a fair idea will be reached of the relative coarseness of grain of the beds passed through at different depths of the same boring. This method is perfectly justifiable, as, for the present purpose, the absolute coarseness or fineness of grain is immaterial, and we merely want to know whether in any part of the bore hole the beds are on the whole coarser or finer grained than those above and below.

Adopting this system of classification, we may make an abstract of the Fort William boring,[1] thus:—

0 ft. to 100 ft.	Sand	0	Clay	100
100 „ „ 200 „	„	11	„	89
200 „ „ 300 „	„	95	„	5
300 „ „ 400 „	„	98	„	2
400 „ „ 481 „	„	81	„	0

The increase in coarseness of grain of the beds passed through is conspicuous enough in this abstract, but the reality is even more striking, for in the sand from 180 feet downwards, some beds of gravel and pebbly sand

[1] For detailed section see *Records*, XIV, 221, (1881); *Calcutta Journ. Nat. Hist.*, I, 324, (1841).

are included, and the boring was finally brought to a standstill in a bed of gravel which it was not found possible to penetrate.

The second boring of importance is that made at Umballa.[1] Adopting the same broad classification of clay and sand, we get the following result:—

0 ft. to 100 ft.	Sand 59	Clay 37	Soil 4
100 " " 200 "	" 56	" 42	Kankar 2
200 " " 300 "	" 58	" 42	
300 " " 400 "	" 54	" 46	
400 " " 500 "	" 46	" 54	
500 " " 600 "	" 37	" 63	
600 " " 701 "	" 6	" 95	

Here we have, as the hypothesis requires, a very distinct increase in coarseness of texture in the upper beds as compared with the lower.

Besides these two borings, one has been put down at Agra, the evidence of which is slightly vitiated by the peculiar local conditions. The abstract of the section is as follows[2] :—

0 ft. to 100 ft.	Sand 65	Clay 35	Kankar 0
100 " " 200 "	" 44	" 56	" 0
200 " " 300 "	" 0	" 97½	" 2½
300 " " 400 "	" 18	" 76½	" 5½
400 " " 481 "	" 7	" 74	" 0

Here there would seem to be an increase in coarseness of texture, both upwards and downwards, from 200 feet. The explanation of this is to be found in the fact that the surface deposits round Agra are largely composed of blown sand, and it is probable that the sand beds found in the uppermost 160 feet of the section are of æolian origin, while below that the beds are alluvial and exhibit the gradual upward increase in fineness of texture required by the hypothesis.

The fourth boring at Lucknow has been sunk to a depth of 1,336 feet. As might be expected from its situation, there is no marked increase or decrease in the coarseness of the beds passed through, but near the bottom of the boring some beds of coarse sand were found, and these may indicate an approach to the base of the alluvium and mark a time when its southern boundary was not far from Lucknow.

To sum up, of the four deep borings which have been made, two are completely in accordance with the hypothesis, the third one is in favour of it, though its evidence is vitiated by peculiar local conditions, while the fourth is so situated as to give no evidence one way or the other till it is carried to a greater depth.

We find then that the inductions from observed facts regarding the southerly advance of the margin of the hills, the nature of the boundary between hill and plain, and the mode of formation and growth of the

[1] *Records*, XIV, 233, (1881).

[2] *Records*, XVIII, 121, (1885).

Gangetic plain, agree in all essential points with the deductions from Fisher's theory of mountain formation and using the one to elucidate or amplify what remains doubtful in the other, we may approach the interesting subject of the date of the commencement of the elevation of the Himálayas.

The occurrence of marine nummulitic beds at a height of many thousand feet on the north face of the main snowy range in Hundes, and at a height of 20,000 feet in Zanskar, shows that the elevation of this part of the Himálayas must have taken place entirely within the tertiary period. Further east we have not the same conclusive evidence, but the upper cretaceous fossils that were brought from north-west of Lhasa show that the elevation of this part of the Tibetan plateau could not have commenced at a much earlier period.

The limitation of the marine mesozoic and palæozoic rocks to the northern flanks of the main snowy range, and their absence, so far as is known, to the south of this, may be due to an original limitation of deposit, or it may well be due to the country over which they (are wanting) having been more rapidly elevated, and consequently exposed to more active denudation. But even if the southern limit of these marine formations represents approximately the recurrent shore lines of a long series of epochs, it is difficult to believe that a mountain range at all comparable to the Himálayas of the present day lay immediately to the south of them. The present geographical and geological connection between the Himálayan range and the Tibetan highland is too close to make it at all probable that the elevation of the latter was altogether posterior to, and independent of, that of the former, and consequently the elevation of the Himálayas as a mountain range cannot have been long in progress, if it had commenced, when the sea flowed over Tibet at the close of the secondary period.

On the southern side of the Himálayas there is not the same direct evidence. The close connection between the older rocks of the Assam range, and the corresponding ones of the Indian Peninsula has already been noticed as indicating that the present limits between the peninsular and extra-peninsular areas had not been established at the time that they were being deposited, and the presence of subaerially formed Gondwána rocks in the eastern Himálayas suggests, though it does not prove, that they were formed in the same land area as those of the Peninsula and that no depression, corresponding to that now occupied by the Gangetic alluvium, was in existence at the commencement of the secondary period.

The complete absence of any known exposure of marine nummulitic rocks between western Garhwál on the one hand, and the Gáro hills on the other, might only mean that the shore line ran south of the present limit of the hills, and that the nummulitic beds are hidden by the Gangetic alluvium, but there is not so close a relationship between the nummulitic faunas, so far as they are known, of the Punjab and of Assam as to necessitate, or even suggest, so direct a communication between the two areas. There is consequently some degree of probability that the Indo-Gangetic depression had not been established at the commencement of the tertiary period and we again get the close of the secondary period as the probable date of the commencement of the elevation of the Himálayas.

The stratigraphical relations, between the nummulitic beds of the north-west portion of the lower Himálayas and the subjacent deposits, point to the same conclusion. There is not only a general parallelism of stratification, which might result from the compression both have been exposed to, but there is a very close resemblance in the nature and degree of disturbance they have undergone, and the nummulitics lie with perfect parallelism of bedding on an eroded surface of the former pretertiary deposits, wherever a section showing the original contact between them is found. Were this merely a local phenomenon observed on one or two isolated sections no importance need have been attached to it, but when it is seen wherever the contact between the two rock series has not been modified by faulting, from the inliers of the Jammu hills on the one side, to the outliers east of the Ganges on the other, it shows that there had been no appreciable disturbance of the older rocks, now forming this part of the Himálayas, when the nummulitics were deposited. In other words, that if the elevation of the Himálayas had already commenced in eocene times, it had not extended into the north-western portion, or was confined to the central portion of the range.

The close connection in structure and distribution of the upper Siwálik conglomerates and the submontane deposits of the present time has already been appealed to as evidence that the Himálayan range existed in pliocene times with very much the same limits and elevation as at the present day, and with the main features of hydrography already marked out. But these coarse conglomerates are confined to the upper Siwaliks. As we descend the section pebbles get smaller in size and less in number till, in the lower part of the Siwáliks proper and throughout the immense thickness of the Náhans, not a pebble is known to occur.[1]

It might be held that this was due to the southerly advance of the foot of the hills, and that we must look for the coarse conglomerates of middle Siwálik and Náhan age in the hills north of the main boundary. It has, however, been shown to be extremely probable that neither the Siwáliks proper nor the Náhan group ever extended, in anything like their full thickness, much to the north of the main boundary, and the absence of any known outlier, though merely negative evidence, cannot be altogether

[1] There is only one recorded instance of a conglomerate supposed to be of Náhan age, and that case is so exceptional and its age given with so much hesitancy, that it may well be neglected,—*Memoirs*, III, pt. ii, p. 135, (1864).

ignored. A more probable explanation is that during the formation of the lower portion of the Siwálik series the hills to the north had not attained anything like their present elevation, and that the gradients of the river beds had not become sufficiently steep to enable them to transport anything coarser than fine sand. If this be the true explanation, as seems extremely probable, the greater part of the elevation of the Himálayas has taken place since the miocene epoch, and it is impossible to date its commencement much further back than the commencement of the tertiary or the close of the secondary period.

Another argument of an entirely different character has been adduced by Dr. Blanford, which curiously confirms the conclusion regarding the date of origin of the Himálayas, arrived at on purely geological grounds.

He points out[1] that the mammalian fauna of Tibet has a proportion of species, and even genera, peculiar to the region which is not exhibited by any other continental area of the same size. Omitting all doubtful forms, and taking no account of varieties or subgeneric types, the known Tibetan fauna consists of forty-three species belonging to twenty-six genera, of which twenty-seven species and four genera are not known outside Tibet. Moreover, by far the largest proportion of species ranging outside of Tibet is exhibited by the carnivora, only four out of nine species of ungulata being known outside Tibet, and two of these are represented in Tibet by well marked varieties, while out of sixteen species of rodents only one is not purely Tibetan.

On the now universally accepted theory of the origin of species by descent and modification so large a proportion of peculiar species indicates a long period of isolation. In the case of island faunas, this isolation is due to the sea barrier which mammals cannot cross or can only cross with difficulty, but in the case of Tibet the isolation must be a climatic one, due to the superior elevation of the region, and after comparing the degree of specialisation of the fauna with that of various islands Dr. Blanford comes to the conclusion that this isolation must have commenced in middle tertiary times. This agrees remarkably with that arrived at on purely geological grounds, and from a study of the relations of the Siwálik to the tertiary faunas of Europe,[2] that the elevation of the Himálayas commenced with the tertiary era and that the range only attained an elevation comparable to that which it now possesses towards the commencement of the pliocene period.

Two views have been propounded, regarding the antiquity of the Himálayas which are antagonistic to that just put forward. The first of these,

[1] *Geol. Mag.*, 3rd dec., IX, 164, (1892).

[2] *Supra*, p. 367.

which regards the mountain chain as much older than the commencement of the tertiary period, requires special notice, as it has been advocated by Mr. C. S. Middlemiss, the author of the most detailed study of any portion of the range yet published. It is supported by arguments derived on the one hand from the special structural features of the southern margin of the Himálayas in Kumáun and Garhwál, and on the other by the degree of disturbance which the rocks of various ages have undergone in the same region.

The special structural features are summed up in, and illustrated by, a section drawn north and south along the Rámgangá and Peláni valleys in E. long. 78° 49′, which is reproduced on the accompanying plate. In this it will be seen that, starting from the plainward margin, we have the Náhan sandstones conformably covered by a great succession of middle Siwálik sand rock, north of which the upper Siwálik conglomerates are brought in by a small fault of no structural importance. The upper Siwálik conglomerates are brought into contact with the lower portion of the Náhan group by a great reversed fault of 11,880 feet thrown along the fault, or 6,380 feet in a vertical direction.[1] From this fault there is again, after some undulation of the strata, an ascending section through the greater part of the sand rock, but before the conglomerates are reached the beds low down in the Náhan group are again brought up, and after a series of anticlinal and synclinal folds the topmost beds of this group are brought into contact with the pretertiary beds of the Himálayas along the main boundary.

The rock in contact with the Náhans at the main boundary is a massive unfossiliferous limestone of unknown age. It is overlaid by the Tál beds, presumably mesozoic, and these again by marine nummulitics of the Subáthu group. North of the nummulitic band there is again a reversed fault and purple slates, and volcanic brecias come in, beyond which a great reversed fault brings in crystalline schists.

It will be seen that there are here five zones, each bounded on the north by a great reversed fault, and each successive one showing an older group as its newest member. In the outermost zone the section ranges from Náhans to the upper Siwálik conglomerates, in the next the sand rock in the newest group seen, in the third this is wanting and the section only ranges up to the upper Náhans. In the succeeding zone an entirely new set of rocks comes in, the newest of which is eocene, while in the last the rocks are of unknown, but certainly at least palæozoic age.

It has already been pointed out that the great reversed faults of the sub-Himálayan zone probably mark the successive positions of the outer margin of the hills, that is the limit between the area of elevation on the one hand and subsidence on the other. Mr. Middlemiss opines, with a

[1] *Memoirs*, XXIV, 124, (1840).

very good show of reason, that in each case the youngest rock seen south of the fault marks approximately the period of its completion, and of the commencement of the one next to the south, so that the first fault counting from the south may be ascribed to the upper Siwálik, the second to the middle Siwálik period, the third to the close of the Náhan period, and the fourth to the eocene.

The argument on which this conclusion is based may be epitomised as follows.[1] It is a common character of these long narrow zones bounded on the north by a reversed fault, that they carry along their northern border a still narrow zone of the newest rock they contain, a zone which has been preserved "because the fold involving that zone, and the reversed fault to the north of it, were the companions of the upheaval of that zone from a condition of deposition; that is to say, the uppermost stratum had only just been deposited when it was folded and faulted, and so wrapped up with the older zone to the north that it was preserved from subaerial denudation." It is argued that if this was not so, if for instance the nummulitics had been covered by the Náhans and Siwáliks, and exposed to denudation before they were folded and faulted, the upper members would have been removed in places and left in others, and that when the faulting subsequently took place the irregular patchwork of strata resulting could not have been formed into the regular zones now observable. This argument, however, assumes that the deposits must have been elevated without disturbance and exposed to great denudation before the faulting took place. This is by no means necessary. The whole thickness might have been deposited over the nummulitics, and if the faulting and folding had gone on *pari passu* with the elevation an arrangement, analogous to that which now obtains, might have resulted from the different degree of elevation, and consequent different intensity of denudation, the different zones had undergone. It has been shown, on quite independent grounds, that this supposition is an improbable one, but there is no reason why the nummulitics of Garhwál may not have once been covered by a great thickness of deposits corresponding to the upper members of the Sirmur series further west, if not by part of the Náhan group. The regularity of width of the outcrop, its narrowness, and the absence of these upper members can be sufficiently explained by the high dip of the strata,[2] the narrow patches remaining having been preserved by their having been elevated to a lesser degree, and consequently less exposed to denudation, than the higher beds.

But even if Mr. Middlemiss' argument be admitted to its fullest extent, it does not throw the elevation of the Himálayas further back than the

[1] For the full statement see *Memoirs*, XXIV, 174 ff.

[2] The line along which the section is drawn crosses the strike at an angle of about 40°. The dip shown is consequently considerably less than the true dip.

commencement of the tertiary era, for to the north of the fault, which bounds the nummulitics, we come to conditions so different, both stratigraphically and structurally, from those which obtain to the south that we can no longer apply the same arguments, and as will be shown presently the question of whether the existence of a land area, immediately north of the nummulitic outcrop at the time when these beds were being deposited, can be admitted as evidence regarding the date at which the elevation of the Himálayas commenced, depends entirely on the exact meaning we attach to these words.

The second argument depends on the different degree of disturbance exhibited by the successive zones. As can be seen from the section, the beds in the outermost zone have undergone least compression, those of the next more, and so on. Mr. Middlemiss argues that this increase in the disturbance the beds of each successive zone have undergone, is the result of the successively greater periods of time during which they have been exposed to the disturbing forces, and that the far more intense compression, to which the rocks within the innermost tertiary zone have been exposed, indicates that they have been exposed to pressure, during a period of time, which would carry back the origin of the Himálayas far beyond the tertiary era. To this it might be answered that, even if the compression of every zone had been contemporaneous, it is natural to expect that its intensity would not be everywhere uniform, but would die out laterally, graduating from the zone in which it was greatest to that in which there had been none.

It is, however, probable that, in the tertiary zones, the different degrees of compression exhibited by the successive bands is, to a large extent, the result of the different periods during which they have been exposed to compression. And when we come to the far more intense compression exhibited by the older brocks, which, as descried by Mr. Middlemiss, have been cleaved and foliated by the intensity of the compression they have undergone, he himself affords an explanation, in the observation that the strike is often transverse to that of the tertiaries, indicating that the compression had not all been in a direction transverse to the course of the range.[1] This diversity of strike is appealed to as showing that the compression of the Himálayas was in part due to other, and older, directions of thrust than those which produced the folding of the sub-Himálayan and many of the Himálayan rocks.

Here we are at once brought face to face with the question of what is meant by the commencement of the elevation of the Himálayas. Seeing that the present state of the range is the result of a long chain of physical causation, each step of which was the inevitable result of all that went before, it is impossible to say what was the first origin of the Himálayas.

[1] *Memoirs*, XXIV, 183, (1890).

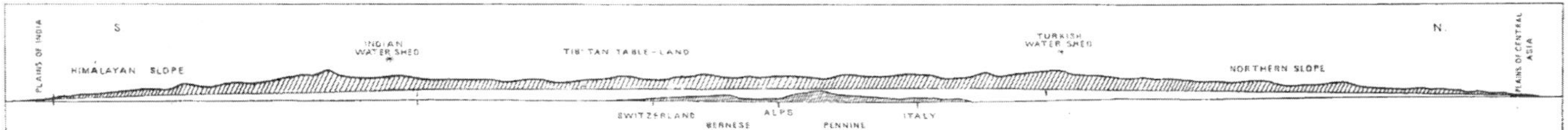

Comparative sections of Alps & Himálayas, after Strachey. Scale 1 inch - 35 miles.

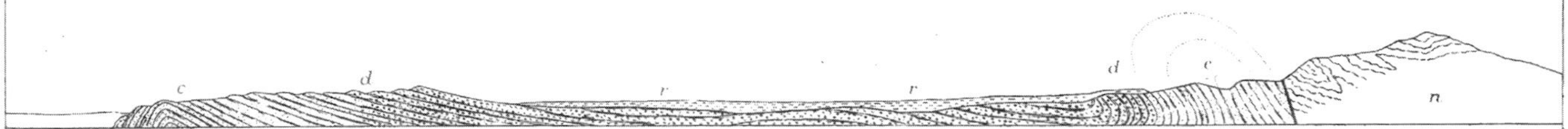

Section across the Dehra Dun, after R. D. Oldham. Scale 1 inch - 2 miles.

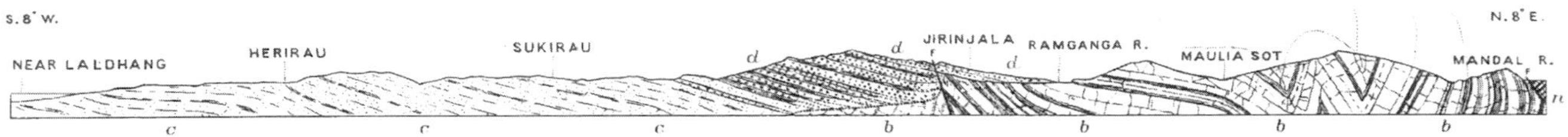

Section across Sub-Himálayas between the Kotah & Pátli Duns, after C. S. Middlemiss. Scale 1 inch = 1 mile.

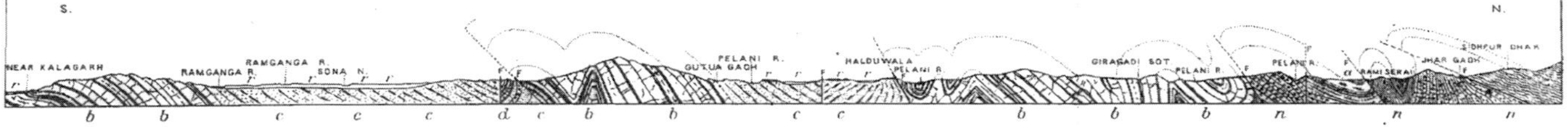

Section across the Sub-Himálayan zone exhibiting its relations to the older zones, after C. S. Middlemiss. Scale 1 inch - 1½ miles.

SECTIONS ACROSS THE HIMÁLAYAS & SUB-HIMÁLAYAN ZONE.

r = recent, d = upper siwalik conglomerate, c = middle siwalik sandstones, b = lower (Nahan) siwaliks, α = nummulitic, n = older rocks of Himálayas.

We may throw it back to the period when the earth first acquired a solid crust, or still further back to that primeval chaos from which, according to one hypothesis, the universe was evolved. But in this place the words are used in a much more limited meaning. If the limitation of the palæozoic and mesozoic deposits along the northern flanks of the snowy range represents at all approximately the general limit of land and sea during their deposition, it might be maintained that the general course of the Himálayan range had been determined in palæozoic times, and yet the elevation of the Himálayas in the sense in which the words are here used, might not have commenced till the dawn of the tertiary era. The further back in time we go the more difficult does it become to follow the sequence of cause and effect, and in speaking of the elevation Himálayas only that final compression is meant, which caused it to rise as a conspicuous mountain range with the same limits and extent as at present, and the antecedents which may or may not have been the direct cause of this result are excluded.

Taking this restricted definition, the transverse strikes mentioned by Mr. Middlemiss, and the systems of compression they indicate would not be connected with the elevation of the Himálayas or belong to the Himálayan system of disturbance.

It would be unnatural to suppose that the great area now occupied by the Himálayas had in no part been exposed to compression, previous to the end of the mesozoic era, and it is noteworthy that the most conspicuous instance of transverse strike quoted by Mr. Middlemiss, where a north and south strike extends for sixty miles, lies on the continuation of the Arávalli range. Now, without assuming, what there is no possibility of proving, that the Arávalli range ever extended so far north, the supposition would explain how there might be an intense crushing of the older rocks, accompanied by a strike transverse to the general direction of the range, which was due to a totally distinct system of disturbance from that which produced the Himálayas.

In this way we see how the crushing of the older rocks of the Himálayas and the divers strikes they exhibit, which Mr. Middlemiss rightly interpreted to indicate successive systems of compression ranging over a long period of time, may have been largely anterior to that final compression to which the elevation of the Himálayas is here restricted.

Another opinion regarding the antiquity of the Himálayas, which requires notice, is that recently propounded by Sir H. H. Howorth,[1] who has gone to the opposite extreme and regards the elevation of these mountains

[1] *Geol. Mag.*, 3rd dec., VIII, 97, 156, 294, (1891); IX, 54, (1892).

as having taken place almost, if not quite, entirely within the pleistocene period. This opinion appears to have been adopted in the first place to satisfy the requirements of his theories regarding the entombment of the mammoth in northern Siberia, but is based principally on the supposed absence of any traces of glaciation in the Himálayas, as also in the Altai, Caucasus and Ural mountains, all of which are supposed to have been elevated within the pleistocene period.

We have already shown that there is evidence of a former great extension of the Himálayan glaciers, evidence which might be amplified to almost any extent, but the only instances, in addition to those already quoted, which need be mentioned here, are the glaciers on the Babeh pass, now barely more than a mile in length, which at one time extended at least fifteen miles and probably more, and the morraine recorded by Colonel McMahon on the southern slopes of the Dháoladhár, at an elevation of only 4,700 feet.[1]

The Himálayan glaciers, it is true, never spread over the low ground in great ice sheets like those of Europe, but there is reason enough for this in the thirty degrees of latitude by which the Himálayas are nearer the equator than the Alps, and in the much greater distance which separates the watersheds from the lowlands. In the Kangra valley, where alone the high mountains rise steeply from the low ground at their foot, there is good reason to suppose that the glaciers once reached to below 2,000 feet above the sea.[2] And the erratics of the Potwár[3] show that ice in large quantities was not unknown there at one time. As it is out of the question to suppose that even in the glacial period these glaciers could have originated at low altitudes, the only possible conclusion is that the mountains must then have had very much their present elevation.

Another argument for a greater age of the Himálayas is the time required for the excavation of the great river valleys. Sir H. H. Howorth avoids this difficulty by denying that the valleys are the work of the rivers that flow through them, but it is impossible for any one who has studied the action of subaerial denudation not to see that the forms of the hills and their intervening valleys are due to the action of rain and rivers, aided by frost. A glance at the photograph so admirably reproduced in the frontispiece of this volume will show to the initiated eye that the shape of the mountain is due to the disintegration of the rock by frost and the removal of the debris from the hollows by streams and glaciers, and not to any disruptive force. Even if we could acknowledge that the courses of the drainage were in the first instance determined by fissures, a long period of time would be required for the opening out of the valleys and the removal of that vast mass which the beds of the Siwálik series tell us was brought down from the Himálayas.

[1] *Records*, XV, 49, (1882).
[2] *Records*, IX, 56, (1876).
[3] *Records*, X, 140, (1877); *Supra*, p. 42.

These arguments would be sufficient to show that the Himálayas must have existed as a mountain range previous to the glacial period at any rate, but it is not necessary to appeal to them, for the evidence of the pliocene sub-Himálayan deposits shows that the range must then have had very much its present elevation, with the main features of the existing drainage system already marked out.

The close agreement in the results attained by the several distinct lines of purely geological reasoning, and that derived from the peculiarities of the living Tibetan fauna, gives a very strong presumption in favour of the correctness of the conclusion arrived at, and discredits at once the hypotheses of an older or a later date for the origin of the Himálayas than here maintained. There seems, however, to be this much truth in Sir H. H. Howorth's supposition of the recent rise of the Himálayas, that their elevation, if not still in progress, has only recently ceased, and that they are probably now somewhat higher than they were during the glacial period.

The evidence pointing to this is of various kinds. There is, firstly, the natural presumption that the mountains which now form the most elevated peaks of the world cannot be in a state of decadence, and as there is no such thing as rest in nature, that they must be still growing. Then there are two recorded cases[1] where a differential movement of the opposite sides of a fault has taken place at so recent a period as to cause interruption of the minor drainage courses, and to exhibit itself as a distinct rise in the surface of the ground, which has hardly been modified at all by denudation. These earth movements show that the Himálayas are still in a state of strain, and we may naturally conclude that this strain is due to the compression which has caused their elevation.

Better evidence is yielded by the sub-recent fossil fauna of the Hundes plain. It was formerly believed that this fauna was tertiary, the presence of a rhinoceros was supposed to indicate that the deposits must have been formed at a very much lower level than that at which they are now found, and that they had subsequently been elevated several thousand feet without any discernible disturbance. The incorrectness of the first supposition has already been shown.[2] The last is one that cannot be granted, and as regards the second, the presence of the peculiar Tibetan genus *Pantholops* outweighs the evidence of the rhinoceros. It is true that a rhinoceros could not exist on the present plains of Tibet, not on account of the cold, for the Tibetan species may well have been protected from that by a thick coat of fur, but on account of the impossibility of its picking up a living from the scanty vegetation of these arid plains. It must not be forgotten, however, that there would almost certainly be shallow lakes and swamps, when these

[1] *Records*, XXI, 158, (1888). | [2] *Supra*, p. 425.

deposits were being formed, and at the present day the river valleys of Tibet, even at a height of over 13,000 feet, can under such circumstances support a growth of grass and shrubs which could easily have given sustenance to the rhinoceros of Hundes. This animal in any case shews that the climate of Hundes must have been somewhat milder than it now is, and as there is little difficulty in supposing that these deposits may have been raised 1,000 to 2,000 feet without any appreciable disturbance, though it is impossible to grant an elevation of 10,000 to 15,000 feet, we may well suppose that this increased inclemency of climate is partly due to the desiccation resulting from the change of condition of the rivers, from deposition to erosion, and partly to an increased altitude of the plains and of the mountains south of them.

The gradual desiccation of the Tibetan lakes points to the same conclusion. There are no data available regarding the rate at which this is taking place, but the fact that some have dried completely up, while others show but little reduction on their original size, indicates that the process is still in progress and that the climate of Tibet was once moister than it now is. There appears to be but one explanation possible of this increased dryness of climate, and that is a rise of the mountains to the south, which has resulted in a more complete cutting off of the moisture from the monsoon winds.

The cause of the origin of these lakes in Tibet is not thoroughly established. Ever since the publication of Mr. Drew's book on the Jammu and Kashmír territories it has been customary to attribute their origin to the damming up of the main valleys by the fans of tributaries, which attained a great development during the glacial periods, when the disintegration of the rocks was more rapid than it now is, while the transporting power of the streams was no greater if so great. The present writer is unable to accept this view in its entirety. In the case of the Pangong lake he believes that its formation is entirely due to differential movements of the surface, which raised a portion of the original river bed at a more rapid rate than the stream was able to erode and dammed back the drainage to produce the present lake. Even in the case of the Tsomoriri in Rupshu, which is accepted as the typical instance of a lake formed by a tributary fan, he has shown[1] that there is reason to believe that this fan could not have caused an interruption of the drainage had there not been an elevation of a portion of the river valley further down its course, and a consequent diminution of the gradient. Whatever may be the cause of origin of these lakes there seems no reason to doubt that the broad shingle plains, which so frequently occur just above where the rivers enter a gorge, are produced by a check in the gradient consequent on a recent elevation of the river bed in the gorge, and consequent checking of the gradient im-

[1] *Records*, XXI, 156, (1888).

mediately above it. A similar action might well, under favourable circumstances, give rise to the formation of an actual lake, while the existence of an exit would depend on the rapidity of the movement, the supply of water, and the nature of the climate. Whether such has actually been the case or not, there have certainly been irregular movements of the beds of the streams and rivers within what is, geologically speaking, a very recent period, and these irregular movements can only be regarded as evidence that the disturbance which caused the elevation of the Himálayas is still in progress.

Thus, from whatever point of view we look at the subject, we see that the decadence of the Himálayas has not yet begun, but whether they have yet reached their maximum development is not so clear. There are no data from which we can decide whether the rate of elevation in the immediate past has been greater than at present or no, but looking to the general indications, throughout the world, that the great earth movements, which caused such profound changes in the form and distribution of land and sea during the tertiary period, have reached their close, and that the present is a period of comparative quiescence in the history of the earth, we may suppose that the chapter devoted to the elevation of the Himálayas is reaching its close and that they soon will enter on their decay.

There remains one more point to be referred to before finally dismissing the subject of the origin of the Himálayas, and that is the supposed connection between mountain ranges and sedimentation. The enormous thickness of sedimentary deposits seen in the sections exposed in mountain ranges has been frequently noticed, and by many observers their accumulation has been regarded as the immediate precedent, and proximate cause, of the mountain ranges. It is, however, doubtful whether in this case cause and effect have not been confused. Sedimentary accumulations of great thickness are known elsewhere than in mountain ranges, but it is only where the beds have been turned up at steep angles and extensively denuded that their thickness becomes conspicuous, and it is only where a great thickness of sediment has been previously accumulated that mountains can be formed of stratified deposits. Otherwise the underlying crystalline and metamorphic rocks will soon be exposed by the denudation which is always much more active in mountain ranges than in more level ground. The subject is, however, of sufficient importance and interest to make it necessary to inquire whether there is any indication of a connection between the present position of the Himálayas and the distribution of the sedimentary deposits which preceded its elevation.

In the north-west Himálayas there is a great series of sedimentary deposits, ranging through the palæozoic and mesozoic eras, which represent long periods of accumulation of sediment in enormous thicknesses of strata. Here there is a distinct temptation to regard the mountain range as the result of this vast accumulation of stratified deposits, but as we trace the range to the eastwards difficulties come in.

The zone of marine deposits found north of the line of highest peaks in that part of the Himálayas which has been accessible to exploration has been referred to, as well as the non-recognition of these beds south of the main range. How far the same distribution holds good further east it is impossible to say with certainty, but we know that jurassic and cretaceous fossils have been obtained from the region north of the hills of Nepál and Sikkim and the discovery of sedimentary strata of unknown, but probably tertiary, age near the Cholamo lakes seems to indicate that the same parallelism between the boundary of the sedimentary deposits and the line of highest peaks prevails at least as far east as Sikkim. Whether the present limit is in the main due to an original limit of deposition or to the effects of disturbance and denudation is for the moment unimportant. The absence from the main range and the hills to the south of them, so far as is known, of any extensive series of sedimentary strata later than older palæozoic or even older, precludes the idea that the elevation of the range was immediately consequent on a great accumulation of strata.

In the eastern Himálayas our difficulties are still very great owing to the scanty observations available. The only sedimentary deposits that could possibly be marine, or that have any great thickness, are certainly long anterior to the carboniferous in age, and these occupy a very small area in comparison with the great expanse of crystalline schists, gneisses and granites. But there are some small patches of coal bearing Damuda rocks, which have been recognised at several spots along the outer edge of the Himálayas, and are important as showing that this region was dry land, at the close of the palæozoic era, when marine formations many thousands of feet in thickness were being deposited in the north-west. It is not possible to say that no marine strata of later date than permian exist in the eastern Himálayas, but it may be taken as tolerably certain that, if present, they cannot be of a very great extent or thickness, and this portion of the Himálayas appears to have been a land area continuous with that of the Peninsula throughout the secondary era, such interruptions of continuity as there may have been, if there were any at all, being of minor importance and only temporary. But though this portion of the Himálayas was a land area, there is no reason for supposing it was a mountain range at these early periods; the great height of the snowy peaks suggests that their upheaval must have been comparatively recent, and the palpable unity of the range as a whole prevents us from ascribing a much earlier date to this portion than to the rest, which it has been shown could not have existed in its present form in the secondary era.

In view of this divergence between the eastern and western portions of the range, it is impossible to attribute the rise of the Himálayas to the sedimentation in what is now its north-western portion, and we must look to some more wide reaching and deep seated cause for its present position and course—a cause which was independent of and able to obliterate long standing structural features and to introduce new lines of separation between areas of elevation and subsidence.

CHAPTER XIX.

GEOLOGICAL HISTORY OF THE INDIAN PENINSULA.

Earliest periods—Origin of Arávallis and East Coast—Mesozoic Indo-African continent—Origin of the West Coast and Western Gháts.

The previous chapters of this book have principally been devoted to the stratigraphical description of the various rock systems of India, and though reference has been made in the course of this description to changes in the distribution of land and sea, and to the earth movements which have marked out the salient features of Indian geography, such references have necessarily been somewhat swamped by other matter. This chapter will consequently be devoted to a brief resumé of the geological history of India, of those changes of land and sea through which it has reached its present form.

The earliest stages of the geological history of India, as of all other history, are wrapped in obscurity. Dimly we can discern an old land surface composed partly of a very ancient granitoid rock, which had even then solidified, been penetrated with quartz veins and trap dykes, and exposed to extensive denudation, and partly of later rocks, themselves the product of the denudation of the granitoid gneiss. From the waste of this land surface the rocks of the Dhárwár system were formed, in a sea where volcanoes poured forth their lavas and ashes, much as at the present time. But whether any living thing was to be found in this sea, or whether the earth was still unfit for the support of either animal or vegetable life, it is impossible to say.

These Dhárwár deposits were in their turn compressed, contorted and exposed to great denudation before the commencement of the Cuddapah epoch, but it is impossible to trace even approximately the changes of distribution of land and sea during this earliest period of the geological history of India.

With the commencement of the Cuddapah epoch, some definite indication of the distribution of land and sea appear. All Southern India, south and west of the Cuddapah and Kaládgi basins appears to have been dry land, while the sea spread out to the east over part of the present Bay of Bengal, and, to the north over what are now the Nizam's dominions and the Central Provinces. The exact limits of this sea cannot be defined with accuracy, western Bengal and Chutiá Nágpur were probably dry land, and this rock area probably stretched to the north-east over the Gangetic delta, to Assam and the eastern Himálayas. In Bundelkhand there was dry land, to the south of which the Bijáwar sea spread to the valleys of the Narbadá and the Son but had probably been obliterated by the time the Cuddapahs were deposited, and at a later period a fresh depression admitted the sea to the north-west of Bundelkhand, in which the beds of the Gwalior system were deposited.

Nothing is known of the early geological history of the great area covered by the Deccan trap, nor of what was going on where the Himálayas now stand, or where the Indus and Ganges rivers have spread their alluvial plains. In fact, what with complete want of information regarding the greater portion of the area, and the incompleteness of that available regarding the rest, the conclusions that can be drawn regarding these earliest periods of the geological history of India are of the most meagre description. This much, however, seems certain that none of the leading features of Indian geography of the present day had been marked out, none of the mountain ranges had arisen, none of the great river valleys had commenced, and the distribution of land and sea was very different to what we now see.

The close of the Cuddapah epoch appears to have witnessed the commencement of the earliest of those earth movements whose effects on the surface contours and geography of India are still prominently noticeable. It was then that the great mountain range, of which the present Arávallis are but the wreck, was raised, and extending far beyond its present limits, stretched across what is now the Gangetic plain, possibly even to the Himálayas. At the same time another zone of contortion was formed running along the south side of the Son and Narbadá valleys, which was probably marked by a range of mountains or hills, not rising to the same height or importance as the Arávallis, and bearing much the same relation to them as the hills west of the Indus alluvial plain do to the Himálayas of the present day.

To the same date must probably be ascribed the zone of contortion which runs along the eastern margin of the Cuddapah basin and can be traced northwards to the Godávari valley.

These three zones of contortion, whose disturbance took place during the Vindhyan epoch, and must once have been marked by mountain ranges much more important in size and elevation than their remnants at the present day, seem to be due to the last great movement of compression which has affected the rocks of the Peninsula. Since then the disturbances have principally taken the form of movements of elevation and subsidence, only to a comparatively minor extent accompanied by compression of the rocks, and it is interesting to note that the earth movements of this period have still their influence on the limitation of the Peninsula.

On the north-west the Arávallis have remained the boundary of the peninsular land area. West of them the great desert of western Rájputána was alternately land and sea through long ages, but the sea never spread east of the barrier of the Arávallis. On the south-east the bend of the east coast north of Madras follows too closely the general course of the Nallamalai range for the connection to be accidental and as we know that from the jurassic period to the present day the position of the coast has been practically where it now lies, we may naturally conclude that its course had been laid down at an even earlier period, contemporaneously with the great Vindhyan epoch of disturbance. In the course of ages there have no doubt been alternate elevations and depressions of the land, at times it has encroached on the sea, at times the sea has flowed over what is now dry land. But the fact that the only marine deposits in this part of India are confined to the neighbourhood of the coast, their small thickness, the manner in which they thin out away from the sea, and the character of the rocks, indicate that when they were formed the shore line could not have been far off, and point to a persistence of the general run and position of this the oldest feature of the geography of India.

The Vindhyan epoch is the age to which the rise of the Arávallis and the demarcation of the east coast has been ascribed, but what this age is in terms of the European sequence there is no means of determining. The upper Vindhyans have been looked upon as devonian, on the strength of their resemblance to the Table Mountain sandstone of South Africa, and though the evidence is insufficient, it is certain that they cannot be much newer than the date indicated, and it seems difficult to make them much older. They may consequently be ascribed to some portion of the middle or latter end of the lower palæozoic, and it is to be regretted that no more exact correlation can be made, for we find that in silurian times the sea flowed over the north-west Punjab and the north-western portion central Himálayas, and over the hills east of the Irawadi valley. No silurian deposits have been found in that small portion of the eastern Himálayas which has been examined, nor in Assam, and it is probable that the land area stretched north eastwards from the Peninsula over these regions in silurian times, as it seems certainly to have done at a later period.

Towards the close of the palæozoic era, at a period corresponding to the upper carboniferous of European chronology, we have some definite information regarding not only the distribution of land and sea, but also the climate. The great Gondwána era opened with a period of exceptional cold. The Peninsula was a land area over which many large lakes were probably scattered, while on land there were glaciers flowing down into these lakes, and into the sea which covered part of the great Indian desert, the north-west Punjab, and a very any large portion, if not the whole, of the area occupied by the Himálayas west of the Ganges valley. The same sea appears to have stretched westwards to the furthest boundary of Afghánistán, and it was continuous in some way with that which flowed over eastern Australia. It is not clear whether this communication was round the south and west of the Peninsula or round the east and north. We know from the evidence of the Salt range fossils that after the glacial period there was an irruption of European forms and a complete change of type of the fauna; this period was one of extensive changes of land and sea when vast areas in South Africa and Australia were converted into dry land, there is consequently a possibility that sea stretched south of the Indian peninsula and the close affinity of the two faunas is more in favour of this direct communication, than of one round by the more circuitous route round the north of the peninsular area, which seems at that time to have extended much further to the north-east than it now does. At the close of the jurassic period the Indian peninsula was still dry land, the east coast was not very far removed from its present position, and on the west the sea flowed over Cutch, the Indian desert and the north-west Punjab and central Himálayas. It is not possible to say whether the north-easterly extension of the peninsular area over Assam was still dry land, but the land connection with Africa was still maintained. Still the presence of some eastern species in the western sea shows that there must have been either a temporary and direct, or more permanent and circuitous, connection between the two. If the first of these explanations is the correct one there may have been a temporary subsidence, by which the land communication between Africa and India may have been severed for a time, and certain forms of life enabled to cross from one marine province to the other. The alternative explanation would be that the form which is common to the two areas, being an abundant and wide ranging one, was endowed with great powers of spreading, and reached the western sea round the north-easterly prolongation of the Indian peninsular land area.

In the cretaceous period the land connection with Africa was still maintained, the eastern coast line of the continent ran not very far from the present east coast of India, across the Ganges delta, and along the south side of the Assam hills. On the west of India a different sea flowed over Arabia and the Arabian Sea and extended inland at least as far as Bárwai on the Narbada. Sea also flowed over the hilly country west of the Indus alluvium and over Tibet.

The close of the cretaceous period witnessed that grea outburst of volcanic activity which buried the whole of western India deep in lavas and ashes, and extended from Sind on the one hand to Rájámahendri on the other. It is not improbable that this great outburst may have been connected, as it was probably contemporaneous, with the commencement of that great series of earth movements which resulted in the elevation of the Himálayas and the extra peninsular mountain ranges generally. But however this may be, the lava flows must have obliterated all the pre-existing surface features and the origin of the main features of the drainage system, of the northern part of the Peninsula at least, cannot be ascribed to an earlier date than the close of the Deccan trap period.

In the tertiary era we find no further evidence of a land connection with Africa; at an early period the west coast was approximately in its present position, and it is probable that at the clcse of the cretaceous or commencement of the eocene period the great Indo-African continent was finally broken up, and all but the remnants in India and South Africa sunk finally beneath the sea.

The eocene sea flowed over western Rájputána and the Indus valley to the west, over a large part of Balúchistán and Afghánistán, and over the whole of the north-west Punjab and the outer Himálayas as far east as the Ganges river. We do not know if this sea stretched eastwards to the north of the Peninsula till it joined that in which the nummulitics of Assam and Burma were deposited, but on the whole it more probably did not. Sea also flowed over the central Himálayas and was probably continuous with that just referred to, across the north-western termination portion of the range.

One of the first effects of the great series of earth movements, which resulted in the formation of the mountain ranges of extra peninsular India, was an encroachment of land on sea, and the driving back of the sea first from the Himálayan and Punjab areas, and finally from Sind and Burma. The same period as witnessed the gradual growth of the Himálayas also saw the rise of the Arakan Yoma, and Manipur and Nágá hills, on the one hand, and the greater part at any rate of the Afghánistán and Balúchistán hills on the other, and as the most important part of this history has been told in the last chapter it will not be necessary to repeat it here.

It would have been in the last degree extraordinary, if such extensive and violent earth-movements all around it had been accompanied by absolute quiescence in the Indian peninsula, but such disturbance as may have taken place in no way took the form of compression, and the only change which can be attributed to this period is the origin of the Western Gháts. Reference was made in the first chapter to the difficulty of accounting for this feature, and its resemblance to a line of sea cliffs modified by subaerial denudation was noticed, as well as the occurrence of a land shell closely allied to a marine form. But though the sea may once have washed the foot of the ghâts, it is impossible to grant that they owe their origin entirely, or even largely, to marine denudation. Marine denudation works slowly on hard rocks, and during the ages that would have been required for the sea to carve the low lands of the Konkan out of the Deccan trap, it is inconceivable that the rivers would not have cut their valleys much further back into the scarp than they have done. It is far more probable that the main features are due to late tertiary earth movements, and the great rock basins of the Narbadá and Tápti valleys show most conclusively that there has been a movement of elevation to the west, which certainly checked and may even for a time have interrupted the flow of those rivers, while the ground along the foot of the ghâts has not been closely enough examined either to prove or disprove the hypothesis.

Whatever may have been the cause of the origin of the Western Ghâts, the present easterly tread of the peninsular drainage must be an ancient one, for had there been any considerable rivers flowing to the west they would have preserved their channels, or if the movement had been sufficiently rapid to reverse the course of the drainage, deep gaps would have been left to mark their former course. There is only one such gap, the Pálghât, north of the Travancore hills, and it is possible that a river may once have flowed westwards through this, whose drainage was reversed by the earth movements which raised the Western Ghâts, leaving the lower part of its course to be occupied by a much smaller stream, while the bulk of the drainage was diverted to the east. With this possible exception it is probable that the main features of the peninsular drainage, the two great westerly flowing rivers to the north, and the series of easterly flowing ones further south, were marked out at the close of the Deccan trap period.

We see then that the origin of the west coast of India dates from the middle of the tertiary epoch or a little earlier, when the dry land which stretched westwards into the Arabian Sea was depressed, and at the same time that to the east was elevated to form the Western Ghâts, the most recent and also, perhaps more correctly therefore, the most conspicuous feature in the geography of the Indian Peninsula.

GEOGRAPHICAL INDEX

OF

INDIAN LOCALITIES.

Latitudes are all north; Longitudes east of Greenwich.

Abbreviations used: —H. Hill or mountain; R, River or stream; I. Island; L. Lake.

In the case of rivers whose names are printed on the accompanying map, the coordinates indicate the position of the name. Where no name is inserted on the map, the coordinates are those of a point on its course which will indicate the locality referred to in the text.

The coordinates of States or Districts are usually those of the capital or head-quarters; where this course could not be followed and in the case of hill ranges, a central point has been selected.

Casual references or the use of the name of one place to indicate the position of another have not been indexed, except where such are the only refreences in the text.

Presidencies and Provinces have not been indexed.

	Latitude.		Longitude.		Page.
	°	′	°	′	
Basí	26	50	76	6	68, 69.
Bassein	16	46	94	48	18, 337, 424.
Bastár	19	6	82	4	4, 91, 375.
Baudwengyee	23	0	97	20	45.
Baxá	26	50	89	36	118.
Bedésar	27	3	70	49	227.
Behar	25	11	85	34	29, 32, 57, 427.
Bela	26	7	78	10	66.
Bela I.	23	52	70	45	218, 220, 321.
Belgáum	15	52	74	34	48, 263, 381.
Bellary	15	9	76	57	24, 35, 40, 48, 375.
Benares	25	19	83	3	427.
Betúl	21	54	77	57	92.
Betwá R.	24	20	78	10	436.
Beypur	11	10	75	51	4.
Beyt I.	22	27	69	10	323.
Bezwáda	16	31	80	39	34.
Bhágalpur	25	15	87	2	31.
Bhágírathí R.	24	0	88	15	440, 444.
Bhagothoro	26	21	67	54	309.
Bhagwa	24	35	79	14	27.
Bhander H.	24	10	80	40	100.
Bhartpur	27	13	77	32	100.
Bhiaura	24	40	85	45	31, 57, 58.
Bhímá R.	17	0	77	59	84, 87.
Bhit H.	26	10	67	40	305.
Bhita	25	19	82	21	94.
Bhopál	23	16	77	26	3, 264, 275.
Bhor Ghát	18	47	73	23	261, 271.
Bhúj	23	15	69	49	216, 220.
Bhusáwal	21	1	75	47	401.
Bhután	27	0	89	30	45, 149, 348.
Bhutra	23	2	78	50	399.
Biána	26	54	77	21	70, 72.
Biás R.	31	45	75	30	450.
Bibinani	29	42	67	26	293, 310.

Q

R

INDEX OF SUBJECTS.

M

N

T